PROCESS ORIENTED ANALYSIS

Design and Optimization of Industrial Production Systems

PROCESS ORIENTED ANALYSIS

Design and Optimization of Industrial Production Systems

URS B. MEYER
SIMONE E. CREUX
ANDREA K. WEBER MARIN

CRC Taylor & Francis
Taylor & Francis Group
Boca Raton London New York

CRC is an imprint of the Taylor & Francis Group,
an informa business

CRC Press
Taylor & Francis Group
6000 Broken Sound Parkway NW, Suite 300
Boca Raton, FL 33487-2742

Printed in the United States of America on acid-free paper
10 9 8 7 6 5 4 3 2 1

International Standard Book Number-10: 0-8493-7494-4 (Hardcover)
International Standard Book Number-13: 978-0-8493-7494-4 (Hardcover)

Library of Congress Cataloging-in-Publication Data

Meyer, Urs B., 1964-
Process oriented analysis : design and optimization of industrial production systems / Urs B. Meyer, Simone E. Creux, and Andrea K. Weber Marin.
p. cm.
Includes bibliographical references and index.
ISBN 0-8493-7494-4 (978-0-8493-7494-4 : alk. paper)
1. Production engineering. 2. Manufacturing processes. 3. Process control. I. Creux, Simone E. II. Weber Marin, Andrea K. III. Title.

TS176.M495 2006
658.5--dc22 2006013504

Visit the Taylor & Francis Web site at
http://www.taylorandfrancis.com

and the CRC Press Web site at
http://www.crcpress.com

PREFACE

Manufacturing started off being an entirely manual process, as the Latin roots of the term indicate. Add the power of machinery, and we arrive at the industry of the times of the Civil War. Arrange processing machinery along the path of the product in making, and we have mass production coming up in the first half of the 20th century, nourishing and sheltering a steadily growing mankind. Put on top computers, closed loop controls, sensors, and servo drives, and we get the full complexity of modern manufacturing plants. Automation is the key, hardware is the base, and software is the tool to realize state-of-the-art manufacturing systems.

What is the art of designing, engineering, and making production machinery? Parts have to match, the nut has to fit to the bolt, the shaft to the seal. Unlike the work of an artist, industrial production relies on a vast number of specifications, common terms, definitions, and methods. In mechanical and electrical engineering, we have a traditional common language in drawings, schematics, measures, and tolerances. With a dictionary for the part names at hand, these documents are understood worldwide. If we go into software, such standards are still missing. We have programming languages, but these are as far off from the final program as ink and paper from a complete drawing. Most of the process control software made today depends on its creator in person to be maintained, adjusted, and improved. And, since manufacturing systems are controlled by software, many of these systems again depend on this single person.

A recent investigation into waste losses in a computer-controlled pasta manufacturing plant revealed that the blend of flour, egg, and water, differs from the recipe by up to 2%. Taking into account the frequent stops and adjustments required to maintain the quality of the product, the impact on the cash flow of the plant is estimated to be around 10%. To avoid this loss, the supervisor of the pasta production line corrects the original recipe based on his own experience and gut feeling. The correct solution, an exchange of the flow rate sensors for better accuracy and an adjustment of the process control software, is no longer feasible because interface documentation is missing.

Successful operation of a manufacturing plant depends entirely on managing interfaces. In planning and setup, the key issue is proper interfacing of the interlinked machinery, robots, conveying, and storage systems. Interfaces again. When running the plant, mutual understanding and teamwork between product design, marketing, order handling, personnel, and maintenance is essential for successful and profitable operation.

For plant management, the allocation of cost is an ever-burning topic interfacing with the accounting department. Thus, we find interfaces in the domains of resources, data, and costing. And, with increasing importance, the interface of plant and product to the environment.

Today, we see two different approaches to plant design and management: First, the method of the bean counter, striving for correctness and precision, keeping a meticulous documentation on everything. He is unable to adjust and get a fit with others outside of his figures. Second, the attitude of the generalist, proud to understand a little of everything, but keeping out of anything in detail. He will soon lose track of the business since general principles and missions, drawn in bold lines in presentation graphics, lead to a disaster when applied in plant engineering.

The method presented here addresses both views. It teaches a method both integral in view and accurate in detail. It tackles the challenge of the interfaces directly, by using them for the specification of the elements of the system. The method is targeted at manufacturing systems in a wide sense of the term. While it makes use of ideas originating in computer science and database design, it does not teach how to make business-related software, but supports a thorough understanding of the manufacturing processes and plants, and prepares a sound basis for real-time and simulation program coding. Most examples in this book are based on the industrial and educational experience of the authors.

After having read this book, you will not tell your superiors how to manage a company, but you will know how to design, run, debug, and optimise modern mass production based on the state-of-the-art, efficient, and easy-to-apply method of Process Oriented Analysis.

New methods must be more efficient than existing ones. Process Oriented Analysis is supported by a computer program to speed up design of diagrams, to set up large models, to check the consistency of the system, and to manage the data base for the objects and their relations. Moreover, this program contains an automatic calculation module for transfer and balancing cost or environmental impact data within a model. Do not hesitate to check in at the download site of CRC Press to get a demonstration of Process Oriented Analysis and further news on applications. (http://www.crcpress.com/e_products/downloads/download.asp?cat_no=7494)

TABLE OF CONTENTS

I INTRODUCTION TO THE PROCESS ORIENTED ANALYSIS

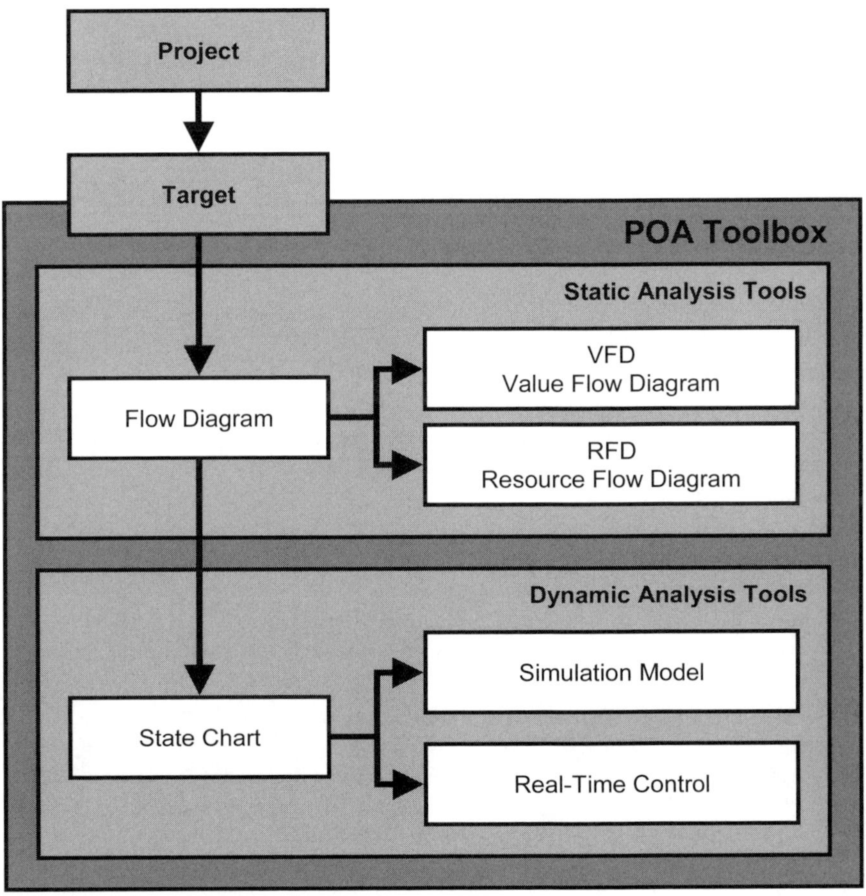

I1 PROCESS ORIENTED ANALYSIS

Key points of the Chapter

- Objectives of this textbook.
- Why, when, and where Process Oriented Analysis is used.
- Purpose and application of Process Oriented Analysis.
- Overview of the different methods in the Process Oriented Analysis toolbox.

Figure I1-1: Ring spinning machines in a spinning mill (Photo: Lauffenmühle GmbH & Co.).

1.1 Introduction

This book introduces a method called Process Oriented Analysis (POA), which has been developed and refined over the last ten years for the analysis of complex production systems. Modern manufacturing machinery and plant automation are becoming more and more complex. As a result, human skills are even more important for the conception, planning, and setup of these systems. In addition, the comprehension required for the interaction of complex manufacturing systems and the behavior of single machines is now a key scientific issue. The goal of POA is to help engineers with conception, optimization, modernization, and maintenance. Existing instruments for the analysis of such systems with advanced technology that include all the technical and operational aspects are few in number and limited in performance.

Application of POA

Process Oriented Analysis can be used for the design of a new production system, for an engineering update, or for the reengineering of manufacturing systems. The main application fields of POA are in:

- System engineering and design for automated production plants
- System analysis of business workflows with regard to profitability
- System evaluation and improvement in safety and sustainability
- Development, modeling and structuring of business processes
- Optimization of static and dynamic behavior in partially and completely integrated systems
- Support and documentation for operation and maintenance of complex production lines
- Improvement of software design and maintenance for machine and process control

POA assists the concurrent engineering of products, manufacturing processes, and production equipment, and helps to improve efficiency for modeling and development. It supports the lifecycle design of products by including technical, financial, and environmental evaluation parameters.

POA contains different graphical tools for the static (time-independent) and the dynamic (time-dependent) analysis of companies, plants, processes, and resources. Financially and environmentally relevant flows are modeled in static diagrams and are applied for the investigation and improvement of industrial systems. In turn, the dynamic system view allows a time-dependent analysis of a production system that enables straightforward programming and documentation. This leads to optimization by simulation as well as to programming of controls for production machinery and processes.

The POA method is very strict in rules and in relationships. The static and dynamic diagrams are absolutely consistent. A step-by-step procedure facilitates the transition from static analysis to dynamic programming.

Objectives of this textbook

The motivation to put this method in print form was the need to compile and record the fundamentals of and the experience with a graphical method used for more than ten years at the Institute of Manufacturing Automation of the Swiss Federal Institute of Technology (ETH) Zurich. This method was born in industrial practice, thereafter developed and applied in education and research. Students working in numerous industry projects in operational plants obtained remarkable results within very short time. Since this method results in an easy to understand depiction of very complex production systems, POA allows an immediate communication with the practitioners in the companies. In recent years, POA has been applied to automated production lines creating almost anything from croissants and pasta to watches and electronic components.

The objectives of this textbook are:

- To define and describe in detail the different diagrams of POA.
- To teach real-world system analysis and design skills in the context of solving realistic problems, and to present practical guidelines and tips.
- To provide a common language for systems analysts and users, management, and other professionals in a typical business organization.
- To offer interesting case studies showing the industrial world application of POA.
- To give an in-depth understanding of how complex systems are modeled in an efficient way.
- To provide a comprehensive POA toolkit that supports the analysis of manufacturing processes, logistics, communications, economics, and energy, as well as code generation for programming.
- To present the subject in a visually appealing, full-color format and in an exciting, easy-to-read style that invites students to learn.
- To create a textbook for graduate courses as well as for self-teaching on the postgraduate level.

Finally, this textbook is intended to improve engineering, both as a science and an art, in order to make life safer and more enjoyable.

This Section I1 of the book outlines the concept and application of POA. It provides a short description of each of the diagrams and methods in the POA toolbox, and explains the setup of an analysis. Section I2 goes into the delimitation of POA compared with well-known system modeling tools.

1.2 Concept of POA

POA investigates production systems on different levels of complexity. The analyzed production system can be an entire manufacturing plant such as a sewing shop for garments, a production line for chocolate, a machine, as for example a packaging machine, or a single process, such as quality control. For analysis and improvement, a system needs to be modeled. This can be done either time-independently (statically) or time-dependently (dynamically). The concept of POA includes both viewpoints of the system, as well as the transition from one to the other. For this, POA comprises two diagram types in order to draw models for the static and the dynamic analysis, as shown in *Figure I1-2*.

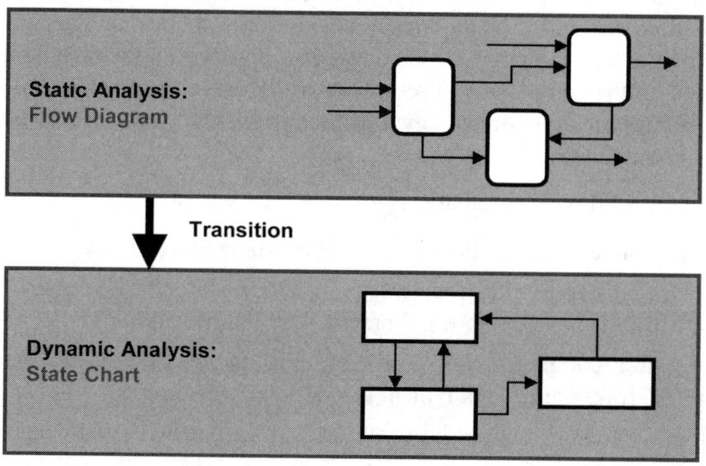

Figure I1-2: Static and dynamic analysis.

Static view

The static view of a production system shows the functional mapping of the system by processes, as well as resource and information flows. The static model is called a *Flow Diagram*. It depicts what happens within a system, which activities are conducted, and specifies where and how what input is transformed into what output. It can be described on several levels of detail in the hierarchy.

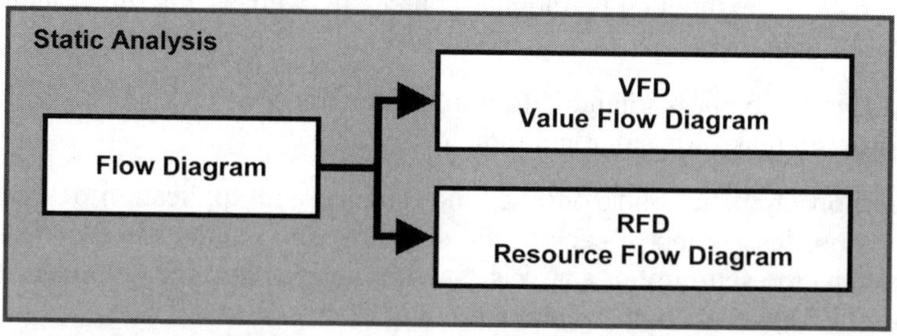

Figure I1-3: Static analysis tools.

In addition to the Flow Diagram, the static analysis tools of the POA toolbox include the *Value Flow Diagram* and the *Resource Flow Diagram* (*Figure I1-3*). Both of these diagrams are based on the Flow Diagram, but incorporate calculations. The Value Flow Diagram contains monetary value calculations of the resource and information flows. The Resource Flow Diagram introduces energy and environmental calculations for the specified resource values.

Transition from system structure to system behavior

In addition to the static views, POA also offers tools for dynamic modeling. The dynamic model is based on the structure of the static model. In POA, the link from the static to the dynamic model is defined with consistent rules. This well-defined link forms one of the main strengths of POA. Only through the consistency and the clear connection of both viewpoints can the static and dynamic analysis be integrated into one model ensuring its correct set-up. In other methods, static and dynamic tools are kept separate. The static tools are mostly used for planning and the dynamic tools primarily for program coding. For supporting state-of-the-art engineering, the two components need to be interlinked. POA specifies the connection between static and dynamic models and provides consistency, which enables to offer an integrated analysis tool.

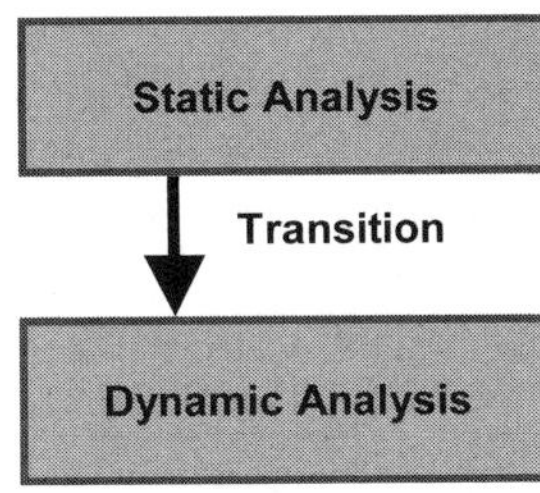

Dynamic view

The basic dynamic diagram is the *State Chart.* It introduces the behavioral view of the function of a production system. It specifies the sequence of states and under which condition the states are entered. To break down a complex system, a State Chart can be detailed hierarchically on different levels.

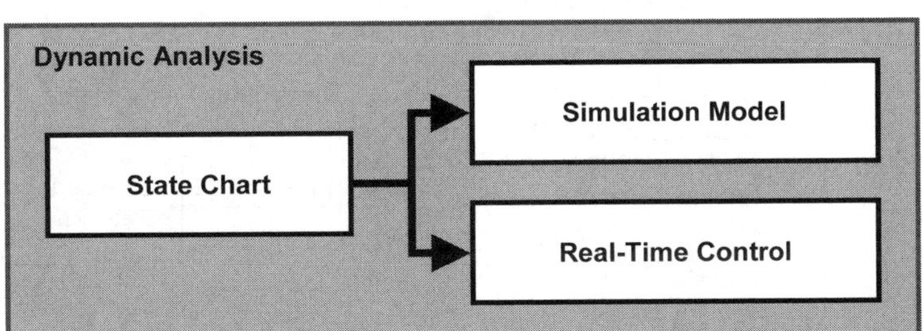

Figure I1-4: Dynamic analysis tools.

The dynamic analysis tool consists of the basic State Chart, the *Simulation Model,* and the *Real-time Control* (*Figure I1-4*). The State Chart can be directly put into code for a simulation program or a real-time control for a machine or a production line. By basing the State Chart on a Flow Diagram, the resulting code is easy to understand, well structured, and maintainable. POA explains the procedure to create computer program code from the State Chart step by step.

1.3 Static Analysis

1.3.1 System Specification

The first step of an analysis is carried out using a static model, the Flow Diagram, which shows a production system or a service enterprise, depicting it by flows and processes. Processes describe the activities of people and machinery and the transformation of material, resources, and data within the system. Flows connect processes, acting as interfaces between them. The purpose of the functional analysis with the Flow Diagram is to get to know the system to be investigated: to specify processes, organize the workflow, find gaps in the information transfer, and streamline the operation. A first optimization takes place by rearranging processes, diverting flows, reducing interfaces, and eliminating redundant processes and flows. The Flow Diagram allows to do all this while keeping the purpose of the system in mind. At the same time, it provides as many details as necessary to fully describe the system while still presenting an overview of it.

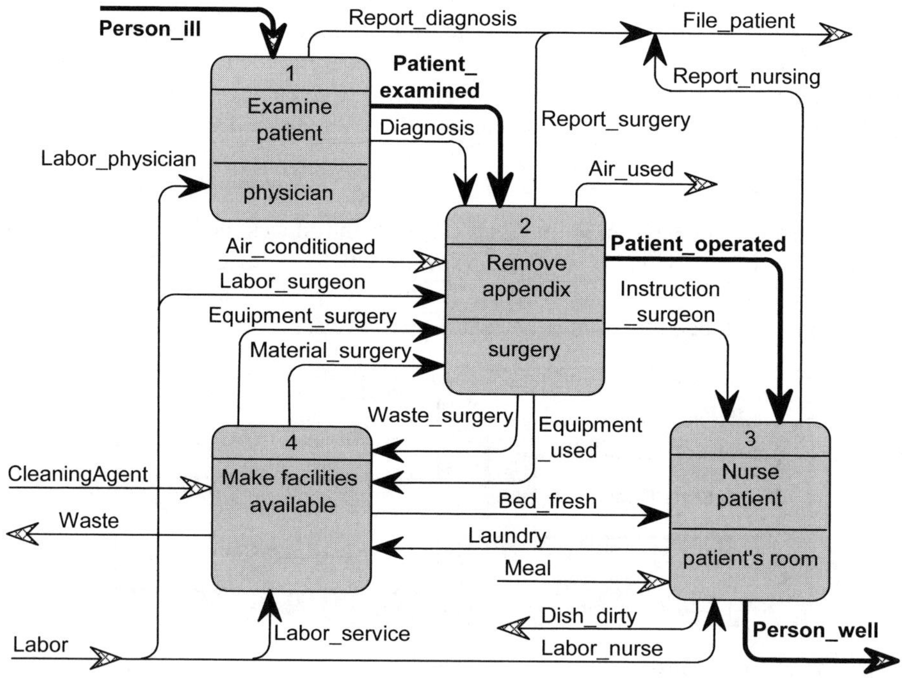

Diagram I1-1: Flow Diagram of surgery in a hospital.

Diagram I1-1 shows an appendix surgery in a hospital, depicted with four processes and several flows. In process 1, the patient is examined, while in process 2, the appendix is removed. In process 3, the patient recovers, and process 4 provides various services that the patient is not aware of, but that enable the function of the hospital.

1.3.2 Economical Analysis

The economical analysis of a system starts by drawing a Flow Diagram. By adding quantities, numbers, and monetary values, the Flow Diagram becomes a Value Flow Diagram. Values expressed in Currency Units (CU) are allocated within the processes and assigned to flows in order to calculate the value of the product. Financial values and costs are connected graphically with the corresponding flows or processes on the diagram. Along the path of the production process chain, the value added is calculated. The value calculation is carried out in order to set the sales price for a product, to evaluate an investment in new production machinery, or to calculate the impact of a change in production parameters. Non-value-adding activities can be identified and eliminated. The value of the product at each step in the production is known. This is helpful for make-or-buy decisions.

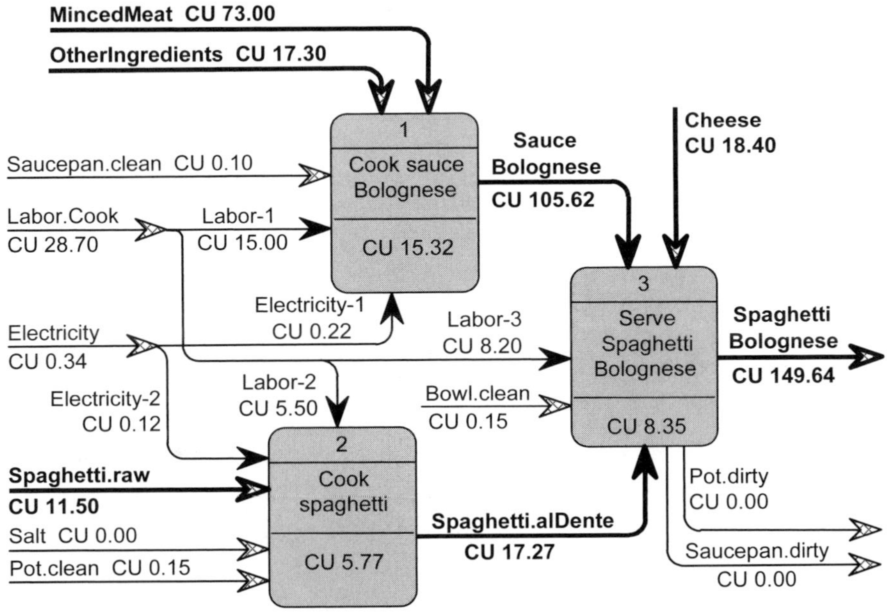

Diagram I1-2: Value Flow Diagram of a spaghetti party.

Diagram I1-2 shows the Value Flow Diagram for the preparation of "Spaghetti Bolognese" for a dinner with 40 guests. The diagram displays directly the costs that are associated with the production of the "Spaghetti al dente" and the "Sauce Bolognese." For the investigation using the Value Flow Diagram, the host's "labor," who cooks himself or herself, is introduced with a value in the diagram. The "Electricity" needed for the cooking is depicted. "Pot" and "Saucepan," and "Bowl" carry values that are incurred by the cleaning and maintenance processes and are charged to the cooking processes. After being used, "Pot" and "Pan" leave process "Serve Spaghetti Bolognese" with the value of 0.00 CU as soon as the spaghetti are transferred onto "Bowl.clean" for serving.

1.3.3 Ecological Analysis

Ecological analysis using the Resource Flow Diagram is based on the Flow Diagram, which is completed by resource flows and their values. The environmentally relevant values are calculated for the resource flows including mass, energy, and embodied energy calculation. Possibilities for recycling and reuse of waste heat and waste material are assessed by in-depth technical energy analysis. The energy consumption necessary for a product during its production and its lifetime is summed up as embodied energy. Calculation and optimization steps result in recommendations for improving the processing system.

The calculation and balancing of the resource flows and their values lead to a complete description of a production system concerning the consumption of resources and the production of waste and heat. At the same time, critical points in the production become visible, such as where resource loops should be closed and where potential for environmental optimization lies.

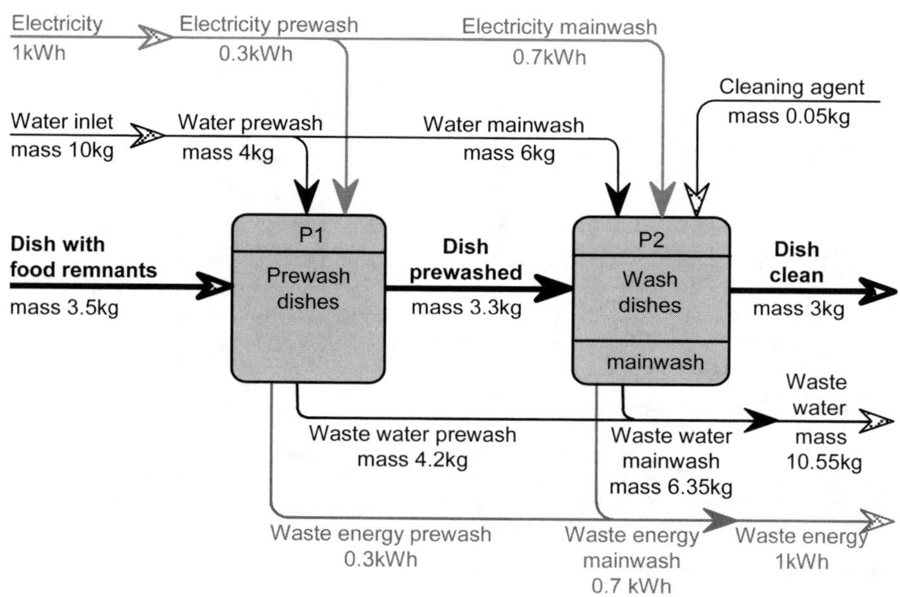

Diagram I1-3: Resource Flow Diagram "Do the dishes."

Diagram I1-3 shows a process with attached flows from a Resource Flow Diagram of a dishwasher. The dishwashing process is organized into the prewashing and the main washing course. Both processes need energy and water. The product flow "Dish with food remnants" changes in the course of the processes. The mass of the product flow decreases because food remnants are washed from the dishes. The food remnants together with the "Cleaning agent" end up in the "Waste water." The consumed "Electricity" leaves the system as "Waste energy" in form of heat.

1.4 Dynamic Analysis

1.4.1 System Behavior

The system behavior analysis is based on a time-dependent, dynamic view of the system, which in POA is depicted by a State Chart. A State Chart may be set up for any process of the Flow Diagram. The State Chart specifies the behavior of the system by defining states and transitions for a given process. It determines the behavior of a process or a system over time in order to model a system dynamically, assess the behavior of a real-time application, and serves as a basis for coding of real-time controls and simulation programs. The description of a system in time must be complete and consistent in order to make sure faultless functioning. This means not only to take into account normal operating states, but also the failure of components or an erroneous input by an operator, resulting in risk or damage.

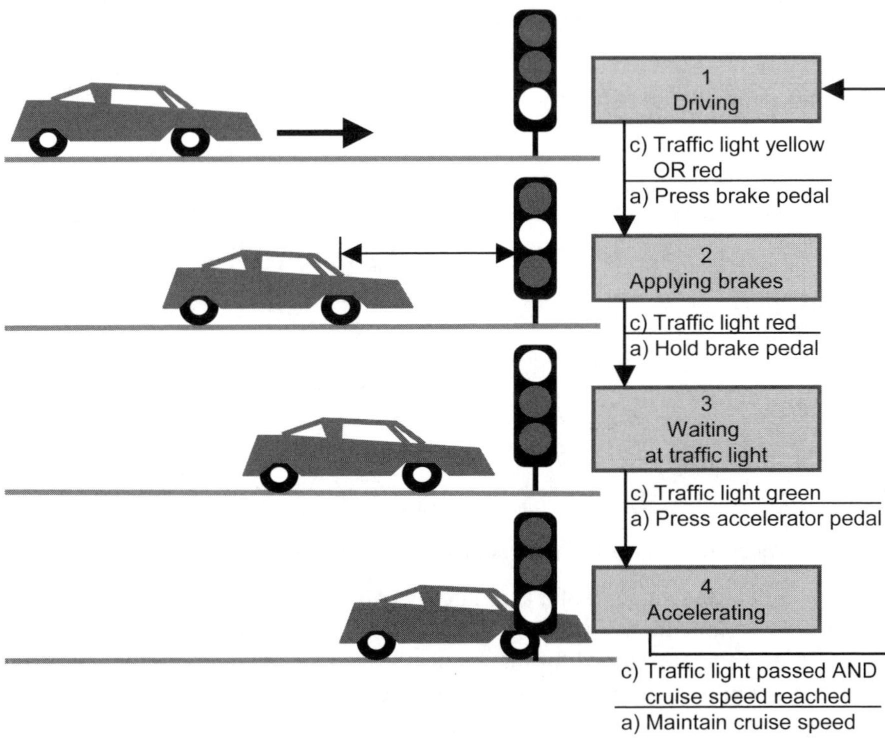

Diagram I1-4: State Chart of "Car approaches traffic light."

Diagram I1-4 depicts the State Chart of the process "Car approaches traffic light." Four states model the behavior of the car approaching the traffic light. Depending on the light that shows (green, yellow, or red), the car is in one of the four states. For the change of one state into another one, the transitions between the states are labeled with a condition (c) and the action (a) to be performed.

1.4.2 Process Simulation

POA supports the transition from the State Chart to design and coding of a computer program. Based on the states and the transitions documented in the State Chart, the program code determines the current System State, depending on the program parameters, and controls the transitions to other states. Resources consumed and products manufactured are calculated in discrete time steps and handed over to the following process by the pertaining flows. The program runs on the computer in exactly the same way as the process works in reality. It may simulate a single machine or an entire production. The simulation model allows to check the function and the performance of a whole system in order to optimize and improve.

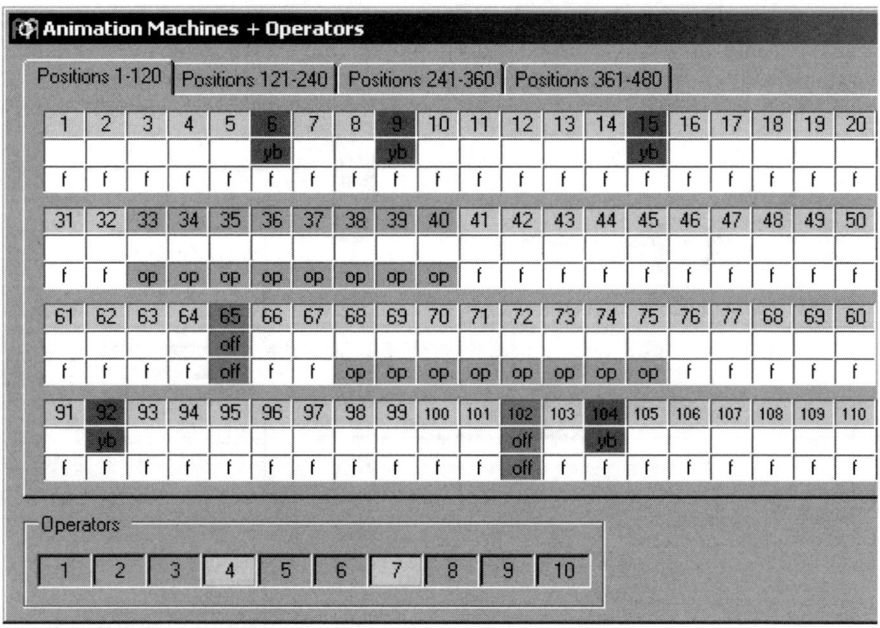

Figure I1-5: Window of a simulation program with animation of a plant.

Figure I1-5 depicts two parts of a user interface of a simulation model of a texturizing plant. This simulation was set up in order to check the function and design of a new machine being developed for this plant. Texturizing means to provide bulk to a synthetic yarn in order to make it elastic and friendly to the skin. *Figure I1-5* shows a window of an ongoing simulation. A simple animation depicts the machines and the operators in the plant by little boxes. These boxes change the color according to the state of a machine or an operator during production. Certain features of the new machine are dependent on the mode of operation, which has an influence on the machine design. Since the new machine is more than twice as fast as the old one, the work organization of the operators may have to be changed. So far, no plant exists with the new machine. To check the feasibility of the machine in actual operation, a simulation model of a texturizing plant was set up.

1.4.3 Machine and Process Control

Based on the Flow Diagram and the State Chart, POA supports the coding of a real-time control for machines or processes. By applying both diagram types of POA, the control software is structured according to the way the hardware components are arranged and named. Software code that is programmed based on the structure of the diagrams is easy to understand and maintain because it reflects the real-world process. With a real-time program, a display panel is used to operate the machine, where certain commands are given by pushing buttons, either on a keyboard or on a touchscreen.

The State Map is a means to determine the System State for a real-time system. In the State Map, states of different processes are drawn against each other to define the operational and the exceptional states of a system.

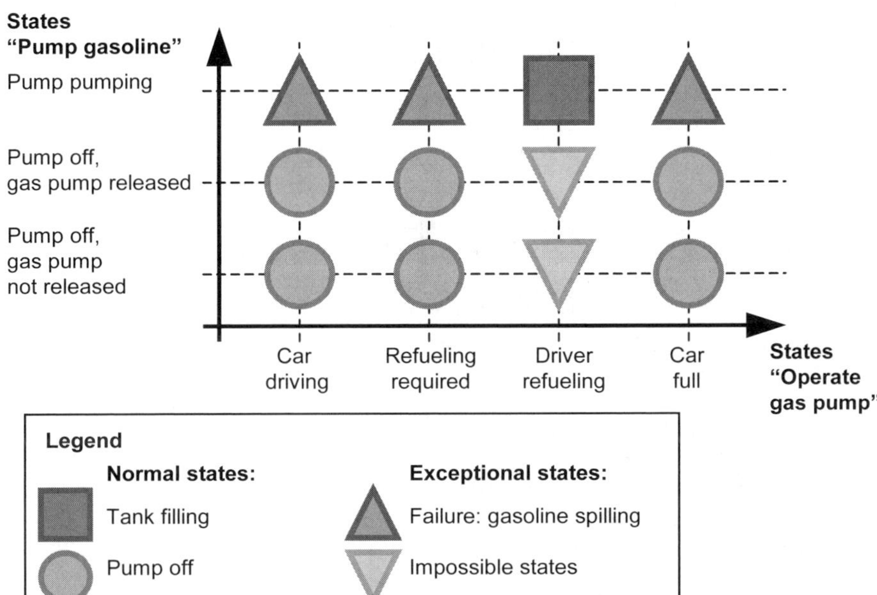

Diagram I1-5: State Map of a gas pump.

Diagram I1-5 shows a State Map of a gas pump. The states of the process "Pump gasoline" are depicted on the y-axis, and the ones of the process "Operate gas pump" on the x-axis. On the State Map, not only the desired System States for the operation of a gas pump can be determined, but also those that lead to failure of the system (e.g. pumping while the nozzle of the gas pump is not in the car tank) or which are even impossible (e.g. filling the tank without pumping). Logical or physical impossible states are an important issue since they point to failure of sensors or inputs to the control. The failure states must be taken into special consideration when programming a real-time control.

1.5 Setup of a Production Analysis

1.5.1 The Real World in a Model

The depiction of a real-world manufacturing system in a model is an abstraction and concentration of essential elements. *Figure I1-6* shows the way in which POA treats the abstraction. The real-world production has to be condensed and described in diagrams that consist of simple symbols with a name and a specification and that are understood by everybody. The extent to which a production can be described by diagrams depends strongly on the understanding of the production itself.

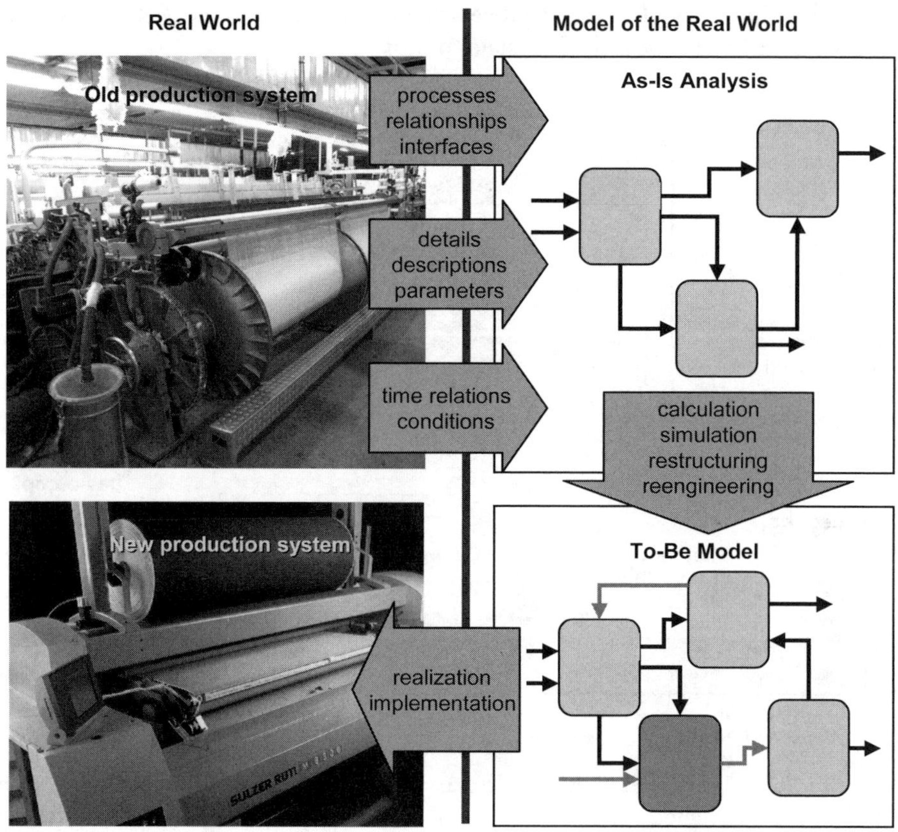

Figure I1-6: Real world and model.

A production system consists of machines, employees and their environment, as well as material, resources, and information that are processed and exchanged. POA translates such a system into a network of processes and flows, which are depicted in the static Flow Diagrams. The State Chart adds the description of the behavior of the system. These static and dynamic investigations start with the As-Is

Analysis, which represents the current system. In the model, alternative solutions and different values for parameters are investigated to gain information for optimization strategies and future scenarios for the To-Be Model, which sets the target to be realized in the real world. The new model is applied thereafter in the real world, and processes and interfaces are adapted step-by-step.

The data used for modeling, calculation, simulation, and reengineering is gained by measurement in the real system, by estimation, by reverse engineering, or by calculation using a generic production system. The calculation results provide a rating of the performance of the production system.

1.5.2 Model Definitions

System

The term "system" is widely used. Within the method of POA, a company, a plant (or parts of it), or a machine (or parts of it) is considered a system. Basically, systems are components of bigger systems and can always be divided into smaller parts. A system is a totality consisting of various elements, including the relations and interactions between these elements. The system is described by the activities of the elements. The dynamics of the relations enable the system to carry out functions, which none of the elements could perform on their own.

Definition: A system is the totality of different elements interacting with each other. Across its boundaries, the system is related to the outside world by the exchange of resources and information.

Model

Generally, a model is a representation of something already existing in reality, or planned to exist. Each description of a system is a model. This model should be easy to understand, so that the reader is hardly aware that he/she is looking at a reproduction of the system, and not at the system itself. The model is used to overview the system, analyze and predict system behavior, and investigate system variables and their impact on system behavior. A model always has a specific goal, and is set up for a specific purpose.

Definition: A model is a representation of a system, created to analyze and examine the function, behavior, and interactions of that system.

Element

The term "element" in this book refers to a single symbol on a diagram. A symbol, or element, can be a process, flow, or external entity in the static analysis toolbox. In the dynamic analysis toolbox, an element is a state or a transition.

1.5.3 Capture a System

As-Is Analysis: The first step in the analysis of an existing system is to capture the actual state. An "As-Is Analysis" is a representation of the system as it is at the present stage. The processes and flows are modeled, as the creator of the model perceives them at the moment of the analysis.

To-Be Model: By analysis and improvement, a better performing and more efficient stage of the production is modeled. A "To-Be Model" is a possible future representation of the system that may be achieved by an optimization step. These improvements can be conceived by reengineering, rearranging, calculation, simulation, or other methods. However, the To-Be Model may also be used as an independent, creative preliminary stage, for example for the design of a new production plant.

The setup of a diagram according to a real-world manufacturing plant can be achieved with different approaches:

Walk through: The chain of production processes is captured along the main material flow, from incoming material to the final product. Additional processes serving to supply resources or dispose of wastes, complete the model. The set-up of models for companies providing services (instead of a tangible good) is done the same way.

Walk around: The system and each process therein are looked at from every side to ascertain that all required input and output flows are introduced. In doing this, all different types of flows have to be considered: material flows, such as raw material or waste, as well as intangible flows, such as information, data, or signals.

Top-down: The analysis starts with the diagram on the top level. The top-down procedure is usually applied if the system is totally unknown and the first step is the setting of the system boundary. The next steps are then detailing and getting to know the system better.

Bottom-up: The analysis is usually done with projects that start on a lower level with a single machine or a machine function. The processes and flows located on the shop floor are drawn on a diagram. This diagram represents a detailed view and is later transferred to a more general view by bundling processes and flows on the next higher level in the model.

1.5.4 Procedure of Setting Up a Model

POA follows a well-defined and systematic workflow to support an efficient analysis. To set up a model, certain steps must be performed. The step-by-step procedure used to set up a model is shown in *Figure I1-7*. The steps of the modeling procedure are not done strictly one after the other from beginning to the end. There are iterative and repetitive loops and simultaneous work.

The modeling procedure begins with the goal definition. The project target states the extent of the study as well as the specific purpose of the project. Based on this target, the system boundaries are set, and the modeling phase begins. The modeling starts with the As-Is Analysis. The creation of the model and the collection of data are intertwined and are conducted concurrently. Data is collected and filled in as it becomes available. If data is not available from the employees of the company, it has to be gained by measuring or estimation. As stated in the previous chapter, the model may be built bottom-up, top-down, or by starting in between. The walk-through procedure ensures the consistency of the flows, while the walk-around procedure checks if all the inputs and outputs of the whole system and of the single processes are included.

As the data collection becomes more complete, the model is continually refined. Based on the diagrams, missing processes and flows, as well as their parameters and values, are identified. It is recommended to discuss the system based on the diagrams with the people involved on different levels in order to get a complete system description.

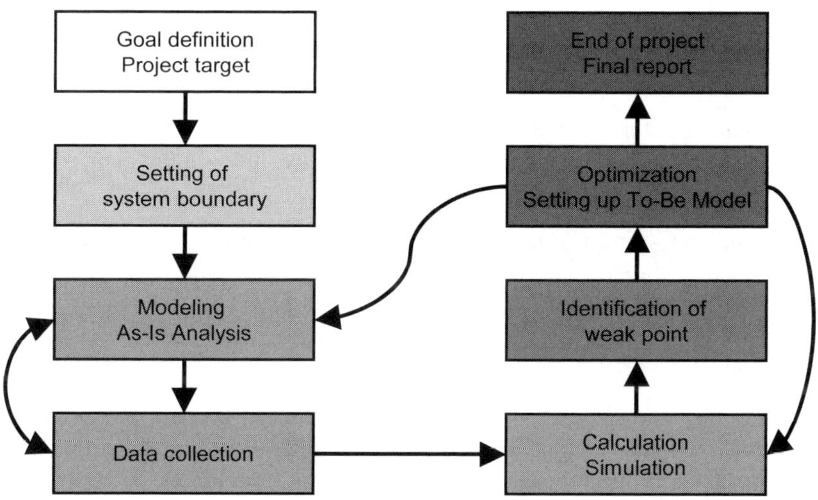

Figure I1-7: Setup of a model.

Calculation and simulation, based on the model and the data collection, give initial results and immediately allow for the identification of weak points and bottlenecks within the production chain, system malfunctions, excessive waste in production, and monetary losses. This uncovers points to optimize and opportunities for improvement in the investigated system.

The optimization requires planning, setting up, and checking of the new system model (To-Be Model). By changing parameters and input data, additional calculations and simulations help to determine the optimal system configuration. The setup of a model ends with the final report that forms a complete documentation of the model (including diagrams and specifications), optimizations, recommendations, improvements, and further steps to be done.

1.6 Projects Using POA Models

Types of projects

In this chapter, different kinds of projects are shortly outlined. To illustrate typical applications of POA and approaches to typical problems, case studies are given in this book. In the following, a short description of these case studies is presented to demonstrate the different kinds of projects carried out.

New system conception

In system design, a top-down approach is recommended according to the aspects of the system. In the case of a new system, processes should be grouped in order to minimize the number of flows. In manufacturing chains, processes should be defined and organized along the path of the product flow. Keeping in mind that flows specify interfaces between processes, a reasonable structure, typical for the specific production plant, should be used. Other flows, such as resources, material, and information, are directed so that the processes along the product flow are supplied optimally.

In the system structure, processes are specified and arranged so that the number of flows on the respective level is minimized. In reality, this is equivalent to avoiding unnecessary interfaces.

In Case Studies C4 and C5, new system concepts are created based on POA. Case Study C4 shows the conception of a new demagnetizing line for TV display tube masks. This process line was planned and realized based on the diagrams and calculations of POA. Also, the machine controls were programmed based on POA State Charts.

Case Study C5 introduces a completely new concept of a texturizing plant with a newly developed type of machine that doubles the speed of production. To obtain the best possible design for the new machine, a simulation model was programmed to determine the workflow and handling of the machine by the operators. Alternatives of automatic or manual operation on different jobs were investigated in order to establish priorities for the subsequent development of automation.

Engineering update

If the objective is to create an engineering update of an existing manufacturing line or plant, a different approach is employed. In this case, the existing structure of processes is broken down from the existing plant and examined in detail (As-Is Analysis). The primary task is to establish the degree of freedom available for alternative solutions. This is not a strictly top-down approach, since the constraints can occur on any level of detail and within any process. Working within a maze of constraints calls for a systematic method with a balanced assessment and comparison of the different solutions. The system model starts with the shop floor processes and is assembled in a bottom-up approach. Finally, the gaps are filled with a top-down approach.

The model of the present system state is changed in the course of the analysis and optimization. By combining processes, changes in the structure and possibilities to reduce interfaces are investigated. This way, a model of the desired system state (To-Be Model) emerges and is implemented step-by-step.

Case Study C6 demonstrates an engineering update based on POA. It introduces the concept for updating an over 50 years old underground cable car system in order to make it useful for the future. The technology controlling the cable car has become obsolete and did no longer comply with regulations. The new machine control allowing automatic, state-of-the-art remote operation was programmed based on POA State Charts, and tested for usability by persons involved in operation before deciding on the final implementation.

Reverse engineering

In the previously mentioned cases, POA was used to support the creative process of system conception with a systematic check for completeness and consistency. In the case of reverse engineering of an existing system, the purpose is to make a complex system easily understandable. Many manufacturing process chains or plants are poorly documented and are kept in operation by relying on the knowledge of a limited number of people within the organization. In this case, it is again preferable to adopt a bottom-up approach, i.e. starting the analysis on the shop floor with a walk along the path of the product. By extending into reworking, and waste-removal, the picture is completed. In addition to the resources, the data transfer to and from the processes is also noted.

Case Studies C2, C3, and C4 show reverse engineering of running productions based on POA. Reverse engineering begins with an analysis of the production line in operation on the example of a typical product or product mix. The pertaining production is modeled as a virtual production facility, using as much information about the original line as possible.

Case Study C2 shows a detailed economic analysis of a woven product. The Value Flow Diagram depicts the processes and flows on several levels of detail to show hidden costs in the system. Values and financial resources are calculated and compared to the real costs.

Case Study C3 conducts an energetic analysis of a croissant production line with the Resource Flow Diagram. Processes and flows are mapped, and energy consumption is calculated. The calculated process efficiency points to processes with a poor efficiency in energy throughput and indicates optimization potential. The ultimate target is saving energy costs.

1.7 Organization of the Book

The six diagrams and methods of POA are organized as a toolbox with static and dynamic analysis tools. The book is structured according to this POA toolbox. As a whole, the book contains four main Parts: introduction, static analysis tools, dynamic analysis tools, and case studies. These Parts are marked with the letters I, S, D and C. Each Part is divided further into Sections.

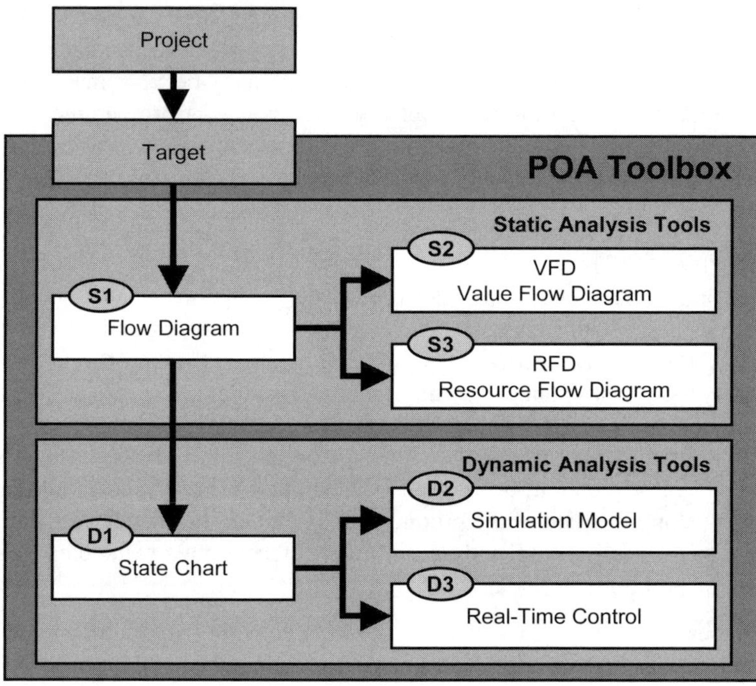

Figure I1-8: POA tools.

The arrows on the POA toolbox in *Figure I1-8* indicate the path of applying the method and diagrams, and guide the reader through the book. Principally, Parts S and D are independent; the reader may read Part D without having read Part S. However when doing this, he/she must have knowledge of the Flow Diagram.

When doing a project, it is strongly recommended to apply POA by following the path indicated by the arrow and beginning with the Flow Diagram in Section S1. That means, to first carry out the static analysis of the system, and then conduct the transition to the dynamic diagram. This transition from the static to the dynamic view is one of the strengths of POA.

Within Parts S and D, the first Sections (S1 and D1) describe the basic diagrams for the static and dynamic models along with the drawing and modeling rules. Following the arrows of the toolbox within Parts S and D shows that Sections 2 and 3 are based on Section 1. Sections 2 and 3 are independent from each other and can be read separately. At the end of each Section, an application example shows how to apply the particular tool. A common application example is used throughout the whole book: a gas station.

Part C is composed of six case studies where the methods of POA introduced in Part S and D are applied. It is organized according to the POA toolbox: there is one case study for each tool in the POA toolbox, as shown in *Figure I1-9*.

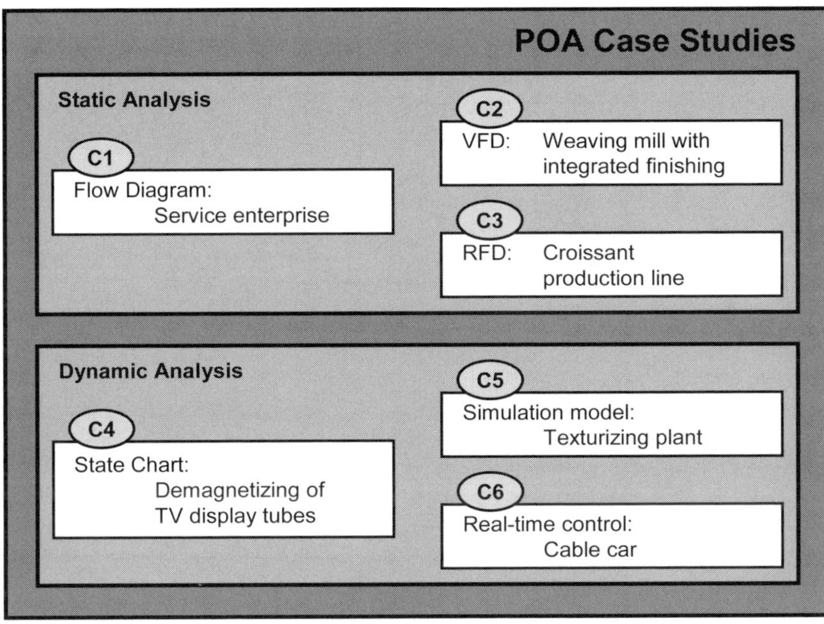

Figure I1-9: Case studies for the POA tools.

Different expressions are in use to describe the Flow Diagrams, the State Charts, and their elements. Throughout this book, consistency is maintained by using one unique expression for each object. Synonyms are found in the glossary.

I2 DELIMITATION OF PROCESS ORIENTED ANALYSIS

Goals of the Chapter

- Become familiarized with the roots of POA and the diagrams of Structured Analysis.
- Know the goals of and the differences between Process Oriented Analysis and Unified Modeling Language.
- Delimit POA to other modelling methods.

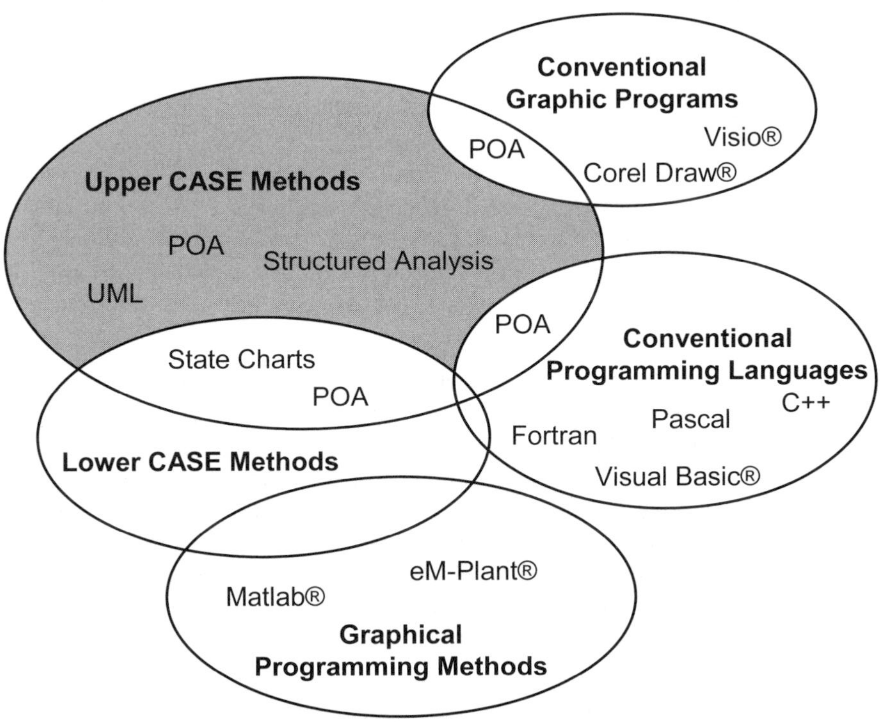

Figure I2-1: Connection of graphical modeling methods.

2.1 Introduction

In the search for a common schematic description of processes and their behavior in complex systems, many methods of software analysis have been proposed in the past. Some of them are presented in this chapter and compared to the method of Process Oriented Analysis (POA).

A well-known method for software engineering is Structured Analysis. As a first generation tool, System Analysis was introduced in the 1970's. It models a system by using three diagrams from different viewpoints: Data Flow Diagram, Entity Relationship Diagram, and State Transition Diagram. Structured Analysis has some strong features, of which POA takes advantage. The basic diagrams of POA emerge from the diagrams of Structured Analysis. POA, in comparison, works with two diagram types: Flow Diagram and State Chart. These are similar to the diagrams of Structured Analysis. However, POA is not limited to data and information modeling but also includes material and resources in the analysis.

A few years later, CASE tools (Computer Aided Software Engineering) were created to support the development of machine controls in logical and time-critical systems. Upper CASE methods are used to analyze and design software. POA counts among these methods and completes this task by offering a tool for integral system analysis and design. In contrast, Lower CASE tools support the step from State Chart to program code by automatically generating source code.

In the endeavor to compile and reunite various object-oriented methodologies for software analysis and design, Unified Modeling Language (UML) was created. UML consists of a range of charts and symbols to represent object-oriented and state-oriented aspects of systems. In order to obtain this universal applicability, the strict axiomatic foundation was sacrificed. There is no defined relation between the various diagram types.

In contrast to the other modeling methods, POA offers a condensed and complete toolbox in the process and time domain. Consistent rules apply for the different models and for the transition from one model to the other. POA is a comprehensive method, which improves communication and supports optimization of productive systems by integrating graphical representation, calculation, and programming methods. The focus of POA is the clearly defined relationship between the two kinds of diagrams Flow Diagram and State Chart, and between diagrams and program code. Thereby, the gaps between visualization, conception, design, and optimization of modern manufacturing systems are closed.

The efficient application of the method of POA relies on computer support. Various commercial tools, the so-called CASE tools, are available, supporting the drawing of the diagrams and supervising the modeling on different levels of detail organized hierarchically.

2.2 Upper and Lower CASE

Computer Aided Software Engineering (CASE) is a programming method that uses computer-generated diagrams and symbols to help systems analysts develop and maintain information systems. System developers distinguish two CASE categories: *Upper CASE tools* and *Lower CASE tools*. Upper CASE tools support the modelling by diagrams with integrated specifications and produce an overview of the information system focussed on its structure. Lower CASE tools make the development and programming process more efficient by automatically generating source code based on low level, detailed diagrams.

The well-known Upper CASE methods are based on Structured Analysis or other object-oriented system representations, UML for example. These methods concentrate on the design and documentation of software programs. Within the Structured Analysis, the modelling by these methods is limited to data and data handling. The advantages of object-oriented methods are used to split up the system into classes, objects, and methods to make possible the inheritance of object specifications.

However, a comprehensive system analysis should start by modelling processing and transfer material and information concurrently. This is done by Upper CASE tools. Upper CASE methods serve generally as analysis tools. They are a tool for the conception and documentation of a system, completely independent of the programming language. They take over the structure, functions, and relationships of the real-world system. Thereby, the application of Upper CASE starts already with the conception of the real-world system, in a typical top-down approach. POA is such an Upper CASE method.

Lower CASE methods, in contrast, focus on the programming job. They offer graphical methods for programming, which help the programmer to automatically generate source code and relieve him from entering line after line of code. They convert graphical representations in form of diagrams into source code text. Higher abstraction levels are not supported. A typical example for a Lower CASE tool is a program running on a PC that is connected to laboratory measuring equipment, using an assembly of icons to build up a measuring chain from sensor to display. A similar programming technique is popular for embedded systems of limited complexity, as for example Programmable Logic Controllers (PLC). Lower CASE methods are not suitable for large and complex systems, as unstructured graphics tend to develop into a maze. In such cases, conventional source code, based on an Upper CASE method for documentation, is easier to develop.

Analysis with POA works with either Upper or Lower CASE methods and may lead to both kind of programming. Independent process controls, which were coded based on a Lower CASE tool, may be integrated into larger systems by POA. As shown in Sections D2 and D3 of this book, even direct programming based on the State Chart is possible. The resulting source code is functionally and conceptually embedded in the real system, and therefore easy to understand and maintain.

2.3 Structured Analysis

2.3.1 Method Description

Structured Analysis (SA) was originally designed in the United States as a software development method for the handling of complex data systems. It evolved in the 1970s. It consists of static and dynamic diagrams for system representation and data organization as shown in *Figure I2-2*. Chris Gane and Trish Sarson [Gane, 1977], Tom DeMarco [DeMarco, 1979], Edward Yourdon [Yourdon, 1988], Keith Edwards [Edwards, 1993], and others have published material on Structured Analysis theory.

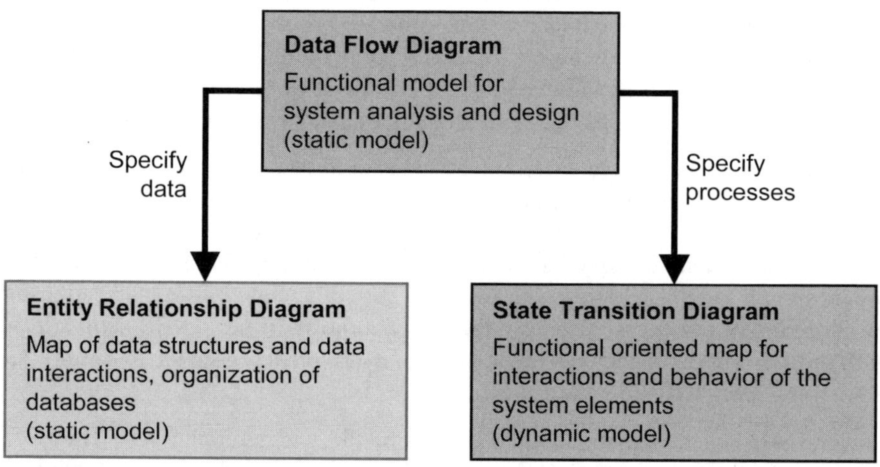

Figure I2-2: Purpose of SA diagram types.

Structured Analysis comprises three different diagram types:

- The *Data Flow Diagram* represents the system and its structure in a static view. It is composed of processes, flows, and stores. During analysis, data and process modeling are used to show how system processes transform input data into output data. It may be structured on several levels of abstraction.
- The *Entity Relationship Diagram* specifies and organizes the data of a system that is modeled in the Data Flow Diagram and serves as a basis for the building of databases.
- The *State Transition Diagram* is used for dynamic modeling. It describes a process by defining its possible states, as well as actions and conditions for transitions from state to state.

The most important features of Structured Analysis are listed in *Table I2-1*. The basics for the Flow Diagram and the State Chart of POA have been derived from

SA to take advantage of the strong features of SA, such as consistency, scalability, and simplicity. However, POA is developed further for system engineering.

Table I2-1: Advantages of Structured Analysis

Feature	Comment
Easy to understand	SA creates a simplified picture of a system, which helps in the understanding of complex systems by revealing their functional structure on several levels of detail, and shows the operational relationships between the elements.
Low user threshold	SA is self-explanatory and improves communication; its graphical nature reveals the interactions between the system elements.
Large system capacity	The hierarchic levels of the system allow the specification of the system in the desired level of detail; quick zoom and pan of the schematic drawings facilitate the overview of the system.
Enforce order, discipline, and consistency	A common object database defines and specifies the elements of the system; automatic rule check and level balance help to keep the database consistent.
Top-down and bottom-up	SA allows working top down, starting with the overall system and developing the detailed processes underneath, or working bottom-up, starting with a part of a production process, proceeding step by step up to an overview of the whole production system.

2.3.2 Delimitation POA to SA

SA and POA apply a similar static model to describe the system from a structural viewpoint. Both methods structure their diagrams and apply level balancing rules. The static models of POA are mainly focused on resource, cost, and information, and therefore do not apply the element data store, as used in the Data Flow Diagram of SA. The storage of material consumes resources, money, and time, and is therefore treated as a process within POA. Regarding the physical real-world implementation, this applies for data storage, too.

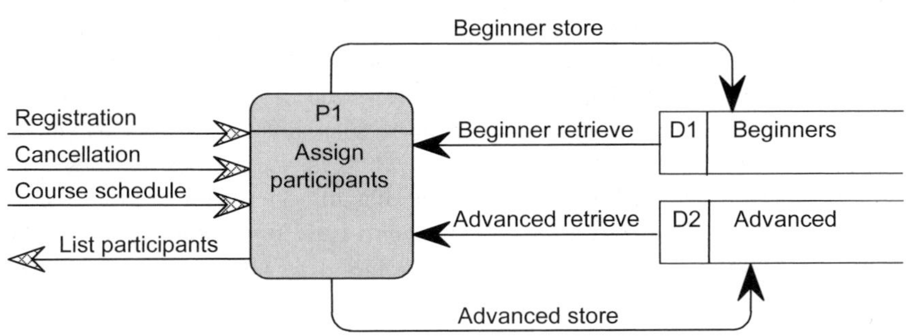

Diagram I2-1: Data Flow Diagram of a course registration.

Diagram I2-1 shows the Data Flow Diagram from SA, using the example of a course registration and the assigning of participants into two groups: beginners and advanced participants. The two data stores D1 and D2 consist of single data sets of the course participants and represent part of the database.

SA is mainly applied today for data modeling and structuring; it defines the structure of databases by Entity Relationship Diagrams. POA does not contain a data organization tool. On one hand, there are already sufficient methods on the market in this field. On the other hand, the structure of processes, flows, and products is imposed by the physical reality in real-world production systems.

Table I2-2: Delimitation POA and SA

	SA	**POA**
Purpose	Software engineering	System engineering
Static modeling	Data Flow Diagram Entity Relationship Diagram	Flow Diagram Value Flow Diagram Resource Flow Diagram
Elements	Flow Process External entity Store	Flow Process External entity
Data and resource modeling	Mainly data modeling	Resource and data modeling
Hierarchical structure	Yes	Yes
Dynamic modeling	State Transition Diagram	State Chart
Connection between static and dynamic models	Two independent models	Consistent relationship between static and dynamic model
Link between modeling and coding	Different concepts for implementation of charts into code	Step-by-step procedure from State Chart to real-time control or simulation program code

Table I2-2 compares SA and POA. Both comprise a dynamic model. In SA, it is called State Transition Diagram. In POA, it is the State Chart. POA goes further than SA by defining a relationship between static and dynamic model with a set of rules. The Resource Flow Diagram and Value Flow Diagram of POA enrich the analysis by including an ecological and economical evaluation. They are extensions of the Flow Diagram used for applications where a special set of rules for the handling of processes and flows applies. The dynamic view is treated by the State Chart, which is closely linked to the generation of code for a simulation model or a real-time program code. POA supports coding based on the State Chart by a step-by-step procedure.

2.4 Unified Modeling Language UML

2.4.1 Method Description

To introduce the Unified Modeling Language (UML), a short summary of the history is given here. Booch, Rumbaugh, and Jacobson introduced the Unified Modeling Language in 1996 with the intention of creating a standardized object-oriented language and modeling method. To achieve this goal, the authors of Object Modeling Technique (OMT), Object Oriented Analysis and Design (OOAD), and Object Oriented Software Engineering (OOSE) united their methods and created a common language. In November 1997, the Object Management Group (OMG) accepted UML as the Standard Object Oriented Modeling Language [Booch, 1998].

UML is a collection of different methods and associated diagrams whose common feature is the object orientation. A single integrated method could not be created due to significant differences between the various original approaches. UML specification consists of a choice of models or diagrams that treat different aspects and overlap. The viewpoints with the pertaining diagrams are listed in the following:

- Static and structural view: Class Diagram
- Behavioral view: State Diagram
- Activity view: Activity Diagram
- Interaction view in time: Sequence Diagram
- Interaction view in system: Collaboration Diagram
- User view: Use Case Diagram
- Physical view: Component Diagram
- Computer hardware/software view: Deployment Diagram

The *Class Diagram* of UML specifies the system's perspective and shows the structural composition of the system. A class is a defined unit that describes a group of objects with common properties. Class Diagrams describe static relationships between classes. These relationships can be defined as follows: generalization/specification, association, aggregation, or dependency.

State Diagrams model possible states of a class or a partial system and the dynamics of system behavior as reactions to external actions. The State Diagram is used as a State Chart in POA in a similar manner.

Activity Diagrams describe procedures in a system by using activities, control flows, and object states. Control flows depict parallel actions (AND) and excluding actions (OR). Activity Diagrams model transmission and reception of signals.

Sequence and Collaboration Diagrams describe partially dynamic aspects of a system by showing specific objects and the succession of interactions between objects over time. The Sequence Diagram concentrates on the time scale, while the Collaboration Diagram concentrates on object interactions.

A *Use Case Diagram* defines a sequence of system actions, initiated by an external actor, providing reactions to the actor or to several actors. It may model different use case scenarios. It specifies interactions between the user and the system, between components in processes, or between components of the system. In POA, this diagram corresponds to the context diagram, which is the top-level diagram of the Flow Diagram. *Diagram I2-2* shows the Use Case Diagram of an environmental evaluation program module called ecoindex that checks the environmental effects of a given substance.

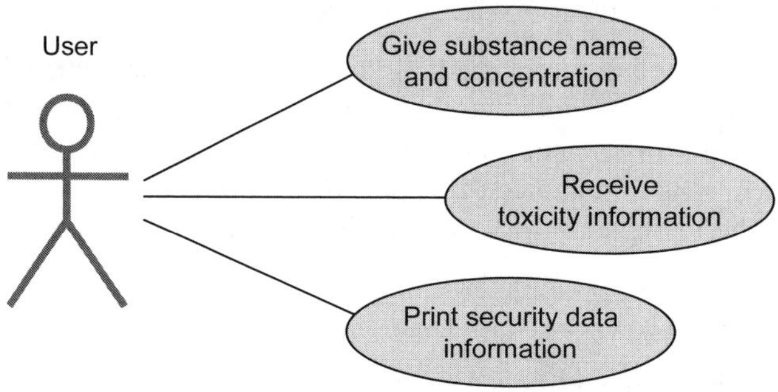

Diagram I2-2: Use Case Diagram of an environmental evaluation program module.

Component Diagrams describe components, modules with their own identity and systems interactions, and their corresponding dependencies. *Deployment Diagrams* model the structure of the hardware and the deployment of software on these hardware components.

2.4.2 Delimitation POA to UML

UML groups its diagram types according to the viewpoints they provide on the system, e.g. behavioral view or functional view. UML offers a variety of diagrams and perspectives but the positioning of theses diagrams is not clearly defined. POA diagrams are assigned to the static and dynamic behavior of a system, while keeping and maintaining a common database for all diagrams. POA uses just two types of diagrams, those being dynamic and static, allowing two fundamental viewpoints of a system. Most UML users apply only two or three diagram types.

Table I2-3 shows the assignment of POA diagrams and diagrams used by UML to a structural (static) view and a behavioral (dynamic) view. The requirements view, represented by its own diagram type in UML (Use Case Diagram), is incorporated into the uppermost level Flow Diagram, the context diagram of POA. Using POA, the process specifications define the requirements both for the general system and for specific cases.

Table I2-3: Dynamic and static diagram types of POA and UML

	UML Diagrams	**POA Diagrams**
Requirements view	Use Case Diagram	Flow Diagram (context diagram)
Structural view	Class Diagram Component and Deployment Diagram	Flow Diagram
Behavioral view	State Diagram Activity Diagram Sequence Diagram	State Chart

The individual diagrams partially correspond to each other. The UML State Diagram corresponds roughly to the State Chart of POA, but it follows different rules. Transitions starting and ending in the same state are allowed in UML, whereas in POA, this is omitted. In UML, there is no equivalent diagram type to the POA Flow Diagram. Therefore, UML does not support the functional, static view of a system, which is required to specify and explain a manufacturing system and introduce a common language for communication. Some UML-users introduce the Data Flow Diagram of Structured Analysis in this respect.

Some other differences between UML and POA are listed in *Table I2-4*. UML is applied specifically for software development, POA however for system engineering and optimization. The diagram types of UML do not have a connection, whereas the POA diagrams are connected by consistent rules and a structural link is given to programming of simulation models and real-time control

Table I2-4: Comparison of POA and UML methodology

Method / **Aspect**	**UML**	**POA**
Purpose	Software engineering	System engineering
Sequence of use of diagrams	Not defined.	Clearly defined.
Connection between diagrams	Connections between diagrams are not fully defined.	Connections between individual diagrams are defined by strict rules.
Coding	Programming is done based on several diagram types. No consistency between diagrams and code.	Structural link is provided through programming based on the State Chart.

The basis of the POA diagrams are rules for representation and balancing, which enable an exact calculation and are checked systematically. Experience shows that the two diagrams of POA with only five types of elements allow a commonly understandable and integral specification of any kind of production and service system.

2.5 Computer Support

2.5.1 CASE Tools

Several commercial software programs, so-called CASE tools, support the drawing of the diagrams as used for the method of Process Oriented Analysis. Beside the drawing of the diagrams itself, they help with the build-up of the model, hierarchical structuring including the numbering, set up and update of the data dictionary, and the consistency check of the model. CASE tools facilitate building and changing of models, creation of alternatives and variants within a model, and adding of subsystems and elements. The update of project documentation can be performed very efficiently.

The drawing of diagrams and the organization of the model with the database may also be done by hand or with a common software program for text editing and drawing if a CASE tool is not available. *Table I2-5* shows the elaboration of a POA model with different tools.

Table I2-5: Tools for the setup of a POA model

Computer support / **Model components**	**Without computer support**	**Common software for text, tables and drawings**	**CASE tool**
Diagram	Drawing by hand	Graphical software	Special software with pre-configured elements
Hierarchical structure	Process number by hand	Process number by hand	Automatic assignment of process number, automatic link within the hierarchy
Element specification	Written by hand	Tables with text-editing software	Specification window
Database, data dictionary	List by hand	Tables with text-editing, spread-sheet	Automatically built and updated
Rule and consistency check	By hand	By hand	Integral check
Report	Written by hand	Text-editing software	Selection of automatically generated reports

Several CASE tools for various purposes are commercially available providing a range of different diagram types. *Table I2-6* lists criteria in form of a checklist that may help to select the CASE tool suited best for the project or company. The diagrams in this book are mainly drawn with System Architect®. While this book is in preparation, POAdesigner is in development to support specifically the POA method.

Table I2-6: Criteria and checklist for CASE tool

Criterion		Requirement
Diagram types		Flow Diagram (equals Data Flow Diagram)
		State Chart (equals State Transition Diagram)
Notation of diagram symbols		There are different notations for the symbols: Rectangle with rounded corners for process (Gane & Sarson)
		Circle for process and state (Yourdon/DeMarco)
		Rectangle for state (Ward & Mellor)
Additional features for diagram and symbols		Split and Merge
		Optional third field for additional information in the process symbol
		Highlighting of loose ends of flows and transitions
		Formatting of line style and text
		Handling of changing of flow and transition paths
		Flow classification
		Automatic numbering of processes and states
		Automatic numbering of the elements in the hierarchy of the model
		Separate fields for condition and action for transitions
		Detailing of a process into a State Chart
Automated rule check for balance and consistency		Balancing rule for check in the hierarchy
		Dangling flows
		Duplicate flow names
		Processes and states without input or output
Element specification		Fields available
		User-defined changes of element specification
Report generation		Automated report generation available
		User-defined adaptations
		Choice of integrated parts
		Export into text-editing software
Program useful for networks		Possibilities for co-workers to work on one project at the same time

The Flow Diagrams in this book are displayed in the notation of Gane and Sarson [Gane, 1977], which uses a rectangle with rounded corners for processes. The State Charts are drawn according to the notation of Ward and Mellor [Ward, 1985] with rectangles.

Note! In the diagrams drawn with System Architect®, the arrowheads change their appearance while drawing. The arrowhead becomes filled, as soon as a flow or a state transition is connected on both ends to processes, external entities, or states. A dashed arrowhead indicates that the flow or the transition is only connected at one end. An arrow with a white arrowhead is not connected at all to an element and should not appear on a diagram, since it is not allowed.

2.5.2 Programming

Sections D2 and D3 focus on the programming of simulation models and real-time control programs. The programming may be done with any commercially available programming language. The examples in this book are programmed with Visual Basic .NET. Visual Basic has been chosen because the code is generally understandable and quickly acquired, even with a minimum of previous experience in programming.

This book does not explain programming fundamentals. Readers of Sections D2 and D3 should have basic knowledge in programming. POA shows how the programming may be modelled in order to apply it easily based on the given diagrams. A comprehensive, maintainable code is generated based on the structure of the State Chart introduced in Section D1. POA completes the program documentation with the diagrams and the database.

S STATIC ANALYSIS TOOLS

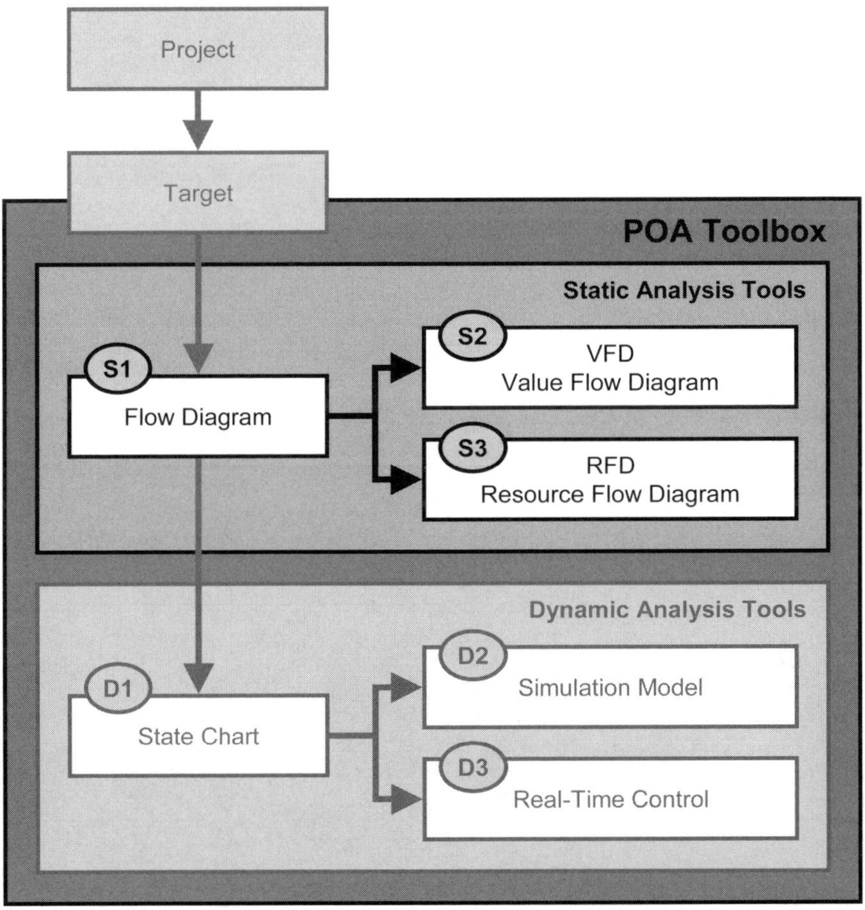

S1 FLOW DIAGRAM

Goals of the Chapter

- Learn to analyze a system graphically using the Flow Diagram.
- Become familiar with the symbols and drawing rules for Flow Diagrams.
- Define a system, set its boundaries, and describe its components.
- Build up the hierarchical structure of a model.
- Balance a set of diagrams.

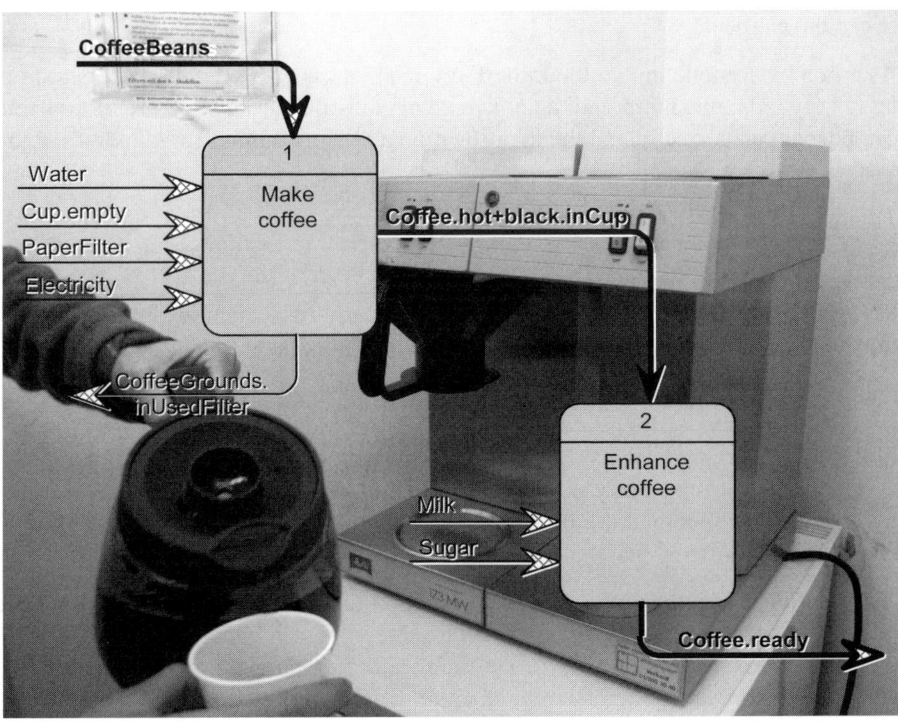

Figure S1-1: Flow Diagram "Make coffee" (Photo: ETH Zurich).

1.1 Introduction

The Flow Diagram is the basic type of diagram used by the static analysis tools of Process Oriented Analysis. Any kind of system can be depicted by a Flow Diagram. It creates a simplified schematic view of a system. The Flow Diagram graphically shows where resources and data come from, where they are going, how output is generated by transformation of input, and what relationships exist between the activities.

A Flow Diagram is a simple, yet powerful tool to structure and order a complex system. Since it is scalable any time into more general or more detailed views, any point within the system may be chosen as the starting point to draw diagrams for a model. A set of diagrams is depicted and built up. It will finally become a picture of the real system and represent a part of the real world, thus looking as similar as possible.

The Flow Diagram helps understand complex production systems by revealing, step-by-step, their underlying structure and the relationships and interactions between the elements.

As a graphical modeling method, the Flow Diagram is easy to grasp. A diagram is less likely to be misinterpreted than its textual equivalent. The structure of a system can be mapped correctly and the identification and elimination of weaknesses in the system setup are easier.

The main focus of the Flow Diagram is on interface-oriented thinking. The flows represent interfaces, relations, and connections. They guide the analysis, hierarchical decomposition, and specification of a system. The interfaces are where most of the system's problems lie. They are a primary cause of difficulties in communication and loss of time within any material and data processing systems.

Starting with the Flow Diagram, it is possible to proceed to cost or energy calculations within POA, as described in Sections S2 and S3. Thereby, the Flow Diagram is the starting point of the analysis and the flows and processes are investigated and calculated concerning costs and energy. One can also continue in Part D with the dynamic analysis tools of the POA toolbox to develop a software code for real-time control or a simulation model.

This Section S1 explains the rules for drawing the Flow Diagram in detail. At the end of the section, there is a chapter with the components of a model and recommendations for the drawing of the diagrams and the optimization of a model. The last chapter contains an application example that models step by step a gas station with a Flow Diagram. Case Study C1 illustrates the analysis of a service enterprise with the example of a bar. The focus of this Case Study is to analyze and model a service as a product flow in a system.

1.2 Flow Diagram: Why?

1.2.1 Purpose

System function: The Flow Diagram of POA is a tool for depicting a system. It is a logical representation, modeling *what* a system does, rather than a physical model showing *where* and *how* it does it. This fact is important for the analysis of a system. The Flow Diagram assists with the analysis and restructuring of existing systems, as well as with the planning and development of modifications, and the conception of new systems.

Communication: Diagrams are a very important means of communication because everybody understands them. The Flow Diagram takes advantage of the descriptive power of pictures instead of words. It does not need further explanations when reading it. The notation is obvious, uncomplicated, and self-explanatory. The concept of processes and flows is immediately understood.

Interface and definition: Within a system, the interfaces and missing definitions cause problems in communication. This can be man-man, man-machine, or machine-machine communication. The Flow Diagram forces the model's designer to use clear, unique names and definitions for the components and interfaces of the system. Consequently, a Flow Diagram model is also a complete documentation of a system.

Completeness: A vast amount of flows are needed to make the activities of a company possible. Many different types of flows exist. With the Flow Diagram, it is principally possible to depict all types of flows, and therefore to show the complexity of a system with all its details.

Hierarchical decomposition: The Flow Diagram model is able to structure the system hierarchically in different levels of detail. This helps to decompose a big, complex system into small, manageable components that are easier to understand. Each process in a diagram may be detailed into child diagrams with subprocesses and subflows.

This hierarchical procedure allows either a top-down or a bottom-up execution of an analysis. Using the top-down analysis approach, the problem is outlined going only into as much depth as is required at the time, but can be further detailed at any time. With the bottom-up analysis, different subsystems are assembled to create a larger system.

System boundary: The diagram on the uppermost level in the model hierarchy clearly defines the system boundary. It defines the elements that belong to the investigated system and those that are part of the external environment. The top-level diagram also shows the interactions or the interfaces of the analyzed system with the external environment.

1.2.2 Application

The Flow Diagram consists of only a few elements: process, flow, and external entity. Using these elements, any type of system can be depicted in a Flow Diagram: Manufacturing a product and providing a serve, for example a car or food manufacturer, a hospital, a bank, a restaurant, or an insurance company. A system can be a whole company or only a part of it, e.g. a department or the functions performed by an employee. It can be a production line, such as a whole pasta line, a single machine, such as a drilling machine, a TV, or a car, or only a part of a machine, e.g. the engine or the control of it.

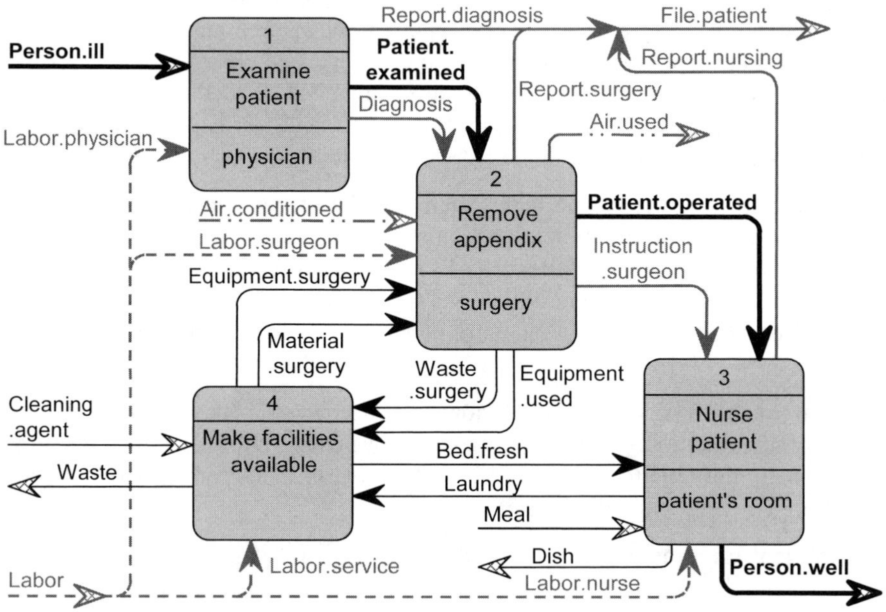

Diagram S1-1: Flow Diagram of surgery in a hospital.

Diagram S1-1 shows the Flow Diagram of a hospital. The patient "Person.ill" enters the hospital and is examined. After the physician diagnoses appendicitis, an operation is scheduled and carried out to remove the appendix. Then, the patient is kept several days for surveillance and recovery before being released. The patient leaves the hospital recovered, shown as the flow "Person.well." In addition to the processes regarding the direct treatment of the patient, other processes are necessary to keep up the operation of the hospital, for example "Make facilities available." This process supplies the departments of the hospital with the necessary facilities, equipment, and resources. This small example shows how complicated a system can be and how a structured procedure helps with the analysis and understanding of this system.

Besides the actual diagrams, the method of the Flow Diagram incorporates two supplementary components: the element specification and the data dictionary. The element specification defines and describes every element used in the model (process, flow, and external entity), whereas the data dictionary is the database for these elements within the diagrams and the hierarchical structure.

1.2.3 Delimitation

Various graphical tools exist for the depiction of manufacturing plants and their activities and for the organization of companies. One of them is the Data Flow Diagram (DFD), which is one of the diagram types of the Structured Analysis method (see delimitation of POA, Chapter I2.3) and was actually developed for software systems.

The Flow Diagram and the DFD use the same main symbols (process and flow) with the same syntax and structuring. Both emphasize the flows as interfaces in a system. While the DFD depicts mainly data flows, the Flow Diagram accepts every kind of flow, for example resource flows like material or energy flows. Using the POA Flow Diagram, any kind of system can be analyzed. It is not limited to the handling of data, as the DFD is, nor is restricted to a specific system, such as a software system. The DFD includes a specific element for the storing of data. The Flow Diagram does not require this element. The saving of data or storing of material is a process that demands activities, needs time and space, and costs money.

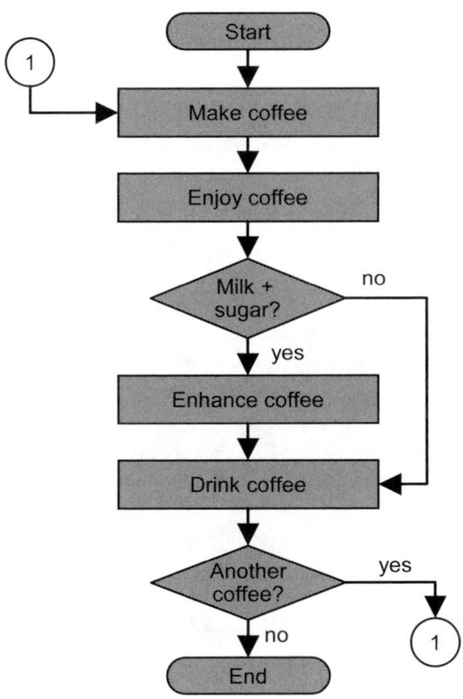

Diagram S1-2: Flow Chart.

A diagram consisting of processes and flows can also be seen as a process diagram. One of the most widely used process diagrams for analysis is the Flow Chart (*Diagram S1-2*). It is mainly used for information processing and depicts the sequence in handling it. The Flow Chart, with rhombuses to split a flow at a decision point, is also applied to decision trees. A Flow Diagram does not normally show decisions and loops in the system.

The conventional Flow Chart does not allow the visualization of a system as a network, and the hierarchical structuring is missing. It is not useful to provide an overview of a system. However, if Flow Charts already exist in a company, they can be used on the lowest level of the hierarchy of a Flow Diagram of POA as a detailed description of a process.

Unfortunately, the conventional Flow Chart mixes up the characteristics of the Flow Diagram, with the process-flow relationship, and of the State Chart, with the time relationship and sequence of activities. This leads to a mix up of processes and states, and therefore, to unclear relations, making the Flow Chart useless. The State Chart of POA, introduced in Section D1, handles the time-related factor of a process.

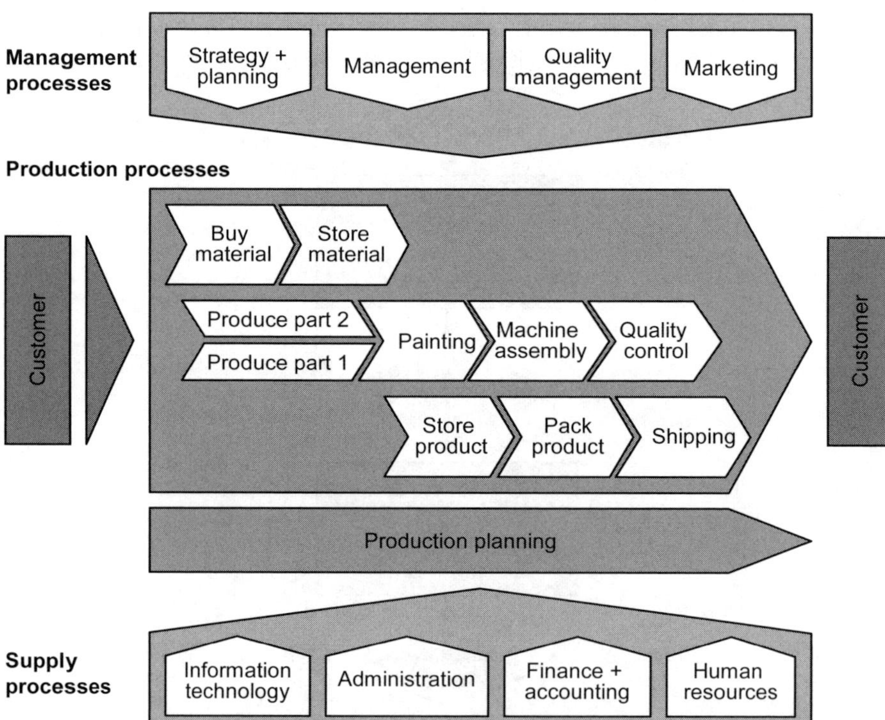

Diagram S1-3: Process diagram.

With still other process methods, only processes are depicted, without showing any flows or connections between them. In such a diagram, one of the most important features of a process diagram gets lost: the representation of the interfaces. The problems imposed by these neglected interfaces are not recognized. Without interfaces, only process chains can be shown, instead of the networks of processes with diverse relationships between them. This is inexact and incomplete because every system is a network.

Diagram S1-3 gives an example of a process diagram without interfaces. It shows the main processes of a company and provides, at the first glance, a good overview of the company. However, as the process naming already demonstrates in this example, it presents the risk that tasks and organization of a company are easily confused. This process diagram is not useful as a tool to handle concrete problems in process design, logistics, or information systems.

The POA Flow Diagram represents the system in the form of a network. It defines what input is processed into what output. It establishes terms for work and communication relating to this network and its elements. The Flow Diagram purposely treats only the stationary state of the system. There is no reference to time, and it shows no conditions or sequences. In other words, it does not serve to plan time-related activities, such as traveling or a production schedule in a factory. For example, one cannot depict waiting queue models or time schedules. The dynamic system behavior is described in POA with the State Chart.

The Flow Diagram does not represent a layout, even if they sometimes look similar. It is function-based, not space-based. Therefore, it cannot be used for plant layout planning. Its purpose is functional analysis and planning.

Table S1-1: Comparison tools

	Flow Chart	**Process Diagram**	**Plant layout**	**Flow Diagram of POA**
Self-explanatory	+	+	+	+
Hierarchical structure	-	-	-	+
Good overview	-	+	-	+
Interfaces	-	-	-	+
Network	-	-	-	+
Functional aspect	+	+	-	+
Space aspect	-	-	+	-
Category of flows	-	-	+	+
Time relation and sequence	+	-	-	-

Table S1-1 compares four graphical tools for drawing diagrams of manufacturing plants and organizations of companies.

1.3 Flow Diagram Elements

1.3.1 Diagram

The Flow Diagram is a graphical analysis method based upon a modeling language with its own vocabulary and syntax. Therefore, the semantics and drawing rules are very important. They obey clearly stated rules that are explained in Chapters S1.3 through S1.5.

A Flow Diagram consists of three elements, each represented by a graphical symbol: process, flow, and external entity. The investigated system is described using processes and flows. The external entities define the external environment of this system.

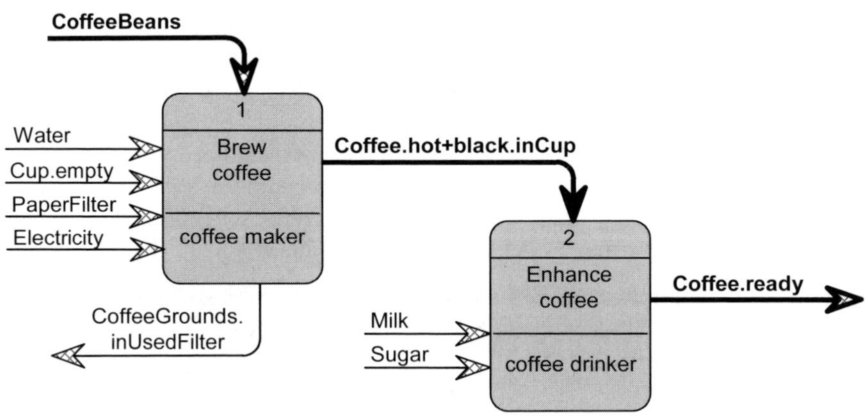

Diagram S1-4: Flow Diagram of "Make coffee" with a coffee maker.

Diagram S1-4 shows a simple Flow Diagram of how to make coffee with two processes and some flows to illustrate the interaction of processes and flows.

Generally, the rules in this book are numbered. In addition, the rules for the Flow Diagram are marked with the prefix F.

Rules S1-1: Flow Diagram

No.	Rule
F1	A Flow Diagram has to contain a minimum of: - one process, - one flow as an input to this process, and - one flow as an output of this process.
F2	Each element (process, flow, external entity) carries an informative name.

1.3.2 Process

To model the manufacturing of a product, the provision of a service, or the functions of a machine, several processes are necessary. These processes can depict the function of a single machine or of an entire company. A process represents a part of a system that carries out an activity or a function. It transforms the process inputs into process outputs. This transformation creates an output of higher value that is more useful than the input.

> **Definition:** Processes are transformations performed on the content of flows. Each change of place, time, state, or value is represented by a process.

Process symbol and marking

The symbol for a process is a rectangle with rounded corners. For the marking, the process symbol is divided by horizontal lines into two or three fields:

- The top field shows the *identification* of the process: the process number.
- The main field in the middle states the *process name*.
- An optional third field on the bottom contains any *additional information*.

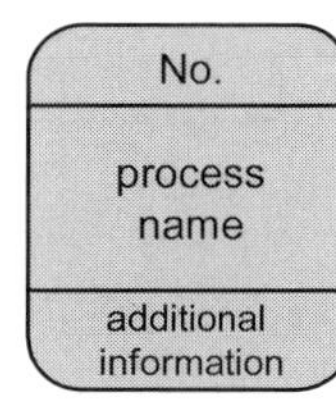

Besides these three information fields, no further information for the process is shown on the diagram.

Every process has an *identification:* the process number. The process number is unambiguous in the Flow Diagram model.

This identification number does not serve as a counting number. The numbers do not represent the sequence of the processes. A system is always a network, and a sequence would not be possible in a network. During the analysis, processes may be divided further and new processes inserted, while other processes are combined or deleted. It, therefore, makes no sense to organize the processes in a certain sequence by the identification numbers.

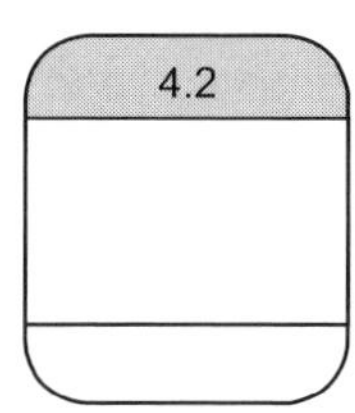

The process number may be preceded by the letter P, meaning "process."

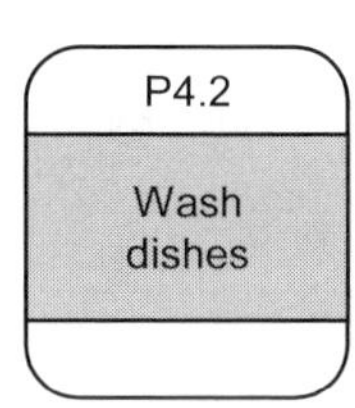

Processes are transformations performed on the content of a flow. The name defines the function that is carried out by a specific process. Therefore, the *name of a process* consists of a single active verb and a concrete object or an objective part of a sentence. It is an imperative sentence; the simpler the better.

The process has a unique name on a diagram. However, the process name can be used again on another diagram. This means, that a process name might appear several times in a model. This is possible because the process is identified by the number.

The process name should make sense relative to its input and output flows. Be sure that the process name is "honest" and describes the complete function carried out by the process. Using a gas station as an example, it is incorrect to name a process "Refuel gas" if it also includes changing the oil, fixing the windshield wipers and changing the tires. The name of such a process should instead be "Service car."

If two verbs are required to name the process, it is an indication that the process should probably be split up in two or more processes for clarification.

One should also avoid meaningless words like "handle" or "process" in the process name. Do not hesitate to change process names, if required, when working on the diagrams to make the name more specific and less ambiguous.

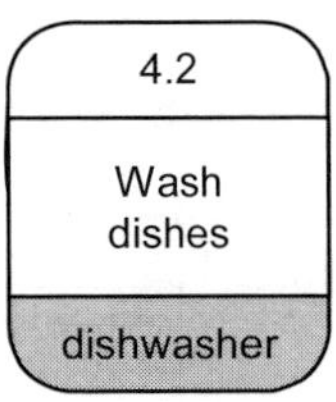

The *additional information* within the process symbol is optional. It can be a physical reference such as where, how, or by whom a process is carried out. It can also be a calculated value, or any other useful information. The only restriction is the size of the field in the symbol.

For the analysis of a system, it is helpful to note the physical circumstances of the process. These can be place, means, position, department, machine, responsibility, etc. If a diagram is to be valid for a longer period of time, avoid filling in the names of individuals because they might change. It is preferable to list the person's function, the workplace, or the machine, i.e. sales department, customer service, assembly line, cash desk, etc.

Process types

In a company, different types of processes are carried out. *Table S1-2* lists the process types with some examples taken from an industrial croissant bakery.

Table S1-2: Examples of processes

Process type	**Explanation**	**Examples**
Production process	Change of state	Bake croissant Mix dough
Storing process	Change, bridging, or passing of time	Store flour Store recipes Store customer data
Transportation process	Change of place	Transport basket of croissants Send bill

Process type	Explanation	Examples
Administration process	Change of state, mainly used for information	Sell croissant Develop croissant with new flavor Answer customer order
Support process	Change of state and/or place	Maintain machine Clean equipment Distribute electricity
Control process	Check state, control parameter	Control oven temperature Check speed of conveyor belt

Note! The storing of material or data is a process. The storing of material requires space and handling, causes costs, and uses energy. The storing of data also involves equipment and is therefore a process.

1.3.3 Flow

Definition and symbol

A flow represents the interface or the crossing of an object that is passed between two processes or between a process and the external world. The flow is an imaginary construct. The object is the content of the flow. It can also be seen as the carrier. This object can be material, as a product, material, or resources, or immaterial, as information, human intervention, energy, etc.

Definition: A flow represents an interface between two processes or between a system process and the external environment.

The symbol used for the flow is an arrow. The location of the head of the flow arrow defines the direction of the flow. It indicates where the flow starts (point of origin) and where it is going (destination). From the viewpoint of the process, the flows on a diagram are referred to differently:

- The *input flow,* also referred to as input, enters a process.
- The *output flow,* also referred to as output, leaves a process.

The flow coming out of one process remains the same flow until it enters another process as an input flow. There is no change in the object of place, time, state, or value between the origin and destination of a flow.

The flow is not a geographical displacement of an object. It is an interface. The displacement demands a process.

For clarity of the diagrams, the arrow ends should touch the outer perimeter of the process box and should not cross into the box. Arrows should attach at the sides of the box, not at the corners.

A *bi-directional* arrow (dialogue flow) is a combination of two flows going in different directions. This is not allowed on the Flow Diagram. The object of a flow is transformed after entering a process. A flow never leaves a process without being changed. But this is what a bi-directional arrow indicates. Even if the returning flow appears to be the same as the incoming one, it is not. If a flow does not undergo any change while in the process, this process is unnecessary for the flow, or the flow is unnecessary for the process.

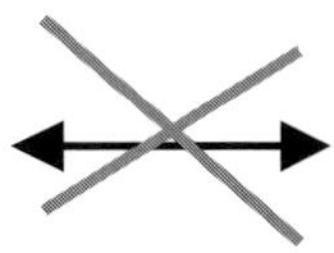

Note that the direction of the flow is defined from the point of view of the two interlinked processes. The arrow gives the relation of direction between the processes. This relation would not be clearly defined using bi-directional arrows.

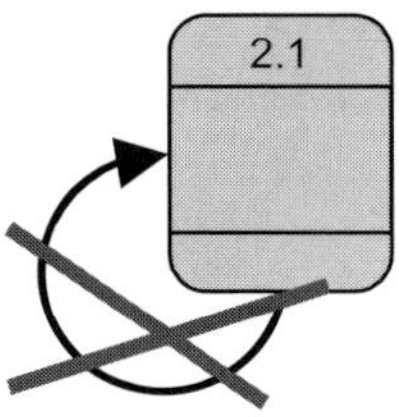

Repetitive flows are not allowed. This means a flow leaving a process cannot enter the same process again to repeat it. Why should a flow leave and return to the same process without undergoing a change outside this process? The process may carry out its activity on the flow as often as needed. It is not possible to depict this as a loop. Instead, the information for the repetition is given in the process specification.

Naming of flows for identification

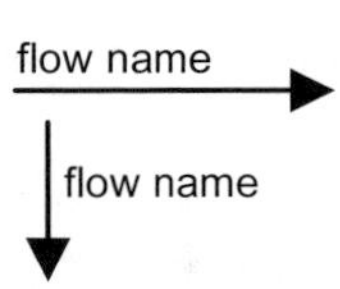

Every flow has a name written along or beside the arrow on the diagram. The name serves as the only identification of the flow in the model. Therefore, it must be unique. In the whole model, two flows never have the same name.

The flow name indicates the contents or object that is represented by the flow. The flow name consists of a noun, preferable in the singular form, and an optional extension. Since the flow is not an activity, no verbs are allowed for the naming.

The extension of the flow names uses an additional word (noun or adjective) or the number of the process, in which the flow enters or from which the flow comes out. This is an easy way to avoid duplicate flow names and to uniquely identify a flow. The name and the extension may be written together or separately using a point, hyphen, or underscore.

A space within a flow name is not practical as a separator. On a diagram with many flows, the names are often written very close together, and it becomes difficult to know which word belongs to which flow name when using spaces. Furthermore, the flow name serves as identification in the data dictionary, and later also as variables for programming. Therefore, they must be compatible for source code. In program coding, spaces within variable names are not allowed because they serve as separators in the source code.

Flow names consisting of two or more nouns are written together with every word beginning with a capital letter, for example: "CustomerOrder," "CoffeeBeans," or "WasteProduction."

Keep the name conventions consistent within the model. *Table S1-3* lists some examples of flow names with different types of extensions.

Table S1-3: Example of flow names with addition

Object	**Type of extension**	**Examples**
Yarn	Noun	Yarn.pallet Yarn.stock Yarn.warp
Electricity	Process number	Electricity-4 Electricity-1.3 Electricity-2.3.2 Electricity-2.3.4
CustomerOrder	Adjective	CustomerOrder_open CustomerOrder_processed CustomerOrder_completed

It is important to make sure that the flow name is straightforward. The name has to represent the entire content of the flow, not just its major component. "Apple" is not sufficient for a flow consisting of mixed fruits. An appropriate name would be "Apple+Pear.inBasket" or "Fruit.inBasket," even if there is only one pear in the basket.

Be careful to group objects into one flow only if there is a relationship between them, for example the same physical characteristics. If they cannot be treated as a whole, they have to be depicted by several flows.

Avoid meaningless names like "data," "information," or "material."

A flow is unnamable if there is no clearly defined idea of the object that is flowing. As the flow represents the interface of this object between two processes, it might be helpful, when not knowing how to name the flow, to look at the origin and the destination processes.

If a flow is not really essential to the main purpose of the analysis by the Flow Diagram, it is probably incorrect or unnecessary to include it. If a diagram describes, for example, the handling and the storing of books in a library, it is unnecessary to introduce the energy flow for the lighting as well.

Resource and information flow

In the Flow Diagram, resource and information flows are distinct. *Table S1-4* summarizes the differences between these flows.

Table S1-4: Differences of resource and information flows

Characteristic	Resource flow	Information flow
Object of the flow	Physical matter: material or energy.	Immaterial: data, information.
Law of preservation of material and energy	Law is applicable: resources cannot vanish and cannot be created out of nothing; they can only be transformed.	Law is not applicable: data can be created and deleted.
Carrier of the flow object	The resource itself is the carrier. It has physical properties to consider.	The carrier of the data (paper, disc, e-mail, etc.) is not important from the viewpoint of the information and does not count as a resource flow. (If it is important, treat it as a resource flow.)
Storing (by a process)	Storing of physical resources needs energy and other resources.	The amount of energy used to store and convey data is mostly negligible.
Examples	Potato chips, a machine, electric power, a person, wrapping material.	Computer file, voice call, keyboard input, alarm sound, customer order.

Note! For the entering and leaving of the resource flows, the process has to obey the rules of the economic balance and the law of nature, such as the conservation of the material. Processes are neither sources nor sinks of resources.

In order to brew a pot of tea, like the output flow "HotTea.Pot" indicates, more than one cup of water for input is needed (*Diagram S1-5*). Keep an eye on the inputs and outputs of a process and do a rough calculation of the quantities. The balance of the mass must be correct. A process can only produce an output if it gets sufficient input. On the other hand, everything going into the process has to come out again, in whatever form. This rule applies also for economical values and energy.

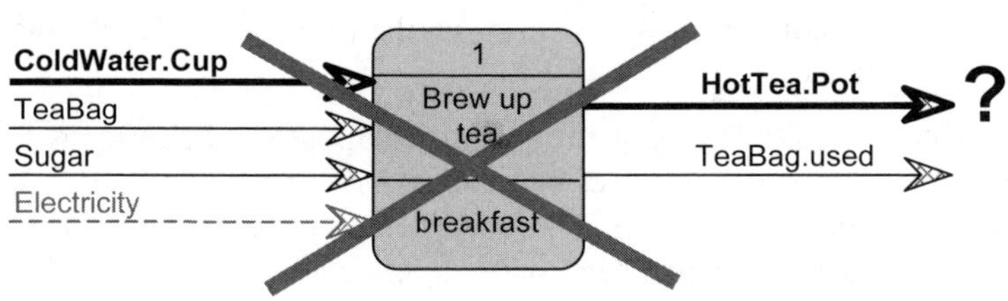

Diagram S1-5: It has to be enough input available.

In contrast to the resource flows, information can be created, transformed, and deleted in a process. A process can create new information. An information output is possible without an information input. In *Diagram S1-6* a sensor measures the temperature of the water and delivers it to another process. No other input information flow is needed to create the output information "Temperature.actual."

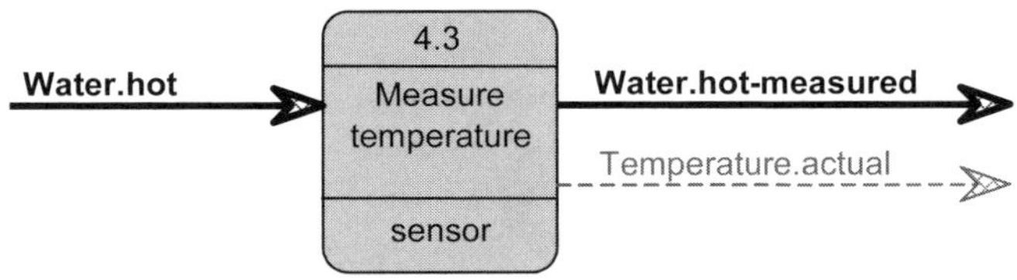

Diagram S1-6: Information output without input.

There can also be an information input without an information output. An example of such a process is illustrated in *Diagram S1-7,* where the supervisor gives an instruction to an employee in a workshop. This instruction is used up during production. It goes into the product as knowledge. There is not a visible information output from this production process.

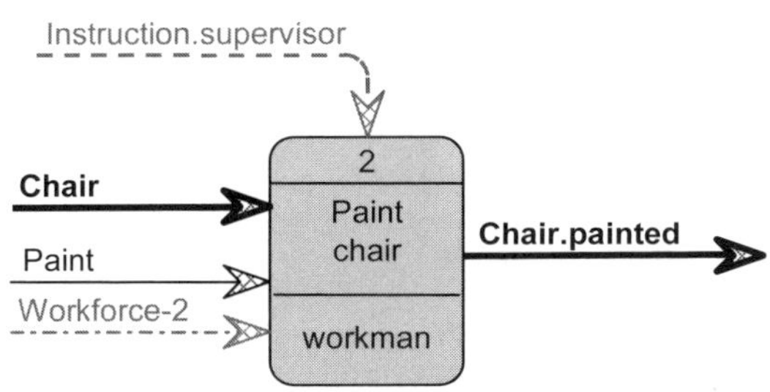

Diagram S1-7: Information input without information output.

Note! Be aware of missing transportation process: A flow represents an interface and not a change of location. Between the point of origin and destination, the content of the flow remains at the same place or position. A change of location demands a process.

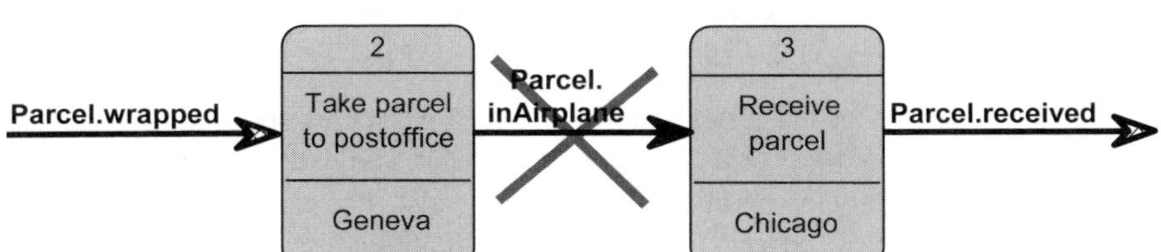

Diagram S1-8: Missing transport process.

Diagram S1-8 and *Diagram S1-9* illustrate the change of location of a parcel. To take the parcel from process 2 at the location of the sender in Geneva to process 3 at the location of the receiver in Chicago, a transport process is needed. The flow "Parcel.inAirplane" cannot be used to indicate the parcel going from Geneva to

Chicago (*Diagram S1-8*). Process 4 "Transport parcel" needs to be introduced, executed by the parcel service that takes the parcel to Chicago.

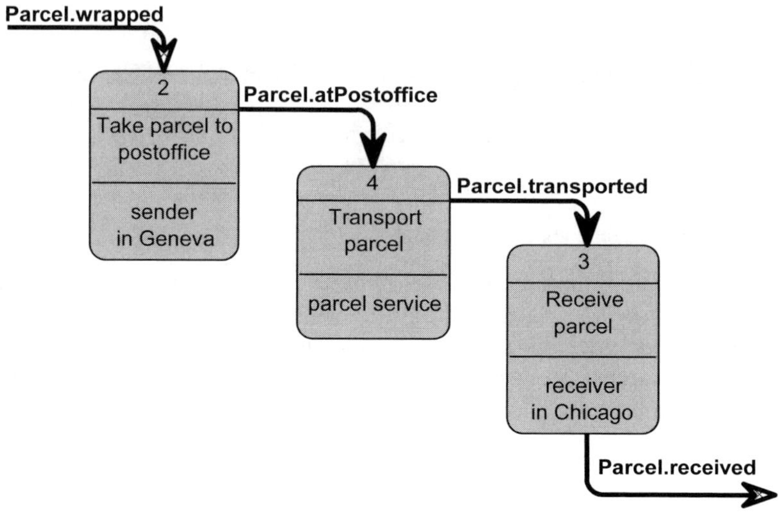

Diagram S1-9: Correct transportation process.

1.3.4 Classification of Flows

In a company, a lot of flows are needed to depict its activity. In order to keep the diagrams readable and easy to handle, a classification of flows is available. The method of the Flow Diagram classifies the flows into flow types and flow groups.

Flow type

The flow types are defined by the designer of the Flow Diagram model. By classification of the flows into flow types, flows with the same physical characteristics are bundled together. If necessary, the flow types can be further classified into subtypes. The flow types are important for hierarchical decomposition. This is done by the split and merge, explained later in Chapter S1.5.5.

The classification into flow types helps to keep track of the flows. For example, it is easier to see with classified flows if a process with a material input also has a material output. The flow arrows may vary in line thickness, style, and color to differentiate between flow types. The flows belonging to the same flow type are drawn identically in the diagram. For example, color is used to distinguish different flow types as product, material, resource, information, etc. It is an advantage to highlight the product flow, as it is done in the diagrams of this book, where the product flow is depicted by a bolt arrow.

Flow group

Flow types can be organized into flow groups. This classification into flow groups is not a functional feature of the Flow Diagram, but an organizational feature. It helps to make the diagrams more readable. There are a vast number of flows within a company. Therefore, a diagram can quickly get overloaded and become difficult to decipher if too many individual flows are entered. For the analysis of a system, all flows are taken into the Flow Diagram model. For a presentation and discussion, depending on the audience, the focus, or the viewpoint of the analysis, flow groups can be made invisible or left out for easier readability of the diagrams.

Flow groups are optional. The designer of the Flow Diagram decides if flow groups are used and defines them.

Table S1-5 gives an example of a flow classification into flow groups, flow types, and subtypes of an industrial bakery. All diagrams in this book are depicted with flow types drawn in different arrow styles.

Table S1-5: Example of flow classification

Flow group	**Flow type Example**	**Flow subtype Example**	**Example**
product flow	product		dough, croissant
resource flow	material	various materials	wrapping
	waste	waste dough	croissant faulty
	water	fresh water	water as ingredient
		sewage	waste water cleaning
	energy	electricity	electricity for lights
		heating electricity	electricity for oven
		waste energy	waste heat of oven
	workforce	operator	operator wrapping
		service	service of oil tank
		supervisor	supervisor of bakery line
	space		space of bakery
information flow	order		customer order
	instruction		instruction of supervisor
	control		form of croissant good/bad

1.3.5 Rules for Processes and Flows

In this chapter, all the rules for processes and flows are listed together because these two elements interact very closely. All flows going into or out of a process have to be considered when applying the rules.

Rules S1-2: Processes and flows

F3	Each element has an identification: - process number for the process - flow name for the flow
F4	Process name = ⟨active verb⟩ + ⟨object⟩
F5	Flow name = ⟨noun⟩ [+⟨extension⟩]
F6	A process name occurs only once on a diagram.
F7	A flow name is unique in the model.
F8	Between the point of origin and destination of a flow, the object of the flow remains the same in place, time, state, or value.
F9	Every flow going into or out of a process has to interact with at least one other flow of this process.

If there is no interaction and the flow does not undergo any change while in the process, this particular flow is not relevant for this process. It should bypass the process.

F10	The process has to comply with the rules of economic balance and the basic laws of nature, such as the preservation of mass and energy.
F11	Each process must have at least one input and at least one output.

An input is needed since it is impossible to create something out of nothing. Without output, the process would be a "black hole" (indefinite sink) that swallows things up.

F12	A product input demands a product output, and a product output demands a product input.
F13	Processes transform input resource flows into output resource flows.
F14	Processes create, transform, and delete data.

1.4 System Boundary

1.4.1 External Entity

The elements of the system not only have contact among themselves, but also with their bordering external environment. The part of the external world that interacts with the investigated system is depicted by the external entities. An external entity is an object outside of the system boundary, drawn on the top level of the model. External entities represent, for example, suppliers or customers of materials, products, or information. These can be partners, industries, compartments, machines, or parts of machines to which the analyzed system is connected.

Definition: The external entity represents the external environment, with which the investigated system interacts.

The *symbol* for the external entity is a shadowed or double-bordered square or rectangle.

The *name* of the external entity is mandatory and serves as the *identification*. It describes the role of the external entity. It must be a noun or a noun with an addition such as an adjective or object. For unique identification, a small letter can be used instead of the name, normally displayed in the upper left corner of the symbol.

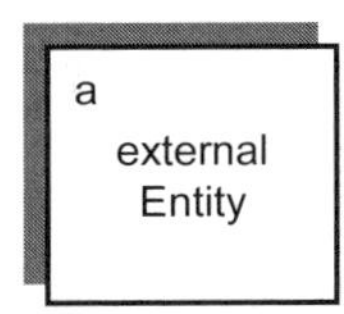

System boundary

The system boundary defines the demarcation between the system and its external environment. The system is separated from the external environment in order to clearly define the analyzed object. The system boundary has to be determined so that it contains those elements that should be investigated in the project and have an influence on the system.

The external entities show the world outside the system boundary. They are specified from the point of view of the investigated system. Because the external entities are only the suppliers and receivers of the system's flows, they have to be assumed as given and unalterable for the system. They cannot be influenced or changed somehow or other by the system. If you try to change an external entity during modeling for a project, it indicates that the system boundary is set incorrectly.

The definition of the system boundary has a decisive influence on the result of the system investigation. The determination of the boundary is not at all easy and should not be underestimated. There are no formal, generally applicable guidelines. It is within the discretion and judgment of the model's designer, his or her viewpoint, and the target of the project. The scope and applicability of the results, as well as the time and labor required to set up the model, have an impact on the system boundary.

1.4.2 Context Diagram

The context diagram is the highest level, the top level, of a model. It is a Flow Diagram that uses also the element external entity. This allows the designer to establish a clear system boundary and to treat the whole underlying structure as a black box. The designer sets on the context diagram what belongs to the investigated system and what is outside of this system. The context diagram defines the model as part of a larger scale.

Definition: The context diagram represents the investigated system as a black box within its surrounding external environment.

The context diagram contains one process, several flows, and one or more external entities. The single process represents the whole system to be investigated. The descriptive process name is general and often the model's name. The flows going in and out of the single process show the interface of the system with its external environment.

The external entities represent the components of the system's surrounding world. The system needs at least one external entity as a supplier or source, and at least one as a customer or receiver for the system's flows. An external entity can be a "supplier" and "customer" at the same time; the context diagram may therefore have only one external entity.

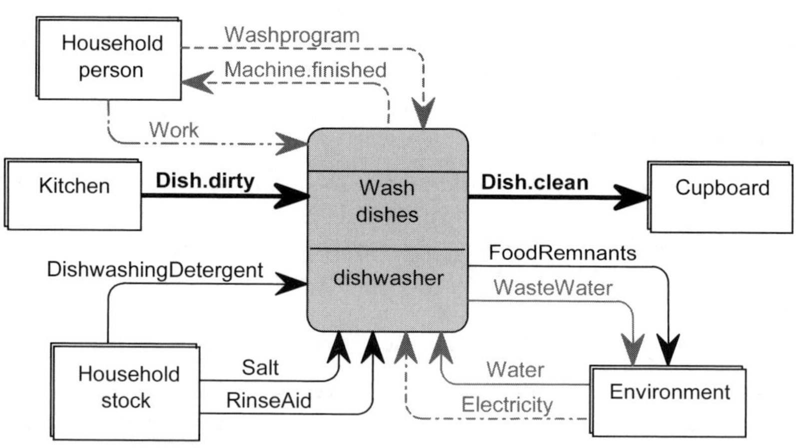

Diagram S1-10: A context diagram.

Diagram S1-10 shows the context diagram of a dishwasher. The single process states the task of the dishwasher: "Wash dishes." The flows going in and out of the process give the interface of the investigated system with the outside world (household and kitchen). They are the dirty and clean dishes, as well as various resources needed to operate the dishwasher. The external entities indicate where the flows are coming from or where they are going.

Note! A flow between two external entities is not allowed.

It is of no interest to the investigated system if two external entities interact. The information flow "Complaint" in *Diagram S1-11* is outside of the investigated system and must not be shown on the diagram. If this complaint is important for the investigation, the system boundary has to be defined differently. The best way to do so, is to integrate the external entity "Customer" or "Bakery" as a process into the system.

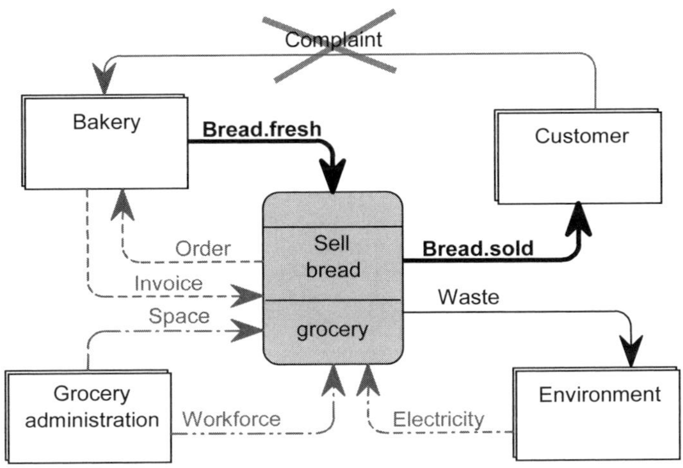

Diagram S1-11: Flow between two external entities.

1.4.3 Rules for External Entity and Context Diagram

This chapter provides a list with the rules for the external entity and the context diagram.

Rules S1-3: External entity and context diagram

F15	At least one external entity is needed to specify the system.
F16	Each external entity has an identification. It is either a unique name or small letter.
F17	Name of external entity = ⟨noun⟩ [+ ⟨adjective⟩]

F18	The system, as defined by the model's designer, has no influence on the external entities and cannot change their function.

The external entities and their behavior toward the investigated system are given for the analysis, and cannot be influenced or changed by the system in any way.

F19	There is only one context diagram for each model.
F20	The context diagram contains the following elements: - a single process, - at least one flow as input of the system, - at least one flow as output of the system, and - at least one external entity.
F21	All external entities are shown in the context diagram. Lower-level diagrams have no additional external entities, only repetitions from the context diagram.
F22	External entities are not hierarchically detailed.
F23	Every flow on the context diagram has to be connected to the single process and to one of the external entities.

On the context diagram, a flow cannot enter the investigated system represented by the process from "nowhere" nor leave the process to "nowhere." It must be connected at one end to an external entity.

F24	Every external entity has to be connected to the system by at least one flow.
F25	The context diagram does not show any relations (flows) between external entities.

Recommendation: The crossing of flow lines should be minimized on the context diagram to simplify it. To achieve this, the same external entity can be drawn more than once on the context diagram. For example, an external entity representing the "Customer" who orders a product can be depicted a second time as the "Customer" receiving the finished product.

1.5 System Structuring in the Hierarchy

1.5.1 System Structuring

The Flow Diagrams of the model are built up hierarchically. A complex system can be broken down into smaller, more understandable subsystems. This detailing into subsystems is done through the processes of the Flow Diagram. Each process in a diagram can be decomposed or broken down into another diagram with its own processes and flows on the next lower level (*Diagram S1-12*). This leads to a hierarchy of processes and flows.

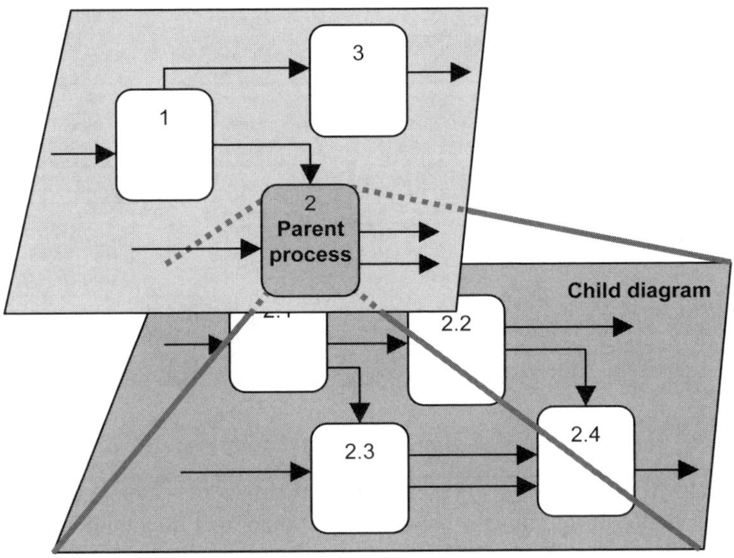

Diagram S1-12: Structure of the diagrams.

The decomposed process is called the *parent process;* the new diagram with its own processes is the *child diagram*. The parent process and the child diagram refer to the same subsystem and deliver corresponding information about it. The parent process appears as a black box for the reader, where inputs, outputs, and general functions are known, but the underlying details are not shown. The child diagram gives the same information in a more detailed manner and reveals the contents and function of the black box.

1.5.2 Numbering of Processes and Diagrams

The process number has two purposes: identification of the process and placing the process in the hierarchy. One must be able to trace back each lower-level process to its parent process. The Flow Diagram uses a logical, consistent numbering convention that also indicates the level of detail shown.

The single process of the context diagram receives the number 0 for identification. The number 0 is generally omitted on the diagram. The processes on the next lower-level diagram are serially numbered 1, 2, 3, etc. The numbers of the processes on the lower levels consist of the number of the parent process and an additional, sequential number, separated by a period: 1.1, 1.2, …, 2.1, 2.2, … 2.2.1, etc., like subsections or chapters in a book (*Diagram S1-13*).

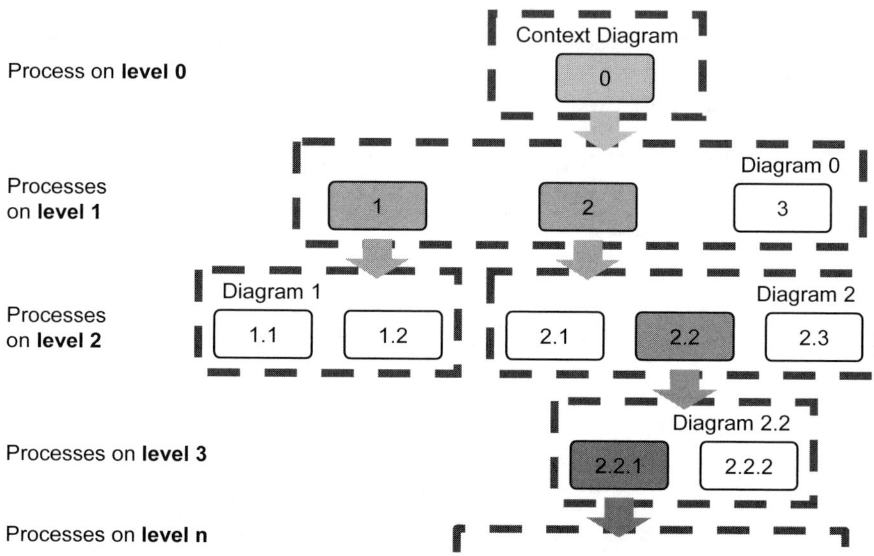

Diagram S1-13: Numbering of the processes in the hierarchy of the leveling.

This hierarchical numbering ensures that two processes never have the same number within a model. If a number were used more than once in a model, the unambiguous identification would get lost.

The identification number of the parent process is at the same time the identification number of the child diagram. To see the details and internal structure of a given process, one only needs to look at the diagram with the same number.

Diagram-set: All the diagrams on all levels of the hierarchy together are referred to as the diagram-set of the model.

1.5.3 Balancing Parent Process and Child Diagram

The balancing rule states that the flows of the parent process and the ones of the child diagram must be balanced. This means that the flows going into and coming out of the parent process must be identical to the flows going into and coming out of the child diagram (*Diagram S1-14*). The parent process and the child diagram represent the same information with different levels of detail. They are consistent.

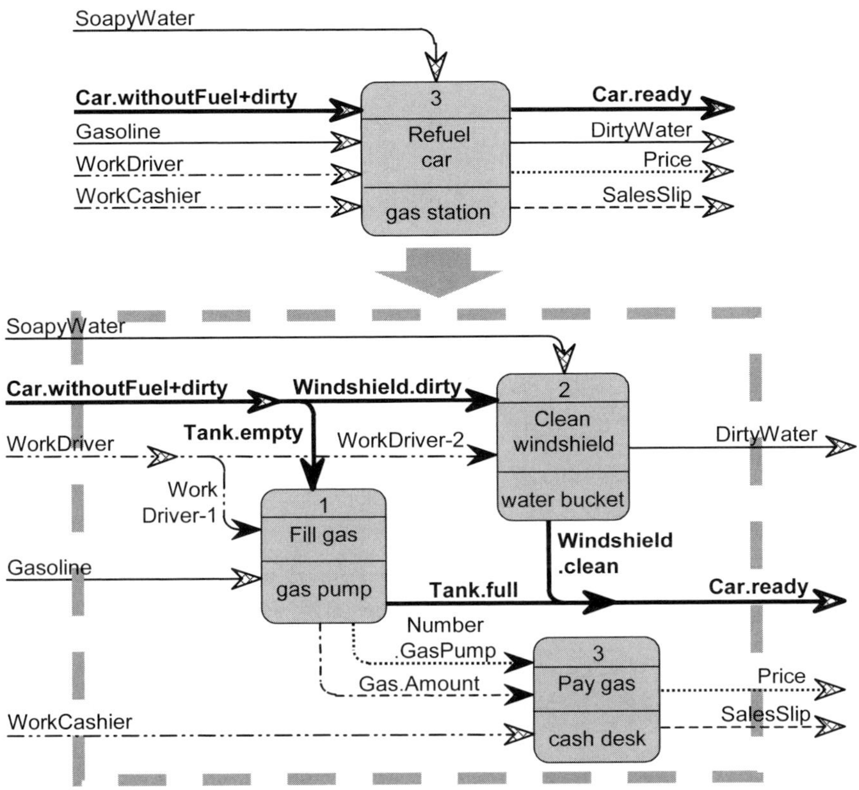

Diagram S1-14: Balancing of parent process and child diagram.

"Refuel car" in *Diagram S1-14* represents the parent process that is detailed into a child diagram with three processes. The flows going into and out of the child diagram are the same as those going into and out of the parent process. The flows between the three processes are added to the child diagram.

1.5.4 Principle of Structuring

Depending on the intention of the model's designer and the purpose of the model, different viewpoints may be adopted that emphasize different aspects of the system. This results in different principles of structuring. Usually, a functional decomposition is chosen. It is also possible to structure the Flow Diagram model from a geographical or personal viewpoint, or in terms of existing cost centers or areas of responsibility. For each model, there is only one viewpoint and, therefore, only one principle of structuring.

Diagram S1-15 illustrates the different principles of structuring on the level 1 diagram with the example of a restaurant serving meals and drinks. This restaurant has two dining rooms; therefore, the decompositions can be done geographically.

Or it might be functional according to the two processes "Handle order" and "Bring meal + serve table." Also, a cost-oriented structure is possible, in order to investigate the difference between drinks and meals because they require different preparations and have different suppliers.

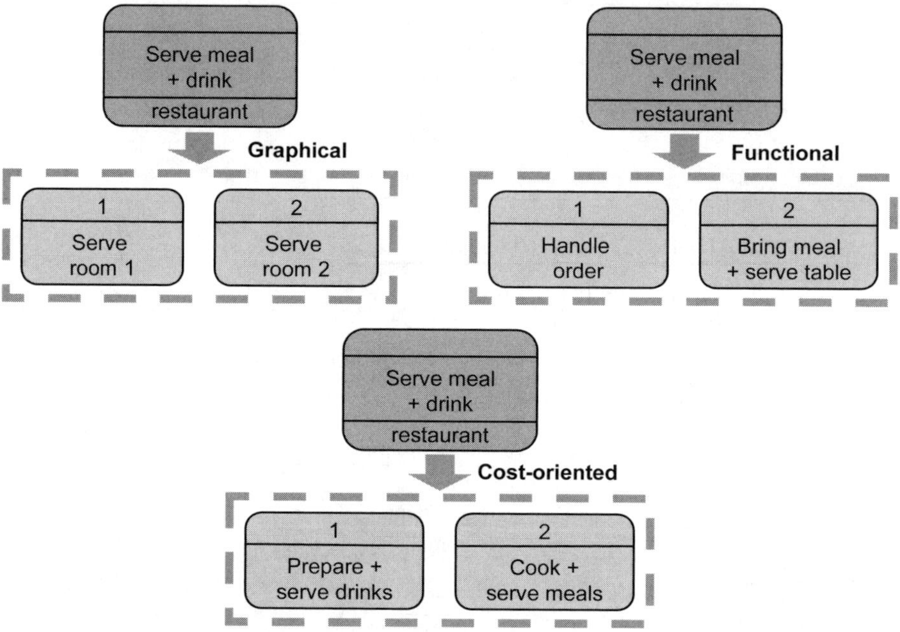

Diagram S1-15: Principle of structuring.

All system functions lie within the single process on the context diagram, which is level 0. The child diagram of the single process on the context diagram is the level 1 diagram. The context diagram gives the outside view of the system. It defines the system interfaces to the external environment. The level 1 diagram depicts the internal view. It is the highest level of the internal structure with the major processes of the model. The terms and structure of the level 1 diagram bind every subsequent level.

The processes on the lowest level of the hierarchy that are not further exploded are the *bottom-level processes*. The composition of all the bottom-level processes results in the complete system.

Top-down and bottom-up procedure

The model can be built top-down or bottom-up. In the top-down approach, one begins with the context diagram or the level 1 diagram and explodes the processes to child diagrams. In the bottom-up method, first the child diagrams are drawn and then condensed into parent process on the higher level diagrams. This process is continued up to the context diagram.

Often the easiest way to begin a model is somewhere in the middle of the hierarchy. From there, the designer proceeds upwards to the context diagram as well as details the process onto lower levels. To determine the level, on which to begin modeling, depends on whether the system to be investigated and its boundaries are already known or if they have yet to be clearly defined. Normally, the system boundaries are concretely set while drawing the diagrams. Beginning the model by defining the system boundary with the context diagram is very difficult.

Number of hierarchical levels

The number of hierarchical levels of a model depends on the details required for the investigation and on the size of the system. A model, where one single function is examined, may consist of the context diagram and the level 1 diagram. For a system depicting a production line, a model with four to five levels is common.

In a model, the number of detailing levels is not necessarily the same for all the processes of the level 1 diagram. Process 1 might be detailed on two further levels, while process 2 has only one underlying level and process 3 has three. However, all the lowest processes in the hierarchy should show the same degree of detail throughout the entire model. For example, all the bottom-level processes represent a machine or a work place.

1.5.5 Hierarchy of Flows by Split and Merge

If a parent process is detailed into a child diagram, the flows can also be detailed on the lower diagram levels. The hierarchical decomposition of the flows is done by split and merge. The flow going into or coming out of the parent process is divided into child flows on the child diagram.

By grouping flows together, the diagrams are lightened. The flows are bundled on the higher-level diagram and are separated on the lower-level diagram. The parent flow is a bundle of the child flows (*Diagram S1-16*); it is the combined flow. The split and merge of flows is merely a graphical element. At the split/merge point, no process is carried out. Therefore, this point receives no name.

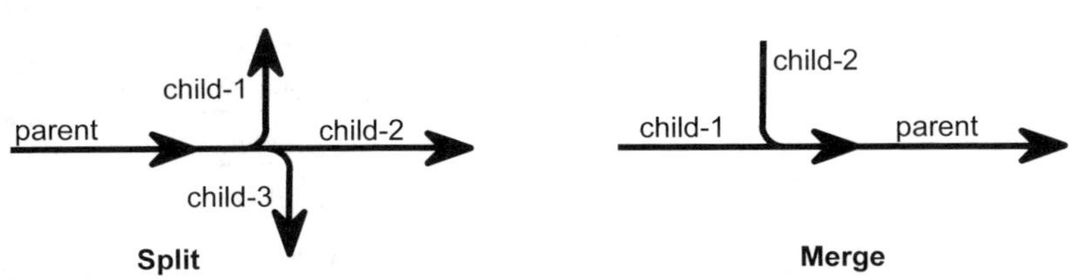

Diagram S1-16: Split and merge.

Diagram S1-14 illustrates with the example of the car, that the split and merge is only a graphical aid, not a process, which involves a physical operation on the flow. The car represented by the flow "Car.withoutFuel+dirty" is not taken physically apart on the child diagram for the two flows "Tank.empty" and "Windshield.dirty." These flows indicate what part of the car is treated in processes 3.1 and 3.2, either concurrently or sequentially.

In Chapter S1.4.4, the flow classification into flow types was introduced. It was stated that flows with the same physical characteristics are organized into flow types. Accordingly, parent and child flows must have the same physical characteristics, meaning only flows from the same flow type can be split and merged.

Note! A flow cannot be split into different flow types. "Material.inBag" from *Diagram S1-17* cannot be split into the flow "Material" and "Bag." Separating material and bag is a process that must be shown on the diagram, for example "Empty bag" or "Take material out of bag."

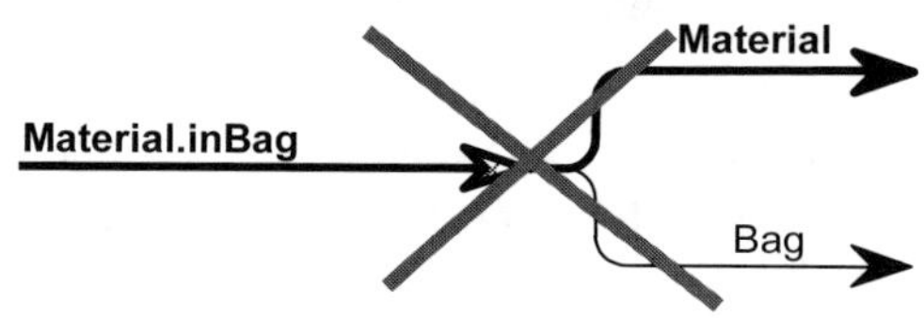

Diagram S1-17: Split in two different flow types.

The split and merge gives the hierarchical relation of the flows. *Diagram S1-18* demonstrates in a specially arranged figure all the hierarchical levels of the flow type "Workforce" at one glance.

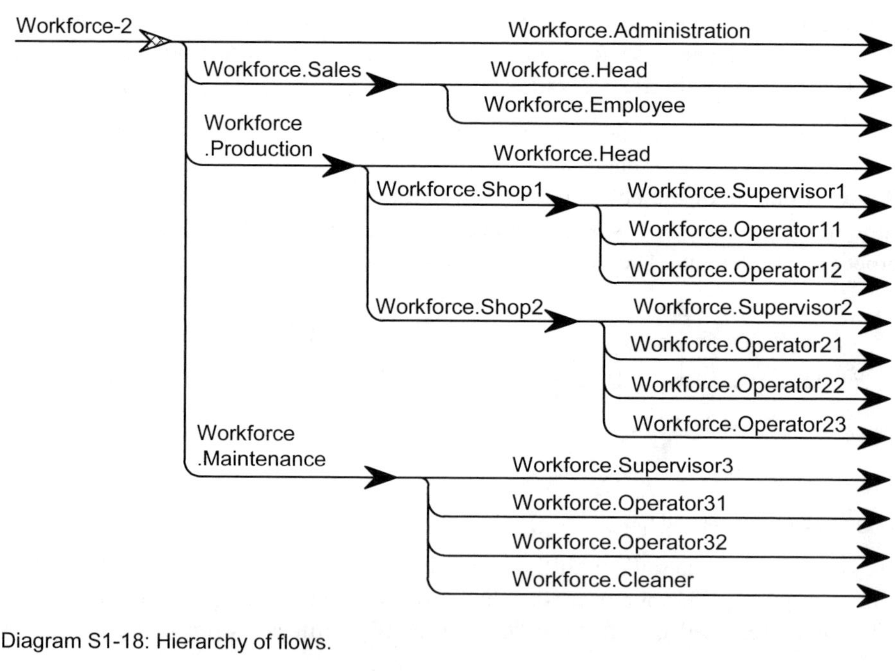

Diagram S1-18: Hierarchy of flows.

Recommendation: The split and merge of flows take place in the child diagram. In the parent diagram, only the parent flow is depicted. This helps to simplify the higher-level diagram.

It is not compulsory to show a split or merge of flows. It is also possible to depict the child flows on the child diagram without a split or merge and without the parent flows. In such a case, the reader of the diagram should be able to assign the child flow to the parent flow by the naming. *Table S1-6* shows examples of such a naming that indicates the connection between parent ant child flow.

Table S1-6: Naming of parent and child flows

Parent flow	Child flow
Electricity-2	Electricity-2.2 Electricity-2.4 Electricity-2.5
CustomerOrder.Original+Copy	CustomerOrder.Original CustomerOrder.Copy

Off-page flow

Flows, going from process to process on the higher level, go on the lower levels from diagram to diagram, which are depicted on different pages. These flows running from one diagram sheet to another are called *off-page flows*. *Diagram S1-19* illustrates how off-page flows are created by exploding processes. The two bold flows A and B between the parent processes 3 and 5 become off-page flows on the child diagrams.

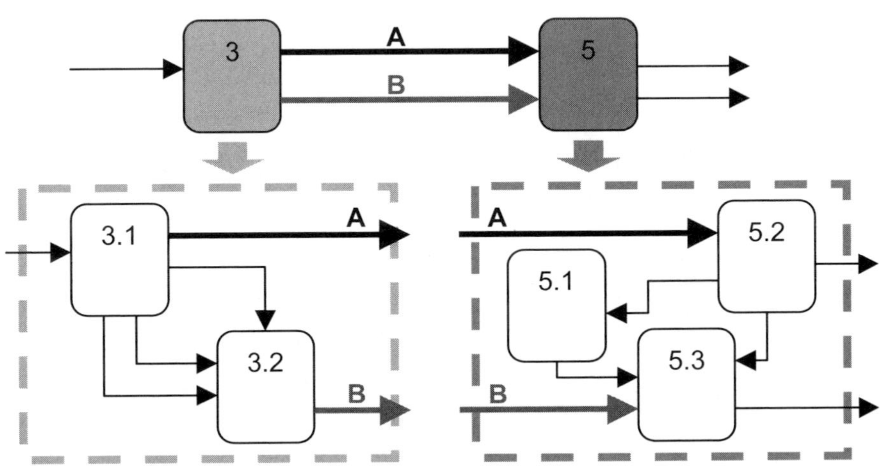

Diagram S1-19: Off-page flow.

Note! A flow must be connected at both ends, to either two processes or to a process and an external entity.

Dangling flow: A flow connected only at one end is called a dangling flow. Make sure while working within the hierarchy that no dangling flows remain or are produced. Off-page flows are especially difficult to trace, because one of the attached processes occurs on another diagram sheet.

1.5.6 Rules for Flow Connections and Hierarchical Structure

This chapter gives the rules for connecting elements and the hierarchical structure, including the numbering of processes and the split and merge of the flows.

The rules for interconnecting elements of a Flow Diagram are summarized and organized in a directional manner, as a matrix, in *Table S1-7*. The table indicates, which connections are possible, that means, from where to where a flow is permitted to go. The caption of the rows gives the point of origin of the flow, and the caption of the columns gives the destination. The permitted connections are illustrated by little elements in the cells. The concerned arrows are given in black. The other connections are not allowed. To make the split and merge functions in the table more obvious, they are indicated by small circles.

Table S1-7: From where to where a flow can go

to / from	Process	External Entity	Off-page	Split	Merge
Process	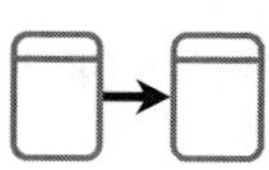	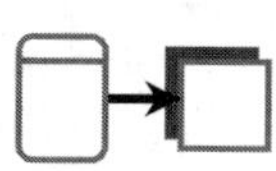	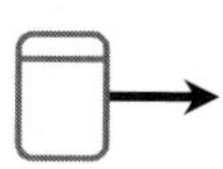	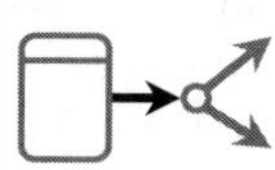	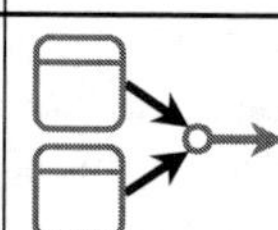
External Entity	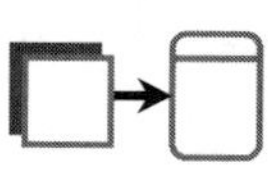				
Off-page	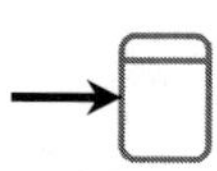			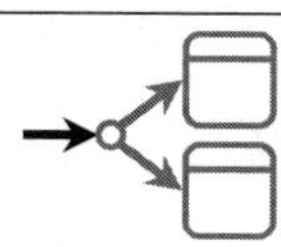	
Split	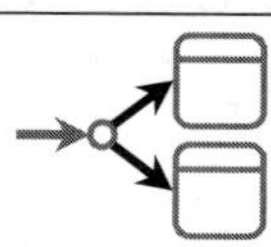				
Merge	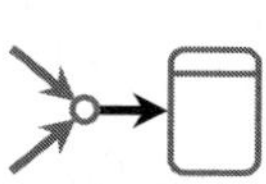		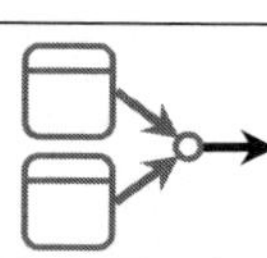		

The only flow connections that must be used are the ones that ensure an exact system boundary and a clear hierarchy. The table indicates, for example, that several flows coming from off-page and going into a merge is not allowed. Such a merge of off-page flows would confuse the strict hierarchical structure of the model. However, as another cell of the table shows, a flow coming from off-page is allowed to enter a split. Flows between external entities and merge or split are not allowed; as they would blur the otherwise clear system boundary.

Rules S1-4: Process numbering

F26	Each process has a number, which is unique in the model, and which indicates the hierarchical placing of the process.
F27	The single process in the context diagram gets the number 0 (zero) for identification.
F28	The child diagram is assigned the number of the parent process.
F29	Every diagram has a name. If it is a child diagram, it receives the name of its parent process.

Rules S1-5: Split and merge, and hierarchical structure

F30	Parent and child flows belong to the same flow type. They have the same physical characteristics.
F31	All child flows as well as their parent flow must be named, and the names must be different and unique in the whole model for every child and parent flow.
F32	On the context diagram, split and merge is not allowed.
F33	At a split and merge, the content of the parent flow is equal to the sum of the contents of all its child flows.
F34	The balancing rule ensures the balance between levels. Flows into and out of a parent process are equivalent to the inputs and outputs of the child diagram.

Note! Make sure that the Flow Diagram model is kept in balance whenever flows are added or deleted at one level. In addition, a change in a flow name must be updated on all the concerned diagrams of the different levels.

Recommendation: Determine if a split or merge can be replaced by a process. This is usually the case in reality. For example, a fuse box carries out the distribution of electricity for different users. In reality, it is therefore an actual process, such as "Distribute electricity."

1.6 Element Specification and Data Dictionary

1.6.1 Element Specification

As seen, the Flow Diagram consists of the three elements process, flow and external entity. Every element used in the model has an element specification (*Diagram S1-20*). While the diagram shows the relationships between the elements of the system, the specification describes and defines the element. The total of all the element specifications in connection with the diagrams results in the complete description of the system.

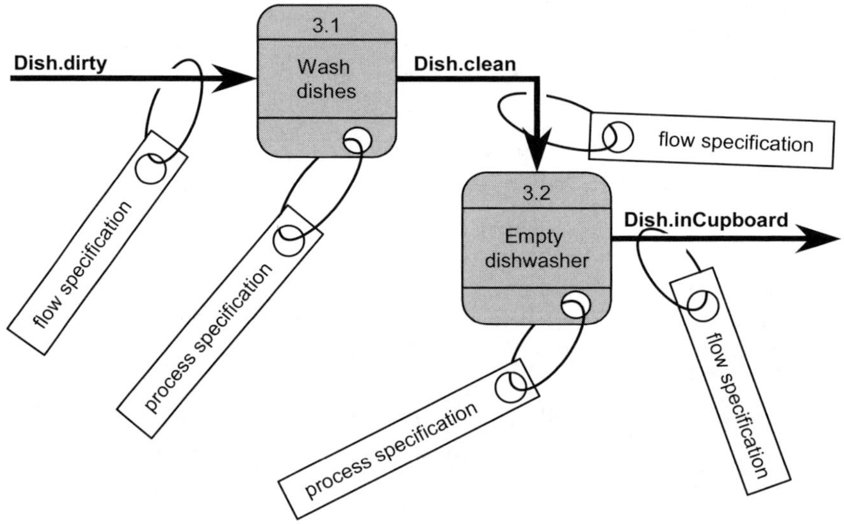

Diagram S1-20: Elements and their specifications.

The specification contains the details of the elements. The description clearly determines the content of the flow or the exact function of a process. A specification should produce a precise definition with no room for ambiguity. The name of an element does not give enough information for the complete description of a system. The specifications especially help people who study or use the diagrams but did not draw them themselves. They are also important for a model that will last a long time and must be clear to others besides the model's designer. The original designer might not be around anymore to give any further information.

An explanation follows for the content of the specification of the different elements. In general, every element specification has a header with the identification and the name of the symbol, and a section for the description of the element. In the description section, anything that specifies and describes the element in detail appears. This might be text, tables, graphics, values, conditions, decisions, equations, or references to other information sources.

The header of the *process specification* contains, in addition to the process name, the process number and, if used, the additional information as shown in the symbol on the diagram. Parameters with quantities can also be stated in the description, for example the time used to carry out the process. Such parameters might be useful for a later calculation or to make possible the step to the State Chart that includes the time-related behavior of a model (described in Sections S2, S3, and D1).

Specification S1-1: Process specification in general

⟨Process No. ⟩	**⟨Process Name⟩**	**⟨Additional Information⟩**
Description	⟨text⟩	

Specification S1-2 gives the details for the process "Cook spaghetti." The header shows the process number 2.2 as identification, the process name "Cook spaghetti" and the additional information "kitchen." The description gives a brief explanation how the spaghetti cooking is handled, as well as information about the cooking time. Specifications of bottom-level processes are longer and contain more detail than the ones of the higher-level processes.

Specification S1-2: Example of a process specification

2.2	**Cook spaghetti**	**kitchen**
Description	Two pots are necessary to cook 3 kg of spaghetti. Do not forget to put salt into the water! Time for boiling the water: 8.5 minutes. Time for cooking the spaghetti: 12 minutes.	

In the *flow specification,* an additional section for the flow type exists besides the flow name and the description. The indication of the flow type is needed for the structuring of the flows.

Specification S1-3: Flow specification in general

⟨Flow Name⟩	
Description	⟨text⟩
Flow type	⟨name of flow type⟩

The details of the flow "Spaghetti.raw" are stated in *Specification S1-4*.

Specification S1-4: Example of flow specification

Spaghetti.raw	
Description	Enough spaghetti for 25 friends at a party. On the basis of 100 g for each person and a surplus for the very hungry, this makes 3 kg of raw spaghetti.
Flow type	product flow

The specification of the *external entity* holds the name as the identification, as it is written in the symbol on the diagram, and the description of the element. If a small letter is used for the identification of the external entity, this is also given in the header of the specification.

Specification S1-5: Specification of external entity in general

[⟨Small letter⟩]	**⟨External Entity Name⟩**
Description	⟨text⟩

Anything that describes the external entity and clarifies its purposes is written in the description section. For example, the external entity "Supplier" can be described by including remarks on his conditions for selling, his delivery time, his quality, or the minimum purchase quantity.

Specification S1-6: Example of external entity specification

Italian grocery store	
Description	It sells the best homemade pasta and spaghetti sauces in town. Not open on Tuesday and Thursday mornings.

Specification S1-6 shows the external entity of an "Italian grocery store" with some comments in the rubric description.

1.6.2 Data Dictionary

A Flow Diagram model consists of several components. The set of Flow Diagrams make up the graphical component of the model. However, the details within those Flow Diagrams are documented separately in a data dictionary (DD), which is also a component of the model. It can be seen as a catalog that lists all the elements used in a Flow Diagram model. The DD, also called a repository, is a central database that organizes all specific facts about the Flow Diagram model. It consists of quantitative specifications and qualitative properties of the elements flow, process, and external entity. All the attributes and properties of any element can be found in the data dictionary.

The objectives of the DD are as follows:

- It stores and provides clear comprehensive information about the flows, processes, and external entities that make up the system.
- It is an ordered set of definitions. The most important role of any dictionary is to give a single place to look up an element and find its definition.
- It helps to establish a common and consistent vocabulary that can be used throughout the model. Aliases and multiple entries with equivalent definitions are avoided.

- It permits a model's designer to take a global view of the project. This may lead to the discovery of commonalities that might otherwise be missed.

Figure S1-2 shows the connection between the diagrams with the elements, the element specifications, and the data dictionary. The elements and the data from the specifications, as well as the organization of the hierarchical structure of the diagrams, are stored in the DD.

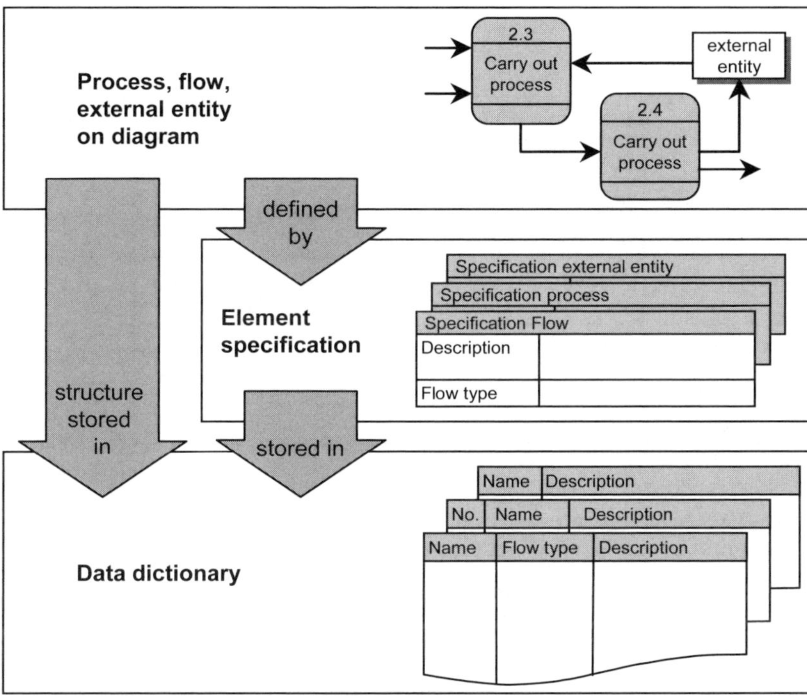

Figure S1-2: Elements and pertaining entries into the DD.

Table S1-8 and *Table S1-9* show an example of a DD for processes and flows, taken from *Diagram S1-1* (Chapter S1.2.2) illustrating the appendix surgery in a hospital.

Table S1-8: DD for processes

No.	Process name	Additional information	Description	Others
1	Examine patient	physician		
2	Remove appendix	surgery	Reservation of operating room necessary	
3	Nurse patient	patient's room		
4	Make facilities available			

The DD for processes lists at least the number, name, and additional information of the process (*Table S1-8*). Depending on the purpose of the model, the DD may be enhanced with further information about the processes.

Table S1-9 gives an extract of the DD for flows of the appendix surgery. The minimum information provided in the DD of the flows is the flow name, which serves as its identification, and the flow type. Again, the DD might be completed with further information about the flows, such as the flow description, the diagram where the flow appears, the origin, and the destination process.

Table S1-9: DD for flows

Flow name	**Flow type**	**Description**	**Others**
Person.ill	product		
Patient.examined	product		
Patient.operated	product		
Person.well	product	Person recovered from surgery and fit enough to leave the hospital	
Report.diagnosis	information		
Work.physician	workforce		
Equipment.surgery	material		
...			

The DD for external entities shows at minimum the name and the description of the elements and, if used, the small letter for the identification.

The rules for the element specification and the data dictionary are listed in the following table.

Rules S1-6: Element specification and data dictionary

F35	Every element from every diagram of the model must have an element specification.
F36	Every single element that appears on a diagram of the model must have an entry in the data dictionary.
F37	Each element from the data dictionary appears on at least one diagram.
F38	Access to the definition in the DD must be possible by the identification of the element.
F39	No redundancy is allowed in the data dictionary.

1.7 Setup of a Model and Recommendations

1.7.1 Components of a Model

A Flow Diagram model is the analysis of a system and, at the same time, the documentation of this analysis. A model consists not only of the diagrams, but of other components as well. The documentation of a Flow Diagram model must be complete. It contains the components listed in *Table S1-10.*

Table S1-10: Components of the model

Component of the model	Short explanation
Model name	Like every project, every model has a proper name.
Target of project	The target of the project should be stated.
Statement of purpose and viewpoint of model	The statement of purpose expresses the reason why the model is created. The viewpoint determines what can be "seen" within the model context, and from which perspective.
Diagrams	All diagrams of every level of the hierarchy together form the complete diagram-set of a Flow Diagram model. A list of the diagrams should be included in the report to give an overview of the diagrams, a "table of contents."
Information text	Besides the elements (process, flow, and external entity), every diagram has additional information text. It contains at least the date of the last update, the name of the diagram designer, and possibly a categorization.
Element specification	The element specification also includes a comment. The specifications form the verbal description of the model, and should be detailed enough to ensure the understanding of the model for any user at any time.
Data dictionary	The data dictionary lists all the elements and all their various properties.
Glossary	A standard terminology is used throughout the whole model to ensure precise communication. The fully defined terms are provided in the glossary, which also includes abbreviations, acronyms, and synonyms.
Comments from model's designer	The model's designer gives comments regarding the entire model. These comments should include what was done in the model, how to proceed, specific explanations, highlights, etc. The organization of the model might also be mentioned.
Evaluation report	The evaluation report is the report on the project, including findings, problems and suggestions.

Viewpoint of model: The context diagram defining the system boundary should present brief statements specifying the model's viewpoint and purpose. The viewpoint determines what can be "seen" within the model context, and from which perspective. Different viewpoints may be adopted that emphasize different aspects of the subject, depending on the model's designer and the audience. Important things in one model may not even appear in another model describing the same system from a different viewpoint.

Statement of purpose: The statement of purpose expresses the reason why the model is created and actually determines the structure of the model. The most important features come first in the hierarchy, as the top-level process is exploded into subprocesses. Those parts are further decomposed until all the relevant details of the viewpoint are adequately exposed.

The viewpoint of the model and the statement of purpose are either additional information on the context diagram or they can be written into the specification of the single process on the context diagram.

Glossary: A standard terminology should be used throughout the whole model to ensure precise communication. This is not easy in a large model. Normally acronyms, abbreviations, key words, or phrases are used for the naming. The fully defined terms should be provided in the glossary. This can be organized together with the data dictionary. It will help to standardize the model and keep order.

A description of the categorization of the flows is also part of the glossary. A legend should describe the flow groups and flow types, and the associated arrow style applied in the diagrams.

Comment of model's designer: The model's designer gives comments for the Flow Diagram model. These comments should include what was done in the model, how to proceed, specific explanations, highlights, etc. The organization of the model should also be mentioned, and an overview given.

Evaluation report: The evaluation report is the final report on the project. It should contain the following information: what was achieved, the findings of the As-Is analysis, the considerations for the To-Be model, what still needs to be done, the benefits, where the problems lie, and proposed solutions for the problems.

1.7.2 Modeling by Hand or CASE Tool

If a model is created by hand, without any assistance of a computer program, a separate list of all the elements should be made and updated as a data dictionary. Initially, it may be sufficient to keep a simple list with all the elements. As the model grows, and with it the dictionary, it becomes necessary to formalize the repository by using a simple database to manage the list or a more focused tool that supports the method directly.

Working with a CASE tool, the DD is created automatically because it is also the basis of the program for the drawing and handling of the diagrams. When a new flow or process is created or an existing element is edited, this change is automatically updated in the DD by the CASE tool. (CASE tool see also Chapter I2.5.1.)

1.7.3 Recommendations and Guidelines for Expedient Procedure

This chapter gives some recommendations and guidelines for the drawing of diagrams that make the procedure of diagram development easier.

Labeling with familiar names: It is important to use names that are generally understood for labeling the elements. Use names that are familiar within the company and that are obvious. The names of the symbols do not necessarily have to correspond to the expressions of production and cost theory.

First an easy case: When beginning to model systems, try using an easy case. Use a well-known product or a well-known process. Concentrate on good practice. Take a standard product or process. Do not indulge in exceptions and special requirements; they should be treated as a secondary priority.

First draft: The first draft of a Flow Diagram is usually done by hand while discussing the problem or project to be investigated. Don't be concerned that the first draft looks like a hopeless tangle. It can be sorted out.

Redraw diagram: Be prepared to start over. To derive the clearest possible diagram, modify or redraw the diagram several times until satisfied.

Verify context diagram: Setting the system boundary by the context diagram is not easy. Therefore, it is recommended to draw only a provisional context diagram if you begin with this. After the lower-level diagrams are drawn and a better understanding of the system is gained, go back and verify the context diagram. Some parts of the system may not be important for the investigation and can be placed outside the system, while others may be within the influence of the system and have to be taken as processes into the system.

Do not change process numbers: When a model reaches a certain stage of completeness and the data dictionary is comprehensively filled, do not change the process numbers anymore. Child diagrams, element specifications, and text descriptions refer to the process by number. In changing the process number, the reference gets lost and the documentation crumbles.

The process number is an identification number and establishes the reference to the other diagrams, the specification, and the data dictionary. It has no meaning as a counting number. Therefore, it does not indicate any kind of order or sequence of the processes. Missing numbers in the sequence may occur when processes are deleted while designing a Flow Diagram.

Cases: Go through cases. It is often very useful to follow a typical flow path all the way through the system and ask: "What needs to happen to this flow next?" "In what form does it leave this process?"

Procedure for setting up a model: For the setting up of a model, refer to the different procedures "walk through" and "walk around" as well as "top-down" and "bottom-up" in Chapter I1.5.3.

1.7.4 Recommendations and Guidelines for Easy Legible Diagrams

This chapter lists recommendations and conventions for the drawing of diagrams that make them easier to read.

Fit on a page: A Flow Diagram should fit comfortably on a page and be easy to read. Avoid complex Flow Diagrams with too many processes, flows, and external entities. Five to eight processes are ideal. This number is also dependent on the amount of flows on a diagram. With many flows, there a less processes; with very few flows, there can be more processes. If a diagram becomes crowded and confused, processes should be grouped into one process and only later detailed on an additional hierarchical level that is introduced into the model.

Flows entering and leaving process: It is easier to follow the production flow if the flows enter the processes in all diagrams systematically from the left or the top and leave the processes to the right or at the bottom. This applies especially for the product flow, as shown in *Diagram S1-21*.

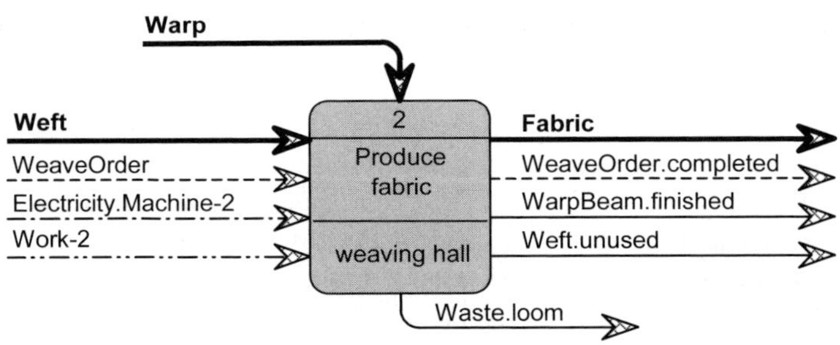

Diagram S1-21: Direction of the flows.

Bundle flows: To relieve higher-level diagrams, bundle flows into a single flow (parent flow) on the parent diagram, and split them into child flows on a lower level.

External entities on lower-level diagrams: Normally, the external entities are only shown in the context diagram and do not appear on the lower levels. However, they can be drawn on the lower levels of the model if it helps the reader to understand the diagram better.

If a Flow Diagram model represents a company with its hierarchy, it might be helpful to show the external entities on the lower-level diagrams. The employees work on different hierarchical levels within the company. Normally, they are interested in the diagram depicting their work and immediate environment. Other diagrams in the hierarchy or even the context diagram may not be of interest to them. It might, therefore, be practical to display the external entities also on lower levels.

Diagonally arranged processes: Avoid placing the processes on the diagram horizontally or vertically in a line. This arrangement can lead to paths of flows that contain many elbows, change of directions, and crossings. They are hard to follow.

With a diagonal arrangement of the processes, the flows become easier to follow and more comfortable to label, making the diagram clearer.

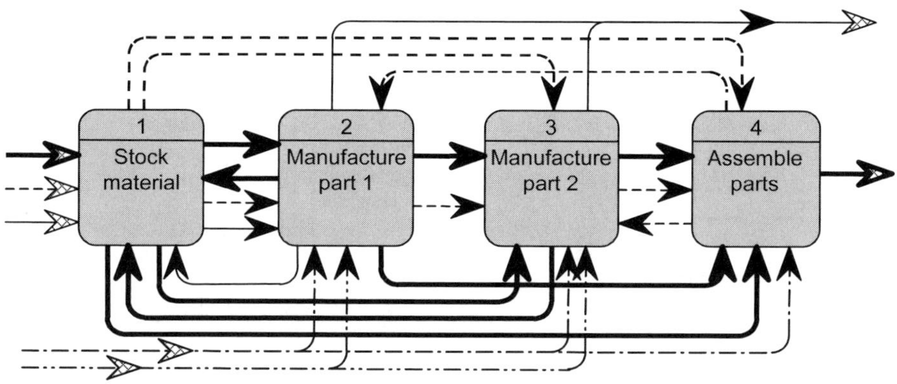

Diagram S1-22: Processes placed horizontally in a line.

Diagram S1-22 shows the same flows as *Diagram S1-23,* but the latter with a diagonal arrangement of the processes. Almost all the flows can be drawn with only one change of direction. Despite some crossing of flows, the paths in *Diagram S1-23* are clearer than those in *Diagram S1-22* with the horizontal arrangement.

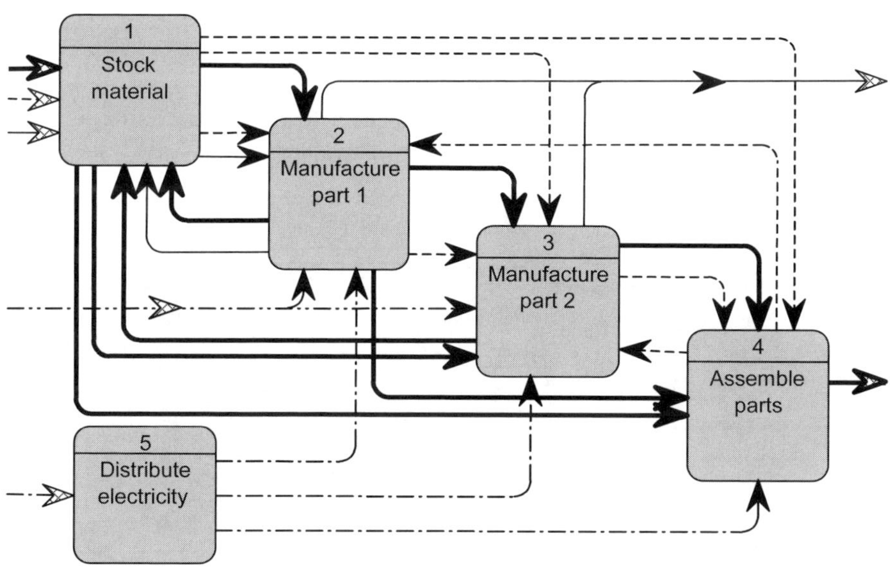

Diagram S1-23: Diagonally arranged processes.

The processes do not have to be strictly in a diagonal line. But with a generally diagonal arrangement, further processes can be more easily added, especially processes that distribute flows to several other processes. This is shown in *Diagram S1-23,* where process 5, which distributes the flows instead of a split, is added.

Labeling of horizontal flow arrows: The labels of several parallel flows are easier to read if written on the horizontal section of an arrow line rather than next to or over the vertical lines. As seen in *Diagram S1-24,* the labels of the flows connected to process 3 are far more difficult to assign to their flow arrows than those attached to process 2. The flow label may be above or below the line.

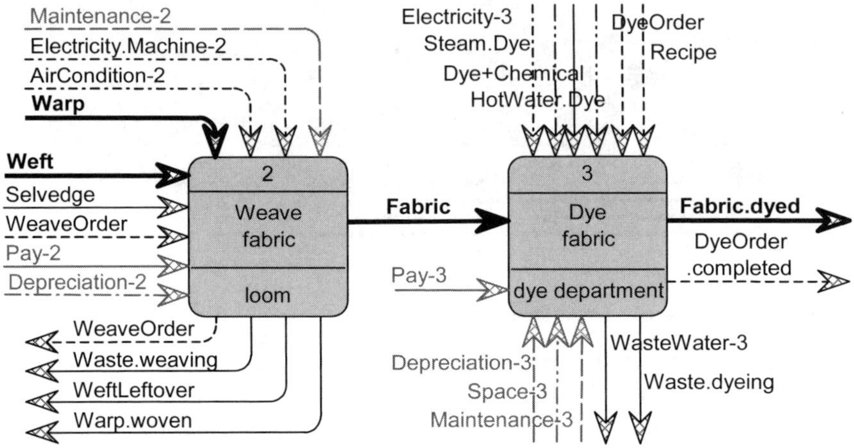

Diagram S1-24: Labeling of vertical and horizontal arrows.

1.7.5 Recommendations for System Optimizations

Some of the following recommendations also help to restructure and optimize a system and, in doing so, to derive the To-Be Model from the As-Is Analysis.

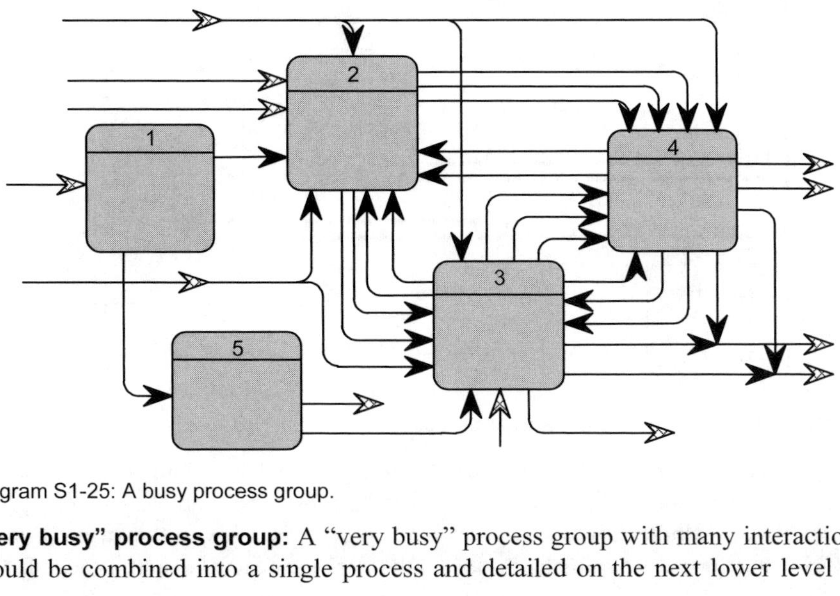

Diagram S1-25: A busy process group.

"Very busy" process group: A "very busy" process group with many interactions should be combined into a single process and detailed on the next lower level on the child diagram. For example, processes 2, 3, and 4 on *Diagram S1-25* are grouped into one process with the number 2, and then detailed into three processes on the next lower level.

Starve the process: The complexity of a process and eventually of the whole system results directly from the number of interfaces, that means the number of incoming and outgoing flows of a process. Therefore, the total number of flows should be minimized. This pertains to the number of flows on a diagram as well as the number of flows connected to a process (this is called "starve the process"). By avoiding interfaces, the complexity of the system is reduced, which is one of the main purposes of optimization.

Bubbling up: A process with many more attached flows than other processes in a diagram is a candidate to be divided into several processes ("bubbling up"). The gray process 1 in *Diagram S1-26* has a lot more flows going in and coming out than the white processes 2 and 3. The diagram has only three processes so far. So the best thing to do is to divide process 1 into several processes.

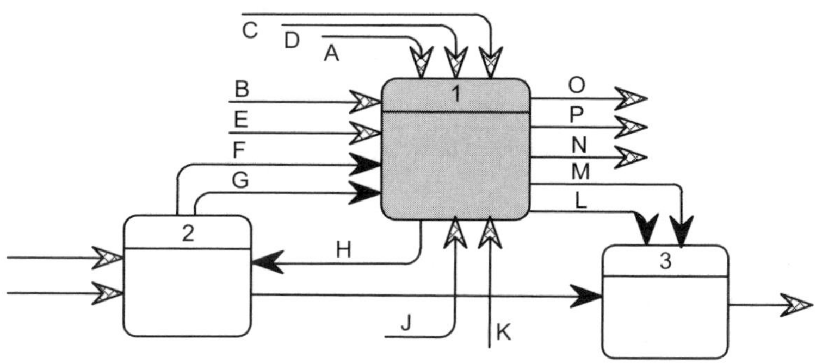

Diagram S1-26: Bubbling up.

Diagram S1-27 shows the original process 1 "bubbled up" into processes 1, 4, and 5. The flows attached to process 1 remain the same, as can be seen by the letters of the flow labeling.

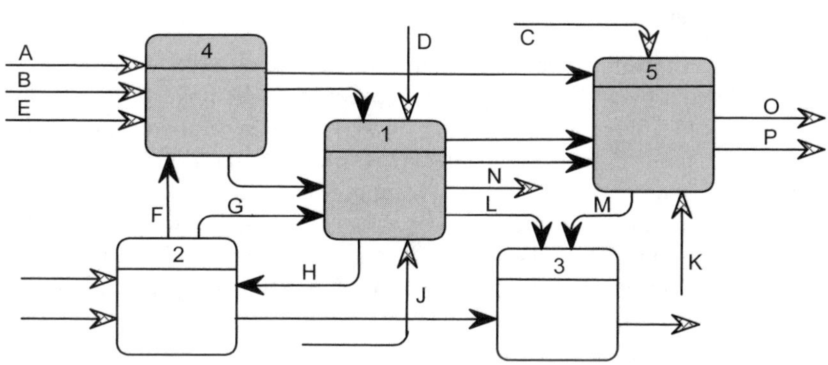

Diagram S1-27: Bubbled up process.

This is not a hierarchical detailing, but a cleaning out of the processes within the same hierarchical level. The "bubbled up" processes represent the same part of the system as the original process. However, the "bubbled up" processes now have, in comparison to the other processes on the diagram, a similar number of flows connected to them. The analysis of the system becomes easier and clearer.

Clustering processes: Clustering of processes is the grouping of two or more processes to form a single process. The goal is to cluster related functions into a single, more general process to eliminate too detailed information on a high level in the hierarchy. Such a process is then detailed into a child diagram.

Looping flows: Looping flows are an indication for potential optimization and are worth investigating more closely. Processes with looping flows are detailed on the lower hierarchical level, where hidden problems may be exposed.

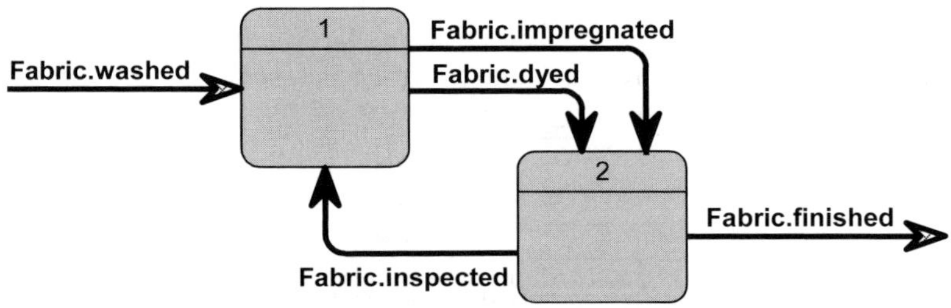

Diagram S1-28: Looping flow "Fabric."

Diagram S1-28 follows the looping product flow "Fabric." The "Fabric.dyed" goes from the floor inspection into process 2 and returns for impregnating to process 1. After the impregnating, it is further treated in process 2.

This going back and forth between the two processes is an indication of the need to investigate the processes on the next lower level. It will probably reveal a muddled and inefficiently conducted product flow. To know why the product flow goes this way, the lower-level diagrams have to be drawn. Based on them, the decision can be made, as to whether a simplification and an optimization of the flows respectively of the sequence of the processes are possible.

Follow up information flows: Due to information being able to be transformed, created, or deleted by a process, special attention must be paid to the information flows. It is best to follow the information flows through several processes. Where is the information created, where is it deleted? Is information created at one place and then used somewhere or is it created for nothing because it is never used again? Are there unnecessary copies of information made? Where and how does the information leave the system? Does the information go to the waste bin or is it stored in archives? Is there informal or verbal information, which is important and should be written down? It is worth answering these questions. A high potential for cost reduction lies in the information flows.

1.8 Application Example: Gas Station

This chapter guides you through the step-by-step modeling of a system. The system we will model is a gas station. The primary service is filling cars with gasoline. In addition, services such as windshield cleaning, car washing, and oil checks are offered, and restrooms are available. Furthermore, there is a small convenience store where one can buy food and newspapers.

Product flow

Begin the Flow Diagram model with the diagram where you best know the processes and flows. This will be one of the upper-level diagrams, probably the first or second level. In this case, it is the future level 1 diagram (*Diagram S1-29*).

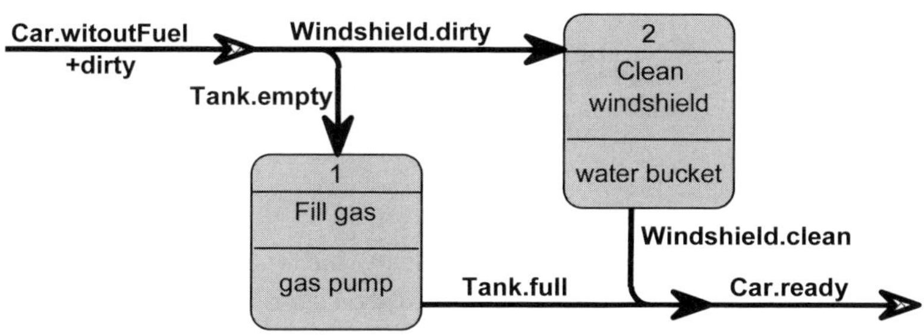

Diagram S1-29: Product flow.

Draw the path of the product through the most important processes. It gives a basic understanding of the processes and of the system as a whole. We "Fill gas" in our car and, during the filling, quickly clean the windshield with the provided water bucket containing soapy water and a sponge.

Resource and information flow

Then add all of the other flows, such as various materials, energy, information, and labor flows.

Some questions that should be considered are: What is done with the process output that is not used anymore? For example, is paper thrown away or is it stored in an archive? In the gas station example, what happens to the sales slip? Do you throw it in a garbage can or does the driver put it into his wallet because he can reclaim a refund for expenses?

Consider all types of waste. For example, what about the dirty water from the windshield cleaning? Should a flow "Water.dirty" be introduced in the model?

Diagram S1-30 depicts the processes we want to investigate in this model. Somebody else might also investigate further processes, such as "Check air pressure in tires" or "Organize shop." For the time being, we will keep our system simple.

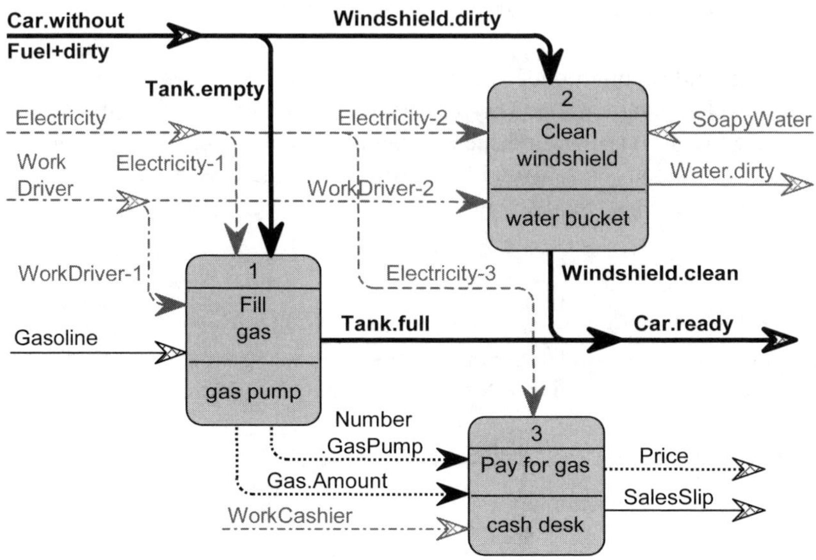

Diagram S1-30: Resource and information flows.

The driver fills the tank with gasoline and cleans the windshield. This is depicted by the flow "WorkDriver" and its two child flows. These flows do not represent the person of the driver, but his work, his invested time. Therefore, the flow can be split. A part of the work is needed for the refueling ("WorkDriver-1") and the other part for the cleaning ("WorkDriver-2").

Now, check the diagram against the rules. *Table S1-11* shows a checklist to help with a self-check for the model. This is a general checklist that can be used for any Flow Diagram model. In this application example, the checklist will continue with the subsequent checkpoints after each step of modeling.

Table S1-11: Checklist Flow Diagram

Question	**Checked**
The model has a name: "Refuel car."	✓
Each element (process, flow, external entity) carries a unique, informative name. Process name: ⟨verb⟩ + ⟨object⟩ Flow name: ⟨noun⟩ [+⟨optional addition⟩] External entity: ⟨noun⟩ [+⟨optional addition⟩]	✓
The process name makes sense for the activity carried out and the transformation performed on the content of the flows. Check the process names against inputs and outputs.	✓
Process names are consistent with the details on the lower levels.	✓
All flows are named.	✓

The flow name reflects the meaning of the represented object of the interface.	✓
Both the child flows and the parent flow are named. The names are different for every child and parent flow.	✓
Parent and child flows belong to the same flow type.	✓
Every process has at least one input and one output flow.	✓
The process fulfills the basic laws of nature such as the conservation of mass and energy, and the rules of economic balance.	✓
Everything going into a process comes out again.	✓
All the flows are connected at both ends at a process or a process and an external entity. There are no dangling flows or unconnected processes.	✓
Every flow going into or out of a process interacts with at least one other flow of this process.	✓
Between the start point and end point of a flow, there is no alteration of the object in place, time, state, or value.	✓
Every change in location is depicted by a transport process.	✓

Context diagram and system boundary

Once there is a basic understanding of the system to be modeled by the Flow Diagram, the context diagram that defines the system boundary is depicted. It determines what is inside the system and what is outside (*Diagram S1-31*).

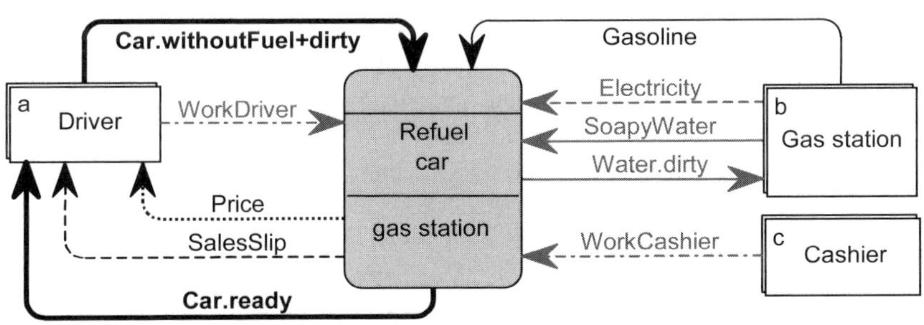

Diagram S1-31: Context diagram.

Drawing the context diagram is not as easy as it seems and requires various reflections. For example, a cashier operates the cash desk, meaning that the person is involved in the process "Pay for gas," but is not part of the system. Only the workforce of this person as a resource flow goes into the system, not the physical presence, or the body.

Setting system boundaries requires some decisions. Is the convenience store that sells candy bars and newspapers or the restroom part of our system? Do we plan to investigate and eventually change their function with regard to an optimization? If yes, they must be included in the model as processes. Otherwise, they are external entities.

The continuation of the checklist for the Flow Diagram model follows.

Table S1-12: Checklist Flow Diagram – continuation

At least one external entity is on the context diagram to specify the surroundings of the system.	✓
Every external entity has a unique identification.	✓
Every external entity is connected to the system by at least one flow.	✓
Every flow on the context diagram is connected to the single process and to one of the external entities.	✓
All external entities are shown in the context diagram. Lower-level diagrams expose no additional external entities.	✓
The process in the context diagram carries the number 0, which is generally not indicated on the diagram.	✓
All the functions to be investigated are within the model, represented by a process.	✓
There are no flows between two external entities.	✓

Detailed level

Detail the processes to the required level. Process 1 "Fill gas" of *Diagram S1-30* is exploded onto the next level, as shown in *Diagram S1-32.* It consists of the two processes "Pump gas," carried out by the gas pump, and "Handle gas pump," performed by the driver. The amount of "Gasoline" filled into the car tank is transmitted automatically to the cash desk, shown by an information flow "Gas.Amount." The number of the gas pump is given verbally by the driver to the cashier, also depicted by an information flow. Also, the flows "Handle.squeezed" and "Handle. released" are information flows created by the car driver in process 1.2.

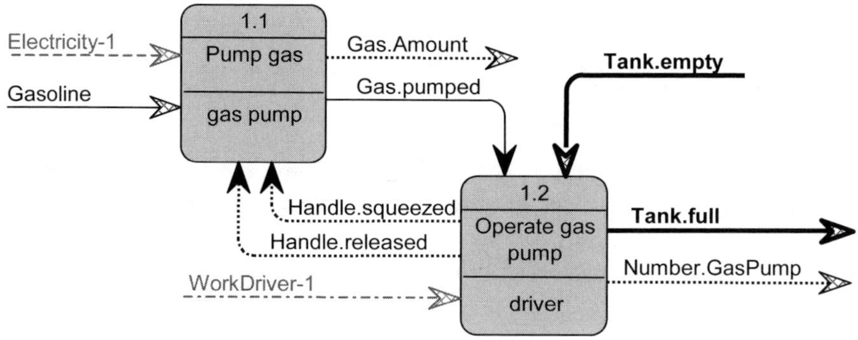

Diagram S1-32: Lower-level diagram.

Check the elements, the diagrams, the balance in the hierarchy, and the structure against the rules.

Table S1-13: Checklist Flow Diagram – continuation

The flows of the parent process and of the child diagram are balanced.	✓
The processes are sufficiently detailed in the hierarchy.	✓
Every process has its unique number indicating the hierarchical level.	✓
There are no dangling flows on a diagram. Every flow is connected at both of its ends to a process. One of these two processes might be on another diagram of the model (off-page flow).	✓

Principle of structuring

Once the model is completed, check to see that the processes are broken up in a useful manner. This is something that cannot be done from the beginning, but only if the model has a certain degree of completeness. *Diagram S1-33* gives the current structure of the gas station model.

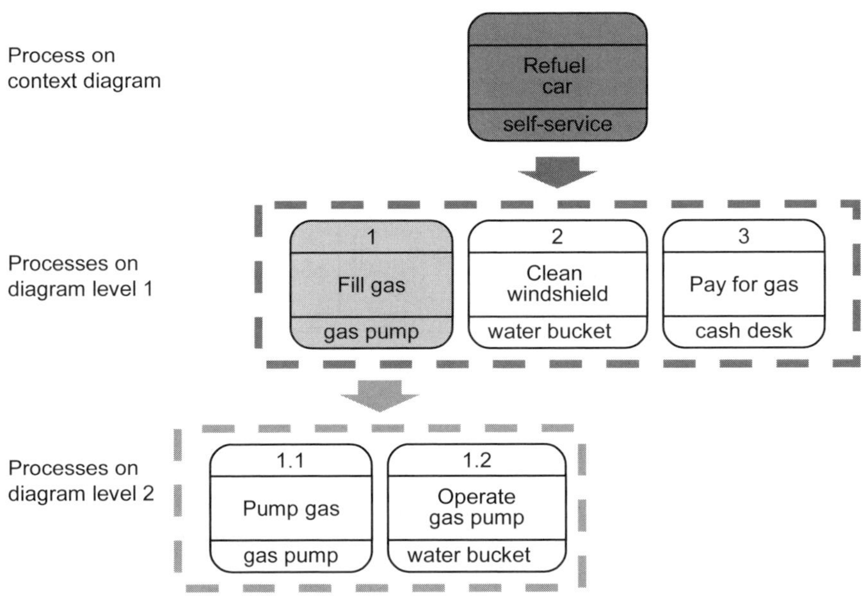

Diagram S1-33: "Refuel car" is the system you want to investigate.

The following are the types of questions that should be asked to evaluate the usefulness of the model structure and whether it needs to be changed:

- What are the goals of the project?
- What is the viewpoint that should be tackled? Is the right viewpoint emerging with the current structuring?
- Does the model include everything that needs to be investigated?

- Check that the structure is consistent with the investigated aspect. Is the functional structuring useful for this investigation? Would a geographical one be better? Was a process division used when a division by product would be more useful?

Imagine a gas station with a large number of served gas pumps. *Diagram S1-33* and *Diagram S1-34* show two different ways to structure the processes of such a large gas station. *Diagram S1-33* gives a functional breakdown, while *Diagram S1-34* shows a geographical breakdown: a service person is responsible for the operating of 5 gas pumps in the old-mannered way of a full-service station. The level 2 diagram then details into the processes "Fill gas," "Clean windshield," and "Collect cash." These processes correspond to the level 1 processes in *Diagram S1-33,* where the model starts with functional structuring, but they are written from the viewpoint of the serving personnel, hence the process names "Collect cash" instead of "Pay for gas."

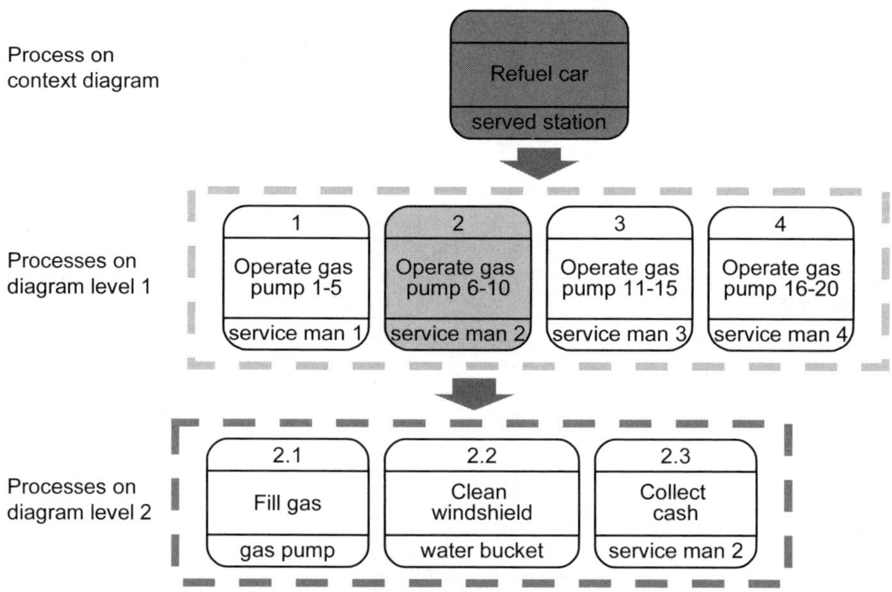

Diagram S1-34: Geographical hierarchy structure.

The modeling process involves deciding what to investigate and look at in detail. Those activities are placed within the boundaries of the system. The designer of the model makes this decision.

Scope of system

Diagram S1-33 and *Diagram S1-35* show two different scopes of the system. The current hierarchy in *Diagram S1-33* shows that "Refuel car" is the process of the context diagram. *Diagram S1-35* now illustrates how the process "Refuel car," highlighted in dark grey, becomes part of a bigger system, such as a service station where you may have the oil checked, tires filled with air, or a car wash. The hierarchy attached to the process "Refuel car" (the child diagram of "Refuel car") can stay the same.

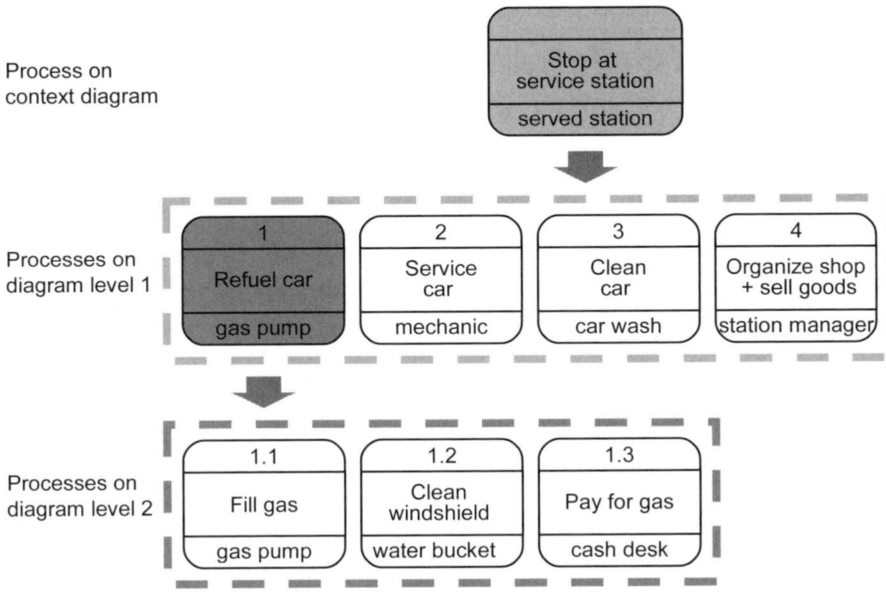

Diagram S1-35: Functional hierarchy structure, view of operating personnel at the gas station.

Change the model to obtain the best possible structure. It is worth the effort because this model will be used for a long time. It will be enhanced by further processes and levels. Different versions will be calculated and simulated. The model will be used to plan and realize the project goals. It will continue to be useful to introduce new employees and instruct the people in the company of the changes that are being made. There are many reasons to build up the best model possible. The effort in cleaning and debugging the model is worth the investment.

A model is typically restructured at least once after it is first created. While restructuring the model, almost all the parts will stay the same; they are just rearranged. The restructuring normally occurs on the first or second level. Such a restructuring demonstrates the advantages of the hierarchical structure.

Element specifications

This section shows how to fill in the element specification. Always keep in mind that the diagram should be readable for everybody. The goal is that the reader does not have to ask any questions about the diagrams to understand them. Every additional comment can be very helpful. Put all available information in the element specifications.

Specification S1-7: Process 0 from the context diagram

0	Refuel gas self-service
Description	Viewpoint: This model shows the processes of a self-service gas station, where the car driver is involved. Statement of purpose: Learn the operation of a gas station.

Besides the general description of process 0 “Refuel car” from the context diagram, *Specification S1-7* also incorporates in the description the viewpoint of the model and the statement of purpose. This information is part of the model and is best put into the specification of process 0. *Specification S1-8* gives another example of a process specification.

Specification S1-8: Process “Clean windshield”

2	Clean windshield water bucket
Description	Driver gets the water bucket and cleans the windshield while the gas pump is filling the tank. Afterwards, the driver brings the water bucket back to its place.

Specification S1-9 describes the flow “Quantity.Gas.” This is an information flow that gives the quantity of gasoline filled in the tank. This quantity is displayed on the gas pump and is also automatically transmitted to the cashier together with the price the driver has to pay.

Specification S1-9: Flow “Quantity.Gas”

Quantity.Gas	
Description	The driver decides how much gasoline he wants to fill in the tank.
Flow type	Information electronic

Specification S1-10 of the external entity “Cashier” also includes the operating hours of the cash desk. In this Flow Diagram model, the external entities are identified by small letters, written in the header of the element specification.

Specification S1-10: External entity “Cashier”

c	Cashier
Description	Cash desk occupied until 10:00 p.m. Afterwards only payment by credit card at the pay station is possible.

Data dictionary

While drawing the diagrams, we keep, of course, the data dictionary updated. In this small model of the gas station, the DD is created by hand, in the form of lists. *Table S1-14* shows the DD of the processes.

Table S1-14: Data dictionary processes

No.	Process name	Additional information
0	Refuel car	driver at gas station
1	Fill gas	gas pump
1.1	Pump gas	gas pump
1.2	Handle gas pump	driver
2	Clean windshield	water bucket
3	Pay for gas	cash desk

The data dictionary of the flows is given in *Table S1-15*. The list is not complete.

Table S1-15: Data dictionary flows

Flow name	Flow type	Subtype
Car.withoutFuel+dirty	product	
Car.ready	product	
Gas.Amount	information	electronic
Gas.pumped	gasoline	
Gasoline	gasoline	
Handle.squeezed	information	command
Handle.released	information	command
Number.GasPump	information	verbal
Price	information	verbal
SalesSlip	information	paper
SoapyWater	water	
Tank.empty	product	
Tank.full	product	
Water.dirty	water	
Windshield.clean	product	
Windshield.dirty	product	
WorkCashier	workforce	cashier
WorkDriver	workforce	driver
WorkDriver-1	workforce	driver
WorkDriver-2	workforce	driver

Table S1-16 lists the external entities. The small letter for the identification of the external entity is also given in the DD.

Table S1-16: Data dictionary external entities

Identification	External entity name
a	Driver
b	Gas station
c	Cashier

Check the data dictionary against the Flow Diagrams to make sure that they are consistent with each other. Apply the checklist for the element specifications and data dictionary.

Table S1-17: Checklist Flow Diagram – continuation

Every element in the model has an element specification.	✓
If possible, a description of the element is given in the specification.	✓
All the elements are defined and registered in the DD.	✓

Next steps

We now have a Flow Diagram model of the actual system. It is ready for optimization and reorganization. For example, flows passing between processes may need to be moved, added, or removed. The hierarchical structure of the model can be rearranged according to the processes.

Such a Flow Diagram model can be further utilized in the future. Within the method of POA, the following possibilities are available, as shown by the example of the gas station (*Table S1-18*).

Table S1-18: Next step in the POA toolbox

Aspect of investigation and optimization	POA-Method
What does filling up a car cost the gas station provider? What does a customer have to pay for the car wash to cover the costs?	→ Go on with the Value Flow Diagram. Section S2
Is there waste? How much energy is used in each process?	→ Use a Resource Flow Diagram. Section S3
Introduce the time-related procedure with its sequences and conditions in order to get a dynamic model. How many cars can be refueled per hour?	→ Do a State Chart. Section D1
Perform a simulation of the gas station to find out how many pumps are needed to plan an optimal gas station?	→ Program a simulation model. Section D3
Use the model to program the real-time control of a gas pump based on a State Chart.	→ Machine control. Section D3

1.9 Apply Your Knowledge

1) Product flow

The easiest way to draw a Flow Diagram is to begin with the product flow going through the main processes on the first-level diagram. For this, the product flow has to be known. *Diagram S1-36* to *Diagram S1-40* give several processes and extracts from diagrams. The product flows are indicated with bold arrows. They are not labeled. Some additional labeled flows are given for better understanding of the diagram. Insert the missing labels of the product flows.

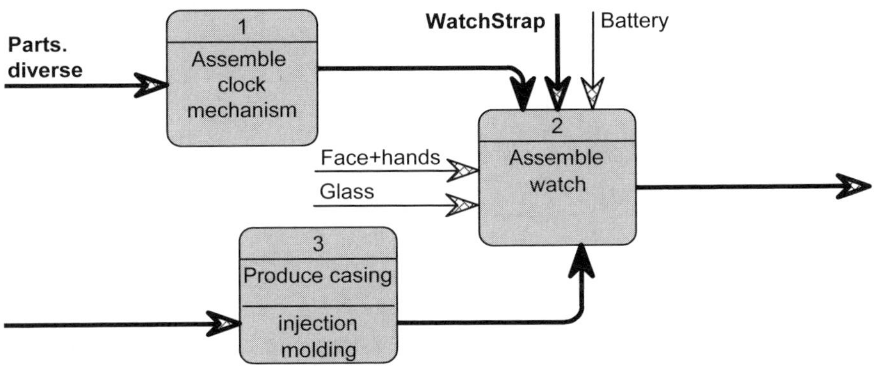

Diagram S1-36: Production of plastic watch.

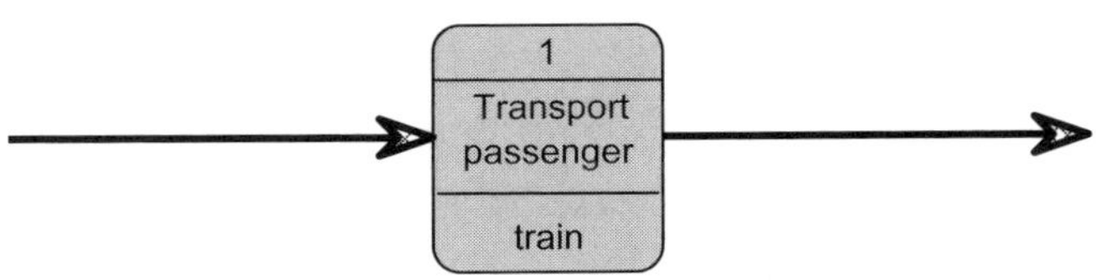

Diagram S1-37: Transportation passengers.

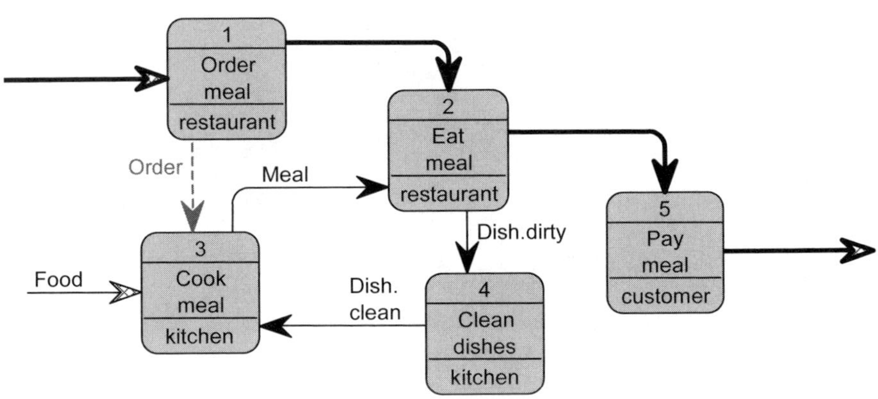

Diagram S1-38: Eating in a restaurant.

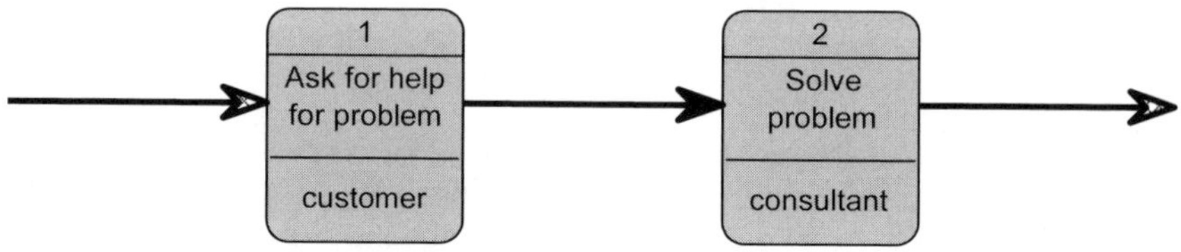

Diagram S1-39: Company hiring an external consultant.

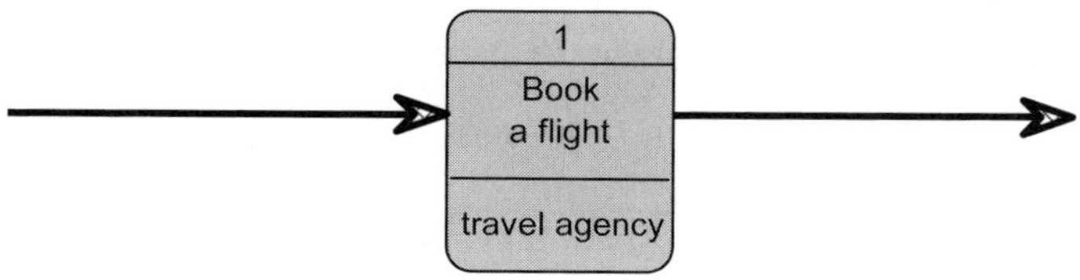

Diagram S1-40: Booking a flight in a travel agency.

2) Flow names

The flow names on the Flow Diagram of doing the laundry in *Diagram S1-41* are incorrect. They violate *Rule F7* (Chapter 1.3.5) that states that each flow name within a model must be unique. Re-label the flows with unique names and find a way to label them in a logical and easy to understand manner.

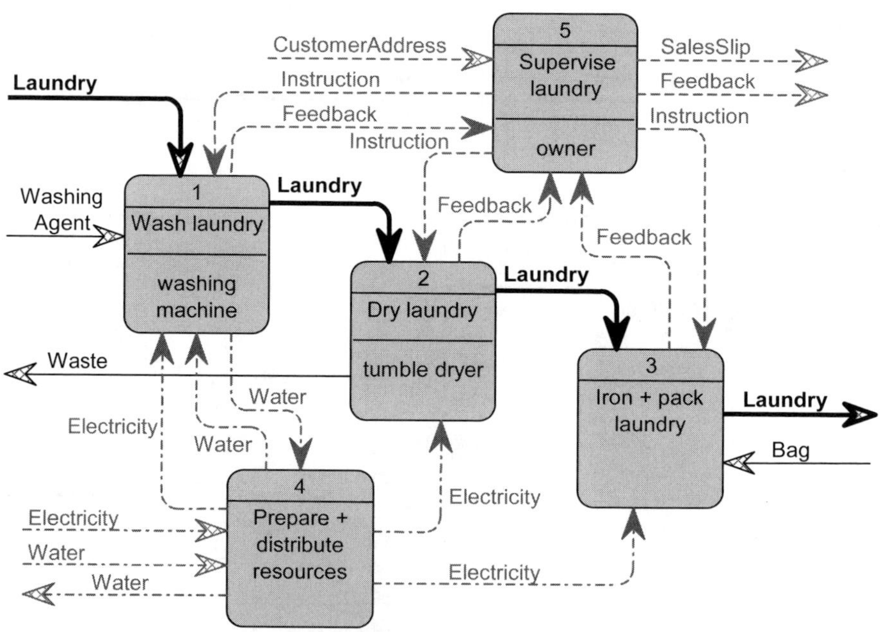

Diagram S1-41: Flow Diagram of a laundry.

3) Unfinished Flow Diagram

The Flow Diagram of the noodle production in *Diagram S1-42* is not yet complete. Finish the Flow Diagram and correct the mistakes. Quote the rules that are violated and give the correct solution.

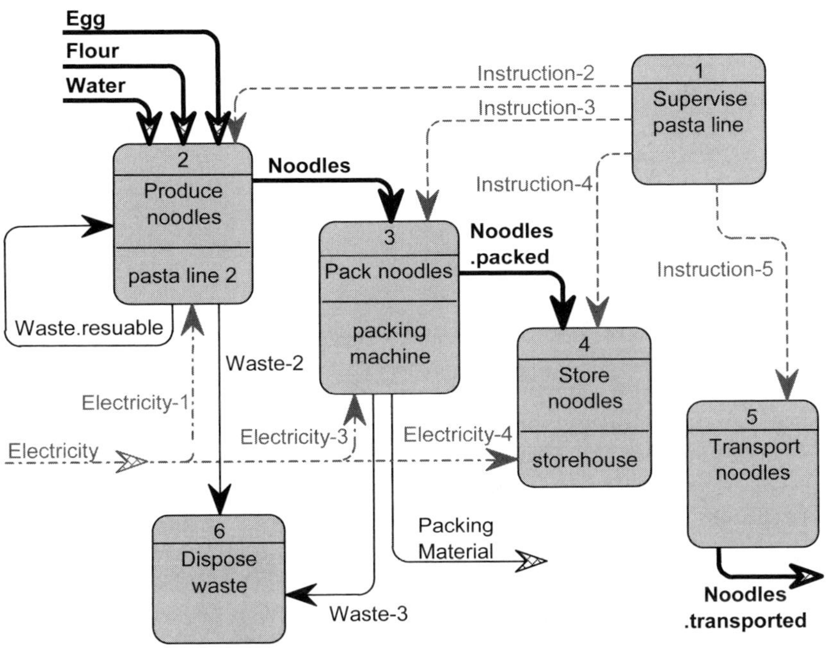

Diagram S1-42: Pasta lines producing noodles.

4) Context diagram

For this question, a system is described in words:

An industrial bakery is to be investigated. The bakery produces different products. The investigation contains only the production line for cookies. The production line of the cookies begins with the mixing of the dough, for which various ingredients are needed, and ends with the finished cookies, packed into a box ready for shipping.

The administration and the sales department of the industrial bakery are not objects of the investigation and should be indicated as external entities.

Draw the context diagram to set the boundaries of the system to be investigated.

5) Balancing rule

Diagram S1-43 describes the production of jeans. The detailing from the parent process to the child diagram is not correct. Look for mistakes in the balancing from the parent process to the child diagram. Put the corrections where it is more feasible; some will be in the parent, others in the child diagram.

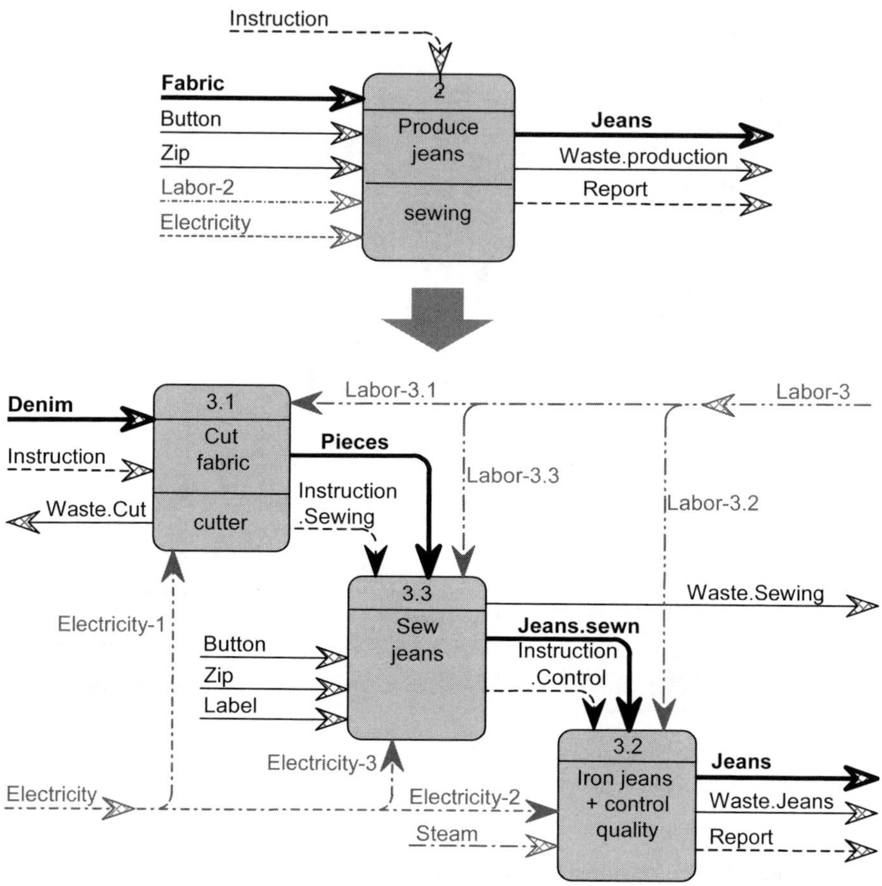

Diagram S1-43: Parent process and child diagram of jeans production.

6) Child Flow Diagram

The context diagram in *Diagram S1-44* shows the entrance to a sports stadium. It depicts the flow exchange with the external entities. The product flow is the “Visitor” entering and exiting the stadium. Draw the child Flow Diagram of the process “Enter sports stadium”.

The ticket check is done electronically when entering. Consider the huge mass of people attending sport events. Do not forget a process to single them out before the revolving door with the ticket check.

How are people treated without a valid ticket?

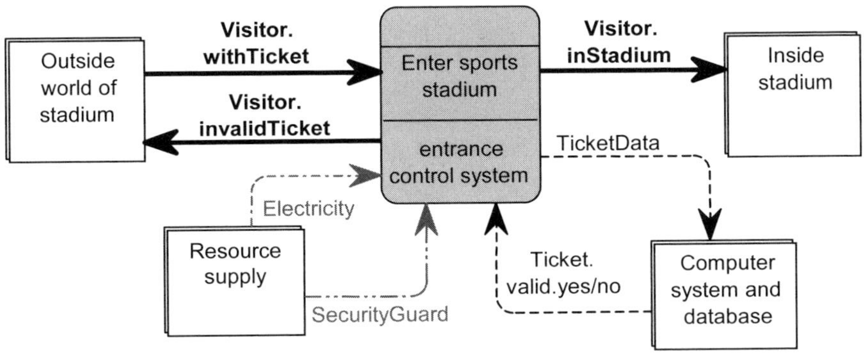

Diagram S1-44: Context diagram of a door for a football stadium.

7) Flow Diagram model

Draw the Flow Diagram for a winery; take into account the 4 to 6 most important processes. For more information on wine production consult the Internet. Depict the context diagram and Flow Diagram of the first level for this winery model. What is the product flow in this diagram?

8) Flow and process specifications

Write the specifications for the product flows "Flour" and "Noodles.packed" and the processes "Produce noodles" and "Store noodles" in *Diagram S1-42*. Besides a description, try to give as many parameters as possible, for the flows as well as for the processes.

S2 VALUE FLOW DIAGRAM

Goals of the Chapter

- Learn how to use a Value Flow Diagram.
- Enhance a Flow Diagram to get a Value Flow Diagram.
- Combine parameters, values, and costs in a company for the monetary value calculation.
- Calculate the value of a product from the value-generating processes.
- Draw a Value Flow Diagram with figures for investment analysis.

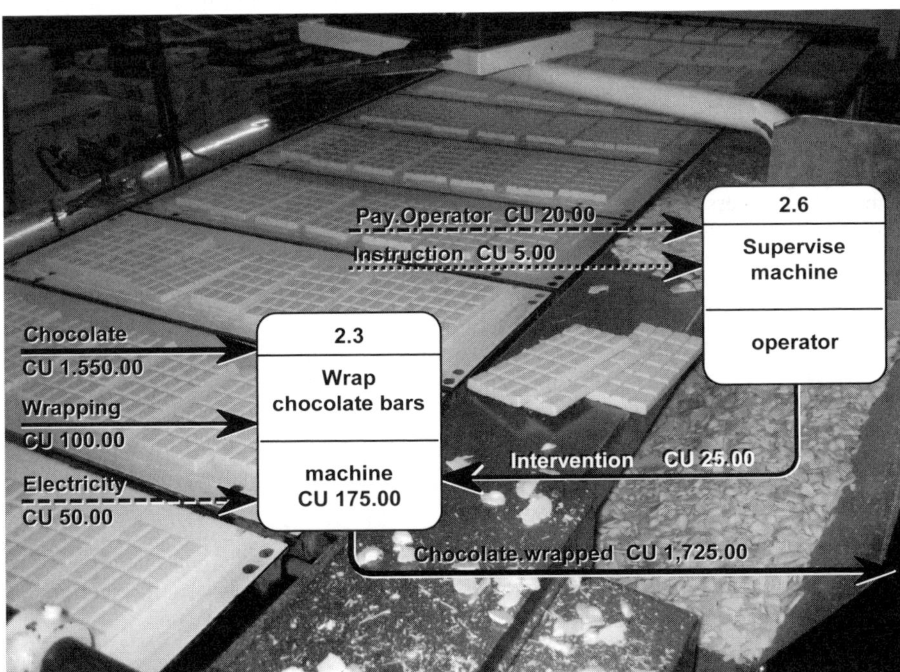

Figure S2-1: Production of chocolate calculated with the Value Flow Diagram (Photo: ETH Zurich).

2.1 Introduction

The Value Flow Diagram (VFD) is based on the Flow Diagram and depicts the manufacturing of a product through processes and flows. Each flow on a diagram, including the product flow, carries a monetary value. The value can be read out from the diagram at any stage in production. This means that the value flows in a company are matched graphically with the production flow. The value added can be followed step-by-step along the chain of production processes, and the origin of the costs is immediately visualized. This makes the VFD a graphical value analysis tool.

In a plant with mainly human workforce and little investment in machinery and equipment, cost calculation used to be simple. Manufacturing costs could be calculated on the basis of raw material consumed and labor hours spent on the product. In a modern and automated production, evaluation of cost is much more complex, since more cost drivers are involved, and issues as efficiency, complexity, and quality have a profound impact on profitability.

Many methods and tools help to model a production system graphically. However, they have rarely been applied for economical use, such as cost analysis or accounting. To date, there is no graphical tool that incorporates a full cost calculation. The VFD is an economic tool that uses graphics to its advantage. These graphics show very clearly if and where a company makes profit, notices losses, or generates wastes, and what apportionment of costs takes place.

The VFD answers many questions, including the following: What does a product cost? Where do the costs come from? Where can costs be saved or avoided? At which point in the production chain does the product have what value? Does an investment in a new plant pay off?

A VFD is created from a Flow Diagram by assigning monetary values and costs to the flows and processes. The specified and calculated values are shown in the diagram. The flows, as carriers of the monetary values, represent the interfaces between the processes related to the value. They carry the value from one process to the next. The process, on the other hand, designates the place of a value change by a new allocation and the value added.

The VFD and the Resource Flow Diagram (RFD) both calculate and display values. The VFD deals with monetary values, while the RFD described in Section S3 investigates energetic and environmental values.

This Section S2 introduces the drawing and calculation rules for the VFD. At the end of the section, the application example of the gas station from Section S1 is further extended to include the monetary calculations of the VFD. Case Study C2 presents a detailed investigation regarding monetary values of a product in a weaving mill.

2.2 Value Flow Diagram: Why?

2.2.1 Purpose

The Value Flow Diagram (VFD) analyses and calculates the monetary value of a specific product.

Detailed cost allocation: The VFD provides detailed cost allocation. The product costs and the value added of every production process are calculated for a given product. The VFD helps with the analysis, as well as the planning and development of products and its processes. An existing product and its processes are investigated for optimization and restructuring. With the VFD, a new product or process is analyzed for design and development, and the future costs are estimated. Costs are cut down by the elimination of non-value-adding activities. Expenses in time, money, or resources that add unnecessary costs to the product and reduce profits become visible and are easily identified.

Graphical context of value and production flow: The creation and increase in value during production is graphically shown in context with the production flow. The valuation of semi-finished or finished goods and products is quick and easy. The VFD shows the value of the product at any stage of the manufacturing. The value can be determined at any time and compared with the cost of purchased or outsourced alternatives. This makes the VFD a helpful tool for make-or-buy decisions: Should a company make a product or parts of a product itself, or purchase it from an external supplier?

Calculation of variants and change of production parameters: Calculation of the cost of product variants and product families is possible. Starting with the calculated cost of an already existing product, the cost of variants and new products of the same product family are calculated very easily by changing a few manufacturing parameters. If parameters are changed, such as the use of a new material or a change in batch size, a new calculation of the product value is performed. This can also be used as a forecast of the consequences of changing processes and production parameters, and it is helpful in evaluating production procedures or offering a special customer requirement.

Estimation and Benchmarking: Estimation and calculation of different options can be carried out. The critical parameters and their impacts on the product value are established. Borderline cases can be calculated and a sensitivity analysis performed. Values can be compared against benchmarks, either benchmarks of known values (for example of competitors) or generic benchmarks. Generic benchmarking is a fairly new procedure applied in POA. A virtual production is set up and calculated, assuming a plant operating in best possible conditions. This provides an overall target for improvement, not depending on the availability of detailed information on competitors, which is often difficult to gain and questionable in content. Compiling and calculation of generic data is treated in Section S3.

2.2.2 Application

A product investigated by the VFD can be a good or a service. With this, the VFD allows calculations for a production plant or service enterprise, such as a restaurant, hospital, or insurance company.

Diagram S2-1 gives a VFD that illustrates the cooking of spaghetti Bolognese. In the diagram, the values of the flows and the value added of the processes are indicated. The values are calculated for a spaghetti party with 40 guests. Besides the product flow, the values of various "Other Ingredients," "Electricity," and "Labor" are also indicated on the diagram.

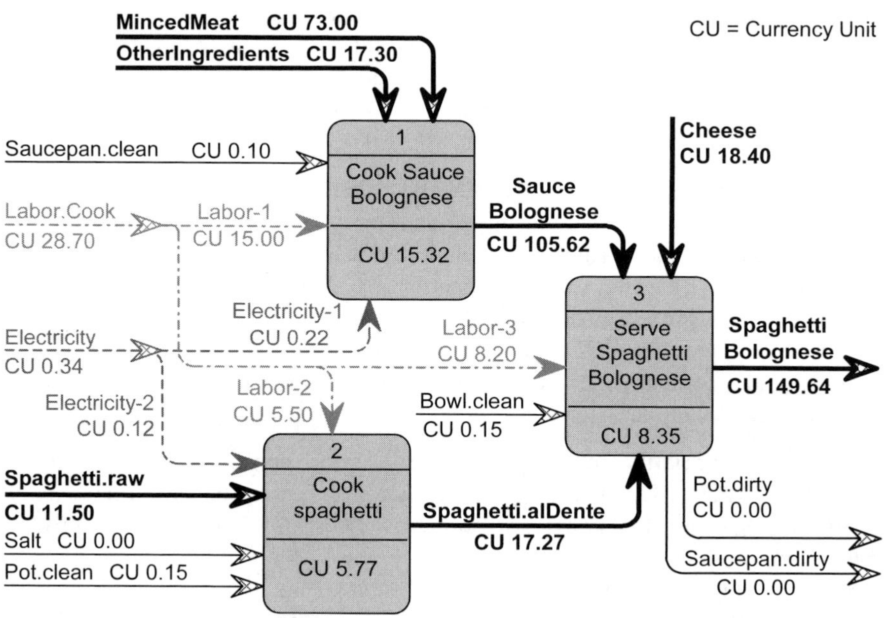

Diagram S2-1: VFD for the cooking of spaghetti.

From the diagram, several things can be noted. The "Sauce Bolognese" and the spaghetti are cooked separately. Authentic Bolognese Sauce has to be cooked for a long time to thicken (3 hours). Therefore, it cooks much longer than the pasta, but at a lower temperature. Including the time needed for the water to boil, the pasta takes 20 minutes to become al dente. This difference in time shows in the different values of "Electricity-1" and "Electricity-2."

The various cooking devices, such as "Pot," "Saucepan," or "Bowl," have a positive value when entering the process and a value of zero when leaving the process. The entering value indicates the work put into cleaning these items outside of the spaghetti cooking system illustrated in *Diagram S2-1*.

2.2.3 Delimitation

Cost accounting

The method of the VFD is not a new method for cost accounting. It uses the same foundation as the existing cost accounting of a company. Every type of cost accounting, such as full absorption costing, direct costing, activity based costing (ABC), or target costing, can be depicted by the VFD in a graphical form.

In general, cost accounting is not subject to a specific form. Structure and content can be freely shapeable and are up to the discretion of the model's designer. The VFD takes advantage of this by doing the cost accounting graphically. Therefore, it can be seen as an independent cost accounting method. By placing the focus on the product, however, the VFD allows a better tracing and allocation of the costs than traditional cost accounting.

Table S2-1 shows the differences between conventional cost accounting and the Value Flow Diagram.

Table S2-1: Comparison conventional costing to VFD

	Conventional cost accounting	**Value Flow Diagram**
Investigated system	Costs of the whole enterprise	Value of a single product
Basis for calculation	A period of time	A produced unit of one product (Standard Unit)
Representation	In tables with columns of figures	Graphical, refers to the production process
Structure	Cost centers; functional-spatial. Only loosely linked, if at all, to the production processes	Hierarchically structured in different levels of detail, directly related to the production processes
Aspect of calculation	Consumption of value and increase in value	Transformation of the values
Allocation of costs	Based on arbitrary chosen cost drivers	Based on production processes
Factors affecting value calculation	Production factors, arbitrarily defined charges and ratios	Value flows
Classification	In cost types	In flow types
Unit, reference	Cost unit	Standard Unit of the product flow

The VFD focuses on the product and the resources needed to make that product. In contrast, conventional cost accounting calculates the costs for all the products in a company, over a period of time. Conventional costing uses the terms "consumption of value" and "increase in value." The VFD is based on the fact that value cannot be consumed, increased, or created; it can only be transformed.

VFD and RFD

The VFD may also be connected to the calculation of energy using the Resource Flow Diagram introduced in Section S3. Similar to the VFD, the RFD is used to calculate and show the values of the energy, exergy, and other resources. The principles of value calculation are similar in both methods. *Table S2-2* shows the delimitation of the VFD in comparison to the RFD.

Table S2-2: Comparison of VFD and RFD

	VFD	**RFD**
Calculated value	Monetary value of information and resource flows.	Physical and energetic value (mass, energy, exergy, embodied energy) of resource flows.
Information flow	Shown on diagram, used for calculation.	Not used for calculation, normally not shown on diagram.
Value calculation of the process	Value added.	Embodied energy added, efficiency of energy and exergy.

Simulation

The VFD is a static method that focuses on the structure of the manufacturing processes, and their cost and value properties. It is a calculation tool, working like a graphical spreadsheet. The calculation is hidden in the process and flow symbols, and the values are displayed at the flow symbol. It is not a simulation tool. However, it can be easily turned into a static simulation. This means, the VFD can serve as a basis for a simulation model that integrates dynamic cost calculation.

If required, the time and logical order of events and conditions may be taken into account, and from there a discrete simulation program can be systematically derived (as described in Section D2 of this book).

2.2.4 Definitions

Value

The term "value" in the VFD method is used as follows:

> **Definition:** "Value" is a generic word for the different terms of cost accounting, such as costs, expenditures, expenses, receipts, income, profits, returns, earnings, performance, output, etc. The value is given in a currency unit as specified for the analysis.

The term "value" implies that a product has a monetary value, and that value is added to the product during the process of making it available to the customer. Each item in production has a monetary value.

The term "cost" relates to charging a resource and has a somewhat negative connotation. In terms of cost accounting, "cost" is just one of the possible monetary values used. "Value" is a generic term, and the most neutral term to represent all the terms of cost accounting listed in the definition box above. Value is an artificially assigned property. Rules for value assignment depend on the accounting principles applied in the investigated system.

Currency Unit

The values in the VFD are given in monetary terms, meaning all legal tender, for example €, $, £, or CHF. In this book, the neutral term "Currency Unit" (CU) is used.

Standard Unit and Reference Unit

The *Standard Unit* and the *Reference Unit* carry out an important function for the calculation in the VFD model and need to be examined more closely. They are necessary for the standardization of units and values, so that they can be calculated together in the process and within the hierarchy of the entire model. Both the Standard Unit and the Reference Unit represent a specified unit of the product that is calculated in the model.

Standard Unit: All values indicated on the diagrams of a model refer to the same quantity: the Standard Unit (SU). It is one unit of the product manufactured in the company and analyzed by the VFD. That means that in a VFD, the given monetary values (CU, €, $) are the values for one SU. The SU quantity is valid for the entire model, including all the different diagrams that build up the model.

> **Definition:** The Standard Unit (SU) is one unit of the product in the investigated system and serves as a common basis for the calculation of all values.

The cooking of spaghetti in *Diagram S2-1* depicts the values in the neutral CU for the Standard Unit of a party with 25 guests. Other quantities used for the SU are the quantities by which the company usually sells its product to the customer, or a multiple of these quantities: 100 pullovers, insurance policy for 1 person, booking for 3 flights, 1 appendix surgery, 1,000 pencils.

Example 1 of weaving mill: SU = 100 m of finished fabric
Example 2 of industrial bakery: SU = 10,000 croissants

From the point of view of conventional cost accounting, the SU corresponds to the so called "cost unit."

Reference Unit: The Reference Unit (RU), like the SU, is one unit of the product investigated in the model. However, the RU is another unit of the product than the SU and is only used in a single process or a specific process group for convenience of calculation. For example, the batch size or the capacity of a machine in a specific process.

For further calculation and display on the diagram, the value per RU is transformed into value per SU.

Example 1 of weaving mill:

in the weaving preparation	RU1 = 1 warp
for the finishing department	RU2 = 1 dye batch

Example 2 of industrial bakery:

in the preparation	RU1 = 400 kg of dough
in the packaging	RU2 = 1 bag with 6 croissants

Conversion of RU to SU

There is only one SU in a model, but there can be many RU. All values for a RU are eventually converted and expressed in relation to the SU. The conversion of each RU to the SU must be defined by the model's designer. *Table S2-3* gives the conversion from RU to SU using the example of the weaving mill.

Table S2-3: Conversion from Reference Unit to Standard Unit

SU, RU	**Quantity**	**Unit**	**Conversion RU → SU**	**Explanation**
SU	100	m fabric		Standardized length of fabric. Valid for all fabrics of the company.
RU 1	1	warp	18.36 SU	For the weaving, the yarn is rolled up parallel on a beam. A warp results in 1,836 m fabric.
RU 2	1	dye batch	1.12 SU	Load of fabric of a dyeing machine.

If an analysis of a system is done with both the VFD and the Resource Flow Diagram, the same SU and RU are used for calculations of mass, monetary values, and energetic values.

SU and RU in the model

Diagram S2-2 illustrates the use of the SU and the RU in the process hierarchy of a model with the examples from *Table S2-3*. Within a single process or a process group, the calculation can be done in the RU: RU1 "warp" is used for the value calculation in two processes 1.1 and 1.2, and RU2 "dye batch" is used in process 3.3. For the calculation in the model as a whole, the process balances, the calculation in the hierarchy, and the values shown on the diagrams (flow value and value added), the SU is used.

Using only the SU to perform calculations in a model is possible, but can be inconvenient. With the Reference Unit, calculations in the model are simplified. The RU is introduced in this method because most of the data and figures in a plant are related to a process-specific unit (i.e. the load of a machine). The people working in a plant know just the process-related parameters. All the internal calculations are done based on the RU.

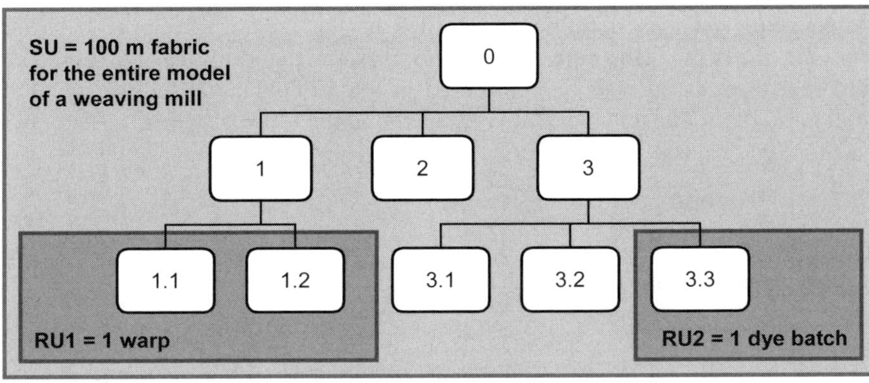

Diagram S2-2: SU and RU in a model.

An example of an RU in a weaving mill is the "warp," as shown above in *Diagram S2-2*. All the declarations, such as time, amount of material, and man-hour, are related to the "warp." The employees of the weaving mill do all the calculations they need for their work with the RU "warp." In these processes, the employees can only provide numbers or quantities for the RU "warp." They cannot provide information directly in terms of the SU "100 m fabric" because the length of the warp is transformed into a fabric of shorter length during the weaving process, and the fabric then shrinks further in washing and in the finishing process. It is only possible, however, to convert the RU "warp" to the SU of "100 m fabric" and to calculate the value of the flows for the SU for the whole weaving mill after having calculated the entire model.

The rules for the VFD are numbered and marked with the prefix V.

Rules S2-1: Standard Unit

No.	Rule
V1	There is only one Standard Unit (SU) in a model. All values indicated on the diagrams refer to this SU.
V2	The final values of the calculation are given in Currency Unit per Standard Unit.

2.3 VFD Elements

2.3.1 From Flow Diagram to VFD

In Chapter S1.4, the elements of the Flow Diagram are introduced and described. The elements and their definitions and rules are the same for both the Flow Diagram and the VFD. However, in order to perform the value calculations, some additional features have to be considered in the VFD. This chapter explains the additions and differences of the VFD compared to the Flow Diagram.

2.3.2 Process

In addition to the general process definition of the Flow Diagram, the value aspect has to be considered for the VFD.

Definition: Each process causes a change of value – each change of value requires a process.

For example, the storage of goods is a process. Storing is a process that helps to pass or bridge time for resources. The space to store goods, the handling of the goods, and the organization of data cost money. Therefore "Storing goods" is a process causing a change in the value of the stored good.

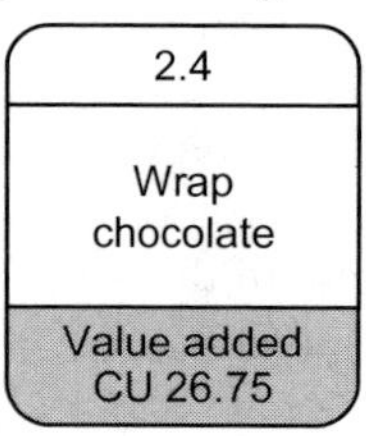

On the VFD, the *value added* of a process can be shown on the diagram in the process box itself. It appears in the third field at the bottom, which is reserved for additional information. The value added of a process expresses the increase of value that the product receives while being treated in this process.

2.3.3 External Entity

External entities represent the external environment. They are the suppliers and receivers – source and destination – of the system and connected to it by the flows going into and coming out of the system. As they do not belong to the investigated system, they are not needed for the value calculation and therefore, do not have values in the symbols on the diagram or in their specifications. The flows, as the interfaces to the external entities, carry all the necessary value information for the calculations.

2.3.4 Value Flow

In the VFD, the flow carries the value; it is therefore called the value flow. The flow, as drawn on a diagram, consists of the object and the value that is carried by the object. The object provides the name and value for the flow. It is also called the flow content or value carrier. The value is written in the monetary unit near the flow label, normally together with the Currency Unit (CU). *Diagram S2-3* shows the flow "Machine" going from the origin process 1 to the destination process 2 with its value of 952.00 CU.

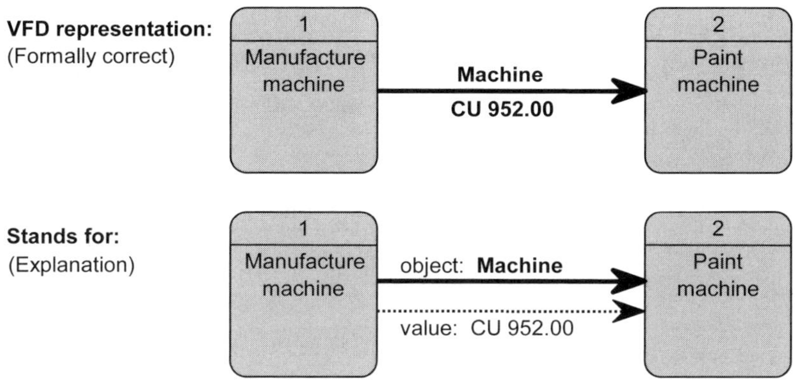

Diagram S2-3: Flow with object and value.

Positive and negative value flows

Positive value flow: The flow direction, relating to the value of the object, is specified by the arrow drawn. For explanatory reasons, the value flow "Machine" on *Diagram S2-3* is taken apart to show how the value sticks to its object "Machine." "Machine" is a positive value flow.

Negative value flow: The value of a flow can also be negative. *Diagram S2-4* explains the flow with a negative value through an abstract example. As with the positive value flow, the negative value sticks to the flow. Either, a negative value goes in the same direction as the object of the flow, or a positive value goes in opposite direction of the flow.

With a positive value flow, the origin process sends the object and value indicated to the destination process, following the direction of the arrow. With a negative value flow, the origin process receives the value for the object that is sent to the destination process.

The positive and negative value flow can also be explained by the vector geometry and graph theory: A positive flow going in one direction becomes negative by reversing the direction of the arrow. The sign of the value is related to the direction of the arrow. Generally expressed: If the arrow of a value flow is reversed, the value is reversed as well; positive becomes negative value, and negative becomes positive value.

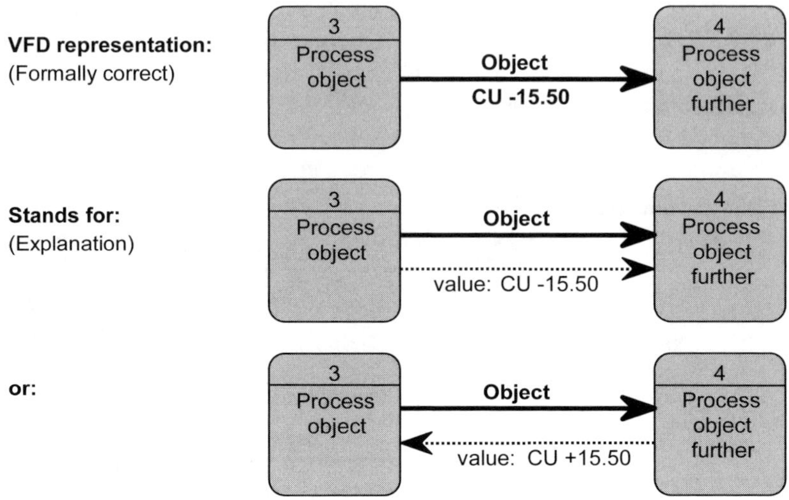

Diagram S2-4: Negative value flow.

Debit and credit by a process

Diagram S2-5 explains the difference between a positive and negative value flow regarding the cost accounting terms "debit" and "credit."

- A positive input value or a negative output value debits the process.
- A positive output value or a negative input value credits the process.

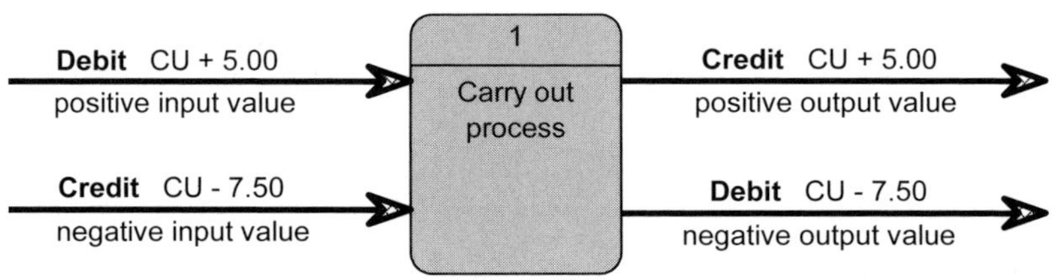

Diagram S2-5: Debit and credit by a process.

The examples in *Diagram S2-6* and *Diagram S2-7* give further explanation of the debit and credit of a product flow.

In *Diagram S2-6*, the positive flow "Transport.Cost" and the negative flow "VAT" (Value Added Taxes) debit the "Product" through the respective process. This means that the values of these flows are charged to the product flow by debiting a process. Debiting a product by a process means the product leaving this process carries with it an additional value. The product has more value, it becomes more expensive.

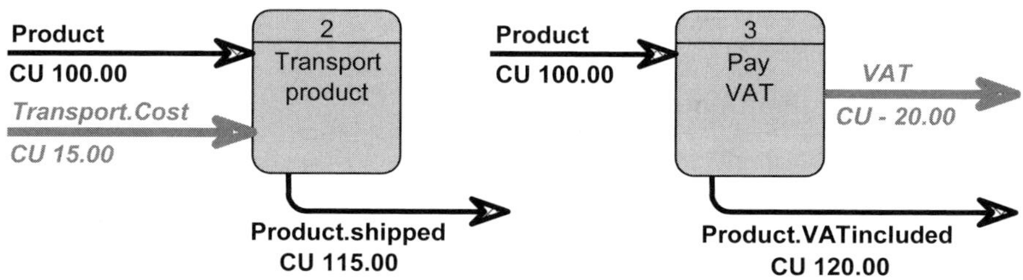

Diagram S2-6: Debit of the product flow.

Diagram S2-7 illustrates the credit. The positive value flow "Cream" credits the product as it leaves process 5 "Skim milk." The value of the product flow "Low-FatMilk" is reduced by the value of the cream as it leaves the process.

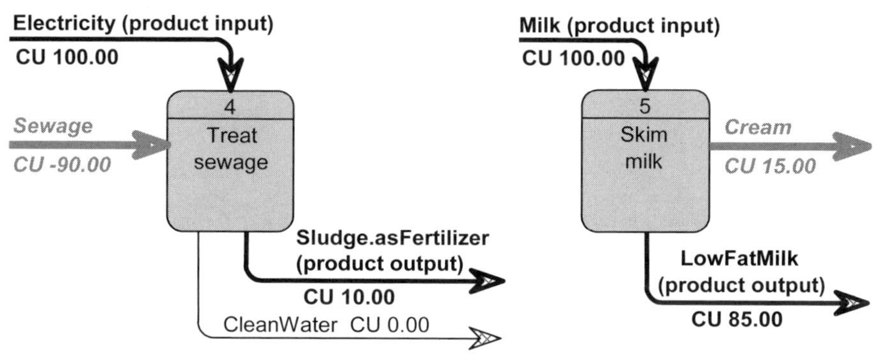

Diagram S2-7: Credit of the product flow.

The credit of the product flow by a negative input flow, is shown here with process 4 "Treat sewage." For the treatment plant, the "Sewage" holds a value, because it generates "Sludge" used as fertilizer out of it. The resulting energy costs "Electricity" are higher than the proceeds for the fertilizer. The people producing "Sewage" (household, company) have to pay for the treatment. This is why the sewage flow comes with a negative value into the process.

Rules S2-2: Value flow

V3	Each flow has a value that is given in a monetary unit. The value may be positive, zero, or negative.
V4	For the calculation, a positive value flow in one direction represents the same value as a negative value flow in the opposite direction.

2.4 Flow Types and Flow Categories

2.4.1 Classification of Flows

In Chapter S1.4.4, the classification of the flows into flow types and flow groups was introduced. In order to enable the value calculation in the VFD method, the classification of the flows needs to be extended by the flow category.

Flow category

Flow categories are necessary to perform different kinds of value calculations in a model. For the calculation of the product value, various operations in cost accounting need to be carried out. Apportion of costs, monetary transactions, or value added are all examples of cost accounting operations. To make these operations in the VFD possible, the value flows are classified into four flow categories:

- resource and information flow
- product flow
- fictitious value flow
- money flow

Besides the basic rules for the flows, each flow category has additional rules for the different value calculations. The flow categories and their rules for the value calculation are defined by the VFD method and their rules explained in the following chapters.

Flow type

The classification of flows into flow types is indispensable for the value calculation in the hierarchy. Flow types are created by bundling flows with the same physical characteristics. The flow types define the flows that are combined by calculation throughout the hierarchy. They ensure that only the flows related to each other are summed up or detailed in the hierarchy. The flow types are defined by the model's designer according to the need and the viewpoint of the system to be investigated. They correspond to the cost types in the cost accounting.

If necessary, the flow types can be further classified into subtypes. The classification into flow types and subtypes is supported extensively by the method of splits and merges. Flows of different subtypes are allowed to be split and merged as long as they are of the same flow type.

Flow group

The flow types have the option of being combined into flow groups to make the diagrams easier to read. This classification into flow groups is an organizational feature to keep diagrams with many flows clearer. It has no influence on the value calculation. For the analysis of a system and calculation of a product, all flows are needed and taken into the VFD model. For discussions, however, depending on the viewpoint of the analysis and the people involved, certain flow groups might be omitted from the diagram for easier reading. Flow groups are optional; the model's designer decides if and how the flow types are grouped together.

Relation of flow classifications

Table S2-4 summarizes the features of the three classifications. The categories are given by the method and are classified into the flow types defined by the model's designer. The flow types may be grouped together.

Table S2-4: Comparison of the flow classifications

Classification is	**Flow categories**	**Flow groups**	**Flow types**
defined by	VFD method	model's designer	model's designer
valid for	every model	one specific model	one specific model
mandatory	yes	no	yes
aimed at	the rules for the value calculation and the different types of calculations in cost accounting.	the clarity of diagrams.	the calculation in the flow hierarchy.

Following, the rules for the flow classification are given.

Rules S2-3: Flow classification

V5	Each flow belongs to a flow category as well as to a flow type.
V6	Parent and child flows of a split or merge belong to the same flow type.
V7	Only flows of the same flow category can be combined into a flow type.
V8	For the calculation in the hierarchy, only flows of the same flow type are allowed to be summed up or split.

Example of a flow classification

Table S2-5 shows the relations of the three classifications flow category, flow group and flow type, and flow subtypes for an exemplary model. The classification into flow types can be emphasized by a different line style and color. The flow classification and the line styles are specified by the designer for one model. Flows belonging to the same flow type are drawn identically.

Table S2-5: Relations of the classifications

Flow categories	Flow groups	Line style (example)	Flow types (example)	Flow subtypes (example)
product flow	product	fat solid blue	product	
resource and information flow	material	solid blue	various materials	
		solid blue	waste	
	resources	solid brown	water	fresh water sewage
		dashed brown	electricity	electricity machines electricity light
		dash-dot brown	thermal energy	oil for steam generating oil for heating propane
	information	solid green	paper	production paper weave order
		dash-dot green	computer	customer order production tracing
		dotted green	verbal	instruction feedback
	monetary resources	solid violet	pay (for labor)	pay operator pay service pay administration
		dashed violet	space	shop floor storage office
		dash-dot violet	maintenance + depreciation	maintenance depreciations + interests
		dotted violet	taxes	
fictitious value flow	fictitious	solid red	none	
money flow	money	dotted black	cash, check	
			bank transfer, bill	

Diagram S2-8 "Produce fabric" illustrates how the flow types and their different arrow styles defined in *Table S2-5* are used on the diagram, here depicted in black, gray, and white. This context diagram shows only part of all the flow types defined in *Table S2-5*. Further flow types and subtypes are used in the underlying child diagrams.

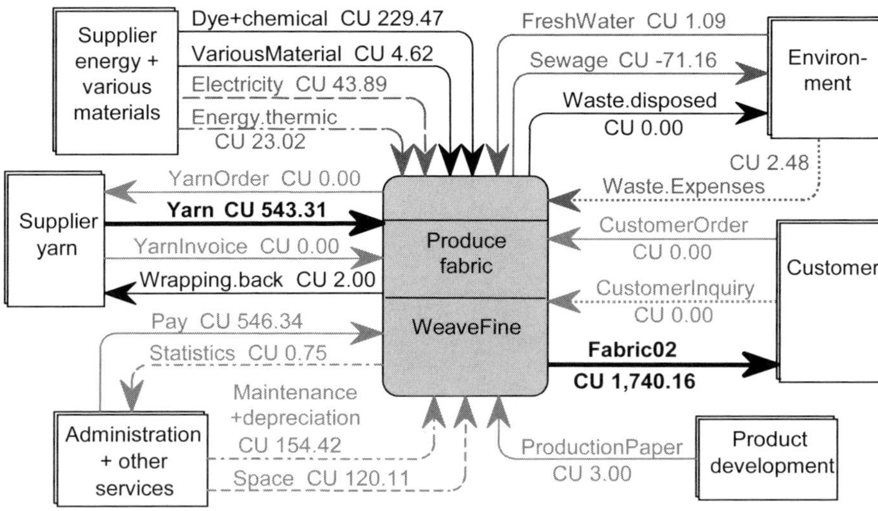

Diagram S2-8: Example of styles of flow types.

Value display on the diagram

It is not always desirable to display all of the values and value flows that where included into a model on the diagram. The model's designer chooses the values that are shown on the diagram. Generally, the interest lies in the increase of value of the product on its way through the production. Depending on the problem, it may be desirable to only look at some flow types and a specific kind of value. Other values that are not of interest at the moment may be suppressed and hidden to improve the clarity of the diagram. For this, the classification of the flows into flow groups is helpful.

View of the investigated system

The values on the diagram are always investigated from the viewpoint of the system. Not everything that has a value for the system has also a value for the external environment. Conversely, something that has a value for the external environment might not have a value for the system. The following are examples from the manufacturing of pasta:

- Incoming wheat flour or final package of pasta have a value for the external environment, as for example a customer. It can be sold.
- Intermediate product dough or processed customer order have no value for the external environment.

Note! With the diagrams in this Section S2, the theory of the VFD is explained and illustrated. Often only part of a diagram is shown, with one or two processes and only some flows.

2.4.2 Flow Category: Resource and Information Flow

The majority of the value flows in a model are resource and information flows. These flows have an object and a value. The object is, for example, information (data), material, resources, or energy. It defines the flow values by its quantity.

Rules S2-4: Resource and information flow

V9	An information flow can be transferred, created, or deleted by a process.
V10	Resource flows are transformed in a process. They cannot be created or deleted. The laws of physics are valid.
V11	A process with a resource flow input must also have a resource flow output. A process with a resource flow output must also have a resource flow input.

2.4.3 Flow Category: Product Flow

The purpose of a company is producing and selling products. A product can be a manufactured good or a provided service. All costs and values occurring in a company are caused by making or providing a product. At the point of sale, all of these costs and expenses must be covered, meaning that they are passed onto the product. The product flow is the equivalent to the cost unit in conventional costing. It is expressed with respect to the Standard Unit (see Chapter S2.2.4). The final output of the enterprise is one or more product flows. A special marking of the product flow in the diagram by color or line style is useful. A bold line facilitates the tracking of the product through the production and helps to orientate on the diagrams.

The product flow is essentially a resource and information flow. It consists of an object and its value. The rules that apply to the resource and information flows are also valid for the product flow. In addition, the product flow obeys other rules for the calculation concerning the value added by a process.

Value added

In conventional cost accounting, a general definition for value added does not exist. The general understanding of value added is the increase in value of the product generated by manufacturing. On principle, any type of value added is calculated by the difference of the outgoing product value minus the incoming product value. This forms the definition of the value added for the VFD.

Definition: The value added of a process in the VFD is the difference of the product value outputs and the product value inputs.

The calculation of the value added in the VFD is possible because the product flow is treated as a separate flow category. *Equation E1* states the general equation for the value added with one product input and one product output. The equation for multiple product inputs and product outputs is given in *Equation E2*.

E1	Value added = po – pi	po value of product output pi value of product input m number of product output flows n number of product input flow
E2	$\text{Value added} = \sum_{i=1}^{m} po_i - \sum_{k=1}^{n} pi_k$	

In *Diagram S2-9,* an example with numbers is shown for the value added of one product input and one product output. "Input.diverse" and "Output.diverse" represent all the other input and output flows that are not product flows.

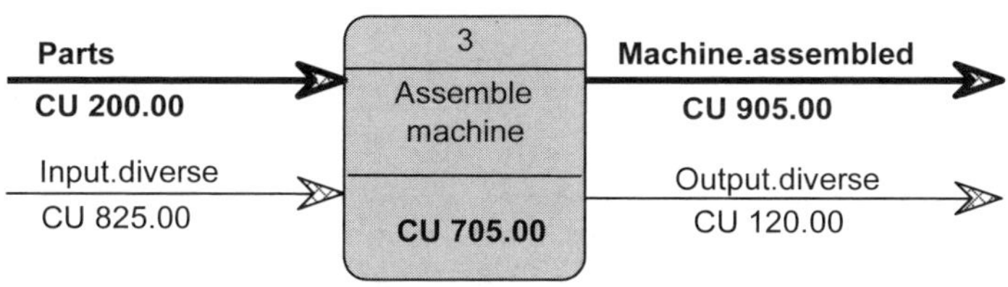

Diagram S2-9: Value added of the product flow.

The calculation of the value added for the process "Assemble machine" is:

Value added	= po	- pi	
	= Machine.assembled = 905.00 CU	- Parts - 200.00 CU	 = 705.00 CU

Value added of several product inputs and outputs

The value added can also be calculated for processes with more than one product input or output. This is used if more than one product flow enters a process, for example to be assembled into one product going further downstream in the production line. Or if a process produces more than one output product.

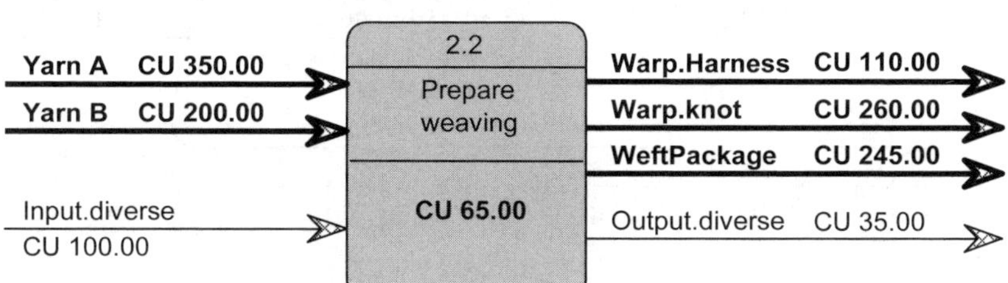

Diagram S2-10: Value added in general.

Diagram S2-10 gives an example from the weaving mill. The value added is calculated and shown for the process "Prepare weaving," which has two different yarns as product inputs and various product outputs. In this case *Equation E2* must be applied. The calculation for *Diagram S2-10* is:

Value added = Σ po				- Σ pi	
	= (Warp.Harness	+ Warp.knot	+ WeftPackage)	- (Yarn A	+ Yarn B)
	= (110.00 CU	+ 260.00 CU	+245.00 CU)	- (350.00 CU	+200.00 CU)
	= 65.00 CU				

Negative value added

The value added of a process can also turn negative. If for example waste leaving a production process has a higher value than the value added of this process, the output product flow has a lower value than the input product flow.

Value added in the hierarchy

Within the hierarchy, the value added of the detailed processes in the child diagram adds up to the value added of the parent process, using *Equation E3*.

E3	$$\text{value added}_{(\text{parent process})} = \sum_{i=1}^{n} \text{value added}_{i\,(\text{child processes})}$$

Diagram S2-11 shows the addition of the value added from the child processes to the parent process. The top process "Produce fabric" of a weaving mill gives the total value added of the product fabrication.

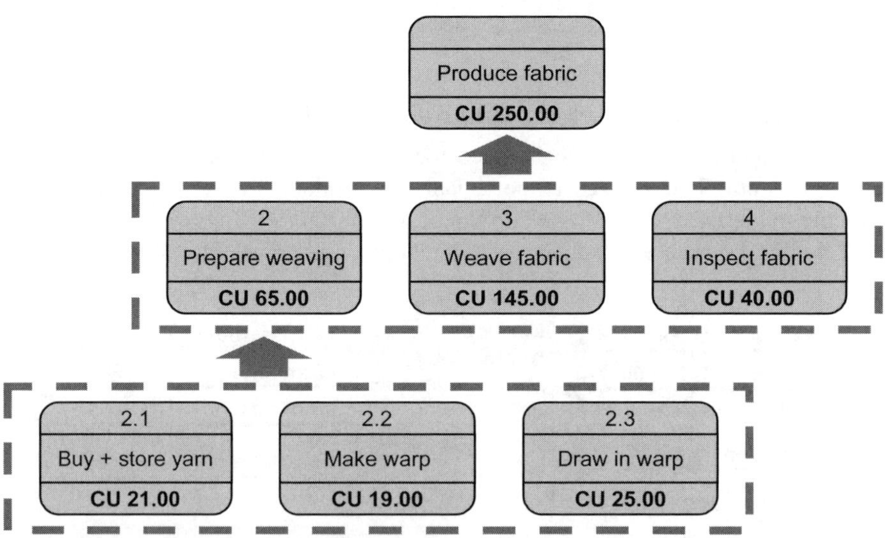

Diagram S2-11: Value added in the hierarchy.

The following rules apply to the flow category "product flow" and the value added of processes.

Rules S2-5: Flow category product flow and value added

V12	Every process involving a product has one or more product input flows and one or more product output flows.
V13	A process with a product input demands a product output, and a process with a product output demands a product input.
V14	The value added is only calculated in processes with a product flow.

2.4.4 Flow Category: Fictitious Value Flow

The fictitious value flow is used to transfer value from one process of a diagram to another process when there is no actual flow nor physical object involved with this task. In the following situations, fictitious value flows are used:

- to pass on costs or to carry out apportionment of costs.
  Costs, such as overhead costs, are passed onto the cost unit, usually the product. This "passing on" or apportion of costs normally takes place in another process rather than in the one that creates the costs.
- to remove value from a process that has no apparent value output.
  Depicting reality in a model always involves simplification of the real world.

The fictitious value flow carries a value but, in contrast to the category resource and information flow, has no object. One can also say that the fictitious value flow is a flow depicted without its object, this is because the object is not apparent or not considered.

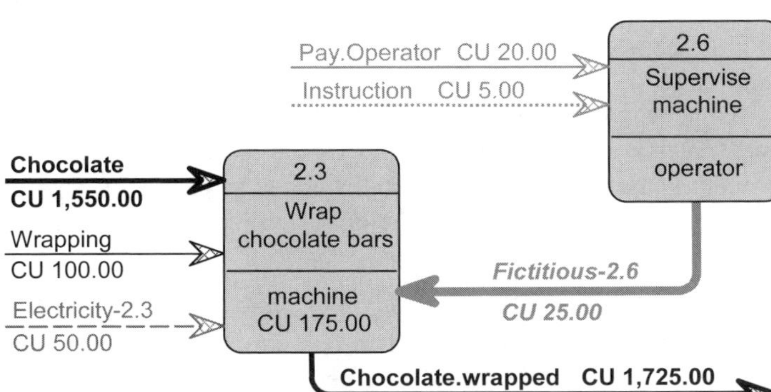

Diagram S2-12: Fictitious value flow.

Diagram S2-12 illustrates a fictitious value flow going from the process "Supervise machine" to the process "Wrap chocolate bars." The process "Supervise machine" is an example of a process without obvious output. Since value must be able to leave a process, the fictitious value flow "Fictitious-2.6" is introduced as output flow of process 2.5.

In reality, the value passed on to the process "Wrap chocolate bars" would be through the time the supervisor spent at the machine, the interactions done while passing the machine, or the experience of the supervisor going into the process "Wrap chocolate bars." When the model's designer knows the system better, he or she might replace the fictitious value flow "Fictitious-2.6" by an information flow, for example "Intervention.Operator."

Apportionment of values and costs

The transfer of value consumption, respectively costs, from one cost center to another cost center is called the apportionment of costs. The costs of all the cost centers that have no direct connection to the product (cost unit) have to be eventually passed on the product. In other words: The indirect costs must be apportioned among the cost centers. This can also be seen as bringing the costs from one place in a company to another.

This happens when

- overhead and indirect costs are passed onto the product,
- profit or loss are indicated to close the balance,
- cross-subsidies are made to support non-profitable products,
- waste costs are passed onto the product.

These operations are well known in cost accounting. Because the apportion of costs has no value flow with a tangible value carrier (such as a resource or a information), the values to be passed on are depicted by fictitious value flows.

For the costs to be passed onto the product, a process is carried out. If there is no suitable process at hand, a "virtual" one must be introduced. This also corresponds to what really happens within a company. *Diagram S2-13* shows the virtual process "Pass on costs." Such a process is usually introduced at the end of the production chain. The product value immediately before the costs are passed on represents the unadulterated costs consumed to produce the product and is therefore of interest as a performance indicator.

Consider the fabrics in the weaving mill on *Diagram S2-13*. "Fabric02" is a long and smoothly running standard product. It is a popular product on the market and sells well. "Fabric99" on the other hand is a very new product. There is not yet anything comparable on the market. This fabric is at the moment the sales hit of the company and has also increased the sales orders of the standard product "Fabric02." However, "Fabric99" is very difficult to produce. The production is not mastered very well, it is still low on the learning curve of production, there are a lot of rejects. The price that would cover the production cost is too high to sell "Fabric99," so the company sells "Fabric99" with a loss and does a cross-subsidy from the profitable "Fabric02." This means the costs of "Fabric99" that are not covered by the sales price are passed onto "Fabric02" by the fictitious value flow "Cross-Subsidy."

Cross-subsidy, unfortunately, is commonplace in companies. The real value of a product is often tampered with. For example, a good selling product covers the costs of a second product that is sold at a very low price in order to stay in the market. The price of the second product does not cover the production or development costs.

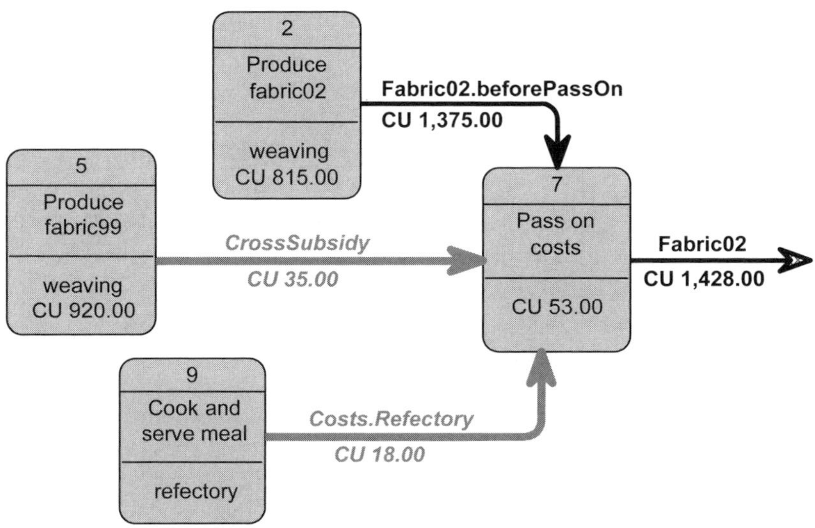

Diagram S2-13: Virtual process "Pass on costs."

Diagram S2-13 also illustrates the passing on of the costs of an activity that is not directly involved with the production. In this case, the costs of the refectory are passed onto the product by the fictitious value flow "Costs.Refectory." This is a common operation in a company where a lot of overhead and indirect costs need to be charged to the product. Examples of such costs are salary of management, costs of human resource department, cleaning of offices, etc.

In general, the fictitious value flows obey the same rules as the resource and information value flows. The following table lists the specific rules for the fictitious value flow.

Rules V16 and *V17* can be summarized by the statement: Fictitious value flows are only allowed to flow directly from one process to another one. They only connect processes.

Rule V17 ensures that the system's boundary is clearly defined.

Rules S2-6: Fictitious value flow

V15	Fictitious value flows are not classified into flow types.
V16	Fictitious value flows are not allowed to split or merge with other fictitious value flows or with flows of other categories.
V17	Fictitious value flows can not exist on the context diagram. Therefore, they cannot flow to or from an external entity.

Recommendation: Instead of drawing many insignificant flows that are not clearly defined, it is easier and faster to use a fictitious value flow. With this, the work involved for a big model may be reduced considerably.

Recommendation: One must always consider whether a fictitious value flow can also be depicted as a normal value flow. This is done by assigning a real carrier, such as data or resources. Wherever possible, a fictitious value flow should be replaced by a resource or information flow. This helps to ensure that the model is as clear as possible, and the true picture of cost in the VFD is not unnecessarily distorted.

2.4.5 Flow Category: Money Flow

Another flow category of the VFD is the money flow. It is used in diagrams with cost accounting, cash flows, and indication of profit and loss. For example, the money flow is used to show the cash transfer when the customer pays for a product in a shop. Another example is the money exchanged between two companies or different subsidiaries of the same company depicted in one VFD model.

The money flow consists of an object and its value. The object may be cash, check, credit card, bank transfer, etc, just any means to transfer money in business between persons or companies. In contrast to resource and information flows, the object of the money flow is irrelevant. That means, the form, in which the money is handed over, is not important.

The money flow is classified in flow types for calculation in the hierarchy. The drawing and calculation rules for the category money flow are the same as for the category resource and information flow.

Diagram S2-14 illustrates the money flows on the context diagram of a sandwich shop. The system “Make sandwiches” buys resources, produces a product, and then sells the product. The flows showing payment for the “Ingredients,” “Resources.diverse” and the product “Sandwich” are the money flows “Cash,” “Credit Card.ingredients,” and “BankTransfer.resources.”

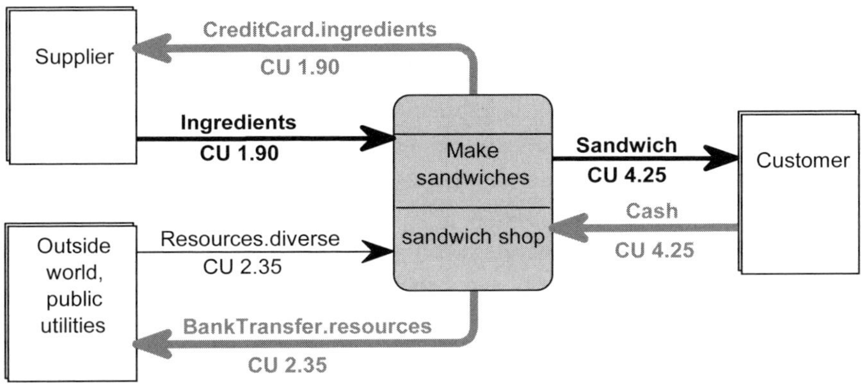

Diagram S2-14: Example of money flows.

The values of the input and output flows of a process and the system must be balanced. (The process balance is explained in detail in Chapter S2.5.5.) A diagram, which displays money flows, has to be balanced in several aspects. These balances are given in *Table S2-6*. The value flows for the examples in the table are from the sandwich shop in *Diagram S2-14*.

Table S2-6: Balancing of diagram with money flows

Balance	**Example from Diagram S2-14**
Balance of resource and information flows: The sum of incoming resources and information flow values must equal the sum of outgoing resources and information flow values.	Ingredients + Resources.diverse = Sandwich 1.90 + 2.35 = 4.25
Balance of money flows: The sum of incoming money flow values must be balanced with the sum of outgoing money flow values.	CreditCard.ingredients + BankTransfer.resources = Cash 1.90 + 2.35 = 4.25
Money paid for input: Every incoming resource or information flow must equal an outgoing money flow.	Ingredients = CreditCard.ingredients 1.90 = 1.90
	Resources.diverse = BankTransfer.resources 2.35 = 2.35
Money received for products and services: Every outgoing resource or information flow must be balanced with an incoming money flow.	Sandwich = Cash 4.25 = 4.25

Double-entry bookkeeping

This multiple balancing of the diagram with money flows can be compared with double-entry bookkeeping in several ways. Double-entry bookkeeping is an accounting technique in which each transaction is recorded as both an asset and a liability. Providing the entries are correct, the sum of the assets equals the sum of the liabilities.

In the VFD, the recorded transaction is represented by a value flow. If the two categories "resource and information flow" and "money flow" are both depicted in the diagram, they represent assets and liabilities for a transaction, as it is done in double-entry bookkeeping.

Refering to *Diagram S2-6*, the representation of debit and credit is different for the money flow than for the resource and information flow:

- A positive money input or a negative money output debit the process.
- A positive money output or a negative money input credit the process.

Consider some of the transactions from *Diagram S2-14*:

- Purchased "Ingredients" have an entry in the bookkeeping as a debit. The material value of the "Ingredients" going into the system represents the use of finances.
- "CreditCard.ingredients" is the money paid for the "Ingredients." It leaves the system and has an entry in the bookkeeping as a credit.
- The "Cash" received from the "Customer" for the sold "Sandwich" goes into the system and has an entry as a debit in the bookkeeping.

Fictitious value flow and money flow

Money flow:	Represents the payment transaction with the external environment and exchange of value between companies. They are expenses or costs, and income and profits. Example: Payment for resources, notice of profit.
Fictitious value flow:	Represents the internal exchange of values. It brings value from one process to another, if no other known flow is going this way. Finally, it is used to charge all the costs, overhead and indirect costs, to the product.

2.5 Calculation of the Value

2.5.1 Procedure of Value Calculation

Table S2-7 summarizes the procedure to calculate the value of a VFD model. The steps are described in different chapters indicated in the list, together with the concerning equation. For the calculation of a whole model with several hierarchical levels, these steps are done iteratively.

Table S2-7: Procedure of value calculation

	Calculation in the VFD	**Chapter / Equation**
Element specification	Enter all known parameters and values into the flow and process specifications.	S2.6.2, S2.6.3
Split and merge of flows	Calculate values in the flow hierarchy. Calculate unknown values within the flow types as they are defined by splits and merges.	S2.5.3 / E4, E5
Flow equation	Set up flow equations. Define all known relations and equations between flows in the flow specification. - Case 1 general: equation with input and output flows. - Case 2 special: sum of several input or output flows. - Case 3 special: output value = input value.	S2.5.4 / E6 E7, E8 E9
Process balance	Do the process balance for each process. It is valid for all input and output flows of one process.	S2.5.5 / E10
Value added	Calculate it in a process with a product flow. It can be declared in the process specification.	S2.4.3 / E1 – E3

If there is not sufficient information in the system for the calculation of all the values, more details must be researched and additional data entered for the value calculation. As a first step, it is very helpful and sometimes unavoidable to use estimated values.

For this Chapter S2.5:

- All monetary values in the diagrams and calculation examples are given in CU per Standard Unit.
- In this chapter, the setup of equations is explained and calculations with parameters are mentioned. They are done in the element specifications, which is described in detail in Chapter S2.6.

2.5.2 Principles of the Value Calculation

The value calculation in the Value Flow Diagram is based on the following two principles:

Rules S2-7: Principles of value calculation

V18	The flow carries the value.
V19	The process transforms input values into output values.

The flow, as value carrier, is the interface of the values between two processes. Every flow has a value, as stated in *Rule V3*. The process, on the other hand, is the place where the values change and the new value allocation takes place respectively. In a process, the sum of the input flow values equals the sum of the output flow values. A process neither creates nor consumes or destroys value. It only transforms input value into output value.

2.5.3 Value Calculation in the Hierarchy

Parent process and child diagram

The hierarchy of the diagrams serves also for the value calculation. The total values of the different flow types are indicated on the context diagram. On the other diagrams of the model, the input and output values of the processes can be read out on the different levels of detail. The hierarchical calculation is done top-down or bottom-up by split and merge of the value flows.

The balancing *Rule F34* as given in Chapter S1.5.6 states that the input and output flows of the parent process are the same as the off-page flows of the child diagram (see *Diagram S2-15*). Consequently, the specifications and the values of these flows are the same. This ensures the correct value declaration and value calculation in the hierarchy of the diagrams.

Split and merge

If a parent process is detailed into a child diagram, the flows can also be hierarchically detailed on the lower diagram levels into child flows by splits and merges. As a result, calculations throughout the hierarchy of the flows, the detailing and addition of the flow values is also done by these splits and merges. The parent flow is a bundle of the corresponding child flows. *Diagram S2-15* shows the split of the parent flow "Resources-1" that is distributed by child flows to the three processes 1.1, 1.2, and 1.3. The merge of the waste outputs are bundled into one single parent flow "Waste." Each split or merge point requires a calculation between parent and child flows. The general *Equation E4* states that the value of the parent flow equals the total value of all the child flows.

E4	$vp = \sum_{i=1}^{n} vc_i$	vp value of parent flow vc value of child flow n number of child flows

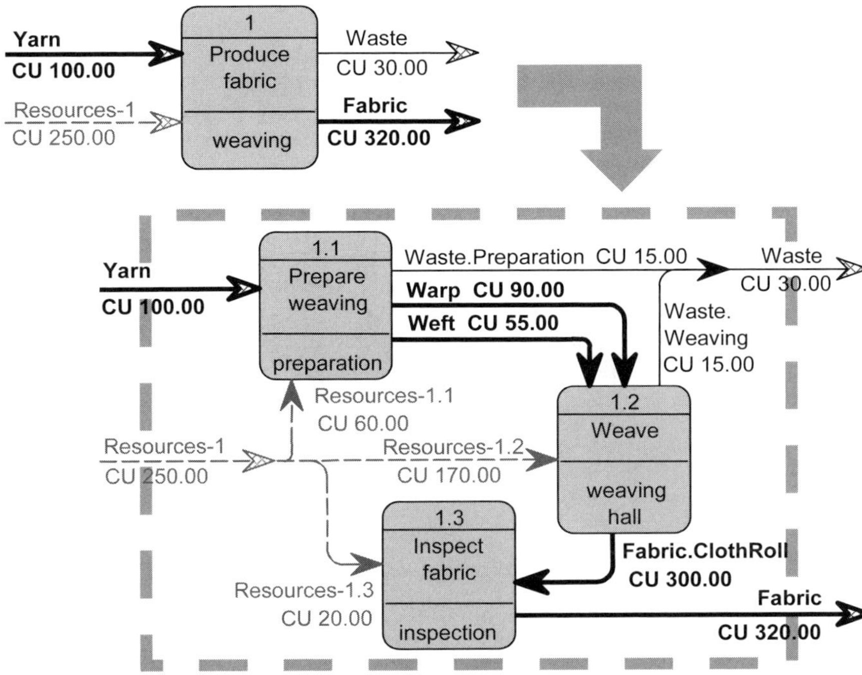

Diagram S2-15: Leveling from parent process to child diagram.

Rule F30 in Section S1 for split and merge states that the parent and child flow belong to the same flow type. Therefore, only flows belonging to the same flow type can be added up by *Equation E4*.

Hierarchically summed up

The value calculation in the flow hierarchy begins on the level where the flow values are known. This can be any level in the model: the top level, the bottom level, or a level in between. From this level, one calculates up or down in the hierarchy. Values that are known on a lower level are added up on the higher level by split and merge from child flow to parent flow. This calculation is best done (and checked) on the diagram itself in order to get a comprehensive picture of the profitability of activities in the company.

Diagram S2-16 shows the calculation for electricity costs. They are calculated separately for all machines on the lower level, and added up on the higher level to obtain the total amount of the electricity used in the process "Produce fabric."

The equation to calculate the parent flow "Electricity-2" from the child flows is determined according to *Equation E4:*

Electricity-2	= Electricity-2.2	+ Electricity-2.3	+ Electricity-2.4	
x	= 20.00	+ 35.00	+ 3.50	= 58.50 CU

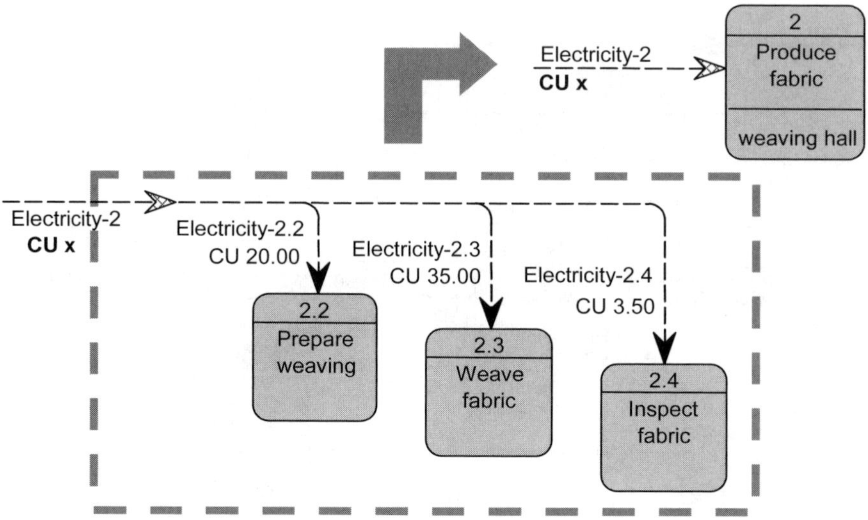

Diagram S2-16: Electricity costs hierarchically added up.

Each flow type is summed up through all the hierarchy levels of the diagrams and ends on the top level, the context diagram. There, the total amount of the resources used to produce a SU of a product shows up.

Hierarchically detailed

The hierarchical detailing distributes the values and costs, known on a certain diagram level, onto the next lower level. On the child diagram, the parent flow values are split up and distributed to the child flows. Both input and output flows are calculated this way in the hierarchy.

By split and merge, an allocation of the parent flow value to the child flows takes place. This allocation may be done using factors or percentages to assign each of the unknown child flows the corresponding part of the value from the parent flow, as indicated in *Equation E5*. The sum of the child flow values equals the value of the parent flow, as previously stated in *Equation E4*.

E5	$vc_x = y_x \cdot vp \quad \text{with} \quad \sum_{x=1}^{n} y_x = 100$	vp value of parent flow vc value of child flow y factor in % n number of child flows

In the example shown in *Diagram S2-17,* the cost of the space occupied by the process “Produce fabric” is assigned on the lower level by several child flows for the different consuming processes. “Space-2” is split to child flows by the percentage of the space occupied by the child processes. Each child flow is expressed as a percentage of the parent flow.

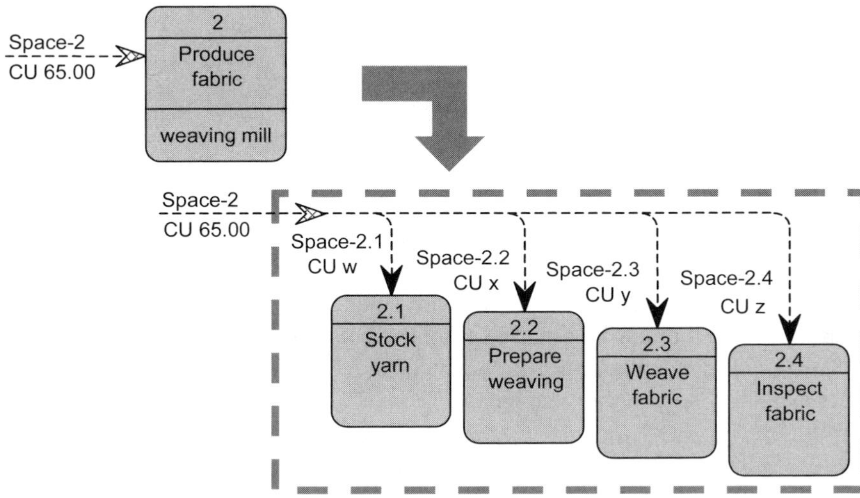

Diagram S2-17: Hierarchically detailed costs of space.

The calculations for the split of the space costs in *Diagram S2-17* are given in the box below. The numbers are assumed for this example. The equations are declared in the flow specifications of the child flows.

Space-2.1 = 10% * Space-2	→	w = 0.1 * 65.00 CU = 6.50 CU
Space-2.2 = 20% * Space-2	→	x = 0.2 * 65.00 CU = 13.00 CU
Space-2.3 = 50% * Space-2	→	y = 0.5 * 65.00 CU = 32.50 CU
Space-2.4 = 20% * Space-2	→	z = 0.2 * 65.00 CU = 13.00 CU

Diagram S2-17 shows the usual way of allocating the overhead costs to the product: They are distributed among the processes according to the rate of consumption. The processes pass the costs onto the product.

2.5.4 Flow Equation

Case 1: General flow equation

The value of an input or output flow can be calculated as a function of one or several other value flows, as given in *Equation E6*. This equation is called the "flow equation." The flow equation must be declared by the model's designer in the flow specification of the unknown value flow (to be calculated). All of the value flows used in the flow equation must be connected as input or output flows to the same process. *Flow Equation E6* allows calculating one value flow by integrating one or more other value flows of the same process.

E6	$v_m = \sum_{n=1}^{m-1} f_n \cdot v_n + \sum_{m+1}^{k} f_n \cdot v_n$	v value of flow f factor k number of all flows of a process

An example for this flow equation is given in *Diagram S2-18*. Process 2.2 illustrates the preparation of the weaving with the outputs "Warp" and "Weft." The value of the "Warp" is calculated with the flow equation. The "Warp" is made of the two yarns using only a certain percentage of "Yarn2." In preparing the "Warp," other inputs are necessary too; these are shown in this example as one flow "Input.diverse," of which the "Warp" needs 60%. The flow equation says:

Warp = 100% * Yarn1	+ 30% * Yarn2	+ 60% * Input.diverse	
Warp = 50.00 CU	+ 24.00 CU	+ 60.00 CU	= 134.00 CU

The dotted arrows within the process box indicate in the example of *Diagram S2-18* the flows that are involved in the flow equation calculation of the value of the "Warp." The calculation for the value of the "Weft" needs the process balance that is explained later in this chapter.

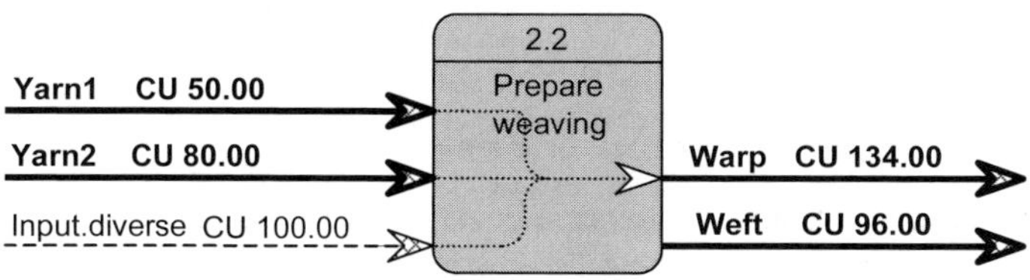

Diagram S2-18: Value input and value output combined by an equation.

Case 2: Sum of input or output flows

A simple, common case of the general *Flow Equation E6* is presented by the two *Flow Equations E7* and *E8*. It is the addition of several input flows of a process to produce one output flow of this process, or the addition of several output flows to produce one input flow.

E7	$vo_y = \sum_{i=1}^{m} vi_x$	vi value of input flow vo value of output flow m number of input flows (1 to all of a process) n number of output flows (1 to all of a process)
E8	$vi_x = \sum_{i=1}^{n} vo_y$	

This calculation is usually done between flows of the same flow type. The designer of the model defines in the flow specification (of the flow to be calculated), which flows are added. This is predetermined by the nature of the process.

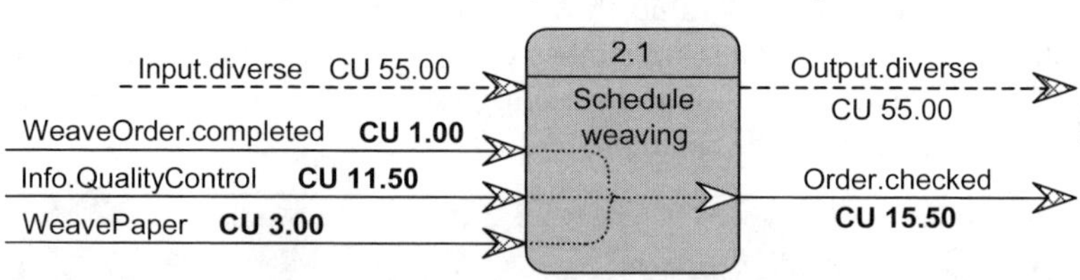

Diagram S2-19: Sum of several output flows.

In the example of *Diagram S2-19,* the foreman of the weaving mill collects all the different papers of a weaving order to check them. The incoming information flows are added up to the outgoing "Order.checked." The dotted arrows within the process box illustrate, which flows are combined in the following flow equation:

$v_{Order.checked} = v_{WeaveOrder.woven} + v_{Info.QualityControl} + v_{WeavePaper}$ $v_{Order.checked} = 1.00\ CU + 11.50\ CU + 3.00\ CU$ $= 15.50\ CU$	v flow value

Case 3: Value output equals value input

As stated in *Rule F9* in Chapter S1.3.5, the output flow is different from the input flow, because the flow is used by the process and interacts at least with one other flow of this process. The process changes state, place, time, or value of the flow. Nevertheless, a value output may equal a value input, as given in *Flow Equation E9*. This equation is a special case of *Flow Equation E6*.

E9	$vo = vi$	vi value of input flow vo value of output flow

In this special case of *Flow Equation E9,* the value of the input flow is the same as the one assigned to the output flow and, therefore, it gives the impression that it is the same flow. The flow seems to run through the process without apparently changing its value. Reasons for this include:

- In reality the value changes, but the change is so small that it is not worth showing it in the model.
  Example: Automatic reading of the bar code of a product unit.
- Certain information flows (example: shop order) accompany the product through several processes along the manufacturing chain. Distributing the value of such a flow is often very tiresome and not worth the labor. Thus, instead of distributing the flow value to every single process, (perhaps even in the right proportion according to the use in the concerned process,) the total amount is passed on in the last process, or in the process with the largest value added.
  In reality the value would change from process to process, but for the sake of simplicity this is not taken into account in the model.

Diagram S2-20 shows how the input value is passed directly to the output flow in the case of a very low cost process. The equation for the flow "WeaveOrder. ready" is:

$v_{WeaveOrder.ready} = v_{WeaveOrder}$ $v_{WeaveOrder.ready} = 13.00$	v flow value

The output flow "WeaveOrder.ready" takes the value from the input flow "WeaveOrder." The dotted arrow in the process box shows the take over of the flow value.

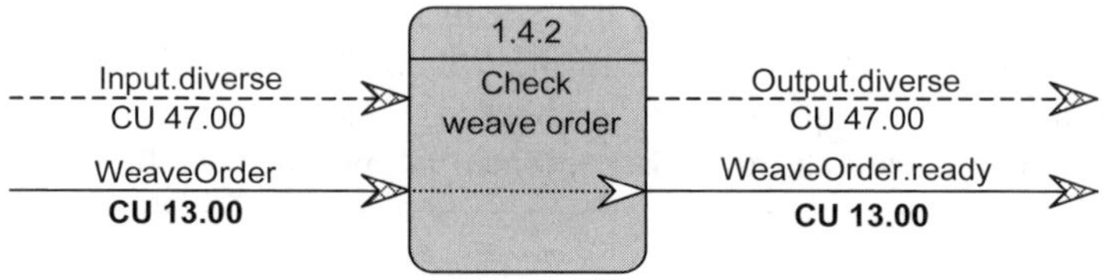

Diagram S2-20: Value input equals value output.

2.5.5 Process Balance

When the flow equations of input and output flows are defined, the process is balanced. The values of the flows, which do not yet have a value assigned by split and merge or flow equations, are calculated by the process balance. The equation of the process balance combines all the incoming and outgoing flows of one process. For each process the process balance as shown in *Equation E10* is:

E10	$$\sum_{i=1}^{m} vi_i = \sum_{k=1}^{n} vo_k$$	vo_k value of output flow vi_i value of input flow m number of all input flows n number of all output flows

The process balance stated in words:

Definition for process balance: For each process, the sum of the value inputs equals the sum of the value outputs.

Note! The calculation of flow values for one process is possible only if the value of one of the flows going into or coming out of this process is unknown. If all flow values are known, the process balance is overdefined.

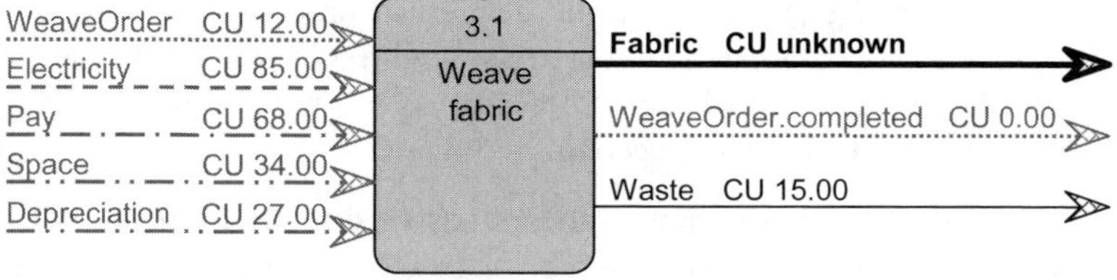

Diagram S2-21: Σ value input = Σ value output.

For a process with a product flow running through it, it is usually the output product flow whose value is unknown in the process equation. For example, the flow with the unknown value in the process balance of *Diagram S2-21* is the "Fabric." The product "Yarn" flows into the process "Weave fabric," and the "Fabric" is the unknown output product flow. The output product value of CU 411.00 is calculated by the process balance.

The input and output flows of the process "Weave fabric" in *Diagram S2-21* result in the process balance in *Table S2-8*. This process balance can be written into the specification of the process.

Table S2-8: Process equation of the process "Weave fabric"

Input	CU		Output	CU
Yarn	200.00		Fabric	411.00
+ WeaveOrder	12.00		+ WeaveOrder.completed	0.00
+ Electricity	85.00	=	+ Waste	15.00
+ Pay (for manpower)	68.00			
+ Space (of building)	34.00			
+ Depreciation (of machine)	27.00			
Input sum	426.00		Output sum	426.00

For a whole model, an equation system can be set up. There is an equation (process balance) for each process. A model can be calculated, if there are as many unknown flow values as processes.

Rules S2-8: Process balance

V20	The process balance is calculated for every process of the model.
V21	Exactly one flow value of the process must be unknown for the calculation of the process balance.

Note! The important point to keep in mind is that the model as a whole is balanced in value as well. For the single process on the context diagram, the process balance applies as well.

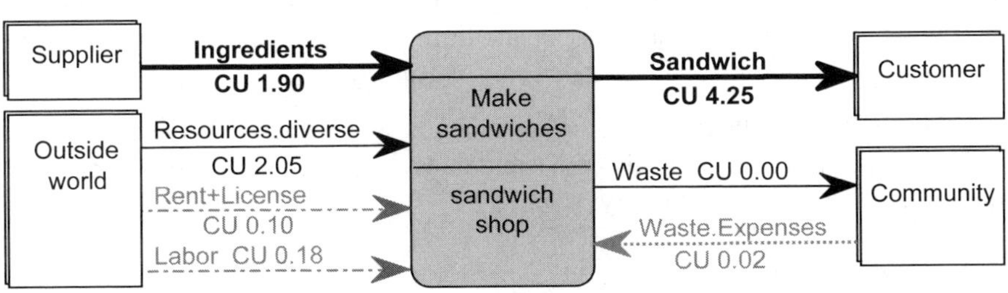

Diagram S2-22: Process balance of the context diagram.

Diagram S2-22 shows an example of a process balance for the context diagram. The balance for the single process is given in the following:

Ingredients + Resources.diverse + (Rent+License) + Labor + Waste.Expenses	= Sandwich + Waste
1.90 CU + 2.05 CU + 0.10 CU + 0.18 CU + 0.02 CU	= 4.25 CU + 0.00 CU

2.6 Element Specification and Calculation

2.6.1 Declaration of Parameters

In the previous chapters, all calculations were made with value flows for which the values were given in the monetary term Currency Unit (CU) per Standard Unit (SU). These values in CU per SU are either already known or have to be calculated using several parameters. The values and the parameters for the calculation are both declared in the flow and process specifications and have to be combined in formulas for the value calculation.

The term "parameter" is applied generally in the method of the VFD. "Parameter" is defined here as a generic term for factors, variables, measured or estimated data that are used for the calculation of the values of processes or flows.

A parameter is given with a name, quantity, and unit. *Table S2-9* shows some examples of parameters.

Table S2-9: Examples of parameters

Parameter name	Quantity	Unit	Source, established by
Yarn price	8.35	CU/kg	statistics, purchasing
Interest of capital invested	3	%	annual report, accounting
Pay-2.3.3	85,000.00	CU/year	estimated, accounting
Electricity machine	75	kW	measured, plant service
Washing time	2.5	hour/warp	timed, analyst
Space-2.1	6.50	CU/100m	measured, accounting
...			

It is helpful to list also the source of the quantity information. Many information sources are available within a company, such as accounting department, sales department, scheduling, production papers, timing of processes, counting quantities on the shop floor, or observation and report of employees.

Each parameter must be defined only once in a model. Otherwise, the system is not consistent. To make sure of this consistency, every parameter used in a specification is listed in the data dictionary.

Rules S2-9: Parameter

V22	Every parameter consists of a name, a quantity, and a unit.
V23	A parameter is defined only once in a model and has an entry in the data dictionary.

2.6.2 Flow Specification

In the VFD, the general element specification of the flow holds sections for the identification of the element, for a remark or description, for the flow type, and for the value calculation. The value calculation needs sections for details in regard to the parameter declaration, the calculation, and the calculated values. A general flow specification is given in *Specification S2-1*.

Specification S2-1: General flow specification

⟨Flow Name⟩			
Description	⟨Text⟩		
Flow type	⟨Flow type⟩		
Parameter	⟨Parameter name⟩	⟨Quantity⟩	⟨Unit⟩
Calculation	⟨Equations E4 to E9⟩		
Flow value	⟨Calculated flow value⟩		

First, the parameters are declared in the section "parameters" of the element specification. Each parameter includes parameter name, quantity, and unit, as previously listed in *Table S2-9. Specification S2-2* shows the flow specification "Light-3.3" with the parameter "Price electricity" as an example.

Specification S2-2: Indication of the flow type in the flow specification

Light-3.3			
Description	Lighting circuit in building 3a.		
Flow type	Electricity light		
Parameter	Price electricity	0.15	CU/kWh
Calculation			
Flow value			

Flow type: Flows belonging to the same flow type often get their values from the same parameters; therefore, these parameters should be declared in the same way all over the model.

Flow category product flow: The specification of the output product flow of a process often has no parameters. This is because the flow value of the product flow is primarily calculated by the process balance and not from parameters (*Equation E10*).

2.6.3 Process Specification

The process specification contains process-specific parameters, i.e. process time or machine data. In detail, the general process specification looks as follows: The heading includes the process number, the process name and the additional information. Thereafter, the process is described. The declaration of the parameters is divided into two parts: "Parameter" and the allocation parameter "key." If preferred, a section for the process balance can be introduced. At the bottom of the specification comes the section with the value added. The general process specification is shown below in *Specification S2-3*.

Specification S2-3: General process specification

⟨Process Number⟩	**⟨Process Name⟩**		**⟨Additional Information⟩**
Description	⟨Text⟩		
Parameter	⟨Parameter name⟩	⟨Quantity⟩	⟨Unit⟩
Key	⟨Parameter name⟩	⟨Quantity⟩	⟨Unit⟩
(Process balance)	⟨Process equation E10⟩		
Value added	⟨Equation E1 to E3 and calculated value⟩		

Allocation parameter "key" and costing

A problem in calculating the value is the distribution of the costs to the different products so that the values are reflected correctly. How are the costs of the building appointed to a given product? How is the salary of the management assigned? How much time is spent on the processing of a customer order? These are all questions that must be considered when assigning the costs.

For cost assignment, an allocation parameter called "key" is used. The key in the VFD is applied the same way as the cost driver in cost accounting. The definition of the key may be difficult because the costs and values of several resources are allocated by this key to the product. The allocation of the costs is normally done by the parameter mass or by time.

In general, the allocation key is a normal parameter. It is defined, like the other parameters, with a name, quantity, and unit. The keys are given in a separate section "key" in the process specification to highlight them and to make the later combination of formulas easier. The keys depend on the process and are needed for the calculation of several value flows connected to this process. Normally, at least one key per process is required for the calculation. However, as many keys can be defined as needed in a process specification.

Specification S2-4 shows the keys with an example of the process "Dye fabric." The assignment of the values of various input and output flows is related to the time (given in hours) needed for the execution of the process. In this case, two keys are required for the calculation: the "Machine hour" for machine related flows, and the "Work hour" for the calculation of the salaries.

Specification S2-4: Example for keys in process "Dye fabric"

3.3.3	**Dye fabric**		**dye machine**
Description	Costs depend on: - Type of dye machine - Type of color (dark, light) - Number of water loads		
Parameter	Number machines	35	machines
	Load of water	2,400	liter/load
	Number of water loads	11	load/dye batch
	Room	295	m^2
Key	Machine hour	5.00	h
	Working hour	2.00	h
Value added		354.98	CU/100m

The key in a process specification often refers to the Reference Unit of this process, for example the time needed to produce one RU.

Considerations for parameter declaration in the specification

When defining the flow and process specification and its parameters, the following must be considered:

- In an enterprise, standardized specifications are essential. The elements and their specifications should be introduced so they can be used generally in the company.
- The calculation of product variants and the treatment of different products of a product family should be as simple as possible. The correct choice of the specification for a parameter can save a lot of time and expense. It is desirable to reuse the model for the calculation of another product with the least changes possible.
- The section "parameter" may be divided alternatively into the two sections "parameter general" and "parameter product specific" (*Specification S2-5* on the next page). This provides an easier handling of the VFD, especially for large models, and saves time when calculating similar products with the same model. The general parameters should be chosen so that they can be used for as many products in a company as possible. The product-specific parameters can only be used for one specific product. For further products, new product-specific parameters must be declared, or at least the quantities of the already existing parameter must be adjusted.
- Parameters can be passed from flow to flow on different levels in the hierarchy. This is possible, if these flows belong to the same flow type. Parameters of the parent flow that also apply to the child flow should be specified in the section "parameter general," for example, the price for electricity.

Specification S2-5: Specification with divided parameter section

2.3	**Weave fabric**		**loom**
Description	Production of the loom depends on - the fabric (width of fabric, weft density, pattern) - the yarn type - the speed of the machine.		
Parameter general	Number of looms	50.00	Machine
	WeaveHour per week	132.00	h/week
	Electricity machine	5.30	kWh
	Room	1,800.00	m^2
	Light weaving hall	2,480.00	kWh/week
	Maintenance per machine	2,100.00	CU/year
	Depreciation + interests per machine	15,600.00	CU/year
Parameter product specific	Efficiency	80.00	%
	WeftDensity	16.00	picks/cm
	WarpLength	3,094.00	m/warp
	Speed	300.00	picks/min
Key	WeaveTime in weeks	2.60	week/warp
Value added			

- A parameter cannot always be allocated unequivocally to one specification. Many parameters can be put into different specifications. Some parameters are used for the calculation of more than one flow. Such parameters are needed for the combination in several formulas of different flow specifications. These parameters can either be placed in one of the flow specifications or in the process specification, where flows to be calculated by these parameters are input or output flows.
- In a specification, parameters may be declared that are not currently used for the calculation of the specific product. This is because, they might be useful in the future for the calculation of another product or variant in a model with the same elements and specifications.

2.6.4 Calculation Based on Equations with Parameters

For the calculation of a flow value, the parameters are combined into formulas. This combination is completed by the model's designer. The formula for the calculation of a flow value is defined in the flow specification in the section "calculation." The parameters to be combined are taken from the specification of the particular flow, but also from the specifications of the two processes connected to this flow. The procedure of combining the parameters in the flow specification is explained by a small example for the value calculation of the flow "Light-2.3" for a model with the SU of 100 m fabric.

Example calculation for value flow "Light-2.3"

The value flow calculation is executed with parameters from the specification of the process "Weave fabric" and of the flow "Light-2.3." The process "Weave fabric" is on *Diagram S2-23,* where only the product flow and the "Light-2.3" are shown.

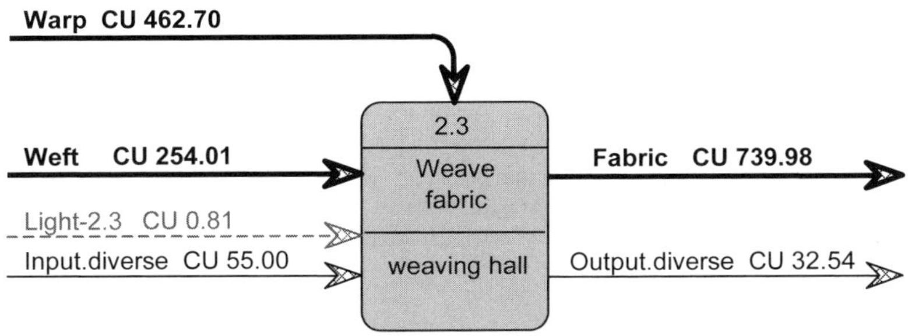

Diagram S2-23: VFD for the calculation light (other flows not shown).

For the value calculation of the flow "Light-2.3," two parameters from the process specification (*Specification S2-6*) are needed: the time ("WeaveTime" in weeks) and the "Number of looms" in the weaving hall. The given quantity of electricity used for light in the process "Weave fabric" applies for the whole weaving hall with 50 looms. The consumption of electricity for the light needs to be broken down and assigned onto a single loom. For this, the parameter "Number of looms" is necessary.

Specification S2-6: Process "Weave fabric" has the key parameter "WeaveTime in weeks"

2.3	**Weave fabric**		**weaving hall**
Description			
Parameter	Number of looms	50.00	machines
Key	WeaveTime in weeks	1.99	week/warp
Value added			

From the flow specification "Light-2.3," the price and consumption of electricity are used (*Specification S2-7*). The formula combining all these parameters is given in the flow specification.

In this case, the calculation is done in two steps. First, the monetary value per RU "warp" is calculated. Then, the value is converted to CU per SU. *Specification S2-7* illustrates this two-step calculation. The formula in step 2 contains the conversion from RU "warp" of 1,836 m to SU of "100 m finished fabric."

To combine formulas, it is often practical to apply a step-by-step approach. For complicated calculations, as many steps as necessary can be introduced. All the formulas are shown in the section "calculation."

Specification S2-7: Flow "Light-2.3," calculation in two steps

Light-2.3			
Description	For the light in the weaving hall (light for weaving), the yearly consumption of electricity is known. This consumption is broken down onto 1 loom for the duration of the weaving of the SU.		
Flow type	Electricity light		
Parameter	Price electricity Consumption electricity light	0.15 2,480.00	CU/kWh kWh/week
Calculation for RU	(Consumption / Number of looms) * WeaveTime * PriceElectricity = (2480k Wh/week / 50 mach) * 1.99 week/warp * 0.15 CU/kWh =	14.81	CU/warp
Conversion into SU	CU/RU / meter on warp * SU = 14.81 CU/warp / 1836 m/warp * 100m =		
Flow value		**0.81**	CU/100m

Combination with flow values

Using values of other flows in the flow calculation formulas is allowed. Such a flow value is treated like a parameter. It is listed with a name, value, and unit in the section "parameter."

Specification S2-8 shows an example for the use of a flow value as a parameter, where the material flow "Waste.loom" is calculated. For this example, the waste is 3% of the quantity of weft yarn, both ends of which are cut off during the weaving process.

The value of the product flow "WeftYarn" is already known with CU 239.00 and is used as a parameter to calculate the value of "Waste.loom."

Specification S2-8: Flow value as parameter

Waste.loom			
Description	Waste of the loom are the ends of weft yarn that are cut off at both edges of the fabric after the weft is inserted into the warp. Quantity: 3% of the weft.		
Flow type	Waste		
Parameter	Percentage cut off	3	%
	Weft yarn	239.00	CU/100m
Calculation	PercentageCutOff * WeftYarn = 0.03 * 239.00 CU/100m =		
Flow value		**7.17**	CU/100m

Some of the combinations go through several processes and flows. For example, the consumption of one flow (e.g. electricity) may depend on the amount of another flow (e.g. material), which is calculated by parameters taken out of several process and flow specifications.

Flow equation

The *Flow Equations E6* to *E9* in Chapter S2.5.4 are also declared in the section "calculation" of the flow specification. Again, the flow values are treated like parameters. Process 2.2 in *Diagram S2-24* illustrates the preparation of the weaving with the outputs "Warp" and "Weft." The value of the "Warp" is calculated with the flow equation.

The "Warp" is made of the two yarns using only a certain percentage of "Yarn2." For the preparation of the "Warp," other inputs are necessary, shown in this example as the bundled flow "Input.diverse," of which the "Warp" needs 60%. The formula is given in the *Specification S2-9*.

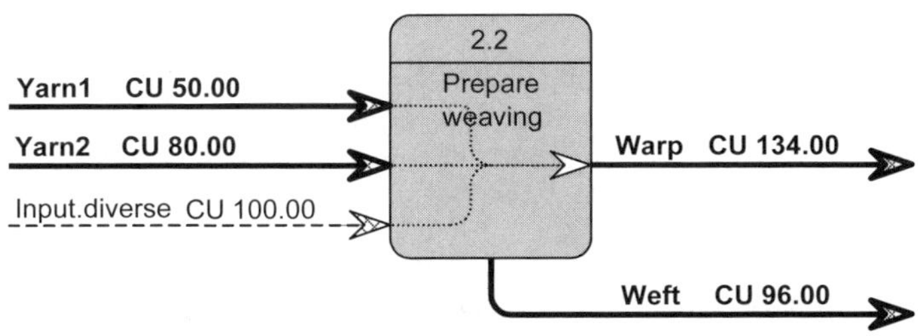

Diagram S2-24: Example of flow equation.

The dotted arrows within the process "Prepare weaving" indicate the flows that are involved in the calculation of the "Warp" value with the flow equation. The "Weft" is calculated by the process balance.

Specification S2-9: Flow values as parameters

Warp				
Description	The warp ends are the yarns that enter the weaving machine in parallel, forming the shed. During the weaving process, the weft ends are interlaced with the warp ends.			
Flow type	Product			
Parameter	Yarn 1 Yarn 2 Input.diverse Percentage Yarn1 Percentage Yarn2 Percentage Input.diverse		50.00 80.00 100.00 100 30 60	CU/100m CU/100m CU/100m % % %
Calculation	100% * Yarn1 1 * 50.00 CU	+ 30% * Yarn2 + 0.3 * 80.00 CU	+ 60% * Input.diverse = + 0.6 * 100.00 CU =	
Flow value			**134.00**	CU/100m

Process balance and value added

Process 2.3 "Weave fabric" from *Diagram S2-23* is used here again (*Diagram S2-25*) to show how the process balance and the value added are indicated in the process specification.

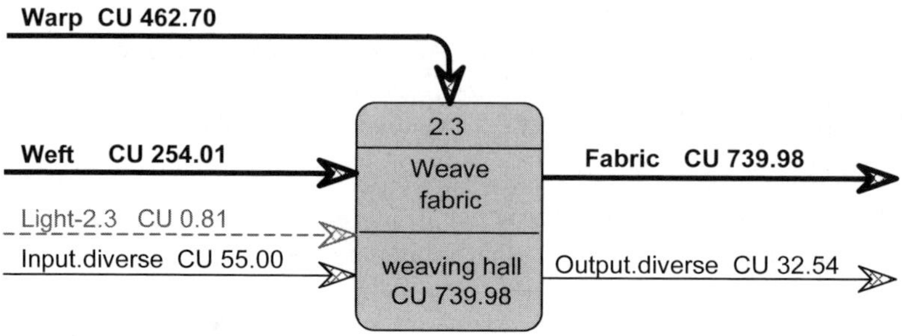

Diagram S2-25: VFD for the calculation value added (other flows not shown).

The process balance *Equation E10* in Chapter S2.5.5 states that the sum of the input flow values equals the sum of the output flow values. With the process balance, the unknown value flow of the entering and leaving flows connected to a process is calculated. In case of the process "Weave fabric," this is the product flow "Fabric."

The process balance can be indicated in the process specification in the pertaining section. In *Specification S2-10,* the process balance is already resolved for the unknown flow "Fabric" that has a value of 739.98 CU for 100 m.

Specification S2-10: Process balance and value added in the process "Weave fabric"

2.3	**Weave fabric**		**weaving hall**
Description			
Parameter	Number of looms	50	machines
Key	WeaveTime in weeks	1.99	week/warp
Process balance	Warp + Weft + Light-2.3 + Input.diverse - Output.diverse = Fabric 462.70 CU + 254.01 CU + 0.81 CU + 55.00 CU - 32.54 CU = 739.98		CU/100m
Value added	Fabric – (Warp + Weft) = 739.98 CU – (462.70 CU + 254.01 CU) =	**138.19**	CU/100m

The value added is calculated from the difference of the values of the input and output product flow, as explained in Chapter S2.4.3. This subtraction can be written in the section "value added" in the process specification.

In the case of the process "Weave fabric," the two input product flows "Warp" and "Weft" are subtracted from the output product flow "Fabric." The value added of the process "Weave fabric" is 138.19 CU for 100 m fabric.

2.7 Special Examples

2.7.1 Exchange of Value with Outside World

The system interacts with the environment and exchanges material and intangible resources with it. In the VFD, all the values are investigated from the perspective of the modeled system. Not all of the flows representing a positive or negative value for the company can be defined with a commercial value from the viewpoint of the outside world. This is the cause when a corresponding market is missing.

There are different possibilities for value exchange of the system with the outside world, depending on the direction, in which the object and the value are flowing. Different cases are discussed in the following paragraphs. Typically, there are four general cases of value exchange of the system with the outside world, as indicated in *Diagram S2-26*:

1) System buys from outside world.
2) System sells to outside world.
3) System disposes into outside world without being paid.
4) System receives from outside world without payment.

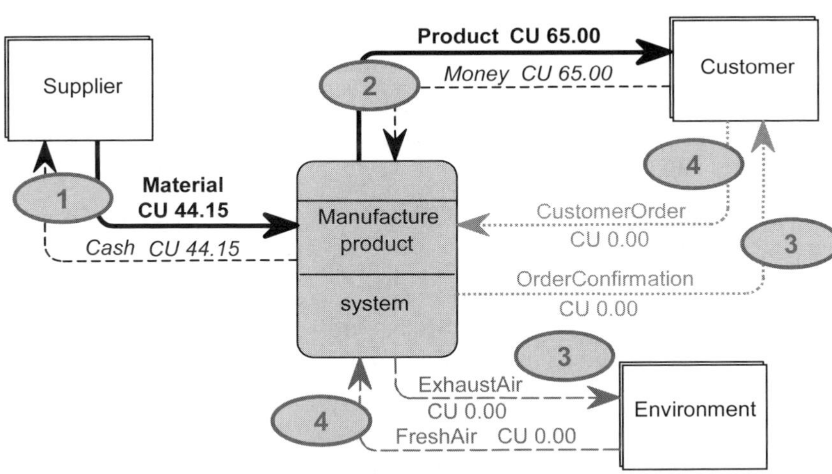

Diagram S2-26: Value exchange with outside world.

Case 1: System buys from outside world

If the system buys from the outside world, an object and its value go into the system by the flows representing the system interface. The system in turn is paying for the object by a money flow. The equivalent value delivered to the outside world by the system in the form of cash, check, or money transfer is an expenditure. The example in *Diagram S2-26* illustrates, how the company buys "Material" from the supplier and in return pays CU 44.15 in "Cash."

Case 2: System sells to outside world

The system sells to the outside world. By a flow, the system passes an object with a value to the outside world and receives the equivalent value as proceeds in exchange for this object. These proceeds are depicted as the money flow "BankTransfer" in *Diagram S2-26*, where the selling of a product for CU 65.00 to the external entity "Customer" is shown.

Case 3: System disposes into outside world without being paid

During the production process, by-products such as waste, heat, or sewage are produced. They represent values or costs for the company because they consumed energy, material, work, etc., but have no value for the outside world. The system has to dispose of these materials. The flows are routed to the environment with the value CU 0.00, because the environment does not pay a monetary value for them. *Diagram S2-26* shows the flows "ExhaustAir" and "OrderConfirmation" leaving the system with the value CU 0.00. No corresponding money flow enters the system.

Waste water is treated according to case 1 because the system has to pay for the disposal of it.

Case 4: System receives from outside world without paying

The environment supplies natural resources (i.e. "FreshAir" in *Diagram S2-26*), while the company (the system) does not pay for them. The flow goes with the value CU 0.00 into the system, even if it is useful or valuable for the company.

Also, flows such as the information "CustomerOrder" enter the system with the value CU 0.00. The value, which these flows represent for the customer, is not of interest for our system and is not charged for it.

Note! The flows of the cases 3 and 4 must cross the system boundary with the value CU 0.00 because no equivalent value flow is received or paid. Otherwise the system would no longer be in balance.

2.7.2 Example of Waste Calculation in a Company

Waste in a company is a very interesting yet often neglected aspect. It is discussed with the example in *Diagram S2-27*. Following the waste flow in a company reveals potential for optimization by indicating poor-running processes and exposes mistakes in the calculation of production values. It can turn out that a substantial amount of resources and, therefore, of value are used to "produce" waste. To investigate the value or cost of waste, the waste flows with values are introduced into the model. To keep *Diagram S2-27* legible product, waste, and money flows are depicted explicitly. All other flows are bundled in the "Resource" flows. For explanatory reason, this diagram includes also the external entities.

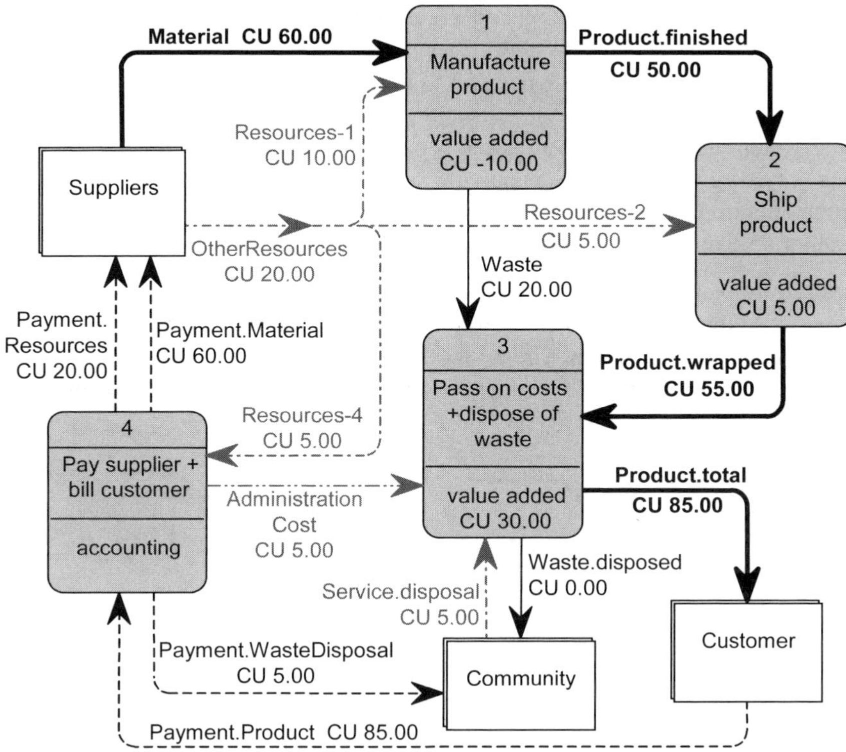

Diagram S2-27: Waste example.

Diagram S2-27 illustrates different aspects of the VFD related to the waste flows:

- The process "Manufacture product" has a negative value added.
  As stated in *Equation E1,* the value added of a process is calculated by the difference of the product output and the product input. The value added may become negative if the product output has less value than the product input. In the example of *Diagram S2-27,* the "Waste" of process "Manufacture product" takes away more value from the product than is added to it by the other input flows. This diagram shows clearly that the process "Manufacture product" should be investigated further because it produces too much waste. A negative value added makes no sense for the company.
  However, it is possible that the company can reuse or sell the waste as material for another product. This could make the value added positive because the process would get a value input from the reuse or selling of the waste. This possibility should be examined when investigating the process further.
- The customer has to pay for the total product costs, including the cost of the waste. For this, the value of the waste has to be passed onto the product.
  Process 3 "Pass on costs + dispose of waste" is introduced into the model. This process exists in every real company, even if you are not always aware of it. It

illustrates how the values from flows created by different processes, such as "Waste" or "AdministrationCost," are charged to the product.

- An important fact to note is that the flow "Waste.disposed" leaves the system with a value of CU 0.00.
  The system cannot give away value because it would be out of balance. Value can only be given away if the equivalent value is received in return from the external environment. The value of flows such as "Waste.disposed" must be passed onto the product, the carrier of all created costs.

To make sure that the flow values are in balance, the money flows going into and coming out of the system are also introduced in this example. It can be clearly seen that the money flow values are in balance. The sum of the input money values of the system equals the sum of the output money values. To see this better, the waste flow is shown again on *Diagram S2-28* as a context diagram.

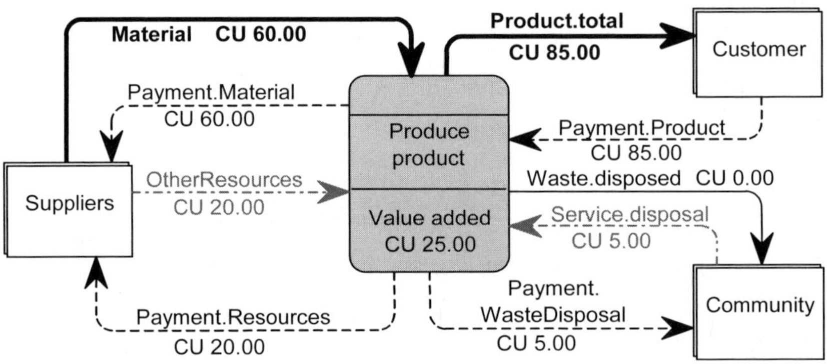

Diagram S2-28: Balance on the context diagram.

2.7.3 Notice of Profit and Loss

The profit or loss from selling a product can be illustrated by the VFD. In the following example, different aspects of depicting profit and loss are discussed. The profit or loss is represented by a money flow going to the external entity "Enterprise financing" by a process. It is calculated as the difference between revenues and product value (total costs of product):

profit (or loss) = revenues – product value

Notice of profit

Diagram S2-29 expands on the sandwich shop example, where the sandwiches are sold for CU 4.40, thus making a profit of CU 0.15. Because "Profit" is a money flow, the other money flows must be introduced in the diagram as well, so that the system is in balance. The money flows are given in dotted arrows with italic names on this diagram.

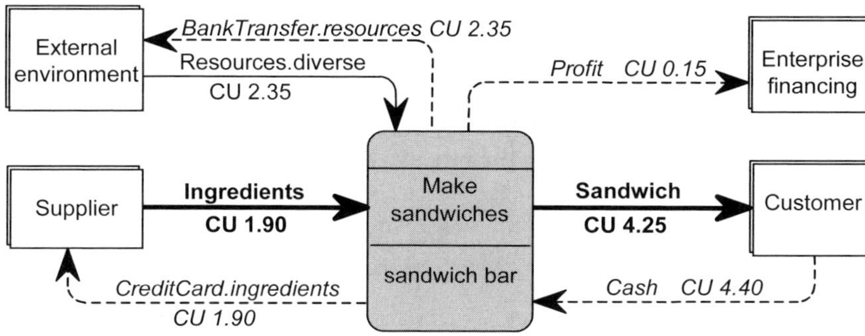

Diagram S2-29: Notice of profit.

The equation for the profit in *Diagram S2-29* for selling one “”Sandwich” to a “Customer” is:

Profit	= Cash	– Sandwich
0.15 CU	= 4.40 CU	– 4.25 CU

Notice of loss

This example of profit is used to illustrate the theoretical explanation of the negative value flow in Chapter S2.3.4 in practical context. Negative value flows do not occur very often in a model and are not easy to explain. Profit and loss situations provide a good example of these flows.

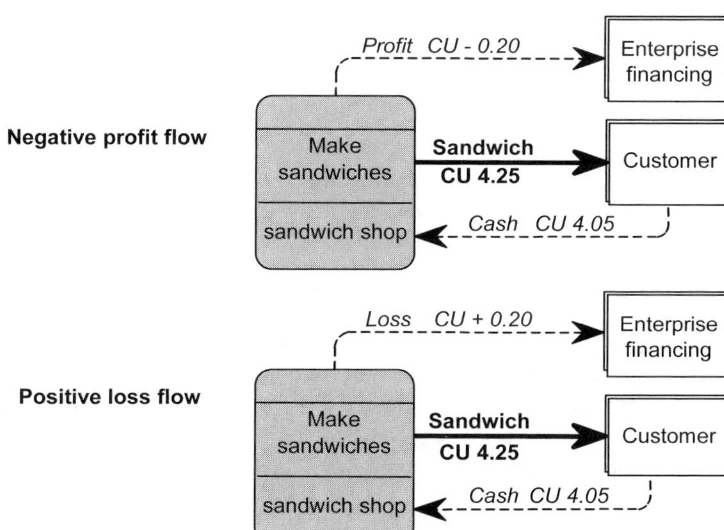

Diagram S2-30: Notice of loss.

Diagram S2-30 shows the loss if a product is sold at a price, which is much too low and does not cover the production costs. It illustrates two possibilities, both depicting the same loss of CU 0.20 for a sandwich:

- Since the loss is actually a negative profit, it can be represented by a negative money flow “Profit” going from the process “Make sandwiches” to the external entity “Enterprise financing.”
- A loss can be shown by the positive value flow “Loss” going from the “Enterprise financing” into the process “Make sandwiches.”

There are several reasons for using either a negative or positive flow. Depending on the aspect that the model's designer chooses to emphasize, either a loss or a profit can be shown. Also, at the beginning of the modeling process it is not yet known if the calculated product produces a profit or loss. As the flow must be drawn prior to obtaining this information, either a profit or loss flow must be chosen. During the lifetime of a product, the values may change; therefore, a loss might turn into a profit or vice versa.

A change from a positive to a negative value flow might occur when the setup of the entire model is completed and the model’s designer starts changing parameters to see how the system behaves.

The two examples in *Diagram S2-30* show very clearly that the change in direction of the flow arrow also inverts the sign of the value.

In general, the following applies for negative value flows:

- A negative value flow pointing in one direction can be replaced by a positive value flow going in the opposite direction. The method allows the model's designer to decide on the direction of the flow. However, it is important that the assignment of a positive or negative value flow is clear for any reader of the diagram.
- The use of a particular direction of an arrow shows the point of view that the model's designer wants to emphasize.

2.7.4 Investment Analysis

The VFD is a very helpful tool in deciding on a potential investment. Investments that do not involve a change in the chain of processes are calculated very easily in the VFD, since they only impact parameters. For example, the replacement of an old machine by a new one requires only a few changes to the parameters and their quantities in some element specifications.

An investment analysis is shown in the example of the weaving mill. An old loom should be replaced. To compare the costs of the old and new loom, only a few changes in the process specification “Weave fabric” are necessary. They are highlighted in *Specification S2-11*.

Specification S2-11: Process "Weave" with the parameters of the old an the new loom

2.3.4	**Weave fabric**	**old loom**	**new loom**
Description	WeavingTime = WeftDensity * WarpLength / Speed * Efficiency Loom is stopped over weekend for 1½ days. Loom in operation for 132 WeaveHour/week.		
	Speed depends on fabric and yarn.	Lower speed	Higher speed
	Normal efficiency is 90%.	Many breakdowns	Better efficiency
Parameter	Number of looms	50.00 Machines	50.00 Machines
	WeaveHour per week	132.00 h/week	132.00 h/week
	Maintenance per machine	2,100.00 CU/year	3,500.00 CU/year
	Depreciation + interests per machine	15,600.00 CU/year	30,000.00 CU/year
	Efficiency	80.00 %	90.00 %
	WeftDensity	16.00 picks/cm	16.00 picks/cm
	WarpLength	3,094.00 m/warp	3,094.00 m/warp
	Speed	300.00 picks/min	330.00 picks/min
Key	WeaveTime in weeks (wk)	2.60 wk/warp	2.10 wk/warp
Value added	Fabric – (WarpEnd+WeftEnd) = 767.80 CU – (393.74 CU – 239.09 CU)	**134.97** CU/100m	**129.41** CU/100m

In the VFD on *Diagram S2-31,* the process "Weave fabric" is shown with all inputs and outputs. For comparison, the old and new values are both written on the diagram. If the value for the new loom differs from the old loom, it is printed in bold. The values of the old loom and the ones that do not change, are printed in standard style.

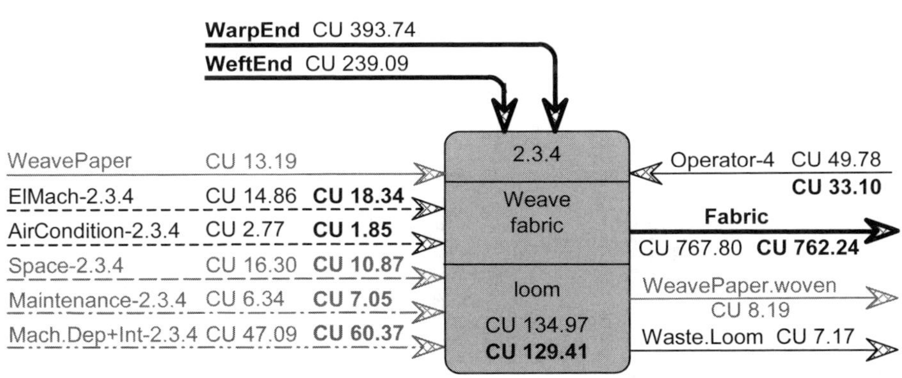

Diagram S2-31: Diagram with the values of the investment calculations.

The comparison of the investment costs can be shown conveniently by placing the parameters of the old and new loom side by side, as demonstrated in *Diagram S2-31* or in the process and flow specifications (*Specification S2-11* and *Specification S2-12*). *Specification S2-12* calculates the value of the electricity consumed by the weaving machine, and again compares the values of the new loom and the old loom.

Specification S2-12: Comparison of Flow "ElMach-2.3.4"

ElMach-2.3.4		old loom	new loom
Description	Time for consumption of electricity taken from the process "Weave fabric."		
Flow Type	Electricity		
Parameter	Consumption ElMachine	5.30 kW	8.10 kW
	PriceElectricity	0.15 CU/kW	0.15 CU/kW
Calculation	Consumption * WeaveTime * WeaveHour * PriceElectricity * (SU/RU) = old: 5.30 kW * 2.30 week/warp * 132 h/week * 0.15 CU/kW * (100 m / 1836 m) = new: 8.10 kW * 2.10 week/warp * 132 h/week * 0.15 CU/kW * (100 m / 1836 m) =		
Flow value		**14.86** CU/100m	**18.34** CU/100m

A comparison like this is performed to help the management to decide about the investment. The diagram shows that the fabric woven on the new loom is lower in cost; its value is CU 762.24 for 100 m. In comparison, the fabric produced on the old loom costs CU 767.80 for 100 m. The calculation done by the VFD clearly shows that the investment in the new machine would be profitable.

2.7.5 Intangible Assets: Labels

A product with a known brand and a good reputation or image can be sold at a higher price than a product with an unknown brand. The increase of the product value by the brand can also be depicted in the VFD. The model's designer decides what type of flow to introduce in this case to bring this increase in product value. For example, the value increase due to product prestige can be added by a fictitious value flow, an information flow, or a resource flow.

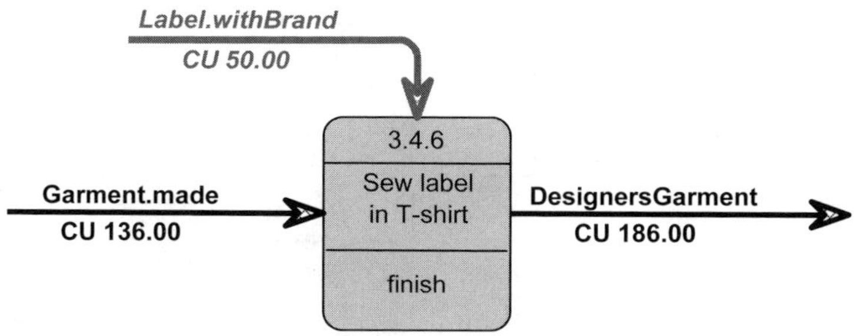

Diagram S2-32: Value of brand.

In a clothing plant, a material flow in the form of the label that bears the brand name carries the value of the brand into the product. The actual value of the material of the label itself is very small and negligible. However, the flow "Label" has a much higher value because it represents not only the actual label material, but also the value of the brand. As soon as the process "Sew label in T-shirt" on *Diagram S2-32* is carried out, the value of the T-shirt is increased significantly, reflecting the prestige associated with the label.

2.8 Application Example: Gas Station

Step-by-step modeling

In this chapter, the Value Flow Diagram of a gas station is set up. The basis for the VFD model is the Flow Diagram of the application example in Section S1. We now learn how to enhance the Flow Diagram to obtain the VFD. A VFD model is created using the following steps, as illustrated in this chapter.

- Step 1: Draw the Flow Diagram.
- Step 2: Fill the element specifications of flows and processes with parameters.
- Step 3: Compose equations in the flow specifications.
- Step 4: Carry out the process balance.
- Step 5: Calculate the flow values in the hierarchy.
- Step 6: Reflections and investigation of impact of different parameter values.

Step 1: Flow Diagram

For this application example with the VFD, we take a slightly different viewpoint than in the Flow Diagram model of Section S1. Instead of cleaning the windshield, we would like to wash the entire car at the car wash. This process is called "Clean car." It is carried out at the same gas station where we get the gasoline.

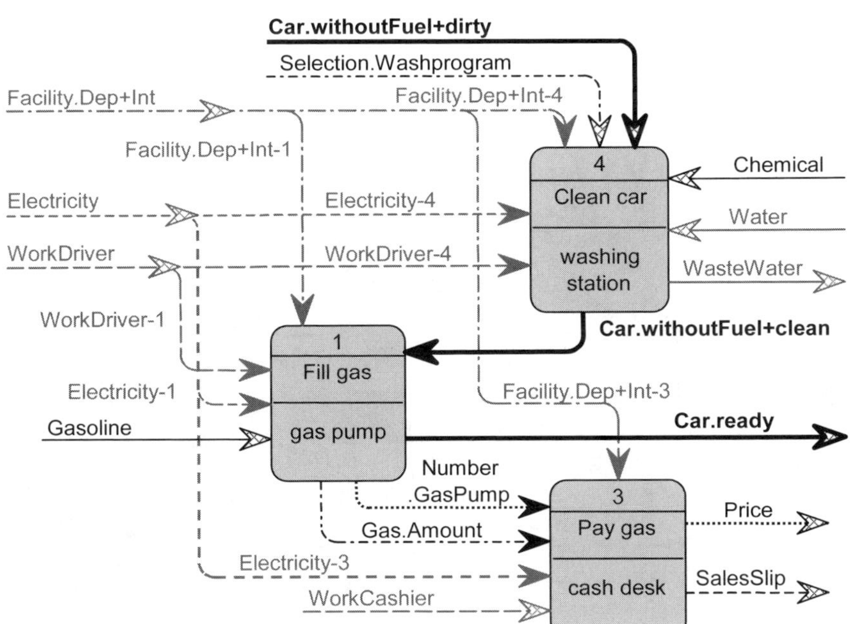

Diagram S2-33: Flow Diagram "Refuel car."

As given in *Diagram S2-33,* the driver first washes the car, then fills up gasoline. Note that the process "Clean car" carries the process number 4, and not the number 2. This number was used in the original application example in Section S1 for the process "Clean windshield." Do not re-number processes if new processes are introduced into an existing model. When both models are compared, two different processes with the same number would show up; that might cause confusion and problems in the data dictionary of the model.

For the VFD, the flows are adapted and more flows are introduced to obtain the real values and costs of the gas station; these are mainly additional resource flows. The colors used for the flow arrows are defined in *Table S2-5* in Chapter S2.4.1.

"Electricity" flows are added for the washing station, gas pump, and cash desk. The costs of the facilities are taken into consideration by the depreciation (Dep) and interests (Int) of the invested capital, as depicted by the resource flows "Facility. Dep+Int." Facilities of the gas station are the buildings, the property, the car wash, the gas pump, the gas tank, and other equipment.

The selection of the wash program, which allows the driver to choose options such as the number of the washing cycles or whether or not to use wax for the finishing, is indicated as an information flow "Selection.Washprogram."

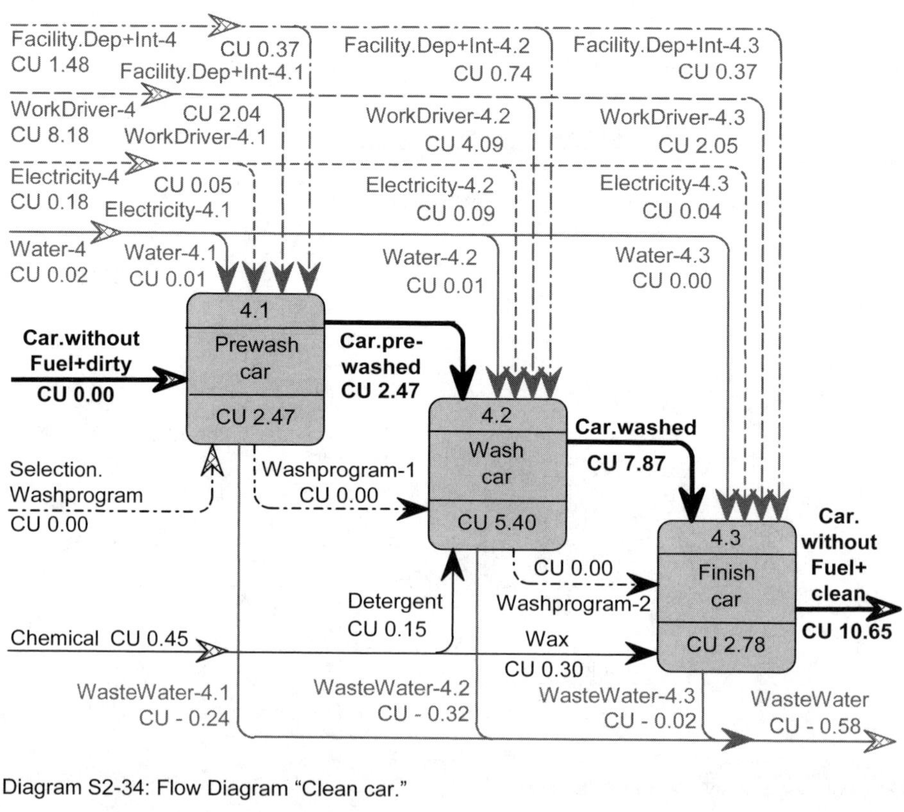

Diagram S2-34: Flow Diagram "Clean car."

In this VFD model, we want to take a closer look at the values and costs of process 4 "Clean car." To do this, we take the next step in the hierarchical detailing and draw the child diagram of process 4, as shown in *Diagram S2-34*. In this model, the car wash is done by hand, instead of driving through an automated carwash. *Diagram S2-34* already shows the values that are calculated later in this chapter.

The car wash is done in three steps. First, the car is prewashed with fresh water to get rid of most of the soil. Second, it is washed with water containing the detergent and then rinsed with fresh water. Last, the wash is completed by applying wax and drying the car. The "Selection.Washprogram," done at the beginning of the wash, automatically goes through the whole procedure from process to process as the information flows "Washprogram."

The *Standard Unit* for the system "Refuel car" is one car. This is what is normally handled by one customer, the driver. He arrives with one car to clean and refuel.

Below in *Table S2-10,* a checklist is provided for the setup of a VFD. The checklist will continue step by step through this application example.

Table S2-10: Checklist VFD

Question	**Checked**
The Standard Unit (SU) of the model is: 1 car	✓
The product flow is specially marked on the diagram.	✓
Waste flows are considered.	✓
Taxes and fees are considered.	✓
Flow types have been defined.	✓
Each flow is assigned to a flow type.	✓
In a split and merge, parent flow and child flow belong to the same flow type.	✓

Step 2 and 3: Element specifications with equations

Numbers, quantities, and parameters are required in order to carry out the value calculations. The element specification of each flow is filled in with parameters. In the next step, the parameters are combined into equations. The calculation procedure in the VFD is an iterative process. In order to keep the application example to a reasonable length, the iterative procedure is not shown here. The following specifications are already completed with the parameters, equations, and calculated values.

Specification S2-13 of the process "Wash car" holds the key "Time washing" that is needed for the calculation of different flow values. The "Time washing" is the time necessary for the driver to perform the car wash process. The value added of "Wash car" will be calculated as soon as the values of the product flows "Car.prewashed" and "Car.washed" are known.

Specification S2-13: Process "Wash car"

4.2	**Wash car**		
Description	The car driver cleans the car with water and detergent, and rinses it with clean water.		
Parameter			
Key	Time washing	**0.2**	h
Value added			

The specifications of all flows going into and coming out of process 4.2 "Wash car" are given in the following. The values of the flows from processes 4.1 and 4.3 are specified and calculated accordingly. *Specification S2-14* gives the details of the product flow "Car.prewashed."

Specification S2-14: Flow "Car.prewashed"

Car.prewashed			
Description	Car rinsed with fresh water.		
Flow type	Product		
Parameter	Price electricity	0.15	CU/kWh
Calculation	(by the balance of process 4.1)		
Flow value		**2.47**	CU/car

The washing time indicated in the process specification "Wash car" is used for the calculation of the value of the flow "WorkDriver-4.2." For the value calculation of the driver's work, it is assumed that an engineer with a salary of 45,000 CU/year washes his car himself (*Specification S2-15*). The introduction of the salary as a value allows the investigation of the carwash from the drivers perspective. The time invested by the driver in such an activity is rarely considered in terms of monetary value. Nevertheless, it can be also very interesting to calculate the costs for work done by yourself.

Specification S2-15: Flow "WorkDriver-4.2"

WorkDriver-4.2			
Description	Owner of the car washes the car himself.		
Flow type	Workforce		
Parameter	Driver salary	45,000.00	CU/year
Calculation	DriverSalary * TimeWashing = 45000 CU/year / 2200 h/year * 0.2 h =		
Flow value		**4.09**	CU/car

The price for the fresh water is 0.6 CU/m^3. This results in very low water flow values (*Specification S2-16*).

Specification S2-16: Flow "Water-4.2"

Water-4.2		
Description	Fresh water.	
Flow type	Water	
Parameter	Water amount Water price	175 l 0.06 CU/m^3
Calculation	WaterAmount * WaterPrice = 175 l * 0.06 CU/m^3 / 1000 l/m^3 =	
Flow value		**0.01** CU/car

The community performs the service of cleaning and disposing of the "Waste Water" for the gas station. In return, the gas station has to pay a waste water fee to the community per cubic meter of water. The "WasteWater" itself goes out of the washing station and leaves the system gas station. The value of it is negative, as shown in *Specification S2-17*. It represents the value of the service received from the community. To get the negative value in the calculation, the parameter "Flow direction" is introduced.

Specification S2-17: Flow "WasteWater-4.2"

WasteWater-4.2		
Description	Water from the washing station goes to the communal sewage plant.	
Flow type	WasteWater	
Parameter	Water amount WasteWater price Flow direction	175 l 1.80 CU/m^3 - 1
Calculation	amount * price * FlowDirection = 175 l * 1.80 CU/m^3 / 1000 l/m^3 * (- 1) =	
Flow value		**- 0.32** CU/car

The value of the "Detergent" in *Specification S2-18* is fixed for one car.

Specification S2-18: Flow "Detergent"

Detergent		
Description		
Flow type	Material	
Parameter		
Calculation	(value fixed)	
Flow value		**0.15** CU/car

The costs for facilities are usually given per year in the cost accounting. To calculate the value of the flow "Facility.Dep+Int-4.2" in the VFD, these costs are broken down by the operating hours per year. Multiplied by the "Time washing" used by process 4.2, the flow value is obtained (*Specification S2-19*). For processes 4.1 and 4.3, the same facilities are utilized.

Specification S2-19: Flow "Facility.Dep+Int-4.2"

Facility.Dep+Int-4.2			
Description	The annual value includes: - Full depreciation within 10 years. - Interest of 2% of capital investment. The washing station is serviced 7 days a week for 18 hours a day. Allowing for breakdowns of the machines, this amounts to 5,400 hours/year. Because the washing station is not always occupied, a percentage for the utilization is introduced.		
Flow type	Depreciation + Interest		
Parameter	Depreciation Interest Facility hours Utilization	5,000 1,000 5,400 30	CU/year CU/year h/year %
Calculation	(Depreciation + Interest) * TimeWashing / FacilityHours / Utilization = (5000 CU/year + 1000 CU/year) * 0.2 h / 3500 h/year / 0.3=		
Flow value		**0.74**	CU/car

Specification S2-20 shows the calculation for the "Electricity." The electricity consumed is given in kW, and is multiplied by the "Price" and the "Time washing" from process 4.2 to obtain the value of energy consumption for one car.

Specification S2-20: Flow "Electricity-4.2"

Electricity-4.2			
Description			
Flow type	Electricity		
Parameter	Electricity price Consumption	0.15 3	CU/kWh kW
Calculation	Price * Consumption * TimeWashing = 0.15 CU/kWh * 3 kW * 0.2 h =		
Flow value		**0.09**	CU/car

The checklist for the VFD is continued below with the points for the element specification.

Table S2-11: Checklist VFD – continuation

Every element has a specification.	✓
The parameters and their quantities are filled in the element specification.	✓
The flow equations are stated in the flow specifications.	✓

Step 4: Process balance

The process balance, according to *Equation E10* and *Rule V19* must be carried out for each process on every diagram in the hierarchy. The process specification "Wash car" is given again in *Specification S2-21*. To illustrate the calculation of the process balance, all input and output flow values are also listed in the process specification. Through the process balance, the yet unknown flow value "Car.washed" is calculated. With this result, the value added by this process can be determined. The originally unknown flow value of this process balance is written in bolt in the specification.

Specification S2-21: Process "Wash car"

4.2	**Wash car**				
Description	The driver cleans the car with water and detergent, and rinses it with clean water.				
Parameter					
Key	Time washing			0.2	h
Value added	Car.washed – Car.prewashed = 21.74 – 7.27 =			**5.30**	CU/Car
Process balance	Input values			Output values	
	Car.prewashed Water-4.2 Electricity-4.2 WorkDriver-4.2 Facility.Dep+Int-4.2 Detergent Washprogram-1	2.47 0.01 0.09 4.09 0.74 0.15 0.00	=	**Car.washed** WasteWater-4.2 Washprogram-2	**7.87** - 0.32 0.00
	Total input	7.55		Total output	7.55

The calculated value of product flow "Car.washed" appears in *Specification S2-22*.

Specification S2-22: Flow "Car.washed"

Car.washed		
Description	Car washed and rinsed.	
Flow type	Product	
Parameter		
Calculation	(by the balance of process 4.2)	
Flow value	**7.87**	CU/car

The calculation for the other value flows and processes is not demonstrated in detail, but the results appear on *Diagram S2-35* and *Diagram S2-36*. *Specification S2-23* gives the specification of the gasoline, including the parameters for the amount and the price.

Specification S2-23: Flow "Gasoline"

Gasoline			
Description	95 octane rating		
Flow type	Resource		
Parameter	Gas amount Gas price	38 0.61	l CU/l
Calculation	GasAmount * GasPrice = 38 l * 0.61 CU/l =		
Flow value		**23.18**	CU/car

Step 5: Calculation in the hierarchy

Flows are calculated hierarchically up and down by split and merge. *Diagram S2-35* depicts the calculated values on level 1 of the gas station model.

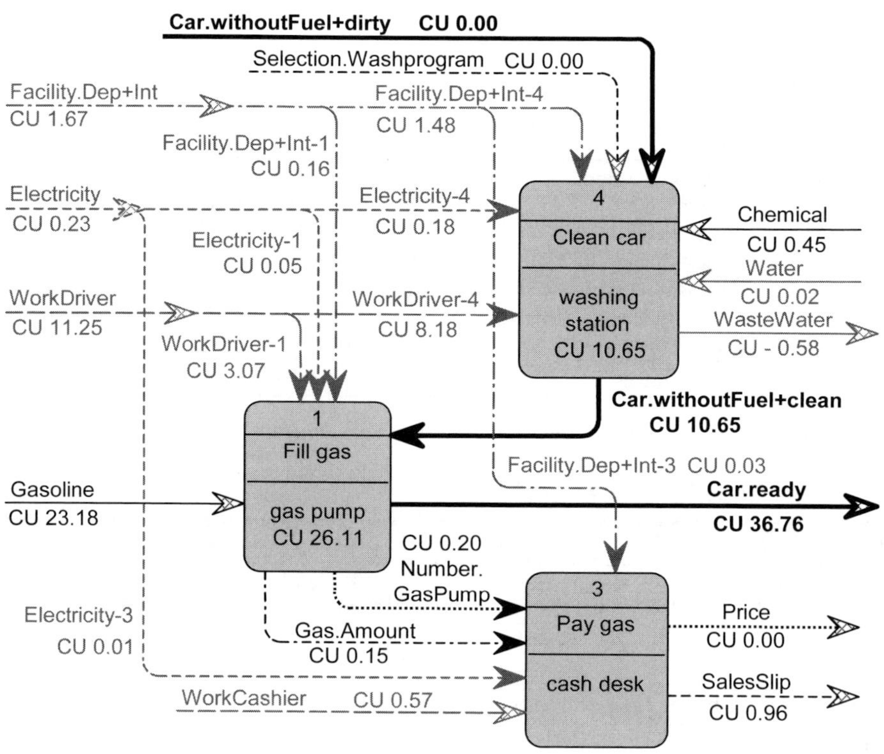

Diagram S2-35: VFD "Clean car."

The resource flow "Facility.Dep+Int-4" in *Specification S2-24* shows an addition of flow values within the hierarchy.

Specification S2-24: Flow "Facility.Dep+Int-4"

Facility.Dep+Int-4	
Description	The "Facility" is added up hierarchically from the child flows.
Flow type	Facility
Parameter	
Calculation	Facility.Dep+Int-4.1 + Facility.Dep+Int-4.2 + Facility.Dep+Int-4.3 = 0.16 CU + 1.48 CU + 0.03 CU =
Flow value	**1.67** CU/car

The checklist for the VFD is continued below.

Table S2-12: Checklist VFD – continuation

A process balance is carried out for every process.	✓
Every process with a product flow has a calculated value added.	✓
A calculation is made for every split and merge.	✓

The following is a reflection on *Diagram S2-35:* Consider that the "WorkDriver" is brought into the process by the car driver. By doing this, we took into consideration the salary of the driver. If the salary of the "WorkDriver" is set to CU 0.00, the value added of process 4 "Wash car" equals the costs that this process causes to the owner of the washing station. The sum of these costs is:

Facility.Dep+Int-4	+ Electricy-4	+ Water-4	– WasteWater-4	+ Chemical	=
1.48 CU	+ 0.18 CU	+ 0.02 CU	– (-0.58) CU	+ 0.45 CU	= 2.71 CU

For a self-service car wash, this is the price that the driver has to put into the machine in the form of coins. The owner of the car wash wants to make profit, Therefore the price is 1.5 to 2 times more expensive than the net price.

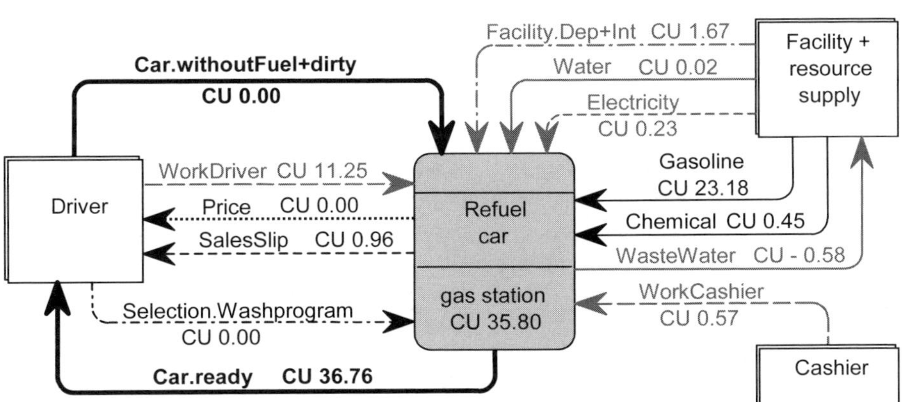

Diagram S2-36: Context diagram.

From the context diagram on *Diagram S2-36,* the total amount of each resource flow can be seen. The value added by the single process on the context diagram "Refuel car" indicates how much fueling and cleaning the car is worth to the driver (including his salary): CU 56.38.

The driver gets no salary for cleaning and refueling his own car. His reward is a shining car with a full tank.

Step 6: Reflections and variations

With the results from the VFD of the gas station, further analysis can be done. If some values had to be estimated to complete the calculation of the VFD, try specifying these values.

Play with the parameters. For example:

- What if you send your daughter or your neighbor's son to wash your car? They still go to school and have no salary. What do you have to pay them as pocket money for this washing job? Insert the value of this pocket money into the parameter "Driver salary" and the VFD will show the new values.
- What if an automatic washing station/line is used? The process times are shorter, but the consumption of water certainly higher. What would the driver have to pay for it? Is it worth it?
- The utilization of the washing station is crucial. How high must the utilization be for the washing station to be profitable?

Next steps

We did a cost analysis of washing a car at a gas station. The following possibilities for further investigations within the method of POA are available.

Table S2-13: Next steps in the POA toolbox

Aspect of investigation and optimization	**POA method**
How much energy is used in each process? Are there energy losses?	Use the RFD. Section S3
Introduce the time-related procedure, with its sequences and conditions, in order to get a dynamic model.	Do a State Chart. Section D1
Perform a simulation of the gas station, to find out how many pumps are needed to keep the waiting of the customer as short as possible.	Program a simulation model. Section D2
Use the model to write the real-time control of a gas pump based on a State Chart.	Code a machine control. Section D3

2.9 Apply Your Knowledge

1) Use of VFD

List some possibilities, where you use the VFD. What problems can you solve that your boss or the management of a company might ask of you?

2) Flow classification

Diagram S2-37 describes the noodle production on pasta line 2. Do the classification for the flows of this production in form of a list, with flow categories, flow groups, and flow types.

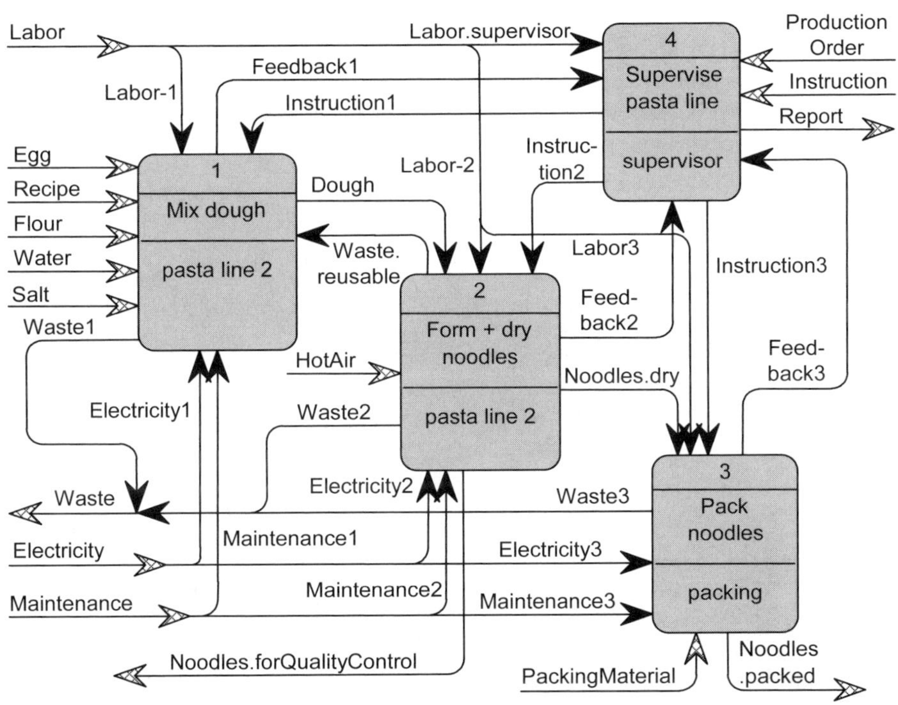

Diagram S2-37: Production of noodles.

3) Process balance and calculation in split and merge

Diagram S2-38 and *Diagram S2-39* describe the production of a business suit. *Diagram S2-39* details process “Produce business suit.” Some flow values are already indicated on the diagrams. Calculate the missing flow values of the parent process and on the child diagram. Set up the process balance for each process. Hint: Split and merge helps calculating the flow values.

The trousers and the jacket of the suit are cut from the same piece of fabric in order to get exactly identical color. After the cutting, the pieces of the trousers and of the jacket are sewn in two different production lines. At the end, the matching trousers and jacket are combined.

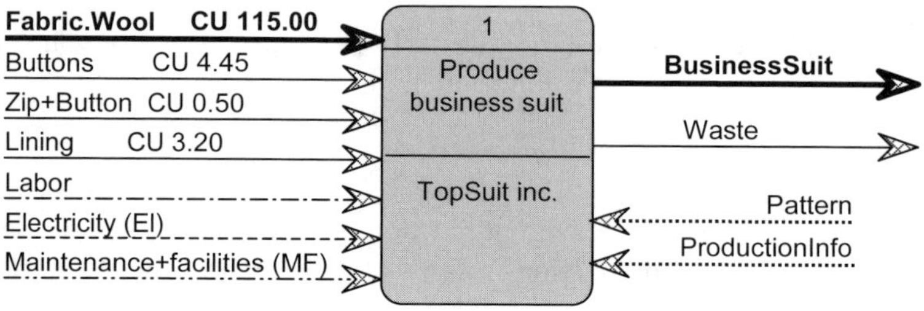

Diagram S2-38: Parent process "Produce business suit."

On the product flows, the length of fabric used is indicated in meter to help allocate the values. The amount of waste is given in percentage of the product flow. The input values of process "Supervise shop floor" are equally distributed to the output flows of this process.

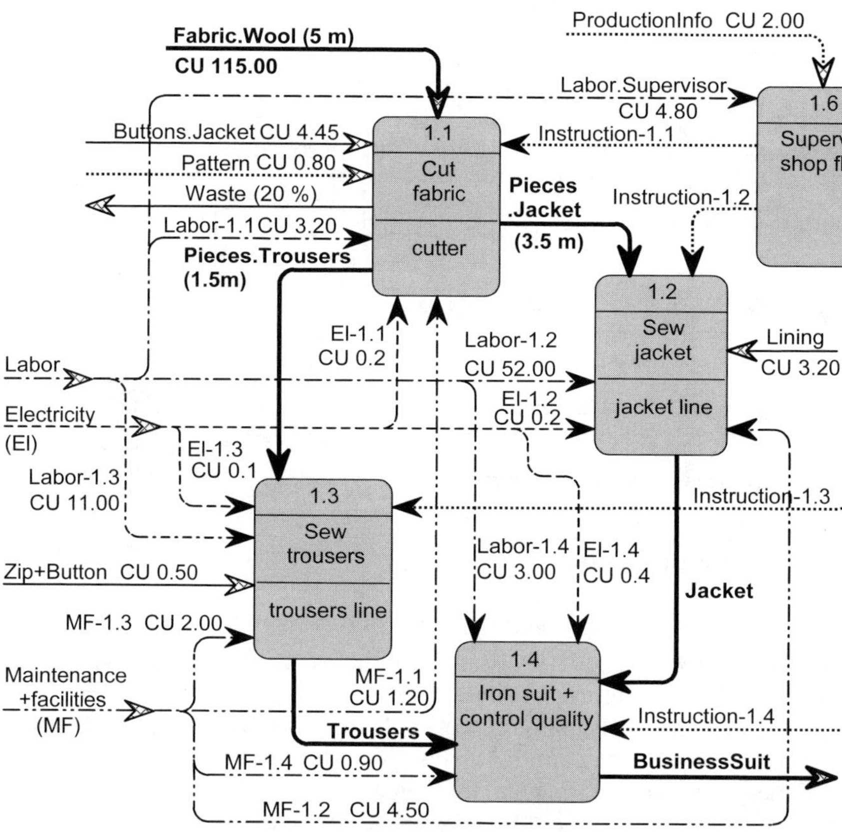

Diagram S2-39: Child diagram of process "Produce business suit."

How much value has the business suit directly after production, before waste, overhead and other indirect costs are passed onto it?

4) Value added

How is the value added defined in POA? Calculate the value added for the processes in *Diagram S2-38* and *Diagram S2-39* of the business suit from question 3. Write the equations for it. Have all processes in this example a value added?

5) Process and flow specification

Model the system for homemade muffins:

- First, draw the Flow Diagrams for baking muffins at home. Insert all necessary flows. You can either mix the dough with all necessary ingredients or take a ready made baking mix for muffins. The model has at least a context diagram and a first detailing level with the main processes.
- Second, set up the flow and process specifications and fill in the necessary parameters for a Value Flow Diagram.

6) Value Flow Diagram

Take the model of question 5 a step further.

- Decide on the Standard Unit for the muffin production. What are the pros and cons for choosing 1 muffin, 1 kg of dough, 1 baking mix, or 1 baking try load?
- Calculate the values for the flows and the value added of the processes for the muffin VFD.

7) Investment analysis

Diagram S2-40 shows process 2 of the production of plastic parts by injection molding: The product is a yogurt cup made of polystyrene. The flow values are indicated on the diagram.

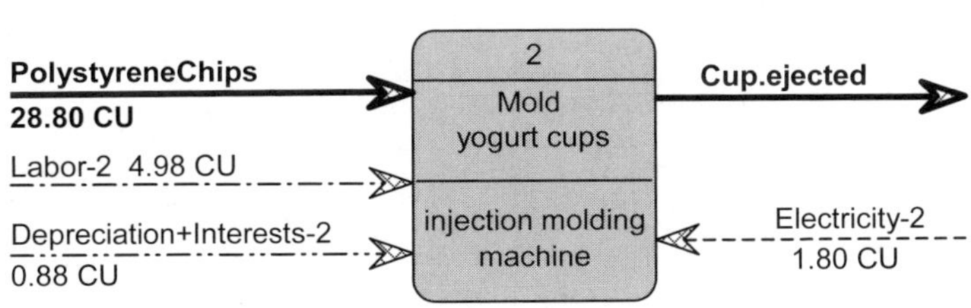

Diagram S2-40: Mold yogurt cups.

This VFD is now used to do an investment analysis. The management would like to decide on buying a new injection molding machine. The one in use was bought second hand and is now 20 years old.

The necessary information about the existing and proposed machine are given in *Diagram S2-40* and in the specifications. The SU is 960 cups. Complete the specification and calculate the values for the new machine. Insert the new values in the VFD. The parameters in the specification that change are highlighted.

In the following, the specifications for the flows and for process 2 "Mold yogurt cups" from *Diagram S2-40* are given (*Specification S2-25* to *Specification S2-30*).

Specification S2-25: Product flow "Plastic chips"

Polystyrene chips		**old machine**	**new machine**
Description	The yogurt cups are made of polystyrene.		
Flow type	Product		
Parameter	Price polystyrene chips	3.00 CU/kg	3.00 CU/kg
	Amount polystyrene	10 g/cup	10 g/cup
	Standard Unit SU	960 cups/SU	960 cups/SU
Calculation	$\frac{\text{Price}\left[\frac{CU}{kg}\right] * \text{Amount}\left[\frac{g}{cup}\right] * 960\left[\frac{cup}{SU}\right]}{1000\left[\frac{g}{kg}\right]} =$		
Flow value		**28.80** CU/SU	CU/SU

Specification S2-26: Process "Mold yogurt cups"

2	**Mold yogurt cups**	**old machine**	**new machine**
Description	The new machine has a shorter injection cycle time. The new molding tool has more cavities, meaning it can produce more parts per injection cycle. The company works 3 shifts. This means 5,800 h/year for the machines.		
Parameter	Machine hour per year	5,800 h/year	5,800 h/year
	Cavities in tool (Number of cups per cycle)	4 cups	16 cups
Key	Injection cycle time (ICT)	12 s	4 s
	Process time for 1 SU	$\text{Time}_{SU}\left[\frac{h}{SU}\right] = \frac{ICT\,[s]}{3600\left[\frac{s}{h}\right]} \cdot \frac{960\left[\frac{cup}{SU}\right]}{\text{Cavities}\,[cup]}$	
		0.8 h/SU	h/SU
Value added		CU/SU	CU/SU

Since the time needed to process 1 SU of cups is used for calculation in several specifications, it is indicated as an equation in the section key in *Specification S2-26* as parameter "Time_{SU}."

Specification S2-27: Flow "Electricity-2"

Electricity-2		**old machine**	**new machine**
Description	Electricity needed to run the injection molding machine. The process time is indicated in the specification of process 2 "Mold yogurt part."		
Flow type	Electricity		
Parameter	Consumption electricity	5 kW	8 kW
	Price electricity	0.15 CU/kWh	0.15 CU/kWh
Calculation	$\text{Consumption}\,[\text{kW}] \cdot \text{Price}\left[\frac{\text{CU}}{\text{kWh}}\right] \cdot \text{Time}_{\text{SU}}\left[\frac{\text{h}}{\text{SU}}\right]$		
Flow value		0.60 CU/SU	CU/SU

Because the calculation in *Specification S2-28* is too complicated to put into one equation, it is done in three steps. First, the time necessary for 1 SU is calculated. "Time supervising" depends on the injection cycle time (ICT). "Time changing mold" and "Time maintenance" take place once in a while after a certain amount of cycles are done. In the third step, the costs for the production time per SU are calculated. The results for the old machine are indicated in the specification.

Specification S2-28: Flow "Labor"

Labor-2		**old machine**	**new machine**
Description	The labor includes the supervising of the machine, the changing of the mold and setting the parameters for a new part, the maintenance of the machine. Since one operator operates more than one machine, the time of the operator supervising is given as a percentage of the processing time (injection cycle time).		
Flow type	Labor		
Parameter	Time supervising TS	2 % ICT	2 % ICT
	Time changing mold TC	2 h	1.5 h
	Time maintenance TM	1 h	1.5 h
	Labor cost per hour	17 CU/hour	17 CU/hour
	Number of cycles per 1 change of mold and maintenance	24,000 cycles	96,000 cycles
Calculation	T1: TS for 1 SU $T1_{\text{old}} = 0.016$ h	$0.02 \cdot \text{Time}_{\text{SU}}\left[\frac{\text{h}}{\text{SU}}\right]$	
	T2: TC and TM for 1 SU $T2_{\text{old}} = 0.3$ h	$\frac{960\left[\frac{\text{cup}}{\text{SU}}\right]}{\text{Cycles} \cdot \text{Cavities}\,[\text{cup}]} \cdot (\text{TC}\,[\text{h}] + \text{TM}\,[\text{h}])$	
	$(T1 + T2)\left[\frac{\text{h}}{\text{SU}}\right] \cdot \text{LaborCost}\left[\frac{\text{CU}}{\text{h}}\right]$		
Flow value		**5.37** CU/SU	

The calculation for the flow "Depreciation+Interests" in *Specification S2-29* is done in two steps. Step 1 evaluates the value for one year. Step 2 calculates the value for the SU. This calculation involves the breaking down of the value per year to the time needed to produce 1 SU.

Specification S2-29: Flow "Depreciation+Interests"

Depreciation+Interests		**old machine**	**new machine**
Description	Depreciation and interests are calculated for the machine and the mold (molding tool). For the new machine, also a new mold is used. The depreciation is linear over 10 years, and the interest are 10% on the invested capital. The depreciation and interest rate is indicated for one year. They must be broken down for the SU. Process time is indicated in process 2.		
Flow type	Depreciation + interest		
Parameter	Value machine V_{mach}	10,000 CU	60,000 CU
	Value mold V_{mold}	10,000 CU	40,000 CU
	Duration depreciation	10 years	10 years
	Interests	10 %	10 %
	MachineHour per year MH	5,800 h/year	5,800 h/year
Calculation per year	$\frac{V_{mach} + V_{mold}}{\text{Duration}} \left[\frac{\text{CU}}{\text{year}}\right] + (V_{mach} + V_{mold})[\text{CU}] \cdot \text{Interest}$		
Calculation per SU	$\frac{\text{Value}_{year} \left[\frac{\text{CU}}{\text{year}}\right]}{\text{MH} \left[\frac{\text{h}}{\text{year}}\right]} \cdot \text{Time}_{SU} \left[\frac{\text{h}}{\text{SU}}\right]$		
Flow value		**0.55** CU/SU	

Specification S 2-30: Product flow "Cup.ejected"

Cup.ejected		**old machine**	**new machine**
Description	The yogurt cups are made of polystyrene.		
Flow type	Product		
Parameter	Standard Unit SU	960 cups/SU	960 cups/SU
Calculation	(calculated by the process balance)		
Flow value		CU/SU	CU/SU

This investment analysis should answer the following questions:

- How much do 960 cups cost with the new machine?
- What is the value added for process 2 "Mold yogurt cups" when producing with the new machine?
- Do you recommend the management to buy a new machine or continue producing on the old one?

S3 RESOURCE FLOW DIAGRAM

Goals of the Chapter

- Learn to draw a Resource Flow Diagram.
- Become familiar with calculating the environmental and energetic values of the flows and processes.
- Execute an energy and/or exergy balance.
- Calculate the energetic and exergetic efficiency of a process.
- Determine the embodied energy of a product and the embodied energy added by the processes.

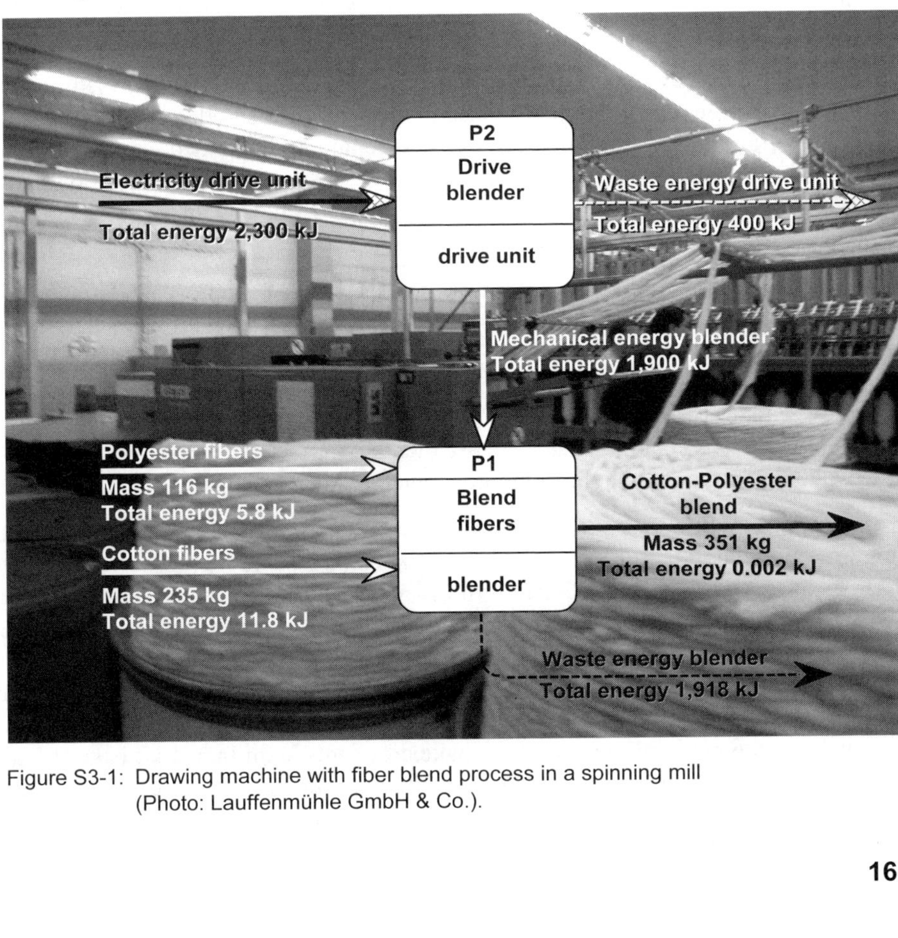

Figure S3-1: Drawing machine with fiber blend process in a spinning mill (Photo: Lauffenmühle GmbH & Co.).

3.1 Introduction

In a manufacturing company, many resources are consumed, and products, emissions, and waste are generated. The Resource Flow Diagram (RFD), an extension of the Flow Diagram, will be introduced in this section of the book. The RFD concentrates on the interfaces of processes and maps resource flows such as energy, material, waste, and manpower. The RFD first analyzes the origin and destination of these flows and then calculates the corresponding energetic and environmentally relevant values. With this analysis, resource and emission management in a company can be optimized and possibilities for recycling and reuse of resources can be analyzed.

The "As-Is Model" of the RFD corresponds to a real system in the actual present day situation. This model also concentrates on the resources and emissions necessary for the manufacturing of the product. The RFD is used to optimize a system, focusing at closed-loop resource flows, reducing resource consumption, emissions, and waste. The RFD is often used as a basis for discussion with project partners and clients because the diagrams and its symbols introduce a common language for the system and show an overview of operations.

The environmental and energy evaluation of the RFD is carried out based on the input data of processes. Resource consumptions are measured or calculated and evaluated with a mass, energy or exergy balance and an embodied energy calculation. The results of the analysis are the quantitative values of the resource flows and the energetic and exergetic efficiencies as well as the embodied energy added of the process.

The RFD analysis generates quantitative results with energetic calculations and evaluations. Energetic values are assessed and evaluated with physical laws; thereby precise results are generated. These results are the basis for technical and logistical improvements within the production chain. In contrast to the Value Flow Diagram, which assesses financial values for the resource and information flows, the RFD focuses on energetic and environmentally associated flows. Nevertheless, the two diagram types could be combined together advantageously.

This Section S3 introduces the RFD, its illustrations and calculation rules. First the application and delimitation is treated, then the elements and their specification are introduced, and finally the calculation rules and results are discussed. At the end of the section, an application example of a gas station will be analyzed to illustrate the use of the RFD. The resource flows of the gas station are assessed in order to calculate the total energy and water consumption. With these results possibilities for reuse of waste energy in the waste water of the car wash are evaluated. Case Study C3 analyzes the production of ready-to-bake croissants in an industrial bakery. The RFD of the croissant line was modeled on site with the help of the employees and filled in with values, which were measured at the machines. The results and optimization possibilities were discussed with the manager of the company.

3.2 Resource Flow Diagram: Why?

3.2.1 Purpose

Analysis and Optimization: The purpose of the RFD is the analysis and optimization of manufacturing systems from an energy point of view. This is achieved by collecting processes and resource flows in a system and mapping them with various levels of detail, according to the rules of the Flow Diagram in Section S1. Furthermore, different values may be calculated for the resource flows, such as the mass, total energy, exergy, and embodied energy. Depending on the goal of the study, one or more of these factors are calculated for the system to reach its best possible performance.

The following points describe the main purposes of the RFD:

- **System analysis with the RFD:** Mapping process interfaces on different levels of detail to get a complete overview of the system, communicate with other partners involved with the object, and calculation of mass and total energy of the flows for an introductory system assessment.
- **Energy evaluation of the system:** Performing an energy and/or exergy balance of the production system in order to determine the energetic performance of the production and optimize energy management, for example concerning the use of waste energy in exhaust air flows.
- **Environmental evaluation of the production system:** Calculating the embodied energy of the product as well as the embodied energy added by processes in order to analyze the environmental performance of the product's supply chain. With this tool, it is possible to compare energy intensities of materials in different phases of production.
- **Weak-point analysis:** Recommendations for technical improvements of processes focusing on closed-loop resource flows, reduction of waste, and the optimization of resource consumption.

Calculation: The RFD assigns resource values to the resource flows. These are mass, exergy, total energy, and embodied energy values. In order to calculate and balance the values, first a mass balance and an energy analysis are performed. In cases where a lot of thermal energy is consumed and a lot of waste energy (heat) is produced, an exergy balance is calculated. The exergy balance evaluates the efficiency of energy management and shows whether other energy resources could be applied instead, or if waste energy flows could be reused. Furthermore, an embodied energy analysis may be performed to better understand the system supply chain.

3.2.2 Application

Any system can be modeled and calculated using the RFD because it allows for the structuring of complex systems on any level of detail. The variety of calculation methods enables a minimal analysis of a system by calculating one or two of the resource values, or a complete analysis including the calculation of all four resource values of the flows (mass, total energy exergy, and embodied energy).

By modeling the system with processes and flows, the resource paths and the process logistics are shown. The product flow is used as a master path through the analysis. The resource values allow a complete assessment of the system. The calculation gives a more detailed insight into the actual resource consumptions and emissions created by the production of the investigated product. Alternative scenarios, like the replacement of a certain resource by another one or the reuse of waste, can be forecasted to indicate opportunities for technical changes and innovations.

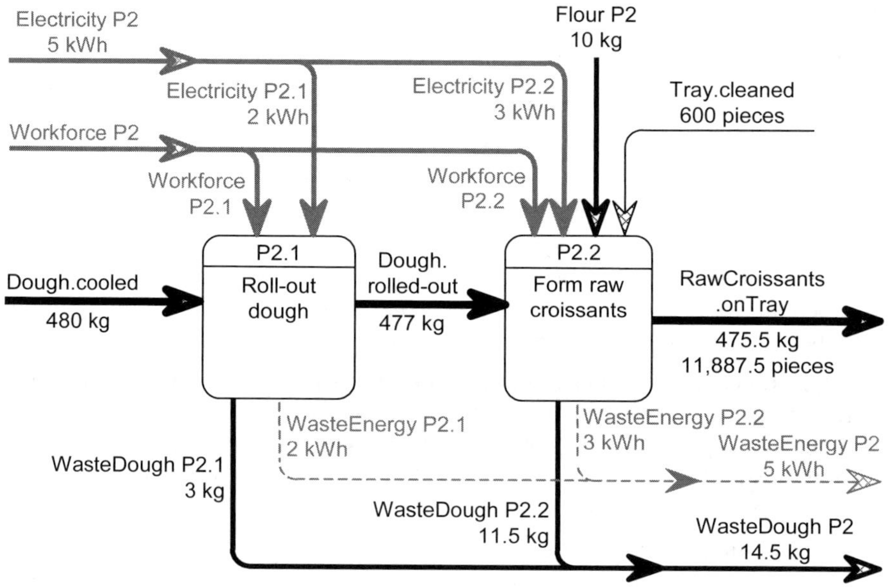

Diagram S3-1: RFD of a croissant production line.

Diagram S3-1 shows the processes and resource flows of a croissant production line. The diagram indicates how much energy is consumed for the production of the raw croissants, and how much waste energy and waste dough result. Additionally, the number of raw croissants made of the dough can be seen. This diagram is part of the case study introduced in Chapter C3.

3.2.3 Delimitation

Life Cycle Assessment

In this chapter, the RFD is compared to Life Cycle Assessment (LCA), which is commonly used for environmental evaluation of systems. Other environmental evaluation methods exist, such as "Input-Output Analysis." LCA is a method that calculates and evaluates the environmental impact of a product or service over its entire life cycle. First, a system analysis is performed, then a data inventory is created, and finally results are evaluated based on various scientific evaluation methods.

The RFD performs an extended energy evaluation of a product and the production. It goes much more into detail using hierarchical system description, graphical modeling and visualization than LCA does. The inventory of LCA is often taken from LCA software tools, where pre-defined datasets are already available, otherwise general literature data is applied. Whereas in the RFD, specific data from companies is used for the models, not every single substance is followed, but only the significant flows. The RFD is able to represent a company, as it is in reality, by an individual model.

The environmental evaluation method of the RFD is based purely on quantitative figures from the calculation of mass, energy, and exergy balances. Additonally, the embodied energy added is calculated and used as indicator for the greenhouse effect. LCA tools give a choice of various scientific valuation methods for the analysis results, which are not compatible and result in technical recommendations.

The LCA starts with data collection and establishing of the database of the inputs and outputs of the system. In general, it is not done graphically. In contrast to the LCA method, the RFD stresses the graphical representation of the system. The graphical model in RFD is made for the life cycle of a product, a part of it, a company, or a suitable machine. The graphics support getting to know the system, understanding the processes, and depicting the system in detail. The analysis starts with the construction of the model.

Table S3-1: Comparison of LCA and RFD

Method / Aspect	Life Cycle Assessment	Resource Flow Diagram
Graphical capabilities	Poor	Powerful
Hierarchical structuring	No	Yes
System boundary	Verbal description	Context diagram
Energy valuation	Energy balance only	Energy and exergy balance and efficiency
Flow classification	No, often only collecting flows or single substances	Different diagram levels, e.g. energy flows only
Cost evaluation	Within LCC (Life Cycle Costing), but no calculation of the value added	By Value Flow Diagram
Transition to dynamic modeling	No	By State Chart

As listed in *Table S3-1,* the primary strength of the RFD is its graphical representation. The LCA method on the other hand, is generally weak in representing a system graphically. A graphical representation is important for setting up the system in cooperation with the company employees, and communicating the results of the analysis to the customer. In contrast to LCA, the RFD shows interactions of the analyzed system with the environment by defining flows and external entities in the context diagram. The diagrams along with the comprehensiveness and consistency of the physical balances, check the resource values of the calculations.

Cost analysis and simulation

The user of the RFD has several options to supplement the analysis, either by a cost evaluation, or by a dynamic modeling of the system. All the methods in POA (see POA toolbox in Chapter Introduction) are connected consistently.

Table S3-2: Comparison RFD and VFD

	RFD	**VFD**
Flows in the model	Resource flows	Resource and information flows
Calculated flow values	Mass, total energy, exergy, embodied energy	Mass, monetary value
Calculated process values	Energetic efficiency, exergetic efficiency, embodied energy added	Value added (monetary)

The RFD and the Value Flow Diagram (VFD), introduced in Section S2, may be combined. *Table S3-2* shows the focus and limitations of both methods. The cost analysis identifies the actual monetary value of a flow at any point in the chain of processes and calculates the value added of the processes. Supplementary dynamic diagrams that lead from process analysis to dynamic state analysis and discrete simulation are given in Part D. A dynamic model based on the RFD serves to evaluate the sustainability of the system in terms of continuous improvement.

3.2.4 Definitions

Resource values

The resource values are calculated for the resource flows of the RFD. Theses resource values are: Mass, total energy, exergy, and embodied energy. Value in general is an artificially assigned property. The rules for the value assignment depend on the system, in which the value is used. The definitions of these resource values are given in the following listing.

- Mass: The weight or mass of product or material flow, for example, the weight of a car, or the mass of 1 liter water correspond 1 kilogram.
- Total energy: The energy amount or content of a product, material, or energy flow, such as the consumed amount of electricity for a process, the combustion value of oil, or the caloric value of a croissant.

- Exergy: The part of the total energy of a product, material, or energy flow, which is completely convertible into any other energy form at ambient conditions, for example, exergy of hot water at 120 °C.
- Embodied energy: Cumulated amount of energy necessary to process the product, material, or energy flow up to the relevant point of the investigation, for example the cumulated energy consumption for coffee growing, harvesting, transporting, and roasting up to the freshly brewed coffee.

Standard Unit

In the RFD, all calculated and displayed resource values are standardized on a certain unit: the Standard Unit (SU). The Standard Unit is defined as the unit of a product that is subject of the analysis. At the beginning of each system analysis, the SU of the product is set. For example, if a company wants to know the analysis results per kilogram of product, then the SU is set to 1 kilogram of the product. All other resource values of the analysis are then calculated relating to the same SU, for example, energy consumption per 1 kilogram of product. By this, the comparison of the values within a model is made possible.

Definition: The Standard Unit (SU) serves as common basis for calculation of all resource values within a model.

Reference Unit

In many production systems, some of the processes along the production chain relate to units different from the SU. Some processes specify their values in other units, called Reference Units (RU). The Reference Unit must ultimately be converted into the SU. The RU, like the SU, is a unit of the product in question. However, the RU is only used in a single process or a group of processes in the production chain.

Definition: The Reference Unit (RU) is an auxiliary reference for the standardization of the resource values. It represents a product unit in a single process or a small group of processes within the production chain.

Example of SU and RU

Using the example of the life cycle analysis of a T-shirt, the SU and RU and their connection are illustrated. The goal is to know the resource consumptions of the whole process line for one cotton T-shirt. In this case, the SU is defined as one T-shirt. The energy consumption is expressed in Joule per one T-shirt, and the water consumption in liters per one T-shirt. During calculation of the single processes along the textile production chain, units other than the SU of 1 T-shirt are used. For example, kilogram of cotton or meters of knitted fabric after the process "Knit fabric." These are Reference Units of the analysis in *Diagram S3-2*. For the hand-over from one process to another one, the RU is transformed into another RU or into the SU. Finally, the RU must be converted into the SU for final calculation and display on the diagram to make all values comparable.

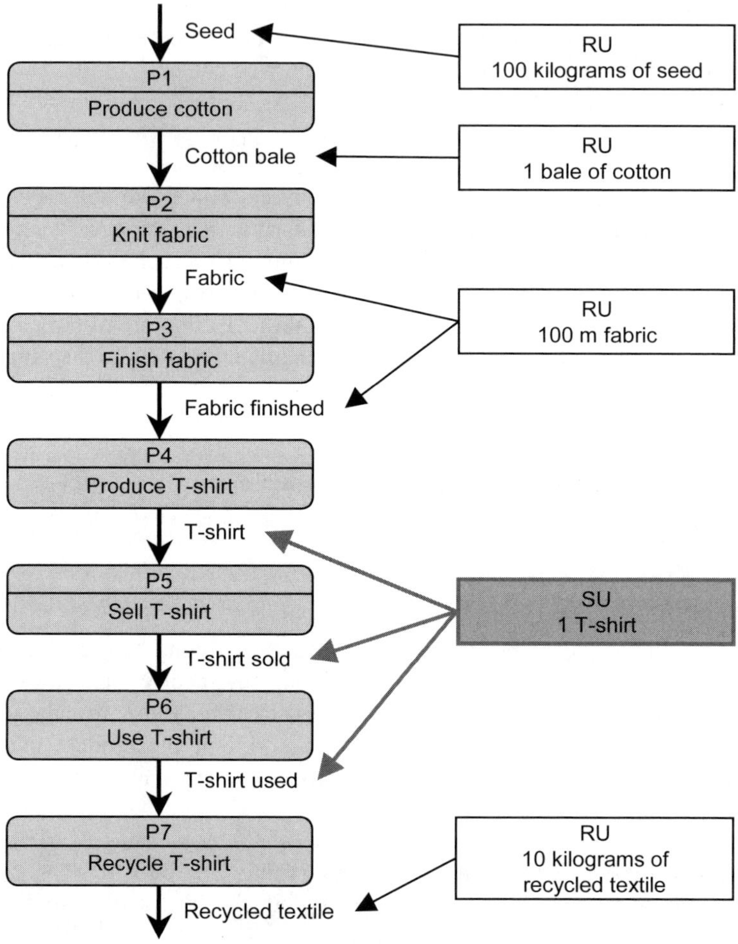

Diagram S3-2: SU and RU for the lifecycle of a T-shirt.

Generic data

The raw data for the RFD may be information taken from the company, measured data from real-life production, or generic data. Data from measurements is documented as comprehensively as possible regarding the process technology, the geographic location, the accuracy of the measurement, and the point in time taken, to make retraceable, for which production situation it applies. If measured data is not available, generic data is preferable to data from literature. Generic data are calculated data, which trace back to generally applicable and recognized data of resources, and processes accessible to general engineering knowledge of the particular production processes. For a first assessment only, data from literature or estimated values may help to build up the model and determine provisional resource values.

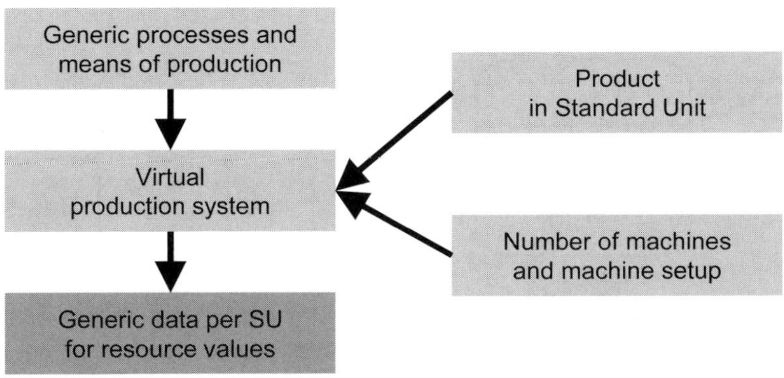

Figure S3-2: Generic data.

Figure S3-2 shows the procedure for establishing generic data. This is commonly determined by assuming the use of state-of-the-art machines to produce the analyzed product. Machine specifications issued by the manufacturer offer data such as production parameters or energy consumption. A virtual production system for a specific product and its Standard Unit with the necessary processes or machines respectively is set up based on the parameters of machine specifications. This allows generic data to be calculated for the production of the product. Using this method, data for input flows are gained, such as energy consumption, compressed air, and other resources. It should be taken into consideration that the generic values are generally valid values for a specific production under optimum conditions, which do not consider local circumstances.

3.2.5 Concept of Energy and Exergy

In this chapter, the concept of energy and exergy for calculation and valuation in the RFD is explained. The RFD enables a complete energy valuation of resource flows. This is most useful for industrial production processes. All energy valuations are based on the first and second laws of thermodynamics. It is therefore not limited to calculating total consumptions, but also includes an assessment of the usefulness of the energy flow based on its exergy content. For example, waste water with a temperature of 100 °C can be useful to heat fresh water through a heat exchanger. On the other hand, it does not make much sense to use waste water with a temperature of 25 °C for the same purpose.

Diagram S3-3 shows the concept of energy and exergy for the RFD. Every resource and energy flow (product, material, or energy) carries total energy and exergy.

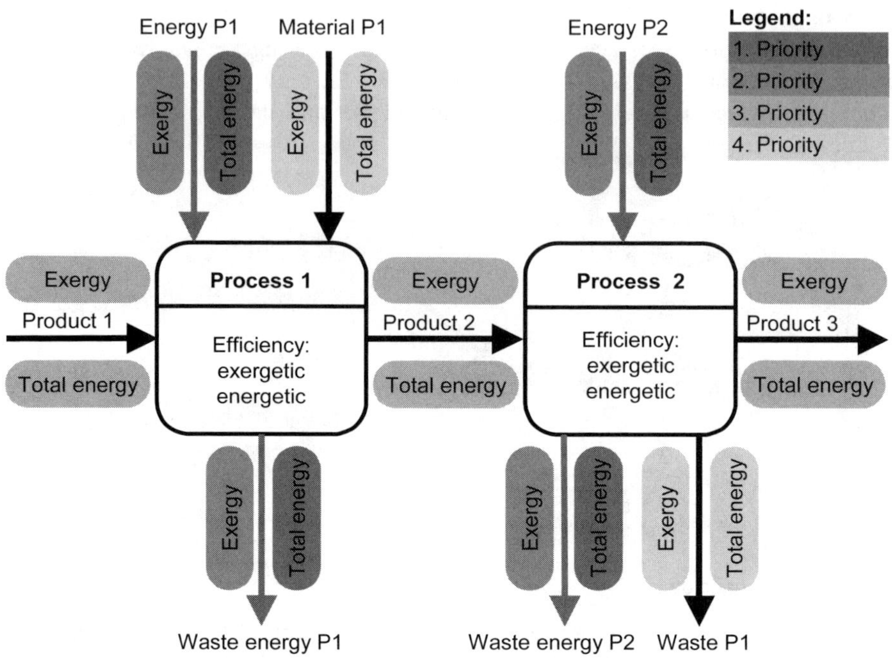

Diagram S3-3: Concept of energy and exergy.

The different color shades of the resource values indicate the calculation priority in a standard analysis for energy and exergy:

1. Assessment of the total energy of the energy flows
2. Calculation of the exergy of the energy flows
3. Calculation of total energy and exergy of product flows
4. Calculation of total energy and exergy of other material flows

1st law of Thermodynamics: The first law of thermodynamics states that energy is conserved in an enclosed system. This law is the basis for all energy transformations and calculations occurring in the manufacturing processes described by the RFD. It allows the calculation of specific energy flows by balancing input and output flows of processes.

2nd law of Thermodynamics: The second law of thermodynamics introduces the concept of exergy and entropy. It states that any transformation of caloric energy into mechanical energy goes along with an exergy loss and an increase of entropy. Entropy is disorder caused in a system by energy conversions, which lead to irreversible loss of useful energy. This implies that no energy form can be converted completely into another energy form.

In the RFD, the emphasis on exergy is useful for defining the value and reusability of an energy flow. Every energy flow comprises an exergy part in a defined environment. Exergy is the portion of energy that can be converted completely into any other energy form, such as work. While the exergy in a system generally decreases, the entropy increases.

Definition: Exergy is the part of energy which is available at ambient conditions and is fully convertible into any other energy form.

Energy flows composed of 100% exergy are entirely convertible. This means that the entire energy amount can physically be converted into any other energy form. Examples for this are electricity, kinetic energy and potential energy. They can be converted, regardless of the temperature level, into another energy form. Restrictedly convertible energy flows have an exergy part that can be converted into another energy form, such as heat.

Entropy: The systems exergy balance quantifies entropy produced by exergy transformations in the system. In technical systems, a minimum of entropy should be produced in the processes and a maximum of exergy should be preserved. Entropy is less decisive for small processes, but for large systems it may become a significant parameter for process setup. In any case, the focus must lie on minimization of produced entropy.

3.3 RFD Elements

3.3.1 From Flow Diagram to RFD

The RFD is based on the Flow Diagram. It supplements the Flow Diagram with a complete set of resource flows and resource values. The RFD consists of three elements: Process, resource flow, and external entity. In this section, some of the rules of the Flow Diagram in Section S1 are repeated and supplemented for the Resource Flow Diagram.

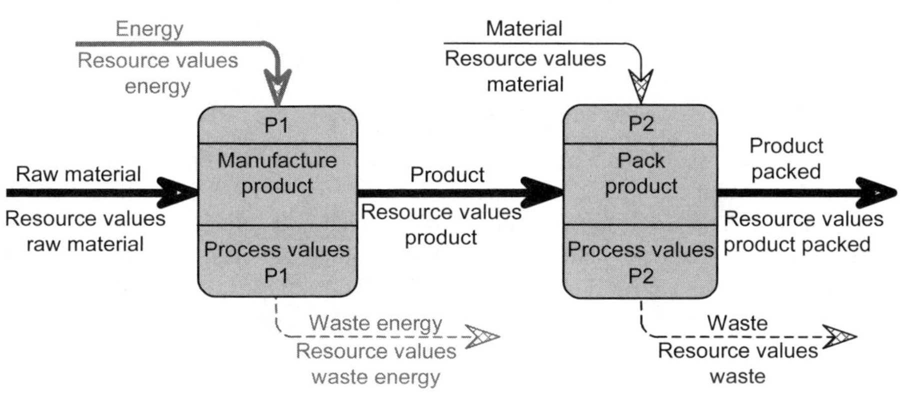

Diagram S3-4: Resource flows and processes.

A basic RFD is given in *Diagram S3-4* including all possible resource and process values. The RFD determines resource values of mass, total energy, exergy, and embodied energy for the resource flows and displays them on the diagram. For the process, values of energetic efficiency, exergetic efficiency, and embodied energy added are shown. The RFD may display only one, a few, or all of these values in the process. Normally, only the values necessary for the analysis of the investigated system are calculated and displayed on the diagram.

3.3.2 Process

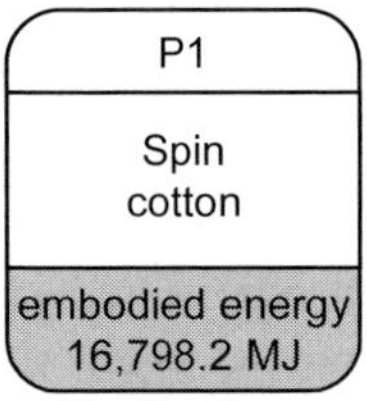

Every transformation of a resource is represented by a process. A process converts incoming resources into outgoing resources. A process performs resource flow transformations based on the laws of energy and mass conservation. The equations for the transformations and all associated data are defined in the process specifications. The calculated values are displayed in the third field, in the lower area of the process symbol that is reserved for additional information. The process "Spin cotton" shows for example the embodied process energy that expresses how much embodied energy is passed on to the product when passing through this process.

3.3.3 Resource flow

The resource flow represents graphically an object specified while passing from one process to the next, as for example a product, material, energy, or manpower. It exits from one activity and enters another activity. A flow defines the interface between two processes or between a process and an external entity (*Diagram S3-5*).

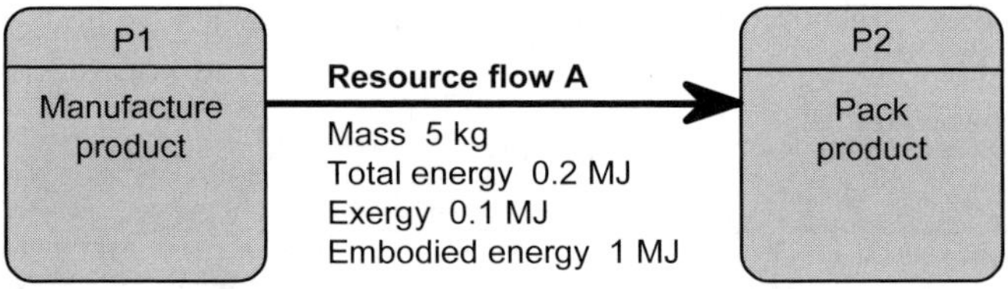

Diagram S3-5: Resource values on the diagram.

The resource flow is depicted by an arrow. A flow must not be regarded as a physical transport of matter as the arrow might suggest. It is a theoretical concept for an object transfer. The flow does not undergo any change of place. It represents an interface and contains only information about an object. Every resource flow is defined by resource values. For this, every flow has a flow specification. The resource values may be displayed directly on the diagram near the arrow.

In the following table, the first rule for the resource flows is given. The rules are numbered and marked with the prefix R for the Resource Flow Diagram.

Rules S3-1: Resource flow

R1	Every resource flow has one or more resource values. These values are higher than or equal to zero.

Note! In general, immaterial flows are not depicted in the RFD because the rules of the mass and energy conservation do not apply. Therefore, no rules for immaterial flows are given in this section. If necessary, immaterial flows may be introduced into the model, for example human workforce for the evaluation of partially automated production lines, or entropy to balance the exergy analysis.

3.3.4 External Entity

The system exchanges resources, products, and information with the outside world. The outside world is represented by the symbol external entity. Such external entities can be for example suppliers or customer of the investigated system. The external entity is displayed only on the context diagram. The external entity is not used for the calculations in the investigated system. Therefore, it has no parameters or calculations in its specification, only its name and description. The flows going to and coming from the external entities carry all necessary information for the calculations in the system.

The context diagram is the top level of the analysis (level 0), setting the system boundary between the analyzed system and the environment. It contains all external entities and one single process. The rules for the context diagram can be found in Section S1.

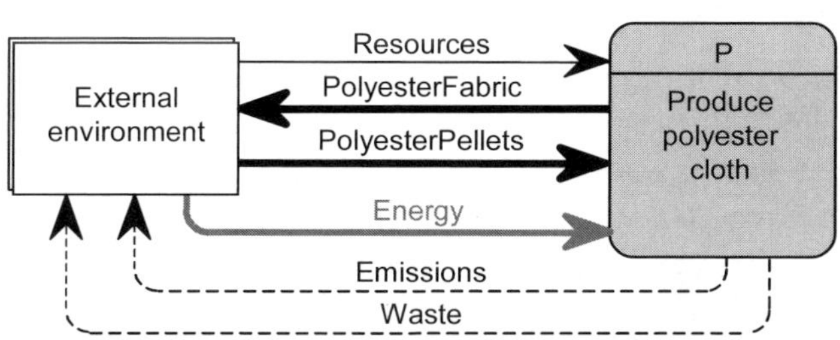

Diagram S3-6: Sample context diagram of a polyester cloth production.

Diagram S3-6 shows the context diagram for the production of a polyester fabric. The investigated process P "Produce polyester cloth" exactly defines the system to be analyzed. Thereby, the process appears as a black box. It is further detailed in the project on different levels of hierarchy. The surrounding of the process is shown on the context diagram. The process "Produce polyester cloth" exchanges resources with the external entity "Environment." From the outside world, the analyzed system gets "Resources," "Polyester pellets," and "Energy." It hands back produced "Polyester fabric," "Emissions," and "Waste" to the "External environment."

3.4 Flow Types and Flow Categories

3.4.1 Flow Classification

Resource flows are organized into flow categories and flow types. This allows an easier overview of the diagram and is necessary for calculations. Resources are classified into five flow categories: Product flow, material flow, energy flow, information flow, and fictitious flow. *Table S3-3* shows the classification of resource flows.

Table S3-3: Classification of resource flows

Flow categories	Flow types (recommendation)
product flow	product
material flow	water
	chemicals and various material
energy flow	electricity
	thermal energy
	energy resources
information flow	workforce
	information
fictitious flow	entropy

3.4.2 Flow Category

Flow categories are mandatory in the POA method and are necessary for applying various calculation rules.

Material flow: The material flow represents all production material, packaging, auxiliary material, water, and other material used for a production. A material flow carries a mass, a total energy, an exergy, and an embodied energy. It is subject to mass balance (mass of the material) and energy or exergy balance (total energy and exergy of the material).

Product flow: The product flow is the main object of calculation in a model. It is also the master path for the analysis of a production system. It runs through several processes and is changed by them. The product flow defines the SU. The product flow of technical systems is mostly a material flow. Therefore, the calculation rules for the material flow apply also for the product flow. There are additional rules for the product flow concerning the calculation of embodied energy. The embodied energy of the system is passed on to the product by the processes.

Energy flow: The energy flow consists of several energy forms, immaterial (e.g. electricity) as well as material (e.g. coal). The immaterial energy flows such as heat and electricity carry no mass. Material energy flows such as fuel to drive machines, heat materials, and condition air, can be treated alternatively as material or energy flows. In the RFD, energy resources are treated either as energy or as material flows. Energy flows are expressed as total energy, exergy, and embodied energy. They are subject to energy balance and exergy balance. The flow specifications of the immaterial energy flows have no record for the mass.

Information flow: The information flow in the RFD is only used occasionally. It is an immaterial flow and does not have a resource value that serves for balancing in the system.

Fictitious flow: The fictitious flow is a theoretical construct of a flow whose object does not exist in reality. (It is also applied in the VFD in Section S2.) The fictitious flow may contain a value for calculation depending on which object it represents. In the RFD, the fictitious flow is used for representing entropy in the exergy calculation. This is a theoretical value that comes off the calculation. The fictitious value flow carries a value and serves for compensation of the process balance.

Rules S 3-2: Flow categories of resource flow

R2	Each flow belongs to one flow category.
R3	Product and material flows are subject to mass balance.
R4	Product and material flows are subject to energy and exergy balance if they carry total energy and exergy respectively.
R5	All energy flows are subject to energy and exergy balance.
R6	Information flows do not carry resource values.
R7	Fictitious flows carry theoretically calculated values, which serve for setup of process balances.

3.4.3 Flow Type

Flow categories are divided into flow types and flow subtypes. The classification into flow types is necessary for the value calculation in the hierarchy of the diagrams. The flow types group flows with identical physical characteristics (from the view point of the investigation). Only flows of the same flow type can be split and merged throughout the hierarchy. The flow types listed in *Table S3-3* are recommendations by the POA method and can be supplemented by the model designer according to the needs of the investigated system. If necessary, flow types can be further divided into flow subtypes.

Definition: Flow types comprise flows of identical physical matter; they allow split and merge in the hierarchy.

The rules for handling of flow types are listed in *Rules R8 and R9.*

Rules S3-3: Flow type

R8	Each flow corresponds to one flow type.
R9	Every child flow belongs to the same flow type as the parent flow.

The RFD is a visual tool and is easily understandable because of the visualization by symbols and symbol styles. The model designer depicts the lines and processes according to a maximum of readability and overview. The product flow is highlighted to give orientation within the production model. Different flow types are distinguished graphically on the diagram by different line styles. Flows of the same flow type have the same line style. *Table S3-4* shows the basic classification into flow types and subtypes, as well as the line style used for each type. The classification in *Table S3-4* is applicable for all diagrams in Section S3.

Table S3-4: Complete classification of resource flows

Flow category	**Flow type (recommendation)**	**Flow subtype (example)**	**Line Style (example)**
product flow	product	product	black solid bold
		waste product	black solid middle
material flow	water (and other natural resources)	input water	black solid fine
		waste water	black dashed
	chemicals and various materials	input chemicals and material	black solid fine
		waste chemical and material	black dashed
energy flow	electricity	electricity (exergy) mechanical energy	gray solid middle
	thermal energy	thermal energy	gray solid middle
		waste energy	gray dashed
	energy resources	energy resources (e.g. oil)	gray solid middle
		exhaust gas and fumes	gray dashed
information flow	manpower		gray solid middle
	information		black dash-dotted
fictitious flow	entropy		black dotted

3.5 Calculation in the Flow and Process Specification

3.5.1 Calculation Procedure

In the RFD model, various calculations are performed. *Table S3-5* summarizes the step-by-step procedure when calculating a RFD. Each step is documented with chapter and equation number, in order to simplify looking up details for individual calculation steps. For an entire model with several detailing levels, these steps are run through iteratively. The first step is the data collection. It aims to find flow values directly or parameters for flow calculations and balancing. If sufficient data is gained, the calculation in the flow specification starts. After calculating all possible flow values, the process balances are set up and calculate remaining not defined flow values. Finally, the process values are assessed.

Table S3-5: Procedure for RFD calculations

Calculation type	**Calculation step**	**Description of calculation**	**Chapter / Equation**
Element specification	Parameter declaration and assessment	All known values of the company or of the literature are supplemented as parameters in the flow specification.	S3.5.2
	Data measuring	Measurable unknown values are measured in the company.	
	Generic data	Unknown values may be determined by a generic system.	S3.2.4
Standard flow calculations	Standardization to the SU	All flow values are calculated per SU.	S3.6.2
	Allocation	The mass or other resource values of input flows are distributed by a transfer coefficient to output flows or vice versa.	S3.6.2 / E2
	Split and merge in the hierarchy	In the hierarchy, the parent flow may be calculated by a merge of child flows. Vice versa, child flows are calculated by a split of the parent flow.	S3.6.2 / E3
Specific flow calculation	Total energy of energy flow	Calculation of the energy content of the energy flow.	S3.7.1
	Total energy material and product flow	Calculation of the energy content of product and material flows by physical equations.	S3.7.1 / E4-E9
	Exergy of energy flow	Calculation of the exergy content of thermal energy flows based on their temperature.	S3.8.1 / E12
	Exergy of product and material flows	Calculation of the exergy content of product and material flows by physical equations.	S3.8.1 / E13

Calculation type	Calculation step	Description of calculation	Chapter / Equation
	Embodied energy calculation	Calculation of the energy consumption for the processing of the resource flow.	S3.9.1 / E16
Process balance	Mass balance	Consistency check of the mass values of input and output resource flows or calculation of a mass value of a resource flow.	S3.6.1 / E1
	Energy balance	Consistency check of the total energy values of input and output resource flows or calculation of a total energy value of a resource flow.	S3.7.2 / E10
	Exergy balance	Calculation of the produced entropy of the process.	S3.8.2 / E14
Process value	Energy efficiency	Efficiency of the process concerning energy consumption and energy generation.	S3.7.3 / E11
	Exergy efficiency	Efficiency of the process concerning exergy consumption and exergy generation.	S3.8.4 / E15
	Embodied energy added	Cumulated embodied energy, which is charged to the product by the process.	S3.9.2 / E17,18

3.5.2 Parameter Declaration and Assessment

Parameters are the quantitative properties of a production system and are necessary to perform the calculations of resource flow and process values. The parameters are declared in the flow and process specifications. For the value calculation, they are related to each other by equations. Parameters are defined by name, quantity, unit, and the data source. *Table S3-6* shows an extract of a parameter list for the model of a bakery.

Table S3-6: Parameter list with examples

Parameter name	Quantity	Unit	Data source
Baking temperature	220	°C	measured
Baking time	50	minutes	recipe
Amount of bread dough	2	kilograms	measured
Number of bread slices	25	pieces	consumer
Electricity conveyor belt	5	kWh	generic
…			

The parameters necessary for calculation are assessed in the company. Usually, the data in company's data base is not directly suitable for use in the RFD and has to be converted or pre-processed.

Energy parameters are often indicated on the machine or available in the machine specifications, for example, nominal machine power consumption. The only reliable and accurate method is to measure parameters directly at machines and installations in operation. This is especially applicable for energy parameters but also for material data. Parameter values may also be calculated based on other known values, for example, by the generic value calculation. (See Chapter S3.2.4.)

If not enough values are available in a model to begin calculation, further information must be made available and additional data gained for the value calculation. In a first step, it is often necessary to assume values or to take them from literature.

The specification of a parameter's data source serves as a reference regarding its reliability. The data source is helpful in the sense to know where the data comes from, to assess its accuracy and applicability in the project. Basically, the most reliable parameters are the ones that are measured or that are retrieved generically from basic physical specifications.

3.5.3 Flow Specification in General

Every element in the RFD has a specification. The flow specification gives momentary picture of the energetic and environmental state of the flow. It is a description with physical, chemical, energetic, and environmental parameters. Generally, it consists of text, parameters, and equations (*Specification S3-1*).

Specification S3-1: General flow specification

⟨Resource flow name⟩			
Description	⟨text⟩		
Flow type	⟨flow type or subtype⟩		
Parameter	⟨parameter name [and data source]⟩	⟨quantity⟩	⟨unit⟩
Mass	calculation	⟨equation to calculate mass⟩	
	resource value	⟨value mass⟩	
Total energy	calculation	⟨equation to calculate total energy⟩	
	resource value	⟨value total energy⟩	
Exergy	calculation	⟨equation to calculate exergy⟩	
	resource value	⟨value exergy⟩	
Embodied energy	calculation	⟨equation to calculate embodied energy⟩	
	resource value	⟨embodied energy⟩	

The flow specification has several sections. The first row describes the flow, i.e. what it represents, which process it comes from, and other comments. The second row specifies the flow type and subtype. The third row contains the parameters that are flow specific and necessary to calculate in the specification. The following sections contain calculations and resource values for mass, total energy, exergy, and embodied energy. Depending on the scope of the analysis, the flow specification is more or less extensive:

- Standard calculation: The resource value mass is calculated for product and material flows and the total energy for energy flows.
- Energy analysis: The resource value total energy is calculated for the product and material flows.
- Evaluation of reuse of waste energy: The resource value exergy of the thermal energy flows is calculated.
- Exergy analysis: The resource value exergy is calculated for the energy, product, and material flows.
- Evaluation of the environmental burden of a product: The resource value embodied energy of energy and material flows is calculated and charged to the product flow.

The complete list of available calculated resource values within the RFD, including their definitions, is given in *Table S3-7*.

Table S3-7: Definition resource values

Resource value	Description	Units
Mass	Mass of material flow	kg
Total energy	Energy amount/content of energy/ material flow	J, kWh, W (if calculated as power)
Exergy	Exergy content of energy and material flows	J, kWh, W (if calculated as power)
Embodied energy	Cumulated amount of energy necessary to produce, maintain, and dispose of the resource up to the point of the analysis	J, kWh, W (if calculated as power)

Note! For a simple system analysis using the RFD, it is usual to calculate the mass for the material flows and the total energy for the energy flows.

3.5.4 Process Specification in General

In contrast to the flow specification, the process specification handles values of incoming resource flows and distributes them to output resource flows. The calculation and checking of flow values is performed by balancing mass, exergy, and energy according to the physical laws of mass and energy conservation. The embodied energy added is calculated by summing up the consumed energy in the system. The process also calculates process-specific parameters to show the performance of the process: Embodied energy added, energetic efficiency, and exergetic efficiency.

A general process specification is given in *Specification S3-2*. The header of the specification contains process number, name, and additional information. The first row of the process specification hosts a description of the activities that are carried out in the process. The second row offers space for the definition of parameters that are used for the value calculations. The third to fifth row contains process equations of mass, energy, and exergy balance. The sixth to eighth section contains calculations for process-specific values such as energetic efficiency, exergetic efficiency, and embodied energy added, and displays the results. The second row of these sections displays the results of process values.

Specification S3-2: General process specification

⟨Process number⟩	**⟨Process name⟩**		**⟨Additional information⟩**
Description	⟨text⟩		
Parameter	⟨parameter name [and data source]⟩	⟨quantity⟩	⟨unit⟩
Mass balance	⟨equation⟩		
Energy balance	⟨equation⟩		
Exergy balance	⟨equation⟩		
Energetic efficiency	calculation	⟨equation to calculate the energetic efficiency⟩	
	process value	⟨value of energetic efficiency⟩	
Exergetic efficiency	calculation	⟨equation to calculate the exergetic efficiency⟩	
	process value	⟨value of exergetic efficiency⟩	
Embodied energy added	calculation	⟨equation to calculate the embodied energy added⟩	
	process value	⟨value of embodied energy added⟩	

The mass, energy, and exergy balance are calculated in the process specifications by equations. The purpose of the process equation is to identify resource values of input or output flows. The results of these calculations are entered as resource values in the specifications of the corresponding flow.

Rules S3-4: Process specification

R10	The process specification indicates the equations used to calculate process values.
R11	In the process specification, resource values are calculated by resolving process balances.

The following chapters show the calculation of flow and process values for mass, total energy, exergy, and embodied energy. Each resource value and connected process value is given in context with calculation rules and examples.

3.6 Mass Analysis in the RFD

3.6.1 Mass Balance

The calculation of the mass of product and material flows is the first step in the analysis of the RFD. There are several reasons for this:

- The mass of the product flow may serve as the SU (and the RU) and is needed for the standardization of the resource values of the resource flows.
- The allocation of resource values may be done by using the mass of the product flow.

The mass of material and product flows is determined by doing the mass balance within the processes and by setting up the general flow calculations. The mass balance quantifies the mass of input and output flows of a process and balances them. The conservation of mass stated in the first law of thermodynamics is the basis for mass balance.

Definition: The mass balance expresses that the sum of the mass of input flows is equal to the sum of the mass of output flows.

The equation for the mass balance is stated in *Equation E1*.

E1	$\sum_{i=1}^{p} mi_i = \sum_{k=1}^{q} mo_k$	mi	mass of input flow
		mo	mass of output flow
		p	number of input resource flows
		q	number of output resource flows

The mass balance is done for every process of the model on its own, as well as for the system as a whole, also for the single process on the context diagram. By this, unknown mass values of resource flows are determined.

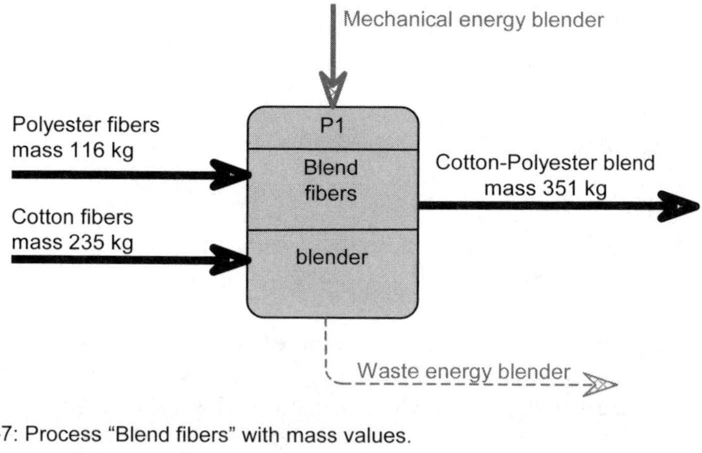

Diagram S3-7: Process "Blend fibers" with mass values.

Diagram S3-7 shows the process “Blend fibers” that is a subprocess within the textile production line of knitting yarn for the fabric of a T-shirt. The two incoming product flows “Cotton fibers” and “Polyester fibers” are blended mechanically in this process and result in the product output flow “Cotton-Polyester blend.”

The mass balance for this process is shown in the process *Specification S3-3* for process P1 “Blend fibers.” Known values for this equation are the masses of “Cotton fibers” and “Polyester fibers.” The mass balance can only be solved if it contains exactly one unknown value, in this case the mass of the mixed fibers “Cotton-Polyester blend.”

Specification S3-3: Process “Blend fiber”

P1	**Blend fibers**
Description	Blending of two fiber types for the production of 1,000 kg/h knitting yarn. Abbreviations used: CO cotton PET polyester
Parameter	Mass cotton fibers 235 kg Mass polyester fibers 116 kg
Mass balance	$\sum$ mass input $= \sum$ mass output $mass_{COfibers} + mass_{PETfibers} = mass_{CO\text{-}PETblend}$ 235 kg + 116 kg = 351 kg

The mass balance as such is set up in the process balance field of the process specification. It serves to balance input and output masses and to calculate missing flow specifications. The process hands over the calculated mass of the resource flow to the corresponding flow specification. The process specification does not contain any results derived from the mass balance.

In the example of the process “Blend fibers,” the mass of the input product flows are summed up to receive the mass of the output product flow according to the mass balance. The calculated result of the mass balance of process P2.1 is given over and stored in *Specification S3-4* of the flow “Cotton-Polyester blend.” It corresponds to 351 kg.

Specification S3-4: Flow “Cotton-Polyester blend”

Cotton-Polyester blend	
Description	Blend of cotton and polyester fibers for knitting yarn by a mechanically driven blender.
Flow type	product
Parameter	
Mass	resource value **351** kg/SU

3.6.2 General Flow Calculations

General flow calculations are independent of physical balances and may be applied to any resource value. The calculation in the hierarchy, for example, is applicable for mass, total energy, exergy, and embodied energy. The resource values are summed up within a specific flow type. General flow calculations feature standardization of resource values with the SU, as well as allocation, and calculation within the hierarchy. These calculation procedures are demonstrated on the example of the resource value mass.

Standardization to the Standard Unit

The standardization of resource values in terms of the SU and RU is done in the flow specification. The resource value of the flow is thereby converted according to the SU to ensure a common basis for the analysis calculations. Finally, all resource values are indicated per SU in the flow specification.

Diagram S3-8 shows the process "Form croissants." This process is a part of an industrial bakery line for croissants that is described in detail in Case Study C3. The external entities also are depicted in this RFD for better explaining of the process that is taken out from its context.

In process "Form croissants," a croissant is formed by cutting out a triangle from the rolled out dough, rolling the triangle, and then bending the rolled piece of dough into a half-moon shape. The croissant, formed and bent, but not yet baked, is the "Raw croissants."

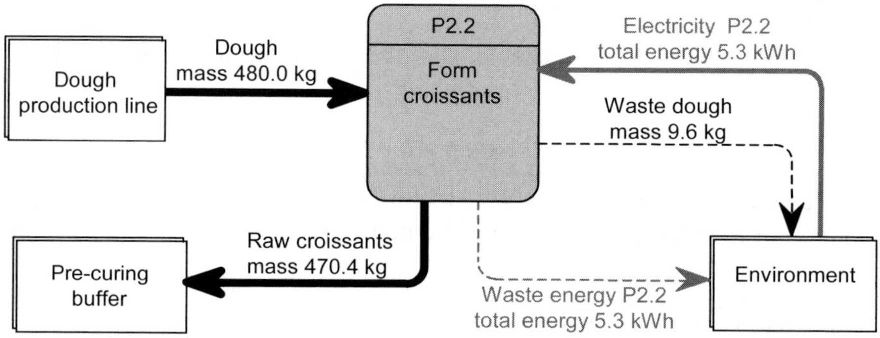

Diagram S3-8: Process "Form croissant."

The SU of the analysis is based on the number of croissants produced per hour, in this case 12,000 croissants. Each resource value of this study has to be standardized by this value.

In order to calculate the mass of the dough needed for the production of 12,000 croissants, the weight of the "Raw croissants" is listed in *Specification S3-5* for the flow "Dough."

Specification S3-5: Product flow "Dough"

Dough				
Description	The ingredients of dough are approx. (percent of weight): 60 % flour, 25 % water, 12 % butter, 3 % other ingredients.			
Flow type	product			
Parameter	Production croissant		12,000	pieces
	Weight croissant		40	g/pieces
Mass	calculation	$\frac{12000\,\text{pieces} \cdot 40\,\text{g/piece}}{1000\,\text{g/kg}}$		
	resource value		**480**	kg /SU

Specification S3-6 of flow "Electricity P2.2" shows the calculation of the energy consumption of process "Form croissants." The calculation is done with the standardized value 480 kg/SU. By this, the result of the "Electricity P2.2" for process 2.2 is also standardized.

Specification S3-6: Flow "Electricity P2.2"

Electricity P2.2				
Description	Electricity used for the process "Form croissant."			
Flow type	electricity			
Parameter	Energy per kg		11	Wh/kg
Total energy	calculation	$11\,\text{Wh/kg} \cdot 480\,\text{kg}$		
	resource value		**5.3**	kWh/SU

Allocation

Allocation describes the assignment of the resource values of the input flows to the output flows. This determines what part of the raw materials are leaving a process as waste, and how much is further processed in the production as intermediate product.

The allocation refers to the rules that apply to the distribution of resource values of one flow into two or more flows by a production process. The allocation in the RFD is done by transfer coefficients (TC), determining the part of an input resource flow to be transferred to output resource flows, or vice versa. The general equation for the allocation of one or several input values to one or several output values is given in *Equation E2*.

E2	$v_{output} = TC \cdot v_{input}$	TC v	transfer coefficient resource value

The TC are parameters given by the process, such as machine settings, or properties of the resource flow itself, i.e. volatility, solubility. Some examples of TC are given in *Table S3-8*.

Table S3-8: Examples of transfer coefficients

Transfer coefficient	Description	Example
Machine TC	Given by the settings of the producing machine	Rate of waste
Energy TC	Waste energy generated, given by the efficiency of the machine	Motor efficiency
Chemical TC	Depends on chemical and physical properties of the resource flows and application recipe	Fixing-grade Impregnating-grade

The calculation of the mass involves often an allocation. The mass of the input flow is distributed to the output flows. Or the mass of the output flow is allocated to several input flows.

Specification S3-7: Flow "Waste dough"

Waste dough			
Description	Dough cut-off and other dough waste of process "Form croissant." It can be reused for the production of dough.		
Flow type	waste product		
Parameter	TC		2 %
Mass	calculation E2	$480\text{kg/SU} \cdot 0.02$	
	resource value		**9.6** kg/SU

In the process "Form croissants," "Dough" is distributed onto the two output flows "Raw croissants" and "Waste dough." The part of "Dough" that is lost as "Waste dough" is empirically known as 2%. This means that 2% of the dough mass is transferred to the flow "Waste dough" (*Specification S3-7*), and the other 98% to the product output flow "Raw croissants" (*Specification S3-8*).

Specification S3-8: Flow "Raw croissant"

Raw croissants			
Description	Formed, unbaked croissant.		
Flow type	product		
Parameter	TC		98 %
Mass	calculation E2	$480\text{kg/SU} \cdot 0.98$	
	resource value		**470.4** kg/SU

Split and merge and calculation in the hierarchy

Within the model hierarchy of the system, flows are split up into child flows, or child flows merged into parent flows. Split and merge in the hierarchy is one of the main calculation methods to determine resource values of flows. The level balancing rule in the diagram states that the child diagram has the same resource input flows and resource output flows as its parent process. The split and merge takes place on the child diagram.

The calculation of split and merge bases on the fact that the total value of child flows must be equal to the value of the parent flow. The general term for calculating the resource valuc of parent flow and child flows is stated in *Equation E3*. This equation may be reversed to calculate the resource value of one of the child flows from the parent flow and the other child flows.

E3	$vp = \sum_{i=1}^{n} vc_i$	vp vc n	resource value of parent flow resource value of child flow number of child flows

This equation is applicable for a calculation in terms of any of the following resource values: mass, total energy, exergy, and embodied energy.

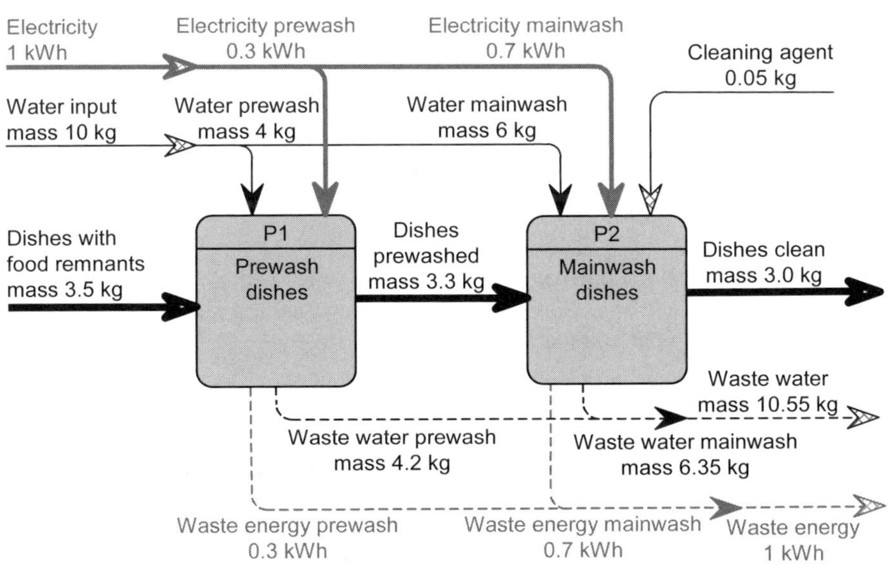

Diagram S3-9: RFD of dishwasher.

Diagram S3-9 shows the RFD of a dishwasher. The diagram shows the split and merge flows of the input resource flows "Water input" and "Electricity" and of the output resource flows "Waste water" and "Waste energy." The parent flow coming from the parent diagram is distributed to the single processes.

Specification S3-9 calculates the mass of the input flow "Water prewash" from *Diagram S3-9,* which happen in a split. We assume the following resource values are known: The mass of the merged flow "Water input" is 10 kg and the mass of the child flow "Water mainwash" is 6 kg. In this case, the mass of the known child flow is subtracted from the mass of the known parent flow, in order to get the unknown mass of the child flow "Water prewash."

Specification S3-9: Flow "Water prewash"

Water prewash			
Description	Water needed for the prewash cycle of the dishwasher		
Flow type	water		
Parameter	Mass water input Mass water mainwash		10 kg 6 kg
Mass	calculation (split and merge)	$mass_{WaterInput} - mass_{WaterMainwash}$ 10 kg - 6 kg	
	resource value		**4** kg/SU

The RFD of the dishwasher shows that the mass of the dishes is reduced by washing, because dirt and food remnants are transferred to the waste water. Furthermore, the input electricity is mainly consumed by the mainwash process, because the water has to be heated up more and the mainwash process lasts longer.

The equation for a split and merge is stated in the flow specification, preferably in the flow specification of the unknown flow. The equation is solved for the unknown resource value, and the result is stored in the flow specification of the concerning resource flow. This may be the flow specification of the parent flow or of one of the child flows.

This hierarchical calculation can be carried out top-down as well as bottom-up. According to *Rule R9,* only flows of the same flow type are allowed to be handled and calculated by a split and merge.

Rules S 3-5: Calculation in flow specification

R12	The flow specification executes allocations and calculations of resource values.
R13	The flow specification uses calculations of resource values in the model hierarchy, i.e. split and merge.
R14	In the flow specification, the resource flow values are standardized by the Standard Unit.

3.7 Energy Analysis in the RFD

3.7.1 Total Energy of Resource Flows

The resource value total energy stands for the energy amount contained in material and energy flows. The assessment of the total energy of energy flows is a standard task for a RFD. The total energy of product or resource flows is generally calculated in studies concentrating on an energy analysis.

In material flows, the correct physical term for this energy is streaming energy.

Total energy of energy flows

The input energy that flows in a technical system is mainly electric energy and thermal energy. Data about electricity consumption is usually available within a company, although this data does not necessarily refer to a production unit, as maybe the SU in the RFD. If the electricity consumption is unknown for a certain machine, the rated power indicated on the machine or found in the machine specification may be used as reference. The power rating has to be multiplied with the standard percentage of machine usage to obtain its electricity consumption as a coarse approximation. If these values are not available, the energy consumption can be read off an electricity counter and then broken down by calculation for the single machine. Another possibility is to measure electric power on a given machine with the assistance of the plant electrician.

Thermal energy is supplied in a company by the purchase of oil or gas. For example, to calculate thermal energy that is necessary to run a gas oven, the combustion value of gas (carrier of the energy) and the efficiency of the burning process are taken into account. To calculate the total energy of a thermal energy stream (e.g. hot water or hot air), the calculation of exergy gives more precise results, as it includes the temperature level in relation to the environment (Chapter S3.7.2).

Output energy flows of technical systems are mainly mechanical energy, which is introduced into a follow-up mechanical process, and waste energy. Mechanical energy is calculated as the physical sum of kinetic and potential energy (*Equations E7* and *E8*). It may also be calculated based on the power rating of a mechanical device and the efficiency of the drive or motor. Waste energy flows are calculated by the energy balance of the input energy flows minus the output energy flows (Chapter S3.7.2).

For any evaluation of energetic performance of systems, measurements at actual production conditions are the most accurate source for data.

Total energy of material and product flows

The total energy of a material or product flow is the energy that is carried by the flow when entering or leaving a process. When material is transferred at ambient temperature, as it is mostly the case, the total energy of a material flow is negligible. However, for product and resource flows moving at high speed carrying a significant amount of heat, the total energy has to be determined for the correct energy balance. For a complete energy evaluation, the total energy of all product flows and of the other material flows must be taken into account.

The first consideration for calculating the total energy of a product or material flow, is to look at the kind of energy that could be contained in the flow and the applicable calculation procedure. Many material flows have an energy content that is expressed by combustion (e.g. energy carriers or combustible waste) or caloric value (e.g. food). These values can be found in literature of basic physics or in the product declaration.

If this data is not available or an additional energy content is present, e.g. by high temperature or any other form of energy stored in the material, physical equations have to be applied for the calculation.

The total energy itself is a combined value consisting of inner energy, kinetic energy, and potential energy. The equations for these energy terms are given in the following list (*Equations E4* to *E9*). This list is helpful for calculating the total energy of material flows. Depending on the available values and the relevant energy form, different equations of this list are used.

E4	Total energy:	$E = e \cdot m$	E total energy of material flow e specific total energy material flow m mass u inner energy ke kinetic energy (also E_{kin}) pe potential energy (also E_{pot}) h specific enthalpy p pressure V volume v velocity g gravity constant z height te heat content c specific heat capacity material flow ΔT temperature difference [K]
E5	Specific total energy:	$e = u + ke + pe$	
E6	Inner energy:	$u = h - p \cdot V$	
E7	Kinetic energy:	$ke = \frac{m}{2} \cdot v^2$	
E8	Potential energy:	$pe = m \cdot g \cdot z$	
E9	Heat content:	$te = c \cdot \Delta T$	

- The total energy is calculated from the specific total energy multiplied by the mass of the flow (*Equation E4*).
- The specific total energy is the sum of the inner energy, the kinetic energy, and the potential energy (*Equation E5*).
- The inner energy is equal to the specific enthalpy minus the energy of volume (*Equation E6*).
- The kinetic energy is the energy of velocity of the mass (*Equation E7*).

- The potential energy is the energy of gravitation (*Equation E8*).
- The heat content is the heat energy of the flow and is calculated with the temperature difference of the object of the flow to the environment and the specific heat capacity (*Equation E9*).

Specification S3-10: Flow "Polyester fibers"

Polyester fibers			
Description	Polyester fibers are introduced to the blender. Because they enter the blender accelerated by air pressure through a tube, they carry kinetic energy. For the calculation, thc vclocity in the center of the transport tube is used.		
Flow type	Product		
Parameter	Velocity channel Mass polyester fibers		10 m/s 116 kg
Total energy	calculation (E7)	$\frac{116\,\text{kg}}{2} \cdot 100\,\text{m/s}^2$	
	resource value		**5.8** kJ/SU

Specification S3-10 determines the total energy of the flow "Polyester fibers," supplied with a certain velocity to transport them to the subsequent process "Blend fibers" from *Diagram S3-10*. In this case, the total energy consists of the kinetic energy of the flow "Polyester fibers." With this equation, the total energy of the other two product flows in this example is defined analogously. For the "Cotton-Polyester blend" however, the velocity for feeding the fibers into the next process is smaller (0.1 m/s).

3.7.2 Energy Balance

The energy balance calculates and balances the total energy of the product, material, and energy flows. It is based on the conservation of energy stated in the first law of thermodynamics. The energy balance sets the total input energy equal to the total output energy. In mechanical systems, the total energy of material and product flows are often neglected, since the transfer of energy is mostly done by stationary equipment, as gears, shafts, or belts.

The simple case of the energy balance includes only the total energy of energy flows. An extensive energy valuation also takes into account the total energy of material and product flows. The energy balance is stated in the process specification in the section of the process balance.

Definition: The energy balance sets the sum of the total energy of input resource flows equal to the sum of the total energy of output resource flows.

The physical equation for the energy balance in the RFD in the stationary state is displayed in *Equation E 10.*

E10	$$0=\sum_{i=l}^{m} E_{i,input}-\sum_{k=l}^{n} E_{k,output}+\sum_{i=l}^{p} Q_{i,input}-\sum_{k=l}^{q} Q_{k,output}+\sum_{i=l}^{r} E_{i,El}-W$$	
	E Total energy material flows Q Total energy thermal energy flows E_{El} Electric energy W Work	m Number of material input flows n Number of material output flows p Number of thermal input flows q Number of thermal output flows r Number of electric input flows

The total energy of material and product input flows (E_{input}), the total energy of thermal energy input flows (Q_{input}), and the total energy of electric energy (E_{El}) equals the total energy of material and product output flows (E_{output}), the total energy of thermal energy output flows (Q_{output}), and the work (W) generated by the system going to the environment.

The example in *Diagram S3-10* shows again process P1 "Blend fibers" (from *Diagram S3-7*) with the additional process P2 for the drive unit of the blender. The blending process is driven by an electric motor that creates waste energy by getting hot and, by this, heating the ambient air. Mechanical energy is handed over to the blender by a V-belt. The total energy values and the mass values of the resource flows are displayed.

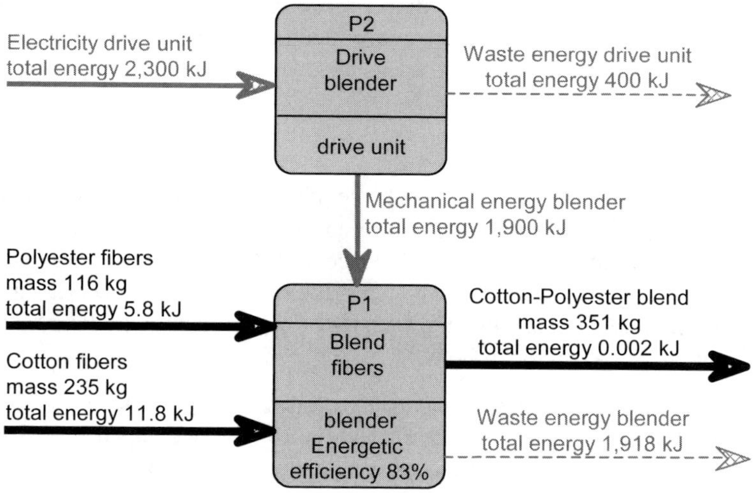

Diagram S3-10: Diagram fiber blender with drive, mass and total energy values.

In *Specification S3-11,* the total energy of the flow "Electricity drive unit" is calculated based on the company's data by standardization of the total amount (electricity consumption per ton of product) to the SU of 351 kg for the analysis.

Specification S3-11: Flow "Electricity drive unit"

Electricity drive unit			
Description	Electricity used for the drive unit of the blender.		
Flow type	electricity		
Parameter	Electricity consumption per 1 ton of product		6,550 kJ
Total energy	calculation	$\frac{6{,}550\,\text{kJ} \cdot 351\,\text{kg}}{1000\,\text{kg}}$	
	resource value		**2,300** kJ/SU

This resource value and the total energy values of the other input and output flows are incorporated into the energy balance for the process "Blend fibers" shown in *Specification S3-12*. The parameters declared in this specification correspond to the resource values of the flows in the diagram. These values of resource flows are already calculated and stored in the pertaining flow specifications. They are not displayed here again.

Specification S3-12: Process "Blend fibers"

P1	**Blend fibers**				
Description	Blending of two fiber types for the production of 1,000 kg/h knitting yarn. Abbreviations used: CO for cotton and PET for polyester.				
Parameter	Total energy "Cotton fibers"				11.8 kJ
	Total energy "Polyester fibers"				5.8 kJ
	Total energy "Mechanical energy blender"				1,900 kJ
	Velocity "Cotton-Polyester blend"				0.1 m/s
	Total energy "Cotton-Polyester blend"				0.002 kJ
Mass balance	$mass_{COfibers}$	$+ mass_{PETfibers}$	$= mass_{CO\text{-}PETblend}$		
	235 kg	+ 116 kg	= 351 kg		
Energy balance	$E_{COfiber}$	$+ E_{PETfiber}$	$+ Q_{electricity}$	$- E_{CO\text{-}PETblend}$	$= Q_{WasteEnergyBlender}$
	11.8 kJ	+ 5.8 kJ	+ 1900 kJ	- 0.002 kJ	= 1918 kJ

The total energy of material flows, including the kinetic energy, is small compared to the electricity consumption of the process. The energy balance in *Specification S3-12* for process "Blend fibers" calculates the waste energy of the blender. Most of the mechanical energy supplied by the motor is wasted by mechanical friction and air drag, in other words converted into heat to ambient air.

3.7.3 Process Value: Energetic Efficiency

In addition to process balances, the process specification consists of specific process values that define the process performance. One of these process values is derived from the energy balance: The energetic efficiency of a process.

Definition: The energetic efficiency expresses the energetic performance of a process.

The energetic efficiency is quantified based on the total energy produced in a process ($energy_{use}$) and the total energy consumed by the process ($energy_{input}$). The total energy produced may be mechanical energy of a drive, structural energy in a product, heat energy or others. The total energy consumed by the process equals the input energy of the process. The equation for the energetic efficiency is stated in *Equation E11*.

E11	$\eta = \frac{energy_{use}}{energy_{input}}$	η $energy_{use}$ $energy_{iput}$	energetic efficiency total energy produced by the process total energy consumed by the process

The calculation of the energetic efficiency is shown in *Specification S3-13* for the process P2 "Drive blender" of *Diagram S3-12*. The calculation takes the energy consumed by the process and compares it to the energy supplied to and used by the drive process. The source energy is the electricity introduced in this process. The energy of use is the mechanical energy needed to drive the cotton blender. The energetic efficiency of the drive results in 83%.

Specification S3-13: Process "Drive cotton blender"

P2	**Drive blender**		**drive**
Description	Converts electricity into mechanical energy to drive the cotton blender.		
Parameter	Total energy "Electricity drive unit" Total energy "Waste energy drive unit"		2.3 MJ 0.4 MJ
Energy balance	2.3 MJ = 0.4 MJ + 1.9 MJ		
Energetic efficiency	calculation	$\frac{1.9\,MJ}{2.3\,MJ}$	
	process value		**83** %

This calculation shows first, how important it is to correctly define flows as interfaces between processes. The power indicated on the name plat of an electric motor, as used here, refers to torque and speed available on the drive shaft of the motor. The energy consumed by the drive is measured by an electricity counter in the supply line. The energetic efficiency in this example is rather poor, pointing to the fact that the motor is oversized and operates at partial load, far from its design rating with optimum efficiency.

3.8 Exergy Analysis

3.8.1 Exergy of Resource Flows

The resource value exergy is calculated to quantify the energy based on its suitability and availability for further use. The exergy determines the part of the total energy of a material or energy flow that is available for further purposes under common ambient conditions. This is necessary to give technical information and recommendations for processes and the possibility of reusing heat within the process. The exergy calculation of material flows is now carried out for an in depth energy analysis. The calculation of the exergy of resource flows is based on the total energy value of the resource flows.

Exergy of energy flows

The calculation of the exergy of energy flows, especially heat flows, is straightforward and very useful in order to identify weak points in the energy management of a system. It especially enables a quick assessment of the value of heat energy, as often encountered in waste energy flows. The exergy of electric energy flows corresponds to the values of their total energy because electricity is a completely usable energy form. Electricity consists of 100% exergy. Heat flows are valuated based on their temperature. The total energy of the heat flow or thermal energy flow is assessed based on the temperature difference to the ambient conditions of the environment. *Equation E12* applies for thermal energy and heat flows.

E12	$E_{x,th.energy} = Q \cdot \left(1 - \frac{T_0}{T_1}\right)$	$E_{x,th.energy}$ Q T_0 T_1	exergy thermal energy flows total energy thermal energy flows ambient temperature [K] temperature of energy flow [K]

The flow "Waste energy P2.2" from *Diagram S3-8* is shown in *Specification S3-14*.

Specification S3-14: Flow "Waste energy P2.2"

Waste energy P2.2				
Description	Heat stream of process "Form croissant" at 60 °C.			
Flow type	waste energy			
Parameter	Temperature T_1 Temperature T_o		333.15 298.15	K K
Total energy	resource value		**5.3**	kWh/SU
kWh/SU	calculation	$5.3\,\text{kWh} \cdot \left[1 - \frac{298.15\,\text{K}}{333.15\,\text{K}}\right]$		
	resource value		**0.6**	kWh/SU

The total energy of flow "Waste energy P2.2" equals the total energy of flow "Electricity P2.2," assuming that the incoming energy is converted into mechanical energy and ultimately into heat. The exergy of "Waste energy," which has a temperature of 60 °C at an ambient temperature of 20 °C, is much lower than the total energy. This is due to the low temperature of the waste heat. This corresponds to practice: the waste energy generated by an electric motor does not always justify a heat recovering system.

Exergy of product and material flows

The value of exergy of product and material flows is negligible in most cases. Only if the material is exposed to heating, evaporation, or condensation, the exergy value may be of importance for the energy evaluation within a system. The exergy for product and material flows consists of the enthalpy difference, the entropy difference, the kinetic energy and the potential energy of the material.

The equation to calculate the exergy for product and material flows is given in *Equation E13*.

E13	$e_x = (h - h_0) - T_0(s - s_0) + \frac{1}{2} v^2 + g \cdot z$	e_x flow exergy of material flows h specific enthalpy s specific entropy v velocity g gravity constant z elevation T temperature [K] X_0 subscript 0: at ambient conditions

Whether or not total energy or exergy of material flows are taken in account has to be decided in every analysis. Generally, the exergy of material flows is only evaluated if the flow has a significant total energy. If it does, this justifies a complete calculation of the resource values exergy and total energy for the resource flows based on the given physical equations. *Equation E13* is applied in Case Study C3 in the example of the hot croissants leaving the baking oven of the production line.

3.8.2 Exergy Balance

The exergy balance is based on the second law of thermodynamics. It assesses the technically available energy and points out possibilities for its reuse, such as for waste energy flows. The calculation of exergy of heat flows or thermal energy flows is based on the temperature difference to the environment.

Definition: The exergy balance calculates and balances the available energy of a flow at ambient conditions.

The equation of the exergy balance for a stationary state is shown in *Equation E14*.

$$0=\sum_{i=1}^{m} E_{x,i,in}-\sum_{k=1}^{n} E_{x,k,out}-W_{0,av}+\sum_{i=1}^{p} Q_{i,in}\cdot\left(1-\frac{T_0}{T_1}\right)-\sum_{k=1}^{q} Q_{k,out}\cdot\left(1-\frac{T_0}{T_i}\right)+\sum_{i=1}^{r} E_{i,El}-T_0\cdot S$$

E14	E_x	exergy material flow	X_{in}	suffix: input
	$W_{0,av}$	work	X_{out}	suffix: output
	Q	total energy thermal energy flow	m	number of input material flows
	E_{El}	electric energy	n	number of output material flows
	S	entropy	p	number of thermal energy input flows
	T_x	temp. of thermal energy flows	q	number of thermal energy output flows
	T_0	ambient temperature	r	number of electric input flows

The equation shows the exergy balance for open systems and time-dependent irreversible processes. It consists of the exergy of input and output material flows and thermal energy flows which are valuated based on their temperature difference to the environment. Furthermore, the equation expresses the work done by the open system. Electricity input is equal to work introduced from the outside; hence the negative prefix. The produced entropy is a measure of the irreversibility produced by the processes of the system, i.e. how much available energy got lost.

The exergy balance is usually not performed in the RFD as a whole, only if the entropy of a system is to be calculated. The calculation of the entropy is mostly done in larger production systems. More often, the exergy of single flows like waste energy flows is determined. Note that the generation of entropy by the system should be as small as possible because it stands for the loss of usable energy.

3.8.3 Example: Exergy Analysis of a Draw Winding Machine

The project analyzed in the following example is an engineering update of an industrial textile yarn processing machine. This process is part of the polyester production chain and consumes a considerable amount of thermal energy for heating. Using an exergy analysis, waste energy flows are assessed and recommendations for technical improvements of the machine are given. *Figure S3-3* shows the physical representation of the polyester drawing process. Polyester (PET) yarn is drawn to increase its strength and make it finer for example for knitting or weaving. The drawing process transforms the input flow "POY" (partially oriented yarn) to the output flow "FDY" (fully drawn yarn).

The Standard Unit for the analysis is 1 kg FDY, with a production time of approximately 1 hour.

The POY, depicted in *Figure S3-3* as bold line that becomes finer, is fed into the first set of godets, which are heated and driven by a motor. Then the yarn runs through the second set of godets, which are not heated and run at a higher speed. This velocity difference between the first and the second set of godets has two effects:

- First, the yarn is mechanically drawn between the pairs of godets 1 and 2.

- Second, “Mechanical energy” is transferred from motor 2 to motor 1 because the second set of godets pulls the first one.

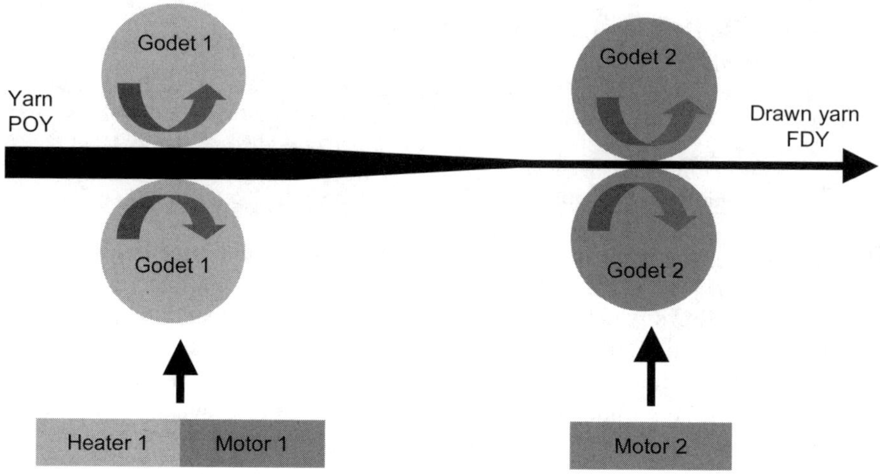

Figure S3-3: PET-drawing layout.

Motor 1 acts as a brake, in fact as a generator, feeding back electrical energy into the mains and covering a part of the consumption of motor 2. In *Diagram S3-11,* “Mechanical energy” is transferred from motor 2 to motor 1 and “Electricity back” is transferred from motor 1 to motor 2.

The drawing process causes a structural change of the yarn by transferring energy to the yarn. This way, the molecules are aligned, as in a crystal, giving the yarn more strength. The total energy of the yarn before the drawing process is negligible, apart from the fuel value of the material. The fuel value is not given here because it is the same after the process. This case deals with the calculation of the thermal and structural fixed energy in the yarn caused by the process.

The RFD on *Diagram S3-11* depicts the second level of detail of the drawing process. Process P1 represents the hot drawing process with heated godets. The yarn in process P1 is restrained and drawn in the hot state by the heated godets. Process P2 represents the mechanical drawing process. The yarn in process P2 is only mechanically drawn, but with a higher velocity than in process P1. The total energies of the resource and product flows are shown.

The mechanical power introduced into the yarn is partially transformed into heat by drawing the yarn and partially returned as electrical power from motor 1 to motor 2. The FDY stores a part of the deformation energy that was introduced into the process. The total energy of the yarn at the beginning of the process is 0 Wh, after drawing the energy equals 12 Wh, as shown on the diagram.

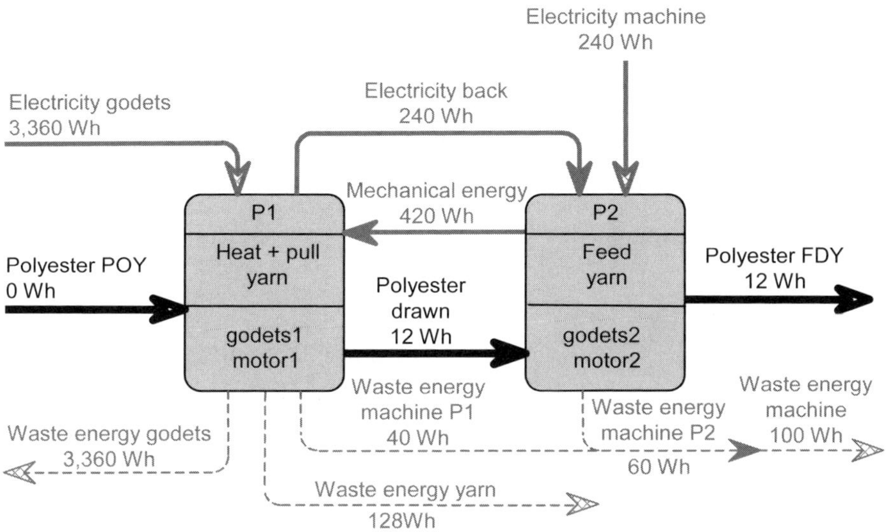

Diagram S3-11: Level 2 polyester drawing.

The energy balance for process P1 and process P2 ensures that the same amount of energy enters and exits the processes. There is a common electric energy supply providing electricity for driving the motors and heating the godets. Waste energy is produced by the motors, the heated godets, and other parts of the machine.

Specification S3-15: Process "Heat and pull yarn"

P1	**Heat and pull yarn**		**godets 1**
Description	Mechanical and thermal drawing of the partially oriented polyester yarn.		
Parameter	Temperature in °C Temperature in K Velocity	115 298.15 800	°C K m/min
Energy balance	3360.0 Wh + 0 Wh + 420.0 Wh = 240.0 Wh + 40.0 Wh + 3360.0 Wh + 12.0 Wh + 128.0 Wh		

The process specifications of P1 and P2 define the parameters of the machine and the necessary calculations. They are shown in *Specification S3-15* and *Specification S3-16*.

Specification S3-16: Process "Feed yarn"

P2	**Feed yarn**		**godets 2**
Description	Mechanical drawing of the partially oriented polyester yarn POY.		
Parameter	Velocity	1,200	m/min
Energy balance	240.0 Wh + 240.0 Wh+ 12.0 Wh = 420.0 Wh + 60.0 Wh + 12.0 Wh		

The parent process of *Diagram S3-11* is shown in *Diagram S3-12*. In addition to the total energy, the exergy values of resource and product flows are shown in the diagram. The energy is summed up from the child diagram and the exergy of the flows is calculated. The value of the total energy of the electricity flows equals the value of the exergy because electricity consists completely of exergy.

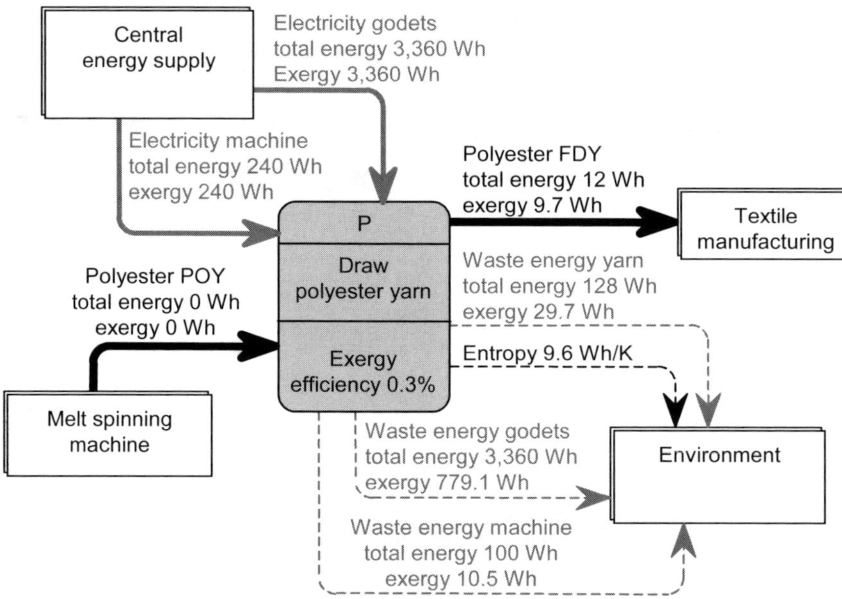

Diagram S3-12: RFD "Draw polyester yarn."

The exergy of the "Polyester POY" before the drawing process equals zero (disregarding the fuel value). After the drawing process, the "Polyester FDY" contains 9.7 Wh of exergy per kilogram. This value was measured and provided by a PET producing company. For the waste energies, the exergy is calculated based on the total energy and their temperature difference to the environment.

Specification S3-17: Flow "Waste energy godets"

Waste energy godets				
Description	Waste energy consists of the introduced energy of the heater and of the drive of the godets.			
Flow type	waste energy			
Parameter	Temperature godets (setting of machine)		388.15	K
Total energy	resource value		**3,360.0**	Wh
Exergy	calculation	$3360\,\text{kWh} \cdot \left[1 - \frac{298.15\,\text{K}}{388.15\,\text{K}}\right]$		
	resource value		**779.1**	Wh/SU

Specification S3-17 shows the calculation of waste exergy of the flow "Waste energy godets." The waste exergy of the machine and the yarn is calculated in the same way. The temperature of the waste exergy of the machine is 333.15 K. The temperature of the waste exergy of the yarn is equal to the one of the godets 2.

With the parameters of *Specification S3-18,* the exergy balance is put together. The entropy expresses the part of energy per Kelvin that is lost to the environment and cannot be recovered. The fictitious flow "Entropy" in *Diagram S3-12* shows the amount of entropy generated by the process.

Specification S3-18: Process "Draw polyester yarn"

P0	**Draw polyester yarn**	**godets 1 + 2**	
Description	Structural change from "Polyester POY" to "Polyester FDY" by mechanical and thermal drawing.		
Parameter	Electricity machine (measured)	240.0	Wh
	Electricity godets (measured)	3,360.0	Wh
	Waste exergy godets (estimated)	779.1	Wh
	Waste energy machine (estimated)	10.5	Wh
	Waste exergy yarn	29.7	Wh
	Exergy "Polyester FDY" (measured)	9.7	Wh
Exergy balance (E14)	$E_{x,PETPOY} - E_{x,PETFDY} + E_{Electricity} - E_{x,WasteEnergy} = T_0 \cdot S$ 0 Wh – 9.7 Wh + 3600.0 Wh - (10.5 Wh + 779.1 Wh + 29.7 Wh) = 288.15 K · 9.6 Wh/K		
Energy balance	240.0 Wh + 3360.0 Wh + 0 Wh = 100.0 Wh + 128.0 Wh + 3360.0 Wh + 12.0 Wh		

The exergy balance shows that the entropy loss of the system to the environment is considerable. It is comparable to the exergy introduced into the polyester yarn by drawing. Furthermore, the high exergy of the waste energy of godets 1 is considerable, but available for reuse in the process.

It becomes obvious that the major portion of the energy used in this process is emitted as waste energy and only a small part is introduced into the yarn. The heated godets are not insulated and the heated surface is very large in comparison to the yarn. This machine has two major options for improvement:

- The electricity consumption could be reduced by using smaller heated surfaces: The ratio between surface of the godet and yarn surface has to be reduced.
- The efficiency of the machine could be increased by enclosing and insulating the heated devices in order to reduce the waste energy production.

These recommendations were implemented in a new generation of machines with a so-called "hot-box." This measure reduced the energy consumption of the machine by 30%.

3.8.4 Process Value: Exergetic Efficiency

The exergetic efficiency is a process value that expresses the consumption and performance of the process in terms of exergy.

Definition: The exergetic efficiency indicates the exergetic performance of a process.

The exergetic efficiency is calculated similarly to the energetic efficiency. It is based on the exergy balance. It is a ratio between intended exergy used (exergy produced by the process) and the total exergy of the energy input (input exergy flows of the process). The exergetic efficiency is stated in *Equation E15*.

E15	$\varepsilon = \dfrac{\text{exergy}_{\text{use}}}{\text{exergy}_{\text{input}}}$	ε	exergetic efficiency
		$\text{exergy}_{\text{use}}$	exergy used for a specific purpose of a process
		$\text{exergy}_{\text{input}}$	exergy consumed by the process

The calculation of the exergetic efficiency is shown in *Specification S3-19* of the process P0 "Draw polyester yarn" in *Diagram S3-12*. The calculation is based on the exergy balance.

Specification S3-19: Process "Draw PET"

P0	**Draw polyester yarn**		**godets 1 + 2**
Description	Structural change from "Polyester POY" to "Polyester FDY" by mechanical and thermal drawing.		
Parameter	Electricity motors (measured)	240.0	Wh
	Electricity godets (measured)	3,360.0	Wh
	Exergy "Polyester FDY" (measured)	9.7	Wh
Exergetic efficiency	calculation	$\dfrac{9.7\ \text{Wh}}{3360.0\ \text{Wh} + 240.0\ \text{Wh}}$	
	process value	**0.3**	%

The purpose of the polyester drawing process is to change the structure of the yarn in order to give it more strength. Therefore, the exergy used is defined as the exergy stored in the yarn after the drawing process. To calculate the exergetic efficiency, the exergy used (9.7 Wh) is divided by the exergy consumed by the process. In this example, the exergetic efficiency is extremely low. Again this shows that a lot of energy is consumed by this process, while only a small portion of this exergy produces a structural change in the yarn. The exergetic efficiency would be significantly higher after insulating the heated godets. With this measure around 30 to 50% of the input energy for heating could be saved.

3.9 Embodied Energy Analysis

3.9.1 Embodied Energy Calculation

The resource value embodied energy is quantified to assess the cumulated energy consumption of a resource or product flow. The energy consumption, which is necessary to process a product from the beginning to the present stage, is cumulated and expressed as an energetic burden of the product flow. The embodied energy is an aggregated value. The calculation of the embodied energy is a theoretical concept. It does not correspond to the physical laws of energy conservation. In the RFD, the embodied energy provides an energetic assessment of the supply chain of a product. It serves to valuate the energy consumption of the product over its lifetime.

The embodied energy is mainly used for comparison of different raw materials and products. Considering product alternatives, the embodied energy may help to choose the less energy-intensive product. In this case, this embodied energy of the process has to be allocated in shares to the embodied energy of the product.

Every input resource and product flow carries an embodied energy. In the process, this energy is charged onto the product. The process itself burdens the product with a part of the embodied energy, which was used to manufacture the equipment and machinery necessary to perform a process.

E16	$Ee = E_{processing} + E_{transport}$	Ee embodied energy E total energy

The embodied energy parameter sums up all energy inputs contributing to the processing of the flow. *Equation E16,* for the calculation of the embodied energy of a flow, sums the total energy of the energy flows necessary to process and transport the material. The embodied energy of a flow can be calculated, estimated, or determined by literature research.

Generally, waste product flows do not carry embodied energy because it was passed on to the product while in the process. An exception is made if the flow is valuable for use in other processes within the same company, for example a by-product. In this case, it carries also an embodied energy value.

Diagram S3-13 shows the parent process of the RFD of the dishwasher in *Diagram S3-9.* Every resource entering the process already carries embodied energy which is passed on to the product in the process. In addition, the total energy of the input energy flows of the process is transferred to the product. The output product flow has a larger embodied energy than the input product flow. Consequently, the output resource flows do not carry embodied energy. In this case, the embodied energy of the dishwasher itself is neglected, since it is used over a long period in time.

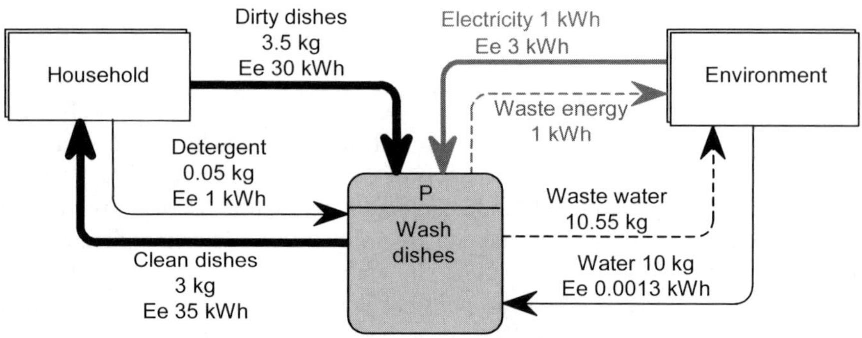

Diagram S3-13: Process "Wash dishes."

Specification S3-20 shows the embodied energy of the flow "Clean dishes." It is calculated based on the process energy (electricity input) and the embodied energies of the input flows "Electricity," "Dirty dishes," and "Detergent." The embodied energy of the flow "Water" and of the process "Wash dishes" is not calculated for this example because it is insignificant. The values for the embodied energy of the flows are estimated.

Specification S3-20: Flow "Clean dishes"

Clean dishes				
Description	Dish after washing.			
Flow type	product			
Parameter	Embodied energy "Dishes with food remnants" (literature)		30	kWh
	Total energy "Electricity" (estimated)		1	kWh
	Embodied energy "Electricity" (literature)		3	kWh
	Embodied energy "Washing detergent" (literature)		1	kWh
	Embodied energy "Water" (calculated)		0.0013	kWh
Embodied energy	calculation	30 kWh + 1 kWh + 3 kWh + 1 kWh + 0.0013 kWh		
	resource value (approx.)		**35**	kWh/SU

3.9.2 Process Value: Embodied Energy Added

The embodied energy of a product shows how much energy was needed to process a product up to its current state. With this analysis, it is possible to evaluate the environmental impacts of the product related to the greenhouse effect. The embodied energy is not balanced by a process equation, but is added and transferred to the output flows. The embodied energy of the input flows and the embodied energy of the process (production and maintenance of machines and installation) are transferred to the product flows.

The process value embodied energy added is the embodied energy that is transferred by the process onto the output product.

Definition: The embodied energy added represents the cumulated energy amount charged to the product flow as energetic burden.

The input product flow carries embodied energy as its energetic burden. The embodied energy added equals the embodied energy of the output product flow minus the embodied energy of the input product flow (*Equation E17*). This corresponds to the definition of the monetary value added in Section S2.

E17	$Ee_A = Ee_{ProductOutput} - Ee_{ProductInput}$	Ee_A Ee	Embodied energy added Embodied energy

The embodied energy added can also be calculated by summing up the embodied energy of the input resource flows (including energy flows), the total energy of input energy flows, and the embodied energy of the machines and installations (*Equation E18*). The process transfers the embodied energy added to the output product flow.

E18	$Ee_A = Ee_{Input} + E_{Input} + Ee_{Mach+Installation}$	Ee_A Ee E	Embodied energy added Embodied energy Total energy

Output resource flows only carry an embodied energy if they are valuable for other production processes. The embodied energy added and the embodied energy of the input product is allocated to the output product and the by-product.

On the diagram, the embodied energy added is indicated in the third field of the process symbol reserved for additional information.

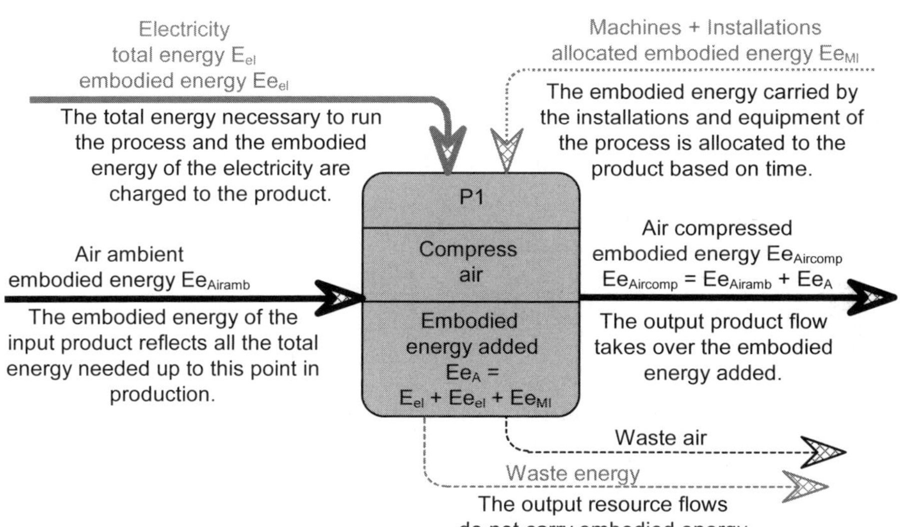

Diagram S3-14: Process "Compress air."

Diagram S3-14 depicts the process "Compress air." This example shows the theoretical calculation of the embodied energy added to the process. The total energy of the energy flow "Electricity," the embodied energy of the resource flow "Air ambient," and the allocated embodied energy of the fictitious flow "Machines + Installations" are summed up to obtain the embodied energy added of the process.

The embodied energy of the process that corresponds to the third term of *Equation E18* (machines and installations) is allocated in shares of the total equipment lifetime. Each product only uses the machine during a specific time. The timely allocation is done by a factor calculated of the production time of the process divided by the total process time of the machine over its lifetime.

The embodied energy of the machines and installations is a theoretical concept, it represents the energy needed to produce and install the machines required for the process. The embodied energy of the equipment and installations is depicted as fictitious flow in order to highlight that it does not really exist. To integrate this concept, the fictitious flow represents the embodied energy as a timely allocated energy investment into the process that is then charged to the product.

3.9.3 Example: Embodied Energy Calculation of a Textile Yarn

This chapter examines the embodied energy of the process and product with the example of a spinning mill producing a yarn. The goal is to calculate the embodied energy added by the process "Spin yarn." This includes the calculation of the embodied energy of machines and equipment (manufacturing and installation of the cleaning machines and spinning machines in the spinning mill), the embodied energy of the input resource flows, and the total energy of the input resource flows.

Step 1: Calculate embodied energy of machines and installation

First, the embodied energy of the machines and installations is calculated and valuated concerning its significance. The calculation begins with a coarse estimation of the embodied energy of the production machines of a spinning mill.

Table S3-9: Machine inventory

Amount	Machine	Mass in 1000 kg
2	Bale opener / Cleaners / Mixer	20
13	Cards	70
12	Drawing frames	18
16	Spinning machines (each with 240 rotors)	160
1	Air-conditioning	20
1	Transporting equipment	6
1	Electricity distribution and switchgear	6
Total mass of machines		300

Table S3-9 lists the number of machines and their weight for a spinning mill (with rotor spinning machines). The number of machines corresponds to the production of 1000 kg of fine yarn (24 tex) per hour. The SU of the analysis is 1 ton of yarn.

The calculation of the embodied energy of the spinning mill is continued in *Table S3-10*. The material composition, weight fraction, and embodied energy are listed for each material. The values for the embodied energy of the basic materials are found in literature. This value contains the energy necessary to mine, transport, and prepare the material, for example aluminum for further processing. Using the weight ratio of the basic materials for the machines, the embodied energy added is summed up.

Table S3-10: Embodied energy of a spinning mill

Material	Weight [%]	Embodied energy [MJ/kg]	Embodied energy total [GJ]
Aluminium	15	210.0	9,540
Brass and copper	5	105.0	1,575
Iron and steel	70	11.5	2,415
Plastics	10	85.0	2,550
		Embodied energy materials	15,990
		Embodied energy assembly (Estimation 15.6 MJ/ton)	4,680
		Embodied energy spinning mill	20,670

After calculating the embodied energy of the materials, the embodied energy of the assembly is added. The last step is the calculation of the timely allocated embodied energy of the spinning mill, as shown in *Specification S3-21*.

Specification S3-21: Flow "Machines + installations spinning mill"

Machines + installations spinning mill			
Description	Embodied energy of the equipment timely allocated to the product.		
Flow type	material		
Parameter	embodied energy spinning mill lifetime spinning mill production time yarn	20,670 20 1,470	GJ years s
Embodied energy	calculation	$20670\,\mathrm{GJ} \cdot 1000 \dfrac{1470\,\mathrm{s}}{20 \cdot 365 \cdot 24 \cdot 3600\,\mathrm{s}}$	
	resource value	**48.2**	MJ/SU

The value of the embodied energy of the machines and installations of the spinning mill is allocated based on the lifetime of the machine and the machine time needed to produce the analyzed product. The calculation is shown in the *Specification S3-21* of flow "Machines and installations spinning mill."

Step 2: Calculate embodied energy product

Now, the embodied energy added of the process has to be calculated. The RFD of the process "Spin yarn" for an open end yarn product is shown in *Diagram S3-15*. The SU for this calculation is 405 kg yarn. The embodied energy of the "Cotton bale" sums up the energy consumption in the supply chain. The embodied energy of the "Electricity spinning" was found in literature. The total energy of "Electricity spinning" was calculated for the SU based on measured values.

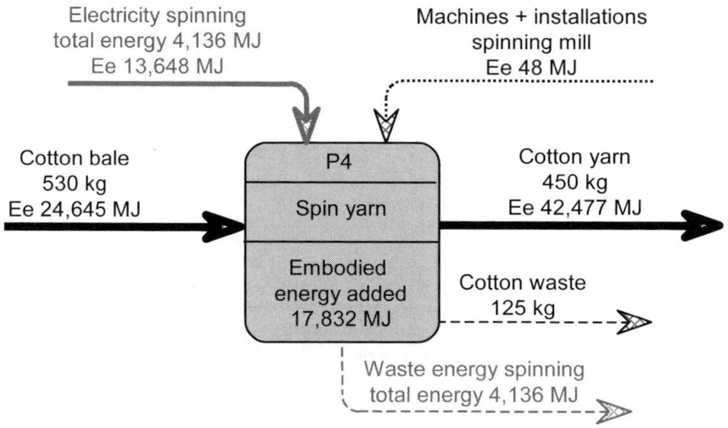

Diagram S3-15: Process "Spin yarn."

The calculation of the embodied energy added is shown in *Specification S3-22* process "Spin yarn." Compared to the embodied energy of the other resources, the embodied energy of "Machines + installations spinning mill" is in fact negligible.

Specification S3-22: Process "Spin product"

P4	**Spin yarn**			
Description	Spin cotton yarn.			
Parameter	Total energy$_{electricity}$ (measured) Embodied energy$_{electricity}$ (literature) Allocated embodied energy "Machines + installations spinning mill" (calculated)		4,136 13,648 48	MJ MJ MJ
Embodied energy added	calculation	4136 MJ + 13648 MJ + 48 MJ		
	process value		**17,832**	MJ/SU

The value of the embodied energy added is summed up to the embodied energy of the input product flow "Cotton bale." The result for 1 SU of the output product flow "Cotton yarn" is the embodied energy of 42,477 MJ (*Diagram S3-15*).

3.10 Application Example: Gas Station

We will analyze a gas station in order to determine its resource consumption and the waste production of its site. Analyzing simple processes which are part of everyday life may reveal unexpected things. All mass and energy values of the resource flows in the gas station will be identified and calculated. The context diagram of the system "Refuel car" is shown in *Diagram S3-16.*

The analyzed system "Refuel car" has interactions with external entities. The resources supply the gas station with "Gasoline," "Electricity.GasStation," "Water.GasStation," and "Chemicals." The "Driver" is an external entity who delivers the empty and dirty car to the system and receives a car that is clean and refueled. The "Environment" is burdened with "GasEmissions," "WasteWater.GasStation," and "WasteEnergy.GasStation." The product flow of the system is depicted with bold arrows. It represents the car entering and exiting the system.

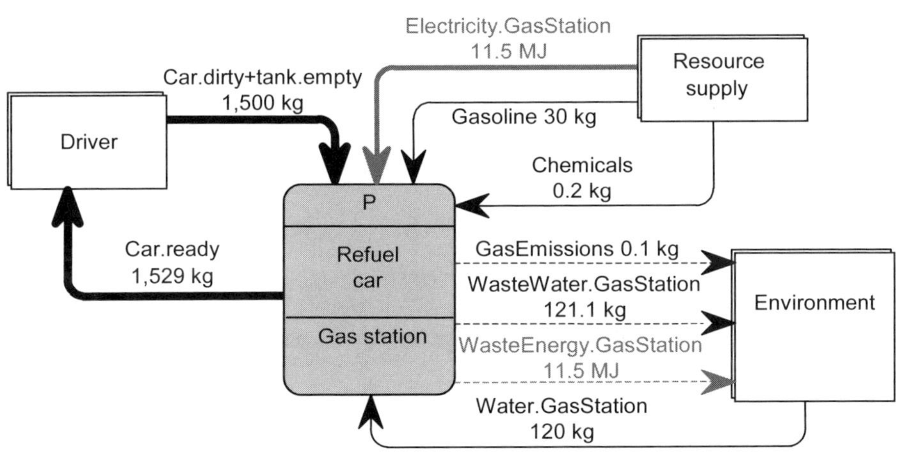

Diagram S3-16: Context diagram of gas station.

Now, we must define what we are going to do at the gas station. Do we just refuel the car, pay, and leave, or do we require further services?

On *Diagram S3-17,* the first level of detail, the child diagram of the process "Refuel car" is shown. In our case, the car is first refueled and then washed in an automatic car wash. There the driver sits inside the car while it is moved through various washing zones. The first level diagram has three processes: "Fill gasoline," "Clean car," and "Pay bill."

The first step is to draw all the input flows. Be careful not to forget less obvious inputs, such as the electricity of the gas pump or the cash register. The output flows can be tricky, too. Remember that every energy input should have an energy output, such as waste energy.

Before starting the calculation, the Standard Unit has to be defined. In this case the SU is one car. All the resource values are calculated for one car.

The RFD in *Diagram S3-17* contains the resource values mass and total energy for the resource and energy flows. The values used for this application example are estimated values. We have some information from the gas station about how much electricity is used in the individual processes, and we know how much gas we need to refuel. With these values, we can perform some initial calculations.

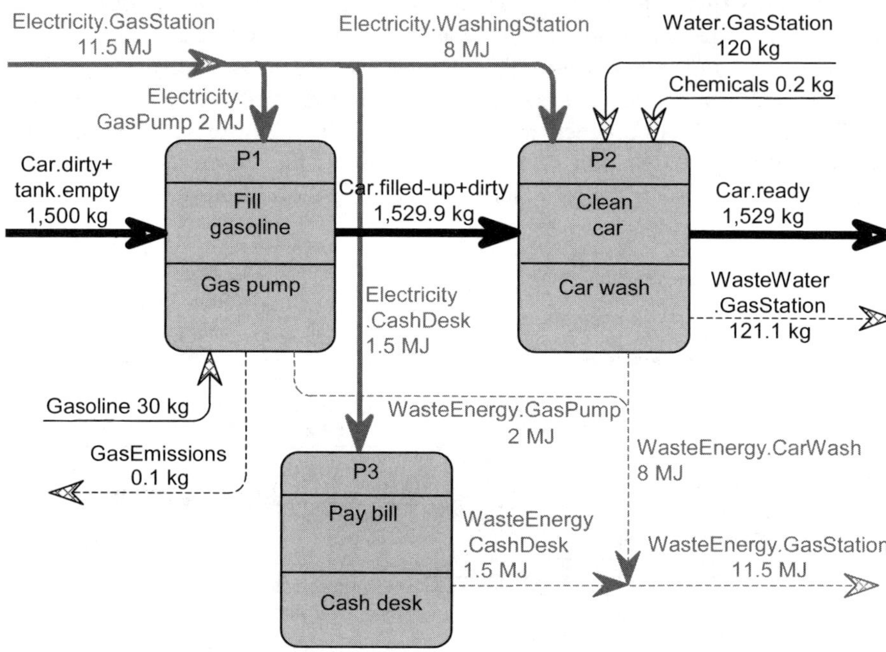

Diagram S3-17: First level of detail of the gas station.

The values for electricity and water consumption of the gas station are available. It is known how much gasoline is filled into the car. Based on this data, the unknown resource values can be calculated by using process balances and flow calculations.

The energy balance for the process "Fill gasoline" tells us that input energy equals output energy. Therefore, we can calculate the total energy in *Specification S3-23*.

Specification S3-23: Process "Fill gasoline"

P1	**Fill gasoline**	
Description	Fill fuel into the car.	
Parameter	total energy$_{\text{electricity gas pump}}$ (estimated)	2 MJ
Energy balance	2 MJ = 2 MJ	

The same applies for the waste energy flows of the process "Clean car" and "Pay bill." Now that we have the values of the three waste energy flows, we can calculate the merged parent flow "WasteEnergy.GasStation" in *Specification S3-24*. The amount of "WasteEnergy.GasStation" is transferred up through the hierarchy to the context diagram.

Specification S3-24: Flow "WasteEnergy.GasStation"

WasteEnergy.GasStation		
Description	Merged flow of "WasteEnergy.GasGump," "WasteEnergy.CarWash," and "WasteEnergy.CashDesk"	
Flow type	waste energy	
Parameter		
Total energy	calculation	2 MJ + 8 MJ + 1.5 MJ = 11.5 MJ
	resource value	**11.5** MJ/car

Now we calculate the product flow. The car entering the system has an empty tank and is dirty; its mass is 1,500 kg. The car is heavier after refueling because of the weight of the gas. *Specification S3-25* shows the mass balance of the process "Fill gasoline." Of the 30 kg gas put into the car, around 100 g evaporated while refueling, as shown in the mass balance. The mass of the flow "Car.filled-up+dirty" equals 1,529.9 kg.

Specification S3-25: Process "Fill gasoline"

P1	**Fill gasoline**		
Description	Fill fuel into the car.		
Parameter	total energy$_{\text{electricity gas pump}}$ (estimated)	2	MJ
	mass$_{\text{car.dirty_tank.empty}}$	1,500	kg
	mass$_{\text{gasoline}}$	30	kg
	mass$_{\text{gasemissions}}$	0.1	kg
Energy balance	2 MJ = 2 MJ		
Mass balance	1500 kg + 30 kg - 0.1 kg = 1529.9 kg		

The refueled car is washed in process P2. The mass is reduced as dirt is washed off. The mass of the flow "Car.clean" can be merged through the hierarchy to the same flow on the context diagram.

The values of the water and chemical flows of the washing station are inspected more in detail. We have the resource values of the water, chemical, and energy consumption of the whole car wash, but not of the single stations. Therefore, we go one level down in the hierarchy to the child diagram of the process "Clean car" in *Diagram S3-18*.

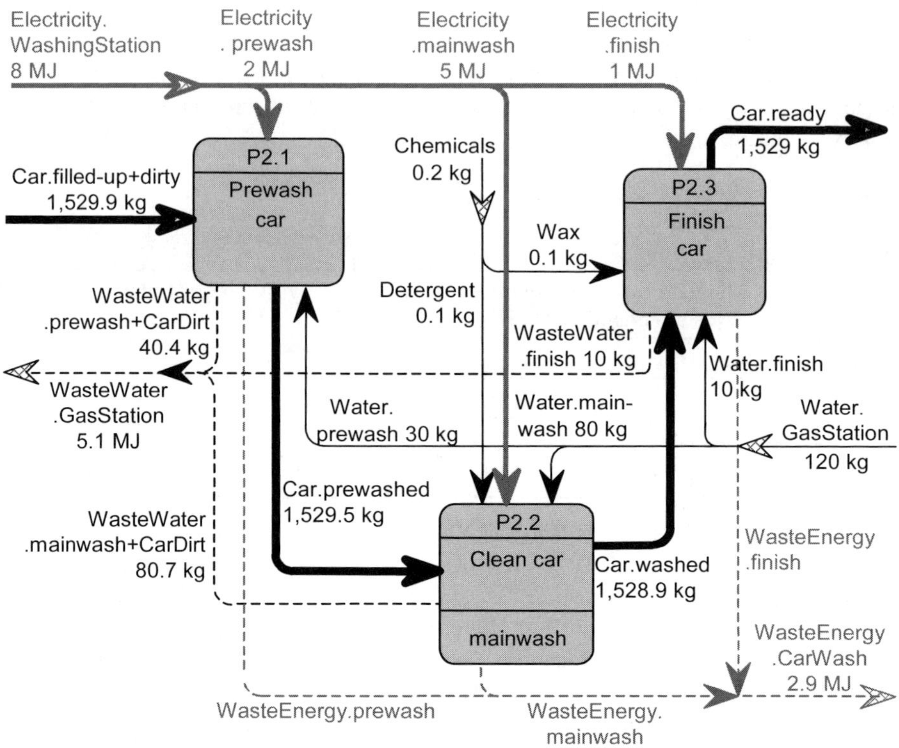

Diagram S3-18: Second level of detail of the car wash.

The car wash analyzed here is an automatic one, where the car is transported on a conveyor belt through three different chambers. The water flows of each washing station can be merged to check the mass of the flow "Water.GasStation" on the parent level. We can calculate the waste water flows by summing up the input water and other liquids for each chamber. The merged flow "Water.GasStation" is calculated in *Specification S3-26*.

Specification S3-26: Flow "Water.GasStation"

Water.GasStation			
Description	Water used for the car wash.		
Flow type	water		
Parameter	$Mass_{Water.prewash}$ (estimated) $Mass_{Water.mainwash}$ (estimated) $Mass_{Water.finish}$ (estimated)		30 kg 80 kg 10 kg
Mass	calculation	30 kg + 80 kg + 10 kg	
	resource value		**120** kg/car

Specification 3-27 shows the calculation of the mass balance of the process "Clean car." The mass of the flow "WasteWater.Mainwash+CarDirt" can be read from the balance; it is the sum of the mass "Water.mainwash," the mass of "Detergent," and the mass of dirt washed from the car. The mass of dirt equals the mass difference between "Car.washed" and "Car.prewashed." As the car is not weighed before and after washing, a value between 0.5 and 1 kg of dirt is used for a slightly dirty car.

Specification S3-27: Process "Clean car"

P2.2	**Clean car**		
Description	Washing car in depth.		
Parameter	$Mass_{Water.mainwash}$ (estimated)	80.0	kg
	$Mass_{Detergent}$ (estimated)	0.1	kg
	$Mass_{Car.washed}$ (estimated)	1528.9	kg
	$Mass_{Car.prewashed}$ (estimated)	1529.5	kg
Mass balance	1529.5 kg + 80.0 kg + 0.1 kg - 1528.9 kg = 80.7 kg		

The flow "WasteWater.WashingStation" is calculated by the merge of each waste-water flow. The mass specification of the flow "WasteWater.WashingStation" is calculated in *Specification S3-28*. The mass specification is transferred up to the first level and the context diagram.

Specification S3-28: Flow "WasteWater.WashingStation"

WasteWater.WashingStation				
Description	Waste water of the process "Clean car."			
Flow type	water			
Parameter	$mass_{WasteWater.prewash+CarDirt}$ $mass_{WasteWater.mainwash+CarDirt}$ $mass_{WasteWater.finish}$ Heat capacity water Temperature difference water		80.7 30.4 10.0 4.18 10	kg kg kg kJ/kgK K
Mass	calculation	80.7 kg+ 30.4 kg +10.0 kg		
	resource value		**121.1**	kg/SU
Mass	calculation	4.18 kJ/kgK · 10 K · 121.1 kg		
	resource value		**5.1**	MJ/car

The mass specification of the flow "Chemicals" can be calculated by merging the flows "Detergent" and "Wax." From the calculation of the total energy of the waste water it becomes obvious that a lot of energy is lost in the hot waste water. Technical possibilities for the reuse of this waste water for instance transferring the heat by a heat exchanger to fresh water should be checked for this car wash. Of the 8 MJ electricity for the car wash, 5.1 MJ are transferred to the waste water and are stored there as total energy. 2.9 MJ are transferred as waste energy to the environment.

We now can fill all the calculated values into the diagrams and check to see that we have defined the system completely. We determined the water, chemical, and energy consumption for this washing station. It would now be interesting to model a RFD for a different type of washing station in order to compare the consumption of resources of different washing methods in gas stations.

Table S3-11 contains a checklist for the RFD.

Table S3-11: RFD Checklist

Question	Checked
The Standard Unit for the study has been defined.	✓
The flow types are fixed and drawn differently on the diagram.	✓
The product flow is highlighted on the diagram.	✓
Every process has an energy input and an energy output.	✓
Every flow is part of a flow type.	✓
Waste flows are included on the diagram.	✓
The total energy of input flows equals the total energy of output flows.	✓
The parameters and their values are declared in the specifications.	✓
The source of the parameter values is indicated.	✓
The sum of mass and total energy of child flows equals the mass and total energy of the parent flow.	✓
Every output of the system is considered.	✓
Split and merge of flows is only done within the same flow group.	✓
For every split and merge, a calculation was performed.	✓
The flow calculations are declared in the flow specifications.	✓
The mass and energy balance is correct for every single process and the whole system.	✓

Next steps

Based on the calculated values, we can perform further analyses, such as determining the energetic or exergetic efficiency of the processes, or perform a further environmental evaluation. In addition, a monetary analysis of the system can be done with the VFD. If we want to analyze the system dynamically in order to optimize its use, we can go over to the State Chart and program a simulation.

3.11 Apply Your Knowledge

1) Values in the RFD

What kind of values are assigned to the flows in the RFD? Give a description of them. Insert the names of the resource values, which can be assigned to the different flows in *Diagram S3-19* that depict the production and packing of dough for later sale in shops.

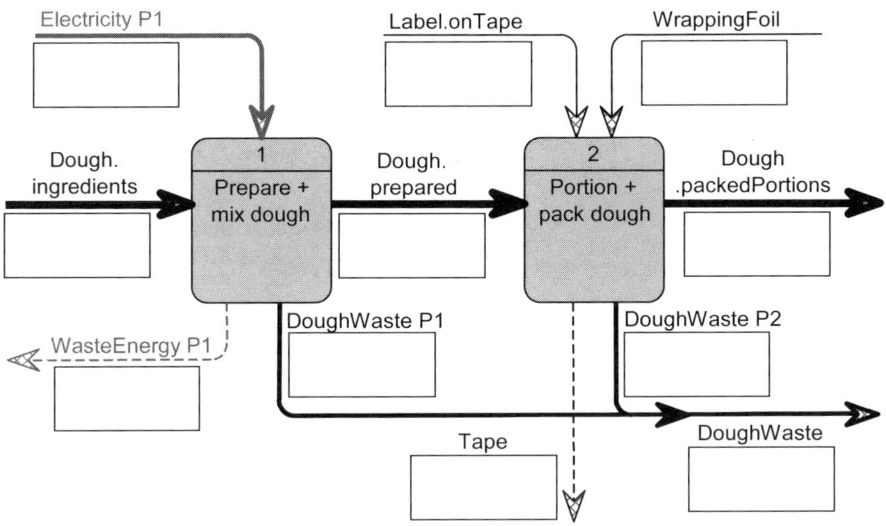

Diagram S3-19: Values in the RFD.

Insert the resource value names into the boxes that are joined to the resource flow names.

2) Mass balance

The calculation of the mass of the product flows in the system is always the first step in calculation. The cornflakes production line with the ingredients necessary for the product and the waste occurring during the production serves as calculation example.

Do a mass calculation and balancing for the production chain of cornflakes in *Diagram S3-20*. The SU is the end product of this production chain that goes to the shops and the customers: 100 boxes of cornflakes frosted with sugar. Each box contains 500 g cornflakes.

Calculate for each product flow between two processes the mass that refers to the SU of 100 boxes at the end of the production chain. The percentages given at the flows give the ratio of several input product flows or the ratio of the output flow (waste, steam, dust) in comparison to the output input flow of the same process.

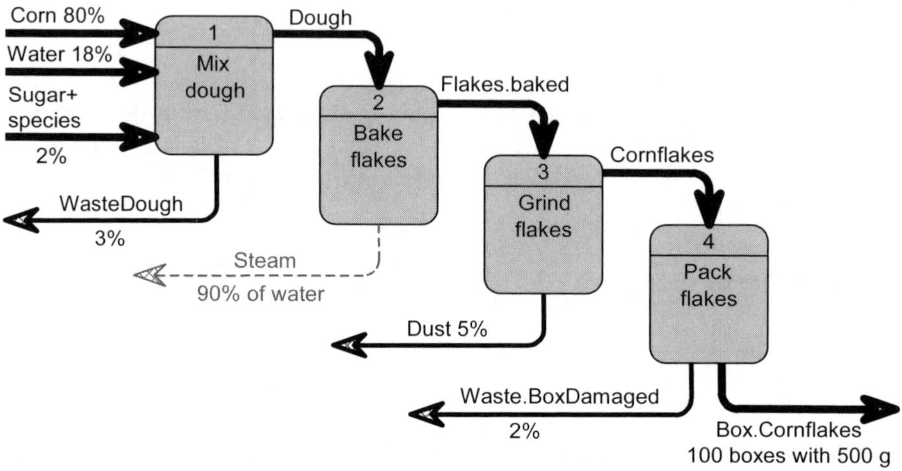

Diagram S3-20: Production chain for cornflakes.

3) Energy balance

Calculate the energy in *Diagram S3-21* that describes the dyeing of T-shirts. Do the energy balance and indicate the missing values on the diagram. Standard unit: 100 T-shirts, assumption: no product waste.

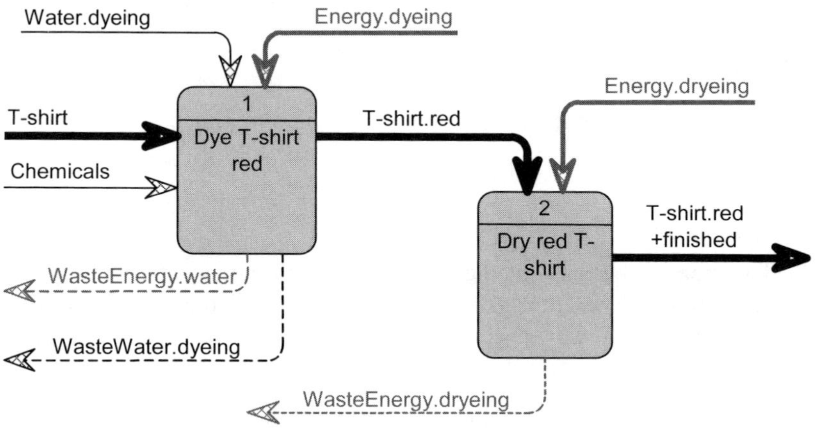

Diagram S3-21: Dyeing T-shirts.

Parameters for process and flow specifications:

weight of 1 T-shirt	300	g
water consumption dyeing	45	l/kg T-shirt
energy consumption dyeing	0.75	MJ/kg T-shirt
energy consumption drying	1.5	MJ/kg T-shirt
chemicals	50	g/kg T-shirt

4) Calculation of exergy

After having performed the energy calculation in the dyeing company example of question 3, the next question arises: Are there flows with sufficient waste energy, which are available for reuse and by this help saving energy and money?

Calculate for the dyeing of T-shirts from question 3 (*Diagram S3-21*) the resource value exergy of the flow "WasteWater" of the dyeing process and the "Waste Energy" of process "Dry red T-shirt." Complete *Specification S3-29* and *Specification S3-30* with the equations and the results.

Specification S3-29: Flow "WasteEnergy.water"

WasteEnergy.water			
Description	Waste energy in water coming from dyeing machine.		
Flow Type	Water		
Parameter	Temperature	80	°C
Calculation exergy			
Flow value exergy			

Specification S3-30: Flow "WasteEnergy.drying"

WasteEnergy.drying			
Description	Waste energy in air released by drying machine.		
Flow Type	Energy		
Parameter	Temperature	150	°C
Calculation			
Flow value			

5) Sources of information

One of the most difficult and time consuming activities in an analysis is the search for data for the calculation of the resource and process values.

Where do you get the information, quantities, and numbers of the system to be investigated and calculated? The numbers, you fill in the flow and process specifications? What do you do, if they cannot be found within the company?

6) Calculation of efficiency

Diagram S3-22 shows the blending of two types of fibers, cotton and polyester, in order to get a mixed yarn. Calculate the energetic efficiency of the process "Blend fibers" for the double of mass processed by the blender, i.e. for an output of 702 kg of "Cotton-PolyesterBlend".

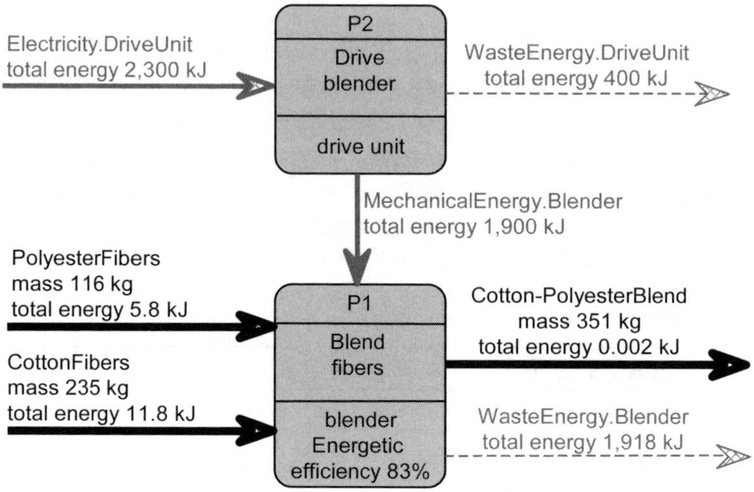

Diagram S3-22: Blend cotton and polyester.

Calculate the mass balance, energy balance, and the energetic efficiency in the *Specification S3-31* below for process "Blend fibers." The missing parameter values can be found in *Diagram S3-22*.

Specification S3-31: "Blend fiber"

P1	**Blend fibers**		
Description	Blending of two fiber types for the production of knitting yarn.		
Parameter	Total energy "Cotton fibers" Total energy "Polyester fibers" Total energy "Mechanical energy blender" Velocity "Cotton-Polyester blend" Total energy "Cotton-Polyester blend"	 0.1 	kJ kJ kJ m/s kJ
Mass balance	$mass_{COfiber} + mass_{PETfiber} = mass_{CO-PETblend}$		
Energy balance	$E_{COfiber} + E_{PETfiber} + Q_{electricity} - E_{CO-PETblend} = Q_{wasteenergy}$		
Energetic efficiency	calculation		
	process value		%

7) Complete system in the hierarchy

Calculate the missing flow values mass and total energy (of energy flows) for a milk bottle production from the values given in the context diagram (*Diagram S3-23*) and the first level diagram (*Diagram S3-24*).

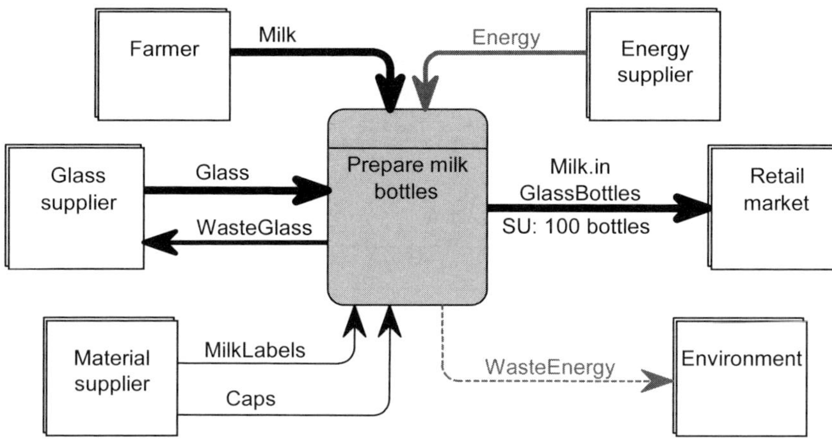

Diagram S3-23: Context diagram.

Some of the resource values are already indicated in the diagram. Some more are listed here:

- For 1 glass bottle: 200 g glass per 500 g milk.
- 2 g label per glass bottle.
- Waste glass: 5% of production.
- 5 g per cap.

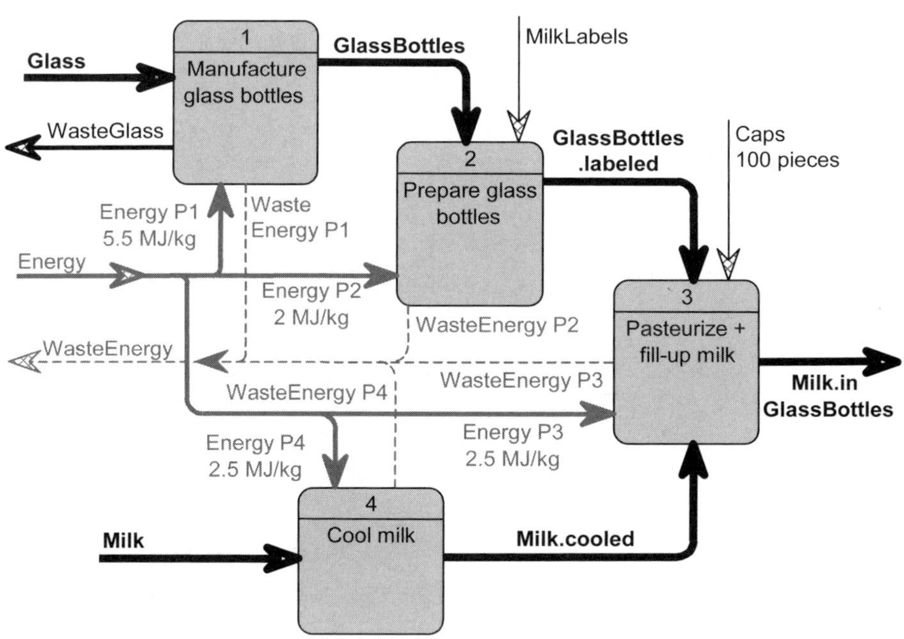

Diagram S3-24: Level 1 diagram “Prepare milk bottles”.

8) Embodied energy

Calculate the embodied energy of the product flow "Milk.inGlassBottles" in the diagram of question 7 (*Diagram S3-24*). Assumption: the embodied process energy is neglected. Calculate the value by using the parameters in of the flow "Milk.in GlassBottle" (*Specification S3-32*).

Specification S3-32: Flow "Milk in glass bottles"

Milk in glass bottles			
Description	Glass bottles filled with half a liter of milk.		
Flow type	Product		
Parameter	Embodied energy glass	26.2	MJ/kg
	Embodied energy electricity	8.2	MJ/kg
	Embodied energy milk	8.0	MJ/kg
	Embodied energy caps (PET)	89.9	MJ/kg
	Embodied energy milk labels	40.0	MJ/kg
Embodied energy	calculation		
	Flow value		

Compare the flow value embodied energy "Milk.inGlassBottles" with the embodied energy value of milk in PET bottles, where the embodied energy of PET bottles is 33.3 MJ/kg and the bottle weight 50 g per 500 g milk.

D DYNAMIC ANALYSIS TOOLS

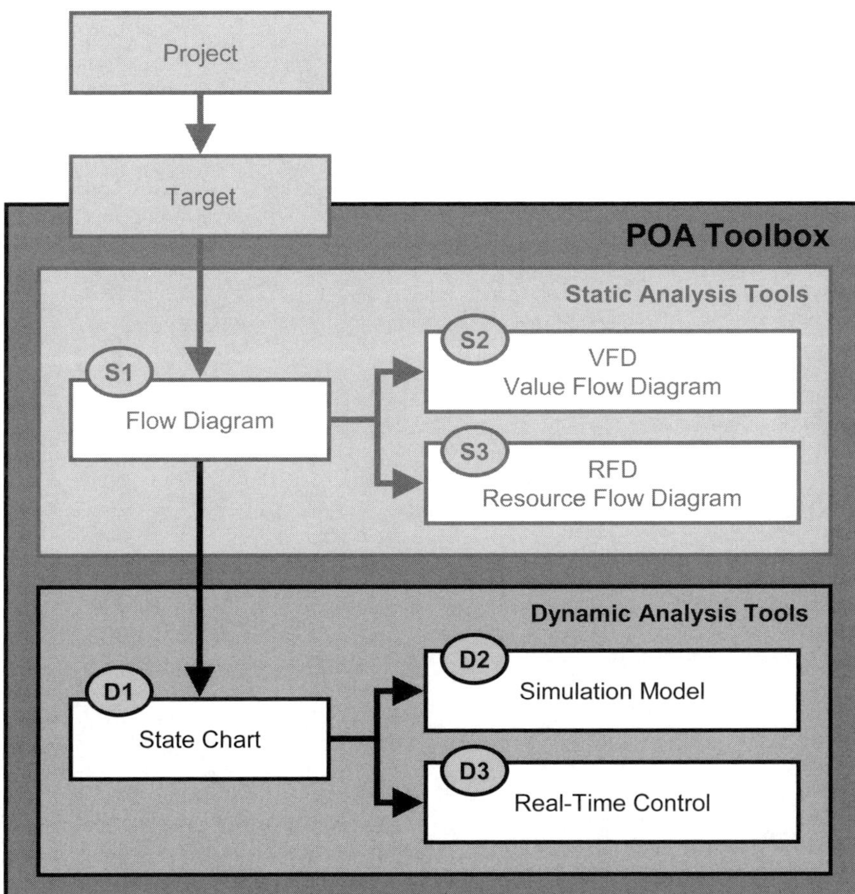

D1 STATE CHART

Goals of the Chapter

- Get to know the elements of the State Chart.
- Describe the time-dependent behavior of a production system using the State Chart.
- Take the step from static to dynamic modeling within the method of the Process Oriented Analysis.
- Draw a State Chart for a production process.

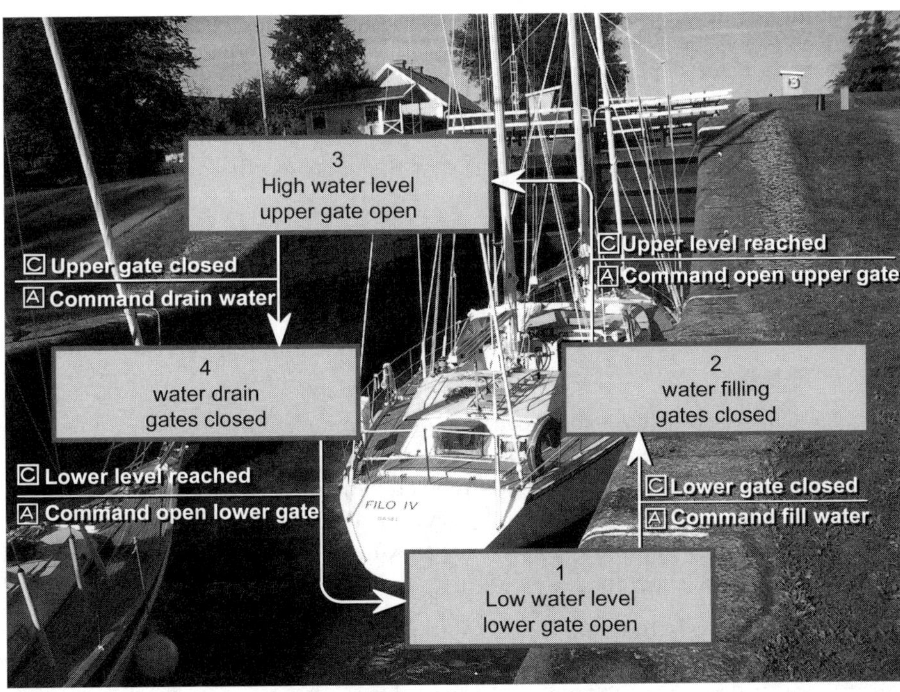

Figure D1-1: State Chart of a canal lock for ships (Photo: ETH Zurich).

1.1 Introduction

The State Chart is popular for event-controlled and time-critical processes such as cyclic machinery operation, interlinked logistic sequences, material handling, and digital data transmission. It is indispensable for the design and specification of non-continuous processes. The State Chart describes the multi-step action of a network of processes and its dynamics. This is often encountered in the control of plants with a high degree of automation.

Through the State Chart, Process Oriented Analysis offers an extension to the static model introduced in Part S. The State Chart investigates the system behavior, which is the time-dependent aspect of a system. The system behavior is described by the states that the system can occupy and the transition to change from one state to another. The State Chart depicts all possible states of a machine or a process, both the intended states as well as the risk and alarm states. The investigation of risk and malfunction states is useful for evaluation and may even be required for documentation of the safety measures implemented in machinery and plants. The sequence of transitions and the conditions for changing the states are determined. The definition and specification of states and transitions complete a State Chart.

This dynamic, time-dependent view of a system is later easily translated into the source code of a program. The State Chart is a specification of a process in time and is used for the analysis and design of simulation models for optimization of plants and real-time controls. It ensures that the source code is clear and understandable.

The starting point for the dynamic state-based view is the Flow Diagram of Part S. A step-by-step procedure goes from Flow Diagram to State Chart to coding. This path is also recommended for any simulation and real-time project because it supports the building of a code that is structured in a way that represents the real world processes.

The State Chart is well known in real-time programming (Lower CASE tools) and has been used for a long time for this purpose. The State Chart alone is not easily understood by someone who is not involved with programming. However, in connection with the Flow Diagram, it becomes a powerful tool for any kind of system engineering. The State Chart adds the timely dimension to the structural analysis of the Flow Diagram.

The dynamic analysis tools of the POA toolbox are introduced in Part D. This Section D1 focuses on the State Chart itself, explaining its elements and the drawing of them. It describes the syntax and the rules for the setup of the diagrams. A gas station is modeled as an application example at the end of the section. The case study in Chapter C4 describes the concept and planning of a new production line for demagnetizing TV picture tubes. The design of the new production line and the integration into the existing plant is done using the Flow Diagram and the State Chart.

1.2 State Chart: Why?

1.2.1 Purpose

State-based and behavioral view: The State Chart is a dynamic model. It represents the state-based and event-based, or behavioral view, of a system in relation to time. On the State Chart, interactions between the elements of a system are shown sequentially. The main purposes of the State Chart are as follows:

- It specifies the behavior of a process in relation to its environment and to time.
- It identifies the relevant states of a process.
- It makes the behavior of a system clear by allowing for a complete overview of its states.
- It discloses lock-up situations that would block the system.
- It helps to design a secure control by identifying dangerous states and introducing transitions to a safe, inactive state that is entered as soon as a fault is detected in the machine or system.

Consistency with Flow Diagrams: The State Chart is fully consistent with the static Flow Diagram. For each process in a Flow Diagram, a State Chart is created and attached as a functional specification. The State Chart can also be drawn independently of a Flow Diagram if it pertains to a single, clearly defined system or process, for example, the behavior of a person at his or her workplace. Nevertheless, it is recommended to set-up the Flow Diagram first in order to grasp the structure of the system correctly and completely. Using the Flow Diagram, the real-world processes are analyzed and their inputs and outputs noted. While defining the behavior of a system or a process in relation to time by setting up a State Chart, the states of the system and the conditions for state changes are determined.

Hierarchical structure: Working on different levels in the hierarchy, the State Chart also enables the modeling of large systems such as entire production plants. The State Chart hierarchy continues the hierarchy of the Flow Diagram. It details the processes on the next level into states.

Simulation model and real-time control: The State Chart serves as the basis for a simulation model and a real-time control program by treating the states as variables in the source code. The operating states, as well as the exceptional states for troubleshooting, are depicted on the State Chart; this allows the programming of a machine control based directly on its State Chart.

1.2.2 Application

A system can be analyzed from different points of view. The following example of a railroad transportation system shows three basic aspects of a system. Each of these aspects is modeled with a different diagram type as shown in *Figure D1-2*.

The map of train stations and lines corresponds to the static view of the system and is represented by the Flow Diagram. It depicts the origin and destination of the trains. The timetable, which shows the time-dependent states and locations of the trains, corresponds to the dynamic view of the system represented by the State Chart. It shows what happens, at which time, and in which sequence. These two diagrams are a part of POA. The typical sequence of the model setup begins with the specification of the railroad network and stations (Flow Diagram). Second, the timetable is set up (State Chart).

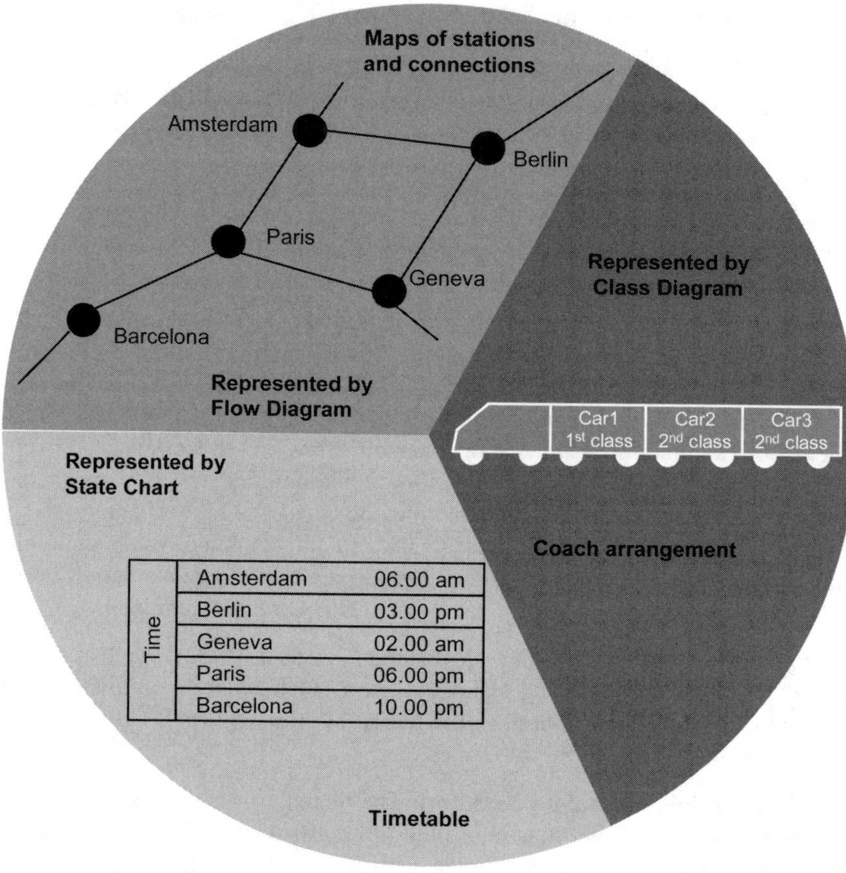

Figure D1-2: Points of view of a transportation system.

Third, if required by the system, the planning of the position of cars and seats in the train is indicated. The arrangement of the individual coaches within the train corresponds to the Class Diagram in the Unified Modeling Language or the Entity Relationship Diagram (see Section I2). These diagrams specify the structure of data within the system; however, this is not subject to POA because the structure and organization of data is already supported by various methods of database modeling.

Figure D1-1 describes the function of a lock for ships in a canal by a State Chart. The states are determined by the parameter "water level" of the canal lock. The State Chart shows, in a very understandable way, which actions are carried out under which conditions in order to get the ships on another water level.

1.2.3 Delimitation

Since the early days of computer programming, programs have been written in text form and arranged in a sequence of lines of code. On a higher level of order, this code is arranged into functional modules that complete a certain task. Graphical tools, which describe the purpose and activity of a program, are a necessity due to the constantly growing size and complexity of program code. This chapter describes the most common graphical diagrams for software design and compares them to the State Chart of POA.

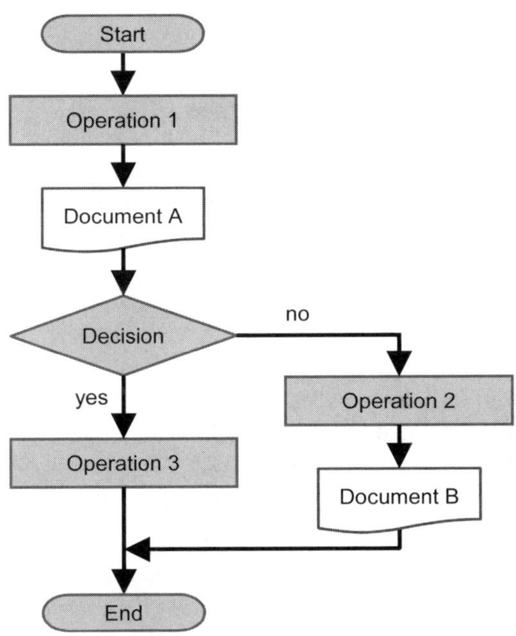

Diagram D1-1: Example of a Flow Chart.

Flow Chart

The conventional Flow Chart belongs to the first generation of such tools. It maps a sequence of activities, actions, and decisions (*Diagram D1-1*). The diagrams are self-explanatory and easy to draw and read. However, all attempts to implement universally acknowledged definitions for the symbols have failed, possibly because the same kind of graphic description is already used for other decision-based methods that do not demand the strict logical structure required to set up a program.

Moreover, this way of mapping does not encourage a modular structure or a hierarchical decomposition.

Nassi Schneidermann

The Nassi Schneidermann diagram is a two-dimensional depiction of the program sequence (*Diagram D1-2*). It is laid out like the layers of an onion. The method includes rules for charting that enforce proper nesting of branching and loops. The increasingly large and complex computer programs resulted in first generation diagrams that sprawl over dozens of pages. Therefore, the complexity of a system may be captured by detailing the system on a single level, but the diagram is not easily understandable. The structuring in hierarchical levels is missing.

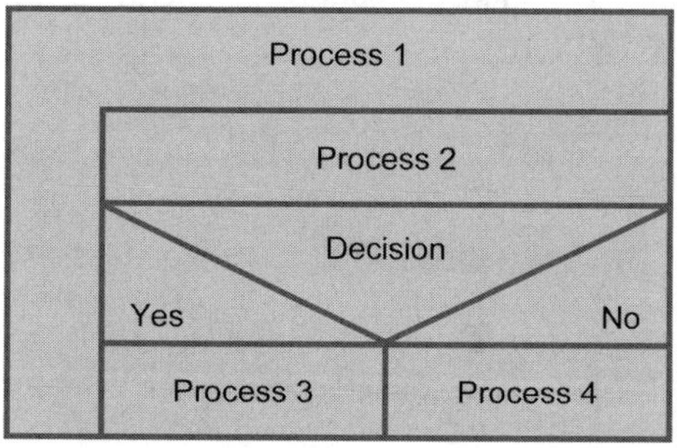

Diagram D1-2: Example of a Nassi Schneidermann diagram.

Petri Net

The complexity of the diagrams led to the development of computer-based charting methods; a strictly defined set of symbols makes their functions and interdependencies compatible with a common database. An example of this second generation of tools is the Petri-Net-based diagrams and graphical programming languages, as used in many state-based analysis jobs.

The original Petri Net depicts a state-oriented and graphical image of the function of a system (*Diagram D1-3*). The actual state of the system is identified by one or more tokens (markers) that jump between the different system states; it is controlled by conditions that trigger the transitions. Based on a single chart, the Petri Net is an excellent tool for analysis and design of functional modules within communication systems. Due to the strict definition of its elements, it is easy to draw and translate into a database. However, the Petri Net is limited by the size of the system since the original concept did not include hierarchical structuring on different levels of detail. The Petri Net is dense, unstructured, and a combination of static and dynamic modeling. It originally served for the graphical display of matrix calculations and is still a convenient way to sketch briefly a limited set of states for low-level real-time programming, as encountered in digital communication and control, e.g. for digital signal processors.

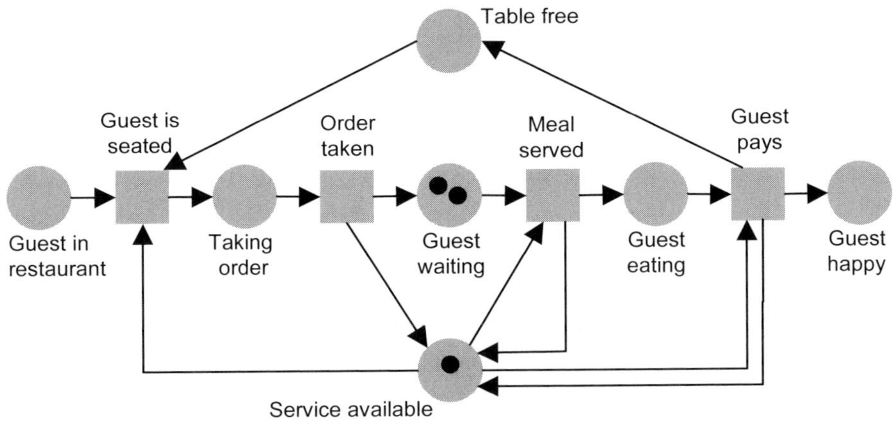

Diagram D1-3: Example of a Petri Net for a restaurant.

Comparison of modeling methods

The comparison between the previously discussed methods is found in *Table D1-1*. In contrast to the other graphical methods, the State Chart of POA supports hierarchical structuring and the transition to programming source code. It is self-explanatory and enables a good overview of the modeled system. The State Chart of POA is viewed one page at a time, and therefore allows for a fast overview. For a complex system, the states are structured hierarchically in new State Charts for further detailing.

Table D1-1: Comparison of dynamic modeling methods

	Flow Chart	**Nassi Schnei-dermann**	**Petri Net**	**State Chart POA**
Self-explanatory	+	+	+	+
Hierarchical structure	-	-	-	+
Good overview	+	-	+	+
Lower CASE compatible	-	-	+	+
Consistent link of static and dynamic view	-	-	-	+

For a better understanding of the system, POA has two different diagrams, one to show the static network of processes (Flow Diagram), and another one to depict the dynamic view of the processes (State Chart). With this, POA offers a complete description of a system. These two diagrams of POA belong in their basic concept to the Upper CASE tools (see Chapter I2.2). Various computer programs are available for the drawing of diagrams and the composition of relevant information (specification of the elements).

1.3 State Chart Elements

1.3.1 Diagram

A State Chart is composed of two graphical symbols: state and transition. The state represents a certain, defined status of a process valid for a period in time. The transitions indicate the changes from one state of a process or system to another one. The minimum number of elements in a State Chart is two states and two transitions.

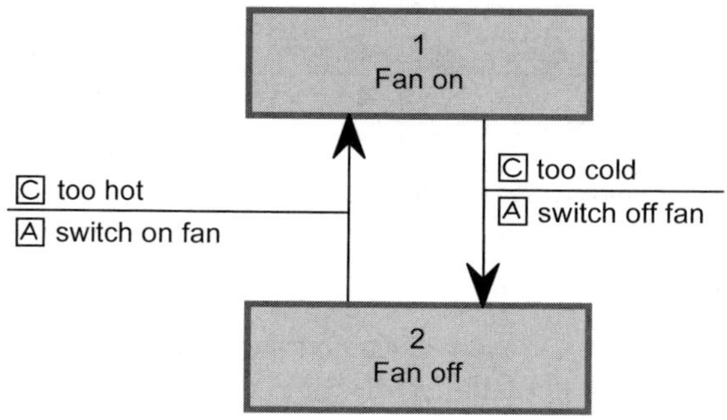

Diagram D1-4: State Chart cooling fan.

Diagram D1-4 depicts a basic State Chart for the system of a cooling fan. The state "Fan on" is active when it is too hot in a room and the cooling fan is switched on. When it becomes too cold, the fan is switched off, and the system enters the state "Fan off."

1.3.2 State

A state represents the properties or the behavior of a process for a given period in time. A state is not an activity, as a process itself, but a momentary situation of a process, e.g. "Machine working."

Definition: A state describes the behavior of a process during a definite period in time.

The complete behavior of a process is given in a State Chart, which is composed of a multitude of states. The behavior of a complete system is described by one or several State Charts. The system and each single process are always in one of the defined states. The system's behavior and characteristics are consistent within each state, so when the system returns to a given state, it behaves exactly as it did while previously in that state.

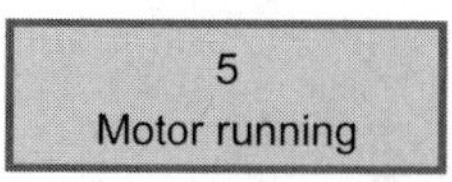

The symbol of a state is a rectangle. Each state is numbered with a unique identification number. On the one

hand, this number identifies the state, and on the other hand, it indicates its position within the model hierarchy (Chapter D1.4.1). Optionally, a prefix may be put before the number: S for state, as for example S2.

Each state is given a name that clearly reflects the particular status of the process or system. Examples of state names are: "Door open," "Engine running," "Stand-by," and "Operating."

State name: The state is named either by a verb in participle form, an adjective, or an adverb, and, optionally, a noun.

Basic states of a system

For every machine or device operating on an external energy supply, there are five basic operating states. These states are either intended or not intended states. *Table D1-2* gives examples for state names for a coffee maker.

Table D1-2: States of a system for a coffee maker

State name	Description	Example coffee maker
De-energized state	Machine inoperable. Considered to be intrinsically safe. Precedes a cold start and terminates a shutdown.	Coffee maker off
Stand-by	Machine is energized, ready and available any time for operation, but not operating. Power saving mode.	Coffee maker on
Target state	Active operation. Represents the system working in its normal operating mode. Purpose of the system.	Brewing coffee
Exceptional state	The system is not following its intended functions, without risking damage. Demands a human intervention.	Water running through filter without coffee grounds
Risk state	Poses a risk to man, machine, and/or environment. Requires immediate attention and intervention to bring the system into a safe state.	Plate heating while empty pot on it

1.3.3 Transition

The states are connected with transitions. A transition only represents a switch of the system into another state, where the system again stays until the next transition occurs. While a system remains for a duration in time in a state, the transition is timeless in comparison to the state. A transition is specified by the origin state and the destination state.

Definition: A transition represents the changeover from one state to another; it takes place momentarily and without delay.

The transition is defined by a condition and an action. The condition describes what has to be fulfilled to trigger the transition. The action is an activity associated with the transition.

The symbol of the transition is an arrow, and the arrowhead points in the direction of the state change from the state of origin to the destination state. The condition and action are written near the arrow. Normally, the condition is placed above the action.

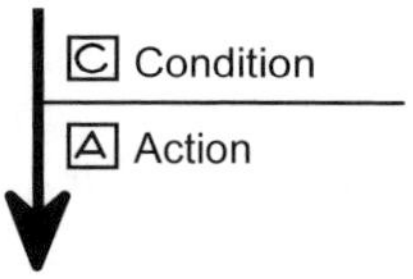

Recommendation for complex State Charts: If condition and action are separated by a horizontal line, the line should preferably touch the transition arrow to make clear, to which transition the condition and action belong.

Action

The action symbol is marked by the letter "A," which is encased, or highlighted by a colon or a closing parenthesis. Actions that are performed during a state transition cause discrete changes within the process. If an action is composed of several components, the action is written as a set of actions.

Condition

The condition stands normally above the action and is optionally separated by a horizontal line from the action. It is marked by the letter "C." To highlight the letter C, it has a frame, a colon, or a closing parenthesis. Transitions are triggered by conditions; as soon as the pertaining condition is fulfilled, the action is momentarily executed, and the system changes into the destination state. Conditions are specified by the concurrent state of other processes of the system or by the actual properties of flows. If a condition is composed of several components, it is expressed as a conditional statement.

Conditions are given in the form of a Boolean expression. Boolean expressions consist of logical variables and operators. The result is again a logical variable, which assumes one of two values, either "True" or "False." When "False," the condition is not fulfilled, preventing the system from making the transition. At the point in time a condition becomes "True," meaning the condition is fulfilled, the associated action will take place, causing the system to change immediately and without delay to the destination state.

Every transition is subject to exactly one condition. This condition can be in form of a *conditional statement* consisting of various individual conditions. These are combined by logical operators AND, OR, or NOT, put in order by parenthesis according to the rules of Boolean algebra. General conditions with examples as written on the transitions in the State Charts are found in *Table D1-3*.

Table D1-3: Examples for conditions

General Expression	**Example of conditions as indicated in the State Chart**
parameter value	C count of bottles = 20
limit value of parameter: $b < a$ $b > a$ $b \leq a$ $b \geq a$	C water temperature < 30 °C C velocity ≥ 60 km/h
negated expression: text NOT text	C heater NOT active
free text for object and attribute, event taking place	C key pressed C part ready at assembly machine C part arrived C car stopped
composed expression: e = w AND v e = w OR u	C Water warm AND clean C Weather rain OR snow
conditional statement: The combination of any of the above conditions.	C (weather nice OR holiday) AND participants ≥ 12

A condition is specified by parameters accessible and events taking place in the investigated system:

- Parameters
  Parameter values, which are measured or calculated, thereafter compared with a threshold, and thus being able to trigger a change in state, e.g. air temperature, mass flow, signal voltage. This comparison provides a Boolean value "True" or "False" in order to formulate a condition.
  For example, for the transition from state 3 to 5, the temperature of the water in *Diagram D1-5* on the next page is compared to the threshold of 36 °C.
- Parameter "time"
  A certain point in time or the elapsing of a certain duration in time may be used to formulate a condition, again by comparison with a given threshold.
- Events
  Any impact from outside, which is detectable, e.g. the flip of a switch, or the arrival of a part in the production line, may be used to formulate a condition. This includes any transition taking place in a different part of the system.

Conditional statement

The conditional statement consists of more than one single condition. It is a combination of several Boolean expressions. The individual expressions or conditions are combined by logical operators, such as AND, OR, or NOT. The last row of *Table D1-3* gives an example of a conditional statement.

Diagram D1-5 shows the states of the process "Take a bath" in a bathtub. Six states describe this process. Several transitions on this State Chart comprise a conditional statement: "Tub empty OR not enough water" on the transition going from state 2 to state 1, "Clean AND relaxed" from state 3 to 6, and the comprehensive conditional statement going from state 2 to 3.

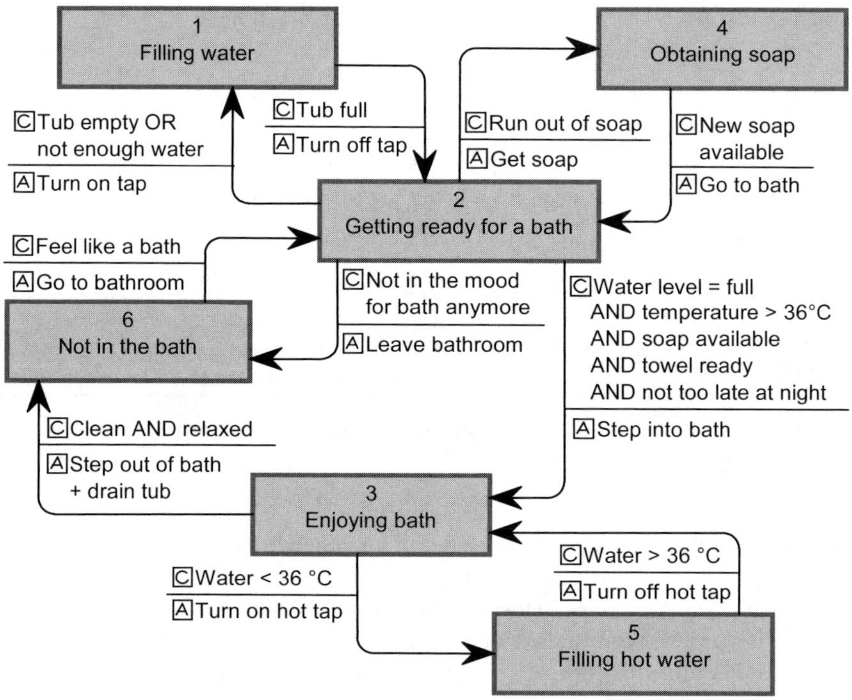

Diagram D1-5: State Chart "Take a bath."

A transition is always triggered by the fulfillment of a condition. This applies also for a conditional statement. Every single condition of the conditional statement must be fulfilled, only then the state change takes place. In the case of the complicated conditional statement in the example of the bathtub going from state 2 to state 3, every single condition must be true. Only then may you step into the bathtub and enjoy it.

At a given point in time, only one condition (or one conditional statement respectively) can be fulfilled on a State Chart. From state 2 "Getting ready for a bath," several transitions lead to different states. Depending on which condition is fullfilled, the process changes into the corresponding target state.

If several transitions leave a state, they are distinguished by their conditions. Depending on the level of detail, it may become difficult to specify exact conditions. However, this is important because the condition decides the transition and therefore the destination state of the system. In such a case, it is necessary to apply conditional statements since one condition may not provide detailed enough distinction between the different transitions. If criteria are missing for setting up unique conditions, more criteria and parameters for the formulation of these conditions must be gained, for example by additional sensors.

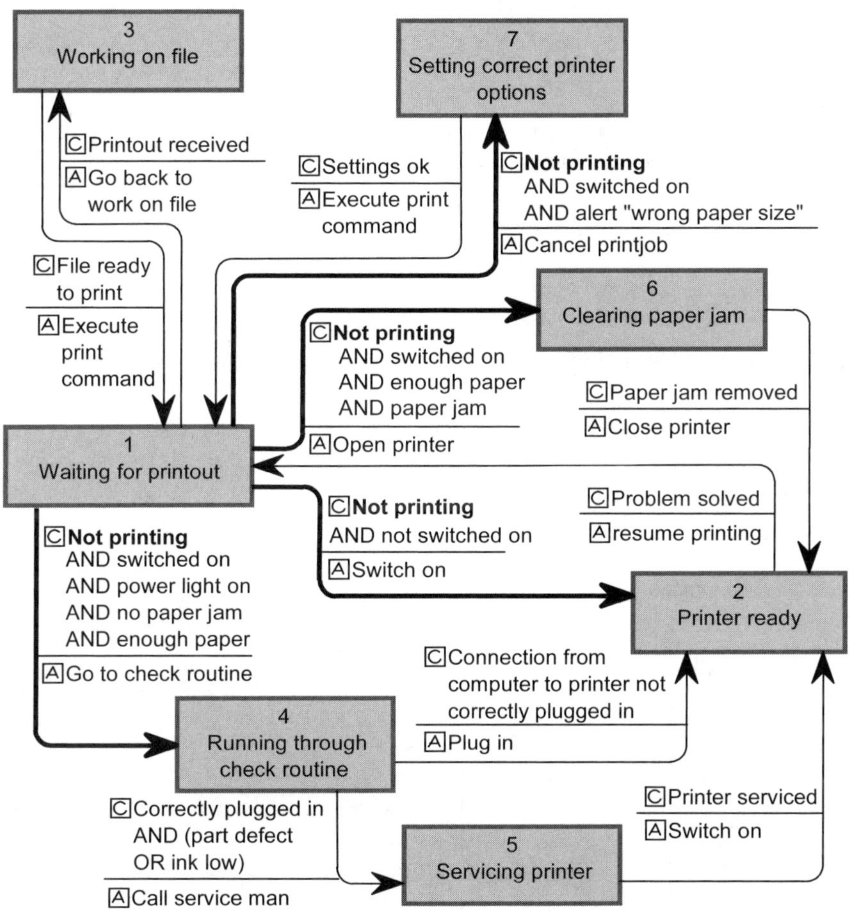

Diagram D1-6: State Chart of printer failing to print.

Diagram D1-6 gives an example for several transitions with the same condition leaving a state. The State Chart describes a printer that does not print. The states are depicted from the viewpoint of the user of the printer, meaning in what state is the user. There may be several reasons why the printer is not printing. Simply to label all the transitions leaving the state 1 "Printer not printing" with the condition "Not printing," indicated in bold, is not allowed. In this case, conditional statements must be used. The condition "Not printing" is extended by further conditional expressions. By the conditional statements, it becomes obvious that there are different reasons for the printer not working are specified in detail.

1.3.4 Rules and Examples for State Charts

In this chapter, rules for states and transitions are given in the following list, arranged by subject. The number of the rules for the State Chart is marked with the prefix SC.

Rules D1-1: State Chart

SC1	A system is always in exactly one state at a given point in time.
SC2	A State Chart contains a minimum of two states and two transitions.
SC3	Each state has a name.
SC4	Each state has a unique identification number.

For the description of a system to be complete, there must be a state to define its behavior at every point in time. Within this description, the system is always in exactly one state at a given point in time (*Rule SC1*).

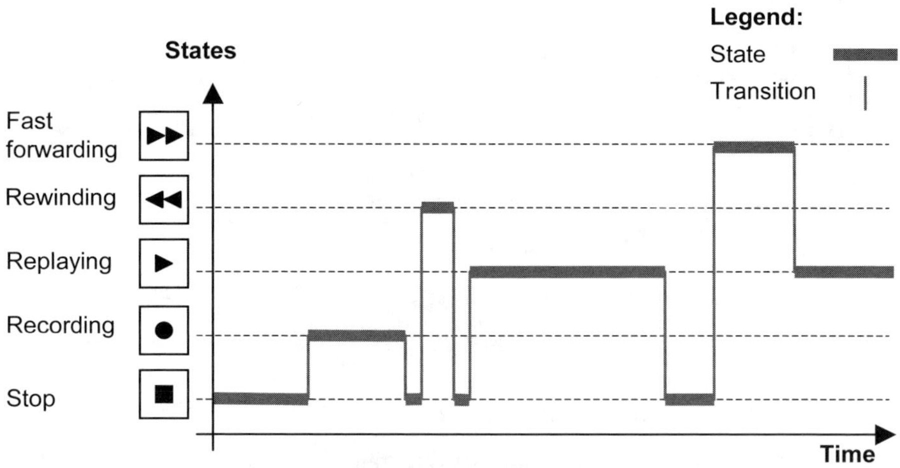

Diagram D1-7: States of a tape recorder.

Diagram D1-7 illustrates the relationship between the system's states and the development in time using the example of a tape recorder. The x-axis shows the time, and the y-axis depicts the different states of a tape recorder. The diagram clearly shows that the device is always in one and only one state, and time is passing while it is in that state. There is never a point in time when the system is not in a state. The transitions between the states are marked with thin vertical lines and are negligible in time (*Rule SC8*).

Note! Avoid confusing processes and states. There are neither flows nor processes on a State Chart. The process is named with an active verb, while the states are labeled with a verb in participle form to indicate the duration of the state.

Rules D1-2: Transition

SC5	A transition is labeled with a condition and, optionally, an action.
SC6	Every state has at least one entering and one exiting transition.
SC7	Every transition connects exactly two states, the origin state and the destination state.
SC8	Transitions are changes of state. Their duration in time is negligible.
SC9	The path, by which a system reaches a certain state, has no influence on the behavior of the system in this state. The system behavior in this state is always the same.
SC10	Every state must be able to be reached from any other state, either directly or indirectly.
SC11	Only one transition can occur between two states in a given direction.
SC12	A transition has only one condition. This condition may be a conditional statement.
SC13	A condition can trigger only one transition from a given state.
SC14	At a given point in time, only one transition can take place on a State Chart.

Rule SC5 states that the action is indicated optionally on the State Chart. Sometimes an action is self-explanatory by the name of the state into which the transition goes. In this case, experienced designers may omit the action to lean the diagram. However, it is not recommended.

A current state may be left by different transitions; each of these transitions is triggered by a different condition. By definition, only one condition may be valid at a given point in time. Therefore, whatever condition occurs first will cause the corresponding transition to occur. Different conditional statements are required to distinguish different transitions leading out of a given state.

Rule SC13 states that a condition of a transition leaving a state is unique for this state. A condition can only cause one transition. There are never two transitions with the same condition.

Example: State Chart for describing a user interface of a program

State Charts are very useful for depicting the user interface of a computer program. In this case, the State Chart is not drawn from the computer's internal view, but from the viewpoint of the user: how he or she works with the program, what he or she sees, and how he or she interacts.

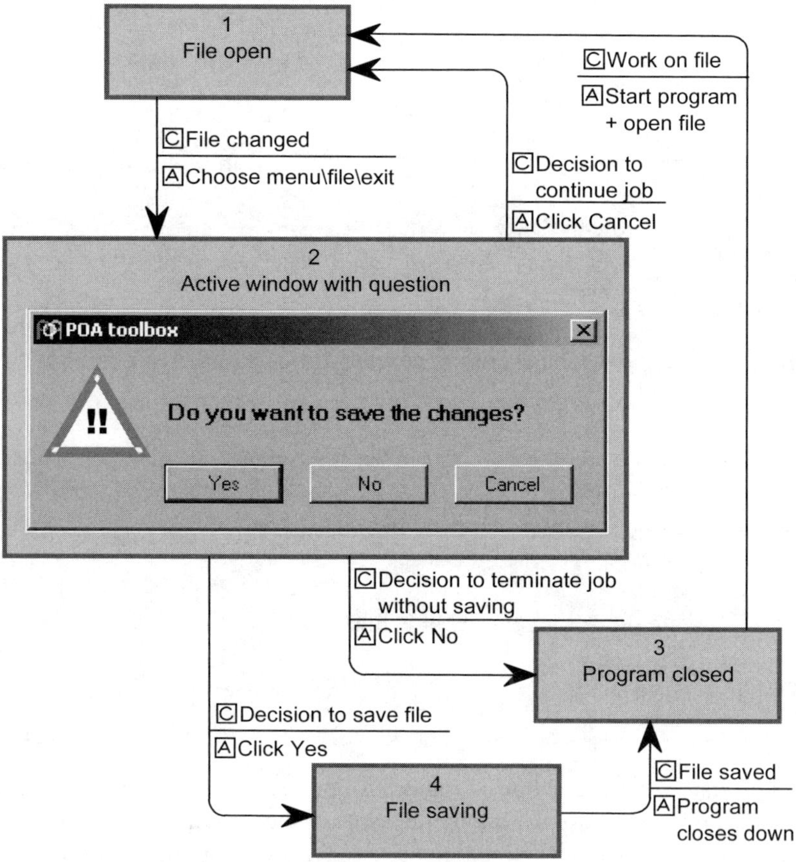

Diagram D1-8: Active window on a user interface for saving a file.

The windows, frames, and dialogues of the user interface are described by states. The active window, or mask, corresponds to the actual state of the system. The choice of a menu, a mouse click on the window, or a keyboard input by the user are all examples of actions of the user that cause a change into another state, respectively onto a new active window of the program. The conditions are given by the choices of the user of the program.

Diagram D1-8 of a State Chart model shows the masks and the activities for the saving of a file in a word processing program. The user has a "File open," while he is working on it. This is illustrated by state 1. Then, the user activates the closing of the program by the action "Choose menu\file\exit" (*Figure D1-3*). Because he made some changes on the file, which he did not save up to now, a dialogue window appears as a warning. This is shown in *Diagram D1-8* by the state 2 "Active window with question." In this dialogue, the user has the choice among three buttons. Depending on the decision of the user (condition), a certain transition is triggered. The action of the transition is the click on the button, and the system changes to a new state, that means to a new active window.

If, for example, the condition "Decision to continue job" is fulfilled first, which means, that the user wants to continue working on his file, he clicks on the "Cancel" button, and the system goes back to the window, namely the state "File open." The user can continue working on his file and make changes.

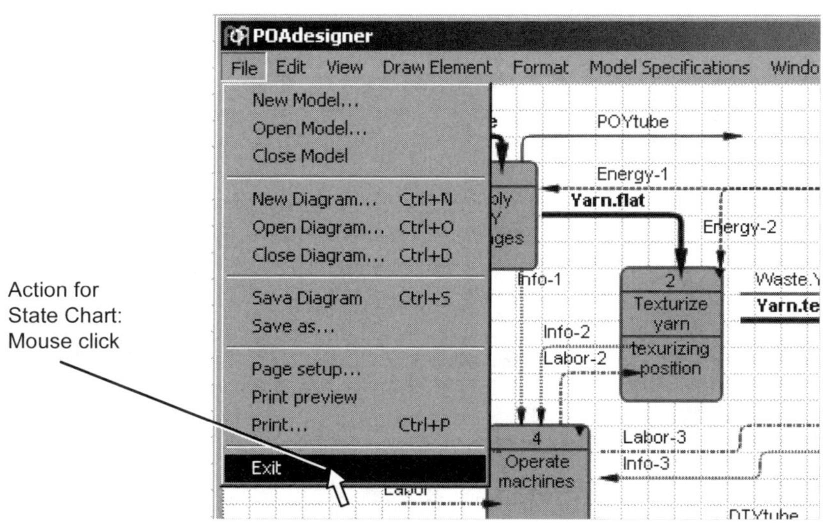

Figure D1-3: Action by a mouse click on the menu "Exit."

According to *Rule SC1,* a process, or a State Chart respectively, is in exactly one state at any given point in time. This rule also applies to the user interface of a computer program, where a state is a window: only one window is active at any given point in time on the user interface. Only the active window accepts input given by mouse or keyboard.

1.4 Model Structure

1.4.1 State Structuring in the Hierarchy

Parent state and child State Chart

Every state of a State Chart can be detailed into a lower-level State Chart. By exploding a state, a child State Chart with its own substates and transitions is created. Usually, this is done if the behavior of certain parts of a production system or a machine needs to be specified. The states of the higher levels represent aggregated states of their lower-level substates.

Simplifying the system by detailing aids in the understanding of its behavior. *Diagram D1-9* shows the structuring of State Charts in the hierarchy. Parent state 2 is detailed into a child State Chart with the two child states 2.1 and 2.2.

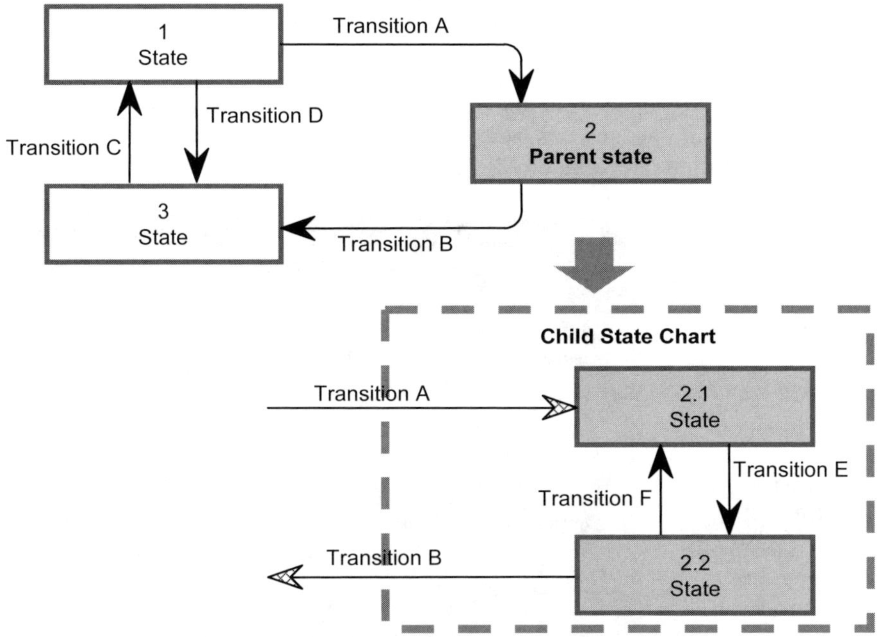

Diagram D1-9: State structuring.

The transitions connecting the parent states on the higher level go off-page on the child State Charts. The child State Chart takes the incoming and outgoing transitions of the parent state onto the lower level where they become off-page transitions. This means that on the child State Chart these off-page transitions connect to State Charts depicted on other pages. Such off-page transitions on *Diagram D1-9* are transition A and transition B that enter and exit the parent state 2 on the top level and appear as off-page transitions on the child diagram.

Numbering in the hierarchy

The numbering of states in the hierarchy is done according to the numbering of processes introduced in Section S1. The states on the child State Chart receive the number of the parent state with an additional, sequential number, separated by a period, as shown in *Diagram D1-10*. This numbering procedure ensures the unequivocal identification of each state in the model. The number of figures divided by a point indicates on which detailing level of the model the state is located. The child diagram carries the number of the parent state.

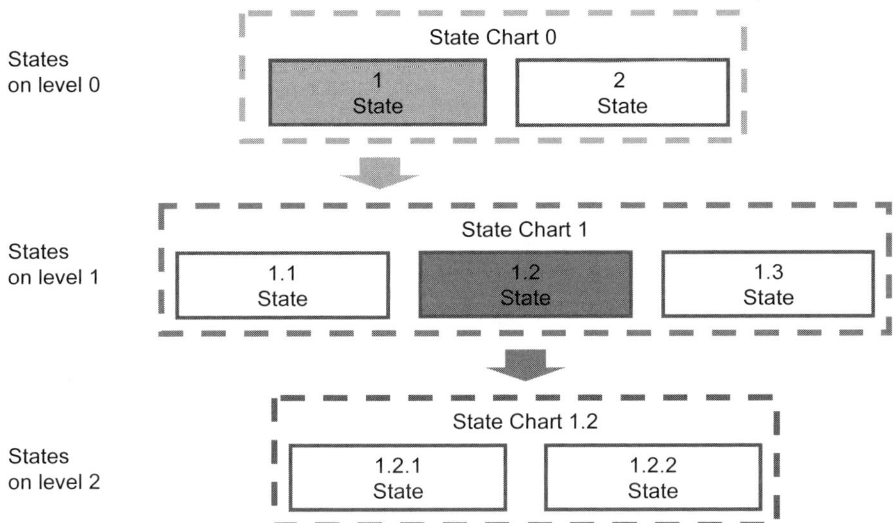

Diagram D1-10: State numbering in the hierarchy.

Regarding the content of the lower-level diagram, there is flexibility in terms of the type and number of states that are set up, as long as the basic rules are followed that each state must be able to be reached from any of the other states within the model, either directly or indirectly. This means a state can also be reached indirectly by states on the parent diagram.

The name of a state has to be unique on a State Chart. However, on another State Chart in the same model, a state name can be repeated.

Example: Canal lock for ships

As an example for the hierarchical structuring, a canal lock for ships, such as the lock of the Panama Canal, is modeled. The behavior of the canal lock is shown as a State Chart in *Diagram D1-11*. There is one state for the filling and one for the emptying of the lock while the lock gates are closed. Two states specify the time when the lock gates are open and the ships can leave and enter the lock: One is on the upper water level and the other one on the lower water level.

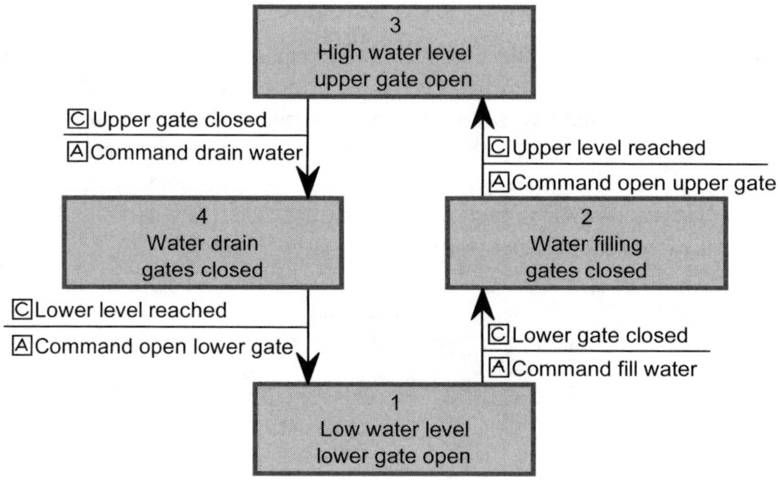

Diagram D1-11: State Chart "Canal lock."

Diagram D1-11 might be detailed by drawing a child State Chart for one of the states. *Diagram D1-12* shows the State Chart for the state 3 "High water level, upper gate open." The ships are only allowed to leave the lock when the lock gate is open. First, all ships have to leave the lock in order for new ships to enter it. All ships have to be moored securely when the lock gate closes.

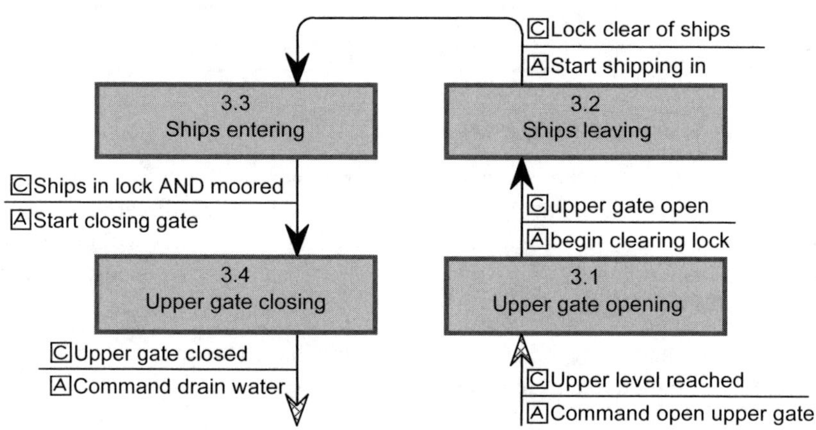

Diagram D1-12: Child State Chart "upper water level, upper gate open."

By hierarchically detailing, one sees how the incoming and outgoing transitions of the parent state become off-page transitions on the lower level State Chart. They depict connections to other states on the parent diagram or on other child diagrams. For example, the condition for the transition from state 3 to state 4 is that the upper lock gate is closed. This condition triggers the action of giving the command to start releasing the water in the canal lock, and therefore the transition to state 4. This transition becomes off-page on the child State Chart in *Diagram D1-12*.

Detailing state with conditional statement into child State Chart

A condition in form of a conditional statement using the logical operator OR indicates an ambiguous situation, which is a hierarchical not enough detailed state. Usually the destination state of the transition with this condition should be further detailed into a child State Chart. *Diagram D1-13* shows the State Chart representing the behavior of a driver noticing that oil or gasoline needs to be refilled, each calling for proceeding to the nearest gas station where the corrective actions are different.

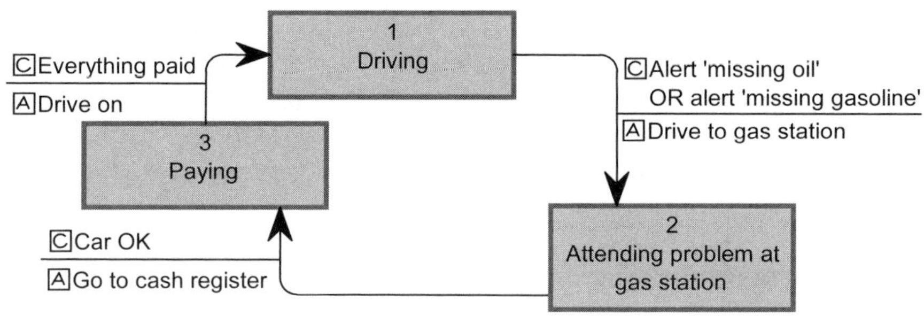

Diagram D1-13: State Chart "Drive car."

Diagram D1-14 depicts the child State Chart of state "Attending problem at gas station." The transition with the conditional statement "Alert 'missing oil' OR alert 'missing gasoline'" is detailed into two transitions on the child State Chart. Every conditional expression of this conditional statement becomes a condition of its own transition.

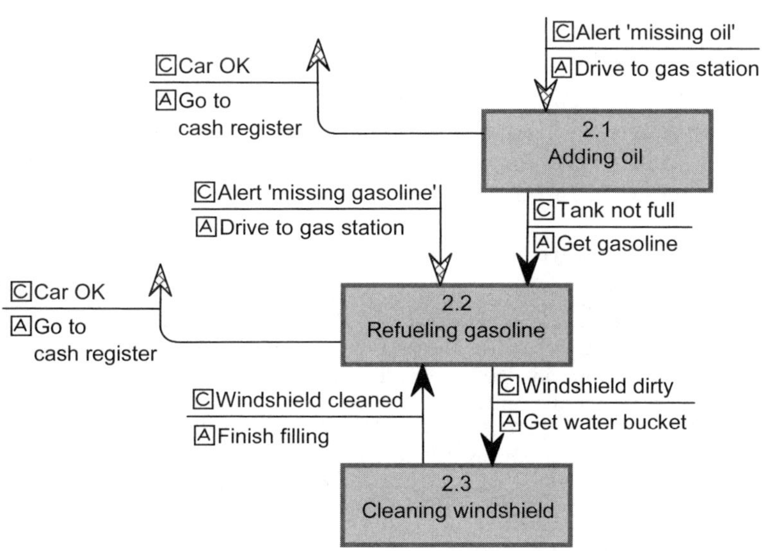

Diagram D1-14: Child State Chart "Attending problem at gas station."

Rules D1-3: State structuring

SC15	Every child State Chart has the name and the number of its parent state.
SC16	The name of a state must be unique on a State Chart.
SC17	The entering and exiting transitions of the parent state must enter and exit the child State Chart as off-page transitions.
SC18	Conditions of different off-page transitions on the child State Chart are combined by the logical operator OR to a conditional statement on the parent State Chart.
SC19	On the top level State Chart, no off-page transitions are allowed.

1.4.2 Element Specification

Every element used in the State Chart, state or transition, has a specification. This element specification contains the parameters and values that are necessary to describe the system states, and later translate it into a program. The total of all specifications in a model results in the complete description of the system. The general element specification of a state is given in *Specification D1-1*.

Specification D1-1: State specification in general

⟨State number⟩	**⟨State name⟩**
Description	⟨Text⟩

The state specification contains the description of the state. The form of the description is principally free. It can be text, table, picture, or graphic. The goal is to make the model available and understandable for as many users as possible. *Specification D1-2* defines state 2 from *Diagram D1-15* "Take a bath."

Specification D1-2: State "Getting ready for a bath"

2	**Getting ready for a bath**
Description	During the preparation for the bath, several things are checked, such as the water level in the bathtub, the temperature of the water, and/or the availability of soap.

The general transition specification is shown below (*Specification D1-3*). The transition is identified by the origin and destination state. This identification is possible, because according to *Rule SC11,* only one transition is allowed between two states in a given direction; therefore, the names or the identification numbers of the origin and destination state appear in the head of the transition specification.

Specification D1-3: Transition specification in general

Transition	from ⟨Origin state⟩	to ⟨Destination state⟩
Condition	⟨text/Boolean expression⟩	
Conditional statement	⟨text/Boolean expressions⟩	
Action	⟨text⟩	
Set of actions	⟨text⟩	

The condition and action of a transition can be simple expressions in text form. These are entered into the section "condition" and "action." The same applies for the "conditional statement" or "set of actions." They are given in the according section. On the diagram, the transition is either displayed by the complete set of conditions or actions, or only by a short label if there is not enough space. In the latter case, the short label is entered in the specification section of the condition and action and the belonging conditional statement or set of actions in the next section.

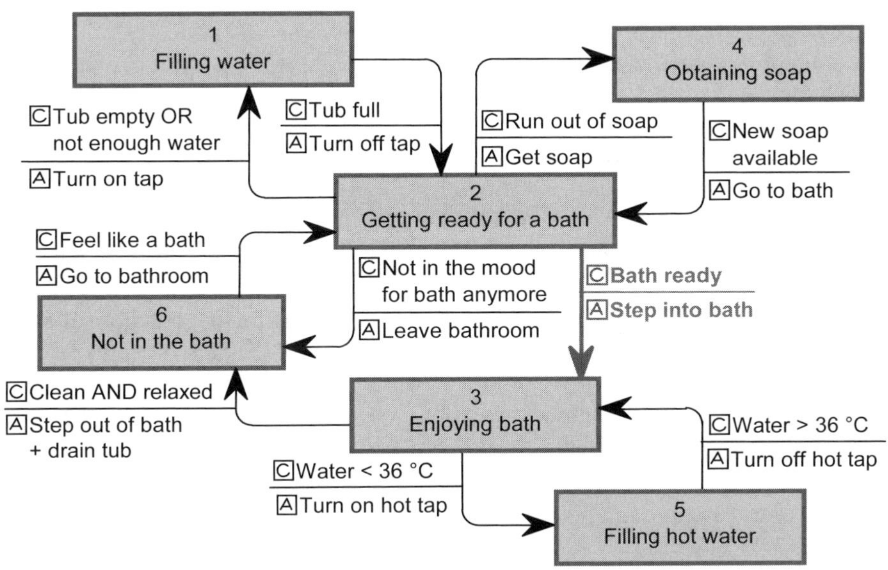

Diagram D1-15: State Chart "Take a bath" with short expression for conditional statement.

Specification D1-4 gives an example for the transition from state 2 "Getting ready for bath" to state 3 "Enjoying bath" in *Diagram D1-15*. Because of space reasons, only the short label "Bath ready" for the condition is indicated on the diagram. This is a tool to make the diagram easily readable. It avoids having large parts of text beside the symbols.

The same applies for a set of actions. In the example of the bath in *Diagram D1-15,* the action "Step into bath" is described by a set of actions in *Specification D1-4*.

Both the conditional statement and the set of action can be shortened and described in detail in the transition specification.

Specification D1-4: Transition from state 2 to state 3 of "Take a bath"

Transition	**from Getting ready for a bath to Enjoying bath**
Condition	Bath ready
Conditional statement	Water level = full AND temperature ≥ 36°C AND soap available AND towel ready AND not too late at night
Action	Step into bath
Set of actions	Undress + put towel next to the tub + step into tub

Rule SC5 says: "A transition is labeled with a condition and, optionally, an action." This still applies when the condition and action is replaced by a short expression on the diagram.

Rules D1-4: Element specification

SC20	Every state and every transition on each State Chart of the model must have an element specification.

Recommendation: The transition with short expressions for the conditional statement and the set of action is highlighted on *Diagram D1-15*. This is useful because it is not clearly visible on the diagram if a stated condition or action is a short expression or not. A highlighted transition refers to the specification where the complete expression is found.

1.4.3 Data Dictionary

As well as the diagrams and the element specifications, the data dictionary (DD) also belongs to the model. The DD contains and organizes all information of a model. The names of the elements, the information contained in their specifications, and the relationships between the elements, including the organization of the hierarchical structure, are stored in the DD. The structure of the DD for a State Chart model and the relationships within the model are shown in *Figure D1-4*.

Within the DD, the information entered in the element specification is listed in common tables for the entire model. These lists are done by hand or by a computer program. Within a CASE tool for example, the individual elements are automatically identified by the identification number and stored in lists with their relevant information.

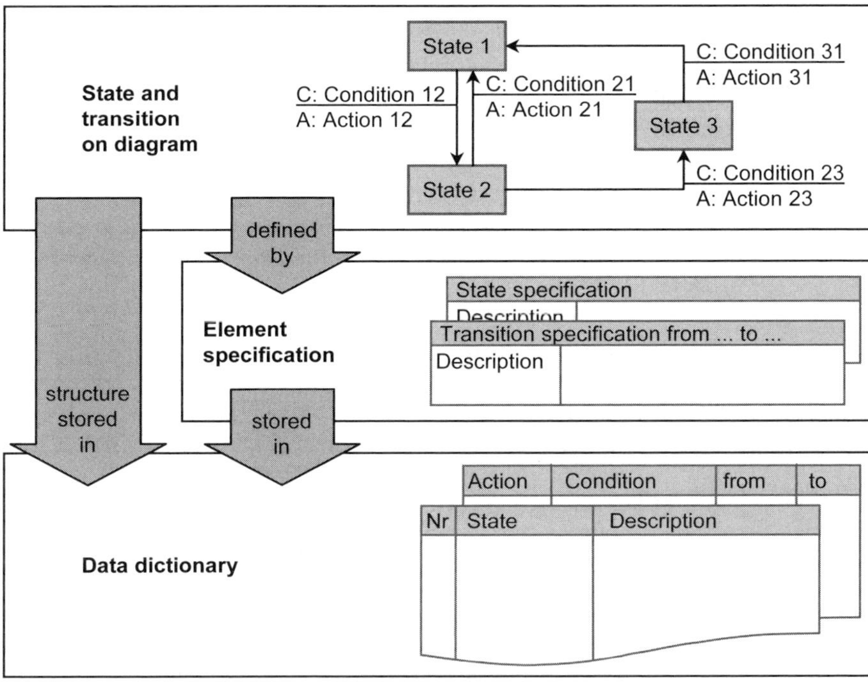

Figure D1-4: Data dictionary of State Chart model.

The data dictionary is an organized set of definitions and fulfills the following tasks:

- It provides a place where elements can be stored, retrieved, and looked up.
- It supplies clear and comprehensive information about the elements, which describe the system.
- It helps to introduce unambiguous terms and definitions and to check the consistency within the model.

For the various states, a state list is set up with at least the state number as identification and the state name. The state list for the State Chart "Take a bath" in *Diagram D1-15* is given in *Table D1-4* The basic list may be enlarged with additional columns if necessary, for example the description of the state.

Table D1-4: State list "Take a bath" of DD

No.	State	Description
1	Filling water	
2	Getting ready for a bath	During the preparation for the bath, several things are checked, such as the water level in the bathtub, the temperature of the water, or the availability of soap.

No.	State	Description
3	Enjoying bath	
4	Obtaining soap	
5	Filling hot water	
6	Not in the bath	

The transitions are compiled in a transition list with at least the origin state, the destination state, and the condition and action. The transition list of the State Chart "Take a bath" is shown in *Table D1-5*.

Table D1-5: Transition list "Take a bath" of DD

Origin state	Destination state	Condition	Action
1	2	Tub full	Turn off tap
2	1	Tub empty OR not enough water	Turn on tap
2	3	Bath ready	Step into bath
2	4	Run out of soap	Get soap
4	2	New soap available	Go to bath
3	6	Clean AND relaxed	Step out of bath + drain tub
3	5	Water < 36°C	Turn on hot tap
5	3	Water > 36°C	Turn off hot tap
6	2	Feel like a bath	Go to bathroom
2	6	Not in the mood for bath anymore	Leave bathroom

The rules for the data dictionary are given below.

Rules D1-5: Data dictionary

SC21	For every single element that appears on a State Chart of a model, an entry in the data dictionary must exist.
SC22	Access to the state must be possible by its identification number.
SC23	A transition is identified by its origin and destination state. These states give access to every transition in the data dictionary.
SC24	No redundancy is allowed in the data dictionary.
SC25	Each element of the data dictionary appears in a diagram.

1.5 From Flow Diagram to State Chart

1.5.1 Hierarchy of Flow Diagram and State Chart

The State Chart is connected to the Flow Diagram within POA. Occasionally, the State Chart is also used separately from the Flow Diagram. However, it is strongly recommended to base it on a Flow Diagram for two reasons. First, the overview of the system and the system structure is developed by a Flow Diagram and interactions of processes displayed therein. To structure a whole system by a State Chart alone without a Flow Diagram leads often to a messed-up system structure. Only if the system is well known to the modeler and not very large, this might be possible. Second, the condition and action of the transitions refer to the flows entering and leaving processes or data accessible within the process. The Flow Diagram makes the intended behavior of the system easy to understand for anybody who is accustomed to a structural view of the system, as e.g. a designer.

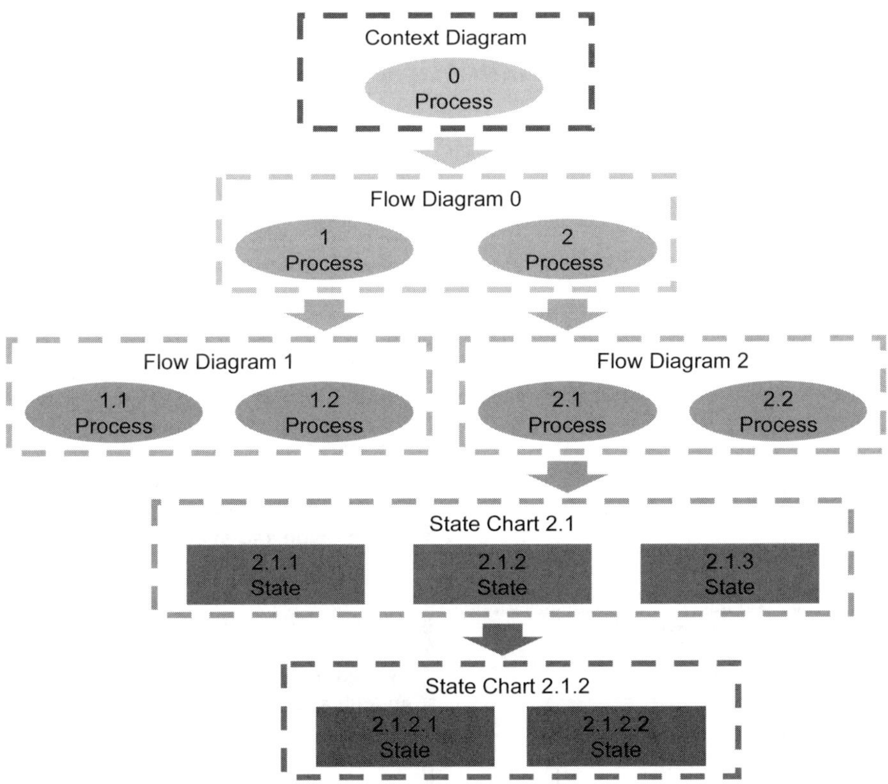

Diagram D1-16: Hierarchy of a model.

This chapter describes how the State Chart is drawn and identified according to the Flow Diagram. By following this procedure, the system will be completely specified in the static and dynamic domain.

Diagram D1-16 shows the hierarchical structure of an abstract model with Flow Diagrams and State Charts. Normally, the State Charts describe the processes on the lowest level of the system. This way, the system behavior in time is completely specified by the states of these processes. The dynamic specification of processes comprises the normal operational behavior as well as the exceptional behavior. The latter describes the system in case of trouble, risk, partial or complete breakdown.

Numbering in the hierarchy

The setup of a State Chart is closely linked to the pertaining Flow Diagram. A State Chart is built for exactly one process of a Flow Diagram. The numbering of the processes clearly indicates the hierarchical link between Flow Diagram and State Charts. The State Chart of a process retains the number of its parent process. The numbering scheme of the states starts with the number of the parent process and adds a new number behind a decimal point as separator. This numbering procedure is the same for processes and states.

A process may be exploded either into a child Flow Diagram or a child State Chart in the system hierarchy, as shown in *Diagram D1-17*.

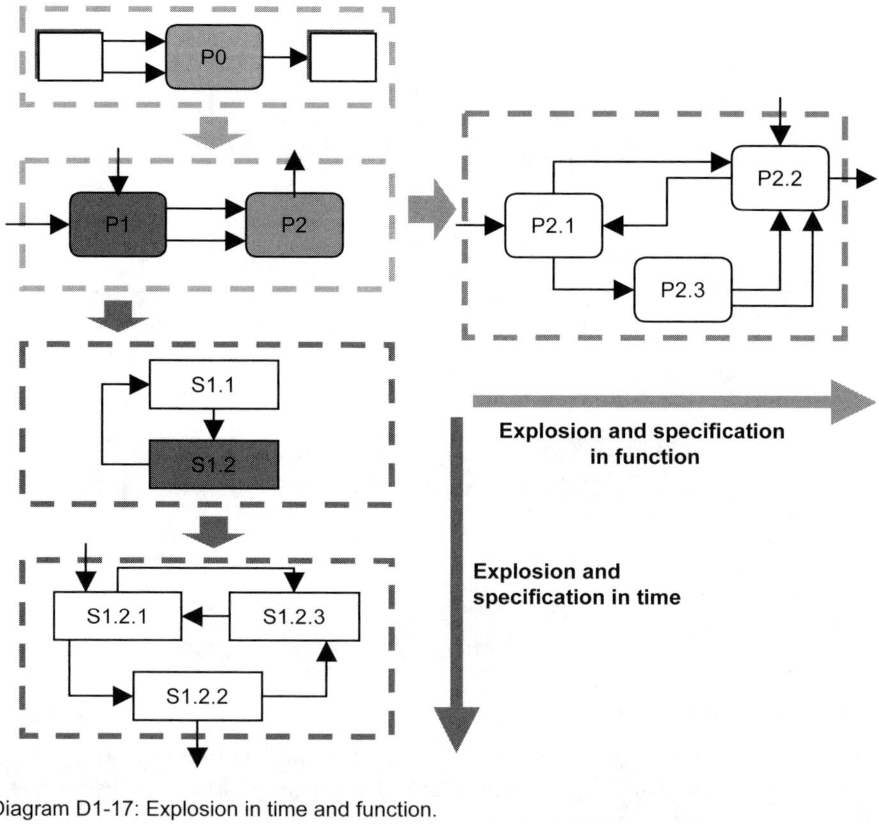

Diagram D1-17: Explosion in time and function.

To analyze the function of the system, the process is exploded into a child Flow Diagram. To analyze the time aspect of the process and to describe the behavior of the system, the process is exploded into a child State Chart. A state may only be exploded to a child State Chart.

Diagram D1-17 also shows clearly that the child State Chart of a process, which is the top State Chart in the hierarchy, has no off-page transitions. Off-page transitions are only possible on a child of a state.

Rules D1-6: Model with Flow Diagram and State Chart

SC26	Every diagram (Flow Diagram or State Chart) has the number of its parent element (process or state).
SC27	The child State Chart of a process has no off-page transitions. It is the top diagram of the State Chart hierarchy.
SC28	A process can have two child diagrams at the same time: a Flow Diagram and a State Chart.

Note! A State Chart is drawn for exactly one process. That means it is the child for a single process, not for several processes. If one tries to do a State Chart for more than one process of a Flow Diagram, it is a sign that the Flow Diagram hierarchy was taken into too much detail. The State Chart should be attached to a process on a higher level in the hierarchy.

1.5.2 Transition from Flow Diagram to State Chart

This chapter explains how to draw a State Chart based on a Flow Diagram and corresponds to the recommended procedure to build a State Chart model. Based on the State Chart a simulation model for optimization or a real-time control may be coded (see also Sections D2 and D3). Following, the transition from Flow Diagram to State Chart is illustrated in general with the example of an electrically powered machine.

Supplement Flow Diagram with further functions and flows

Beginning by a Flow Diagram, the process describing the normal operation and the material flows are drawn. Before detailing the processes into child State Charts, the Flow Diagram is first completed concerning the later purpose of the State Chart. For a real-time control, additional control processes and data and control flows are needed. The data and control flows connect the control process to the other processes.

Diagram D1-18 depicts the use of a machine, operated by a user. In order to draw the State Chart, the control process 2 “Control machine” is added to the process P1

"Drive machine." The incoming data flows, drawn in dashed lines, are commands that the user gives to the machine, such as "StartCommand" or "StopCommand," or information from the control of the machine to the user, such as "Display" or "Alarm." The control process switches the "Electricity" to the motor of the machine on and off.

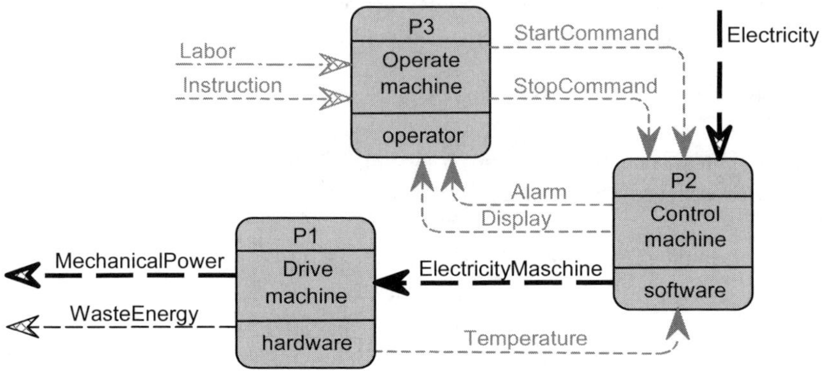

Diagram D1-18: Complete Flow Diagram.

To completely describe a material or data processing system, different types of processes are necessary to specify different aspects of the analyzed system. There are physical and control processes for the machine and operator's processes for human beings (employees, operators, and users).

Check that the Flow Diagram is complete regarding all of these aspects, especially if control processes and control flows are supplemented. Depending on the structure of the model and the type of machine, some of the functions may be combined into one process.

Specify behavior of process

In the next step, the behavior of the processes is specified by the State Chart. The State Chart for the process "Control machine" is shown in *Diagram D1-19*. It has two states: "Machine off" and "Machine on;" the two states are connected by transitions.

The following points explain how the flows from the Flow Diagram are transferred into conditions and actions on the State Chart:

- Input flows of the process may become conditions on the State Chart.
- Output flows may be turned into actions on the State Chart.
- Several input flows of one process may be combined into one condition in form of a conditional statement.
- Several output flows may also be taken together into one action to form a set of actions.

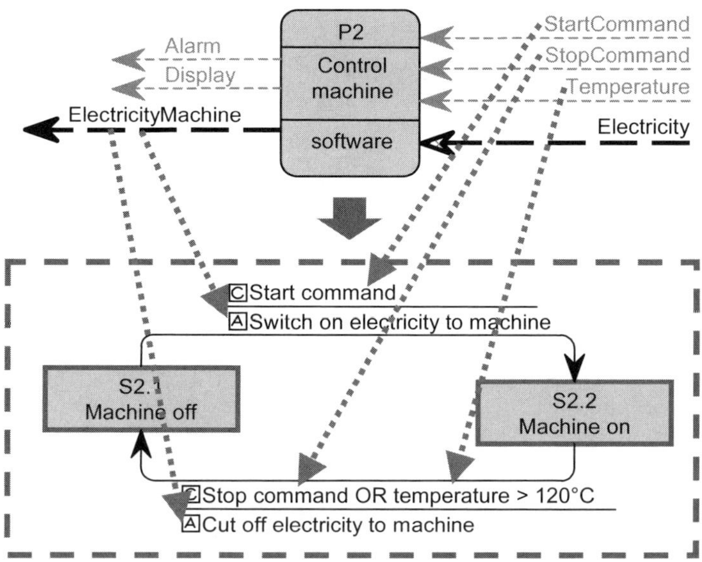

Diagram D1-19: Parent process "Control machine" and child State Chart.

The condition for the transition from the state S2.1 "Machine off" to the state S2.2 "Machine on" is the "Start command." This means that the user has to press the start button to get the machine into the powered up state. The condition on the State Chart corresponds to the input data flow of the control process "Control machine" on the Flow Diagram.

The condition for the transition from the state S2.2 "Machine on" to S2.1 "Machine off" is a conditional statement combining the "Stop command" and the "Temperature." It is a combination of two input flows.

Both actions on this diagram correspond to the output flow "ElectricityMachine," which is switched on or cut off.

The child State Chart of process P2 "Control machine" is the top State Chart in the hierarchy of State Charts and the second level of detail in the entire model. (The context diagram is not given here.) Therefore, there are no off-page transitions on this State Chart; off-page transitions only occur in child diagrams of states. The State Chart normally specifies the lowest-level processes in a Flow Diagram hierarchy.

There may be an interaction between different child State Charts of processes. The type of interaction is given by the flows that are exchanged between the processes. For example: if one process sends a signal to another process (control or data flow), this signal may initiate a certain behavior of the process. The arrival or departure of such a signal corresponds to an action or a condition of a transition on the child State Chart.

To completely describe the behavior of the system in order to design a real-time control, all three aspects of the machine or plant described by a process have to be detailed into State Charts. This is explained in detail in Section D3 of the real-time control.

The steps from Flow Diagram to a State Chart used to describe a system with the purpose of setting up a simulation program are explained in Section D2.

Rules D1-7: From process to State Chart

SC29	For each process to specify its behavior in time, there is one State Chart.
SC30	The conditions in a State Chart are defined by the flows and parameters of the pertaining process and by time.
SC31	The actions on one State Chart may serve to define for conditions used in the State Chart of another process. On the pertaining Flow Diagram, they are indicated by a flow leading from the origin process to the destination process.

In the hierarchy of the Flow Diagram, the State Chart specifies processes of the lowest levels. The State Chart is attached as a child to a process of the Flow Diagram. The amount of detailing levels and the kind of specification of the processes depend on the purpose of the analysis and the kind of system.

Recommendation: After the State Chart is drawn, go back to the Flow Diagram and update flows according to the conditions and actions that were drawn in the State Chart. Check if the naming of the flows and of the corresponding conditions and actions are recognizable as belonging together.

1.5.3 When to Begin with the State Chart in the Hierarchy

This chapter addresses the questions: How far to detail the Flow Diagram, and where to begin with the State Chart? On which level of the hierarchy within the model is it useful to introduce a State Chart? What is the parent process of a State Chart?

The level in the Flow Diagram hierarchy chosen to begin with the detailing into a State Chart depends on the investigated system and on the purpose of the investigation. Is it an investigation of a whole production line in order to discover points of improvements? This is done by setting up a simulation model. Or is it the analysis and description of a machine to be designed that needs a real-time control? In this case, the State Charts help to establish the different states of the machine that need to be covered in the real-time control.

Table D1-6 describes the scope and purpose of the system to be investigated and, within this, locates the hierarchical level of the parent process on which to start with the State Chart. Thereby, the simulation and real-time control are looked at separately.

Table D1-6: Parent process of State Chart

	Scope and purpose of investigation	**Detailing level and parent process for State Chart**		**Examples of states for the underlined parent processes**
		Scope of process for State Chart	**Examples of parent processes**	
Simulation	**Shop floor**: optimization by simulation, plant monitoring system	A production line, a workshop, a group of identical machines	Prepare weaving Weave fabric in plant Finish fabric	Production line working Production line stand still Production line in service
Simulation	**Production line**: simulation of production line (or coding a production line process controller)	A single machine or working place within a production line or production hall	Weave fabric on loom Supervise loom Change fabric	Operator at work Operator not at work Operator supervising Operator dealing with problem
Real-time control	**Machine**: designing machine with real-time control, machine controller	Single function of machine, one aspect of a machine	Carry out function of loom Control loom Operate loom	Loom de-energized Loom stand-by Loom weaving Loom with problem
Real-time control	**Machine component**: designing machine with real-time control for complex machine, individual function controller	Single functions of machine component	Display instructions Move shed Deliver warp yarn Insert weft yarn	Drive OFF Drive ON Drive with malfunction Drive in slow motion

Note! During the analysis using the Flow Diagram, the model's designer often takes the Flow Diagram one step too far in its detailing. This is normally discovered when the naming of the processes gets difficult and process names begin to sound like a state name, or when the processes are arranged into a strict sequence instead as a network with processes working concurrently.

If this level of detail is reached, go back up a level and attach to these processes a State Chart. Perhaps a control process needs to be introduced that is also detailed into a State Chart. Do not panic if you have a child Flow Diagram and a child State Chart of one and the same process. According to Chapter D1.5.1 and *Diagram D1-17,* this is perfectly possible. Often, it is even very practical for better understanding of the system.

1.6 Recommendations and Guidelines

1.6.1 Recommendations for State Charts

Procedure in the hierarchy: The hierarchical structure may be done either top-down or bottom-up. If the analysis was made top-down, the bottom-up approach should be taken to check the analysis, or vice versa. To derive the clearest possible State Chart, it may be necessary to modify and redraw it several times, rearranging states and cleaning up and rerouting transitions.

Fit on a page: A State Chart should fit comfortably on a page and should be easy to read; it should not be overcrowded. Complex State Charts with too many states (e.g. more than 10) should be avoided. The limit in number of states also depends on the number of transitions. The number of transitions is a good measure for the complexity of a system.

Complexity: A State Chart must contain a minimum of two states. If there are very few states in a child State Chart, the best way is to reorganize the parent level diagram and integrate the child states (see *Diagram D1-20*).

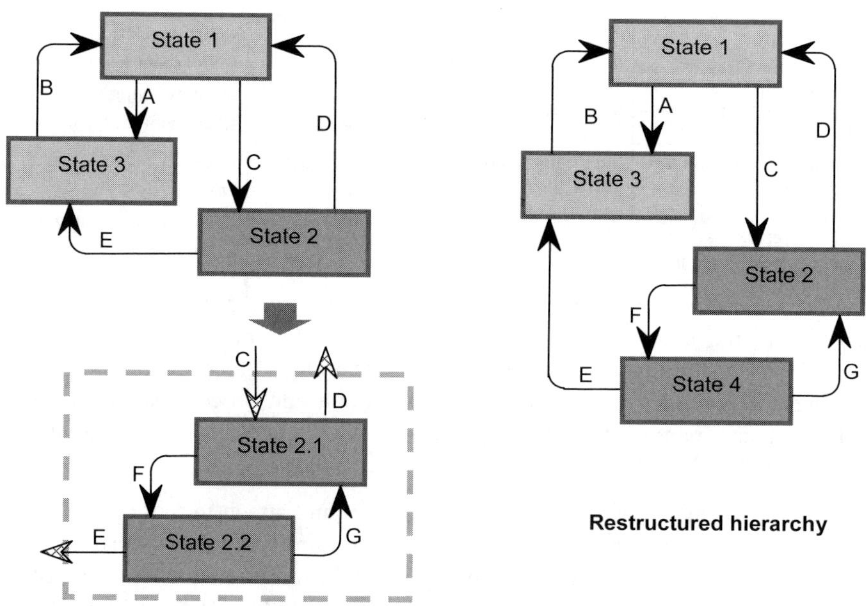

Diagram D1-20: Restructuring hierarchy for a State Chart with only a few states.

If a State Chart is very complex with many states and transitions, reconsider the hierarchical structure. There are two options:

- Several states may be grouped and condensed into one state, which is then detailed into a child State Chart.

- The process structure of the Flow Diagram may be redone. Go back to the parent process on the Flow Diagram. The parent process is refined into a child Flow Diagram. The former child State Chart is split up into several State Charts that are attached to the relevant processes of the child Flow Diagram. This leads to leaner State Charts, as shown in *Diagram D1-21*.

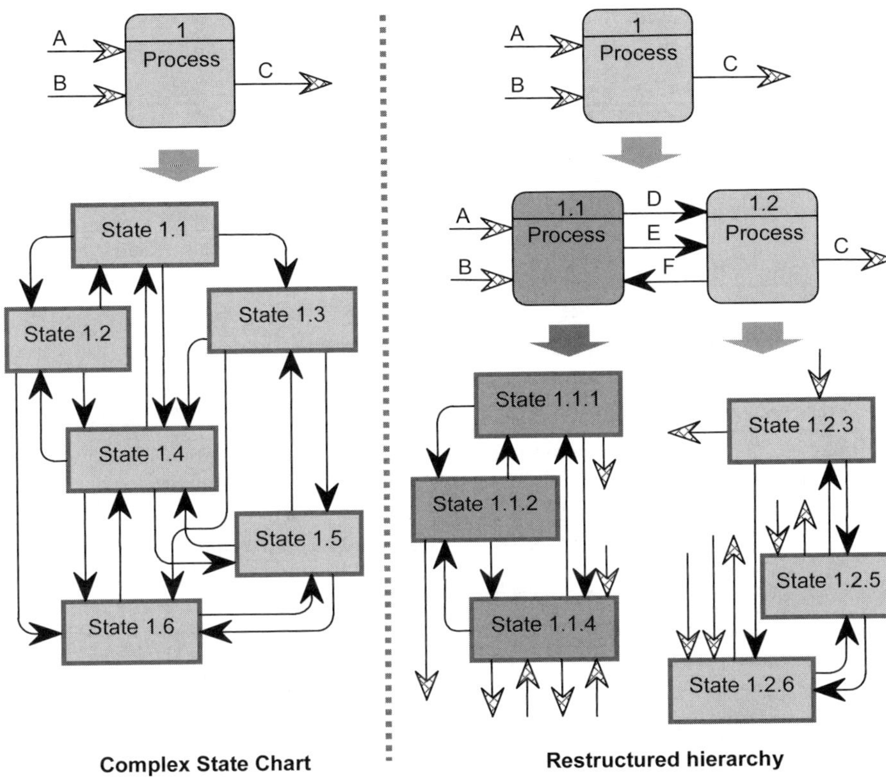

Diagram D1-21: Restructuring of hierarchy to reduce complexity.

Naming: When labeling the elements, it is important to use names that are generally understood and obvious. Preference is given to names that are familiar within the group of users of the model or designers in the project.

The labeling of an action pertaining to a transition is optional in the case where it is self-explanatory, consisting only in the change of state, or where it can be derived from the destination state. The labeling of the action is mandatory if it impacts other processes and State Charts.

Begin with the Flow Diagram: The modeling of a system may begin directly with its dynamic description by a State Chart. However, it is strongly recommended to draw first the pertaining Flow Diagram. This ensures a good structuring of the model and a clear definition of the system boundary and interfaces to its neighbors.

1.6.2 Bottom-Up Approach

When drawing a State Chart because of the reengineering of an existing plant, a real-time control, or a simulation program, the modeling is done bottom-up. It often starts with the lowest-level State Chart. In a first step, the model's designer does not bother structuring the model, but just draws a State Chart. This State Chart may become very large, with many states. In a next step of modeling, several states are condensed into parent states, and the hierarchical structure of the model emerges.

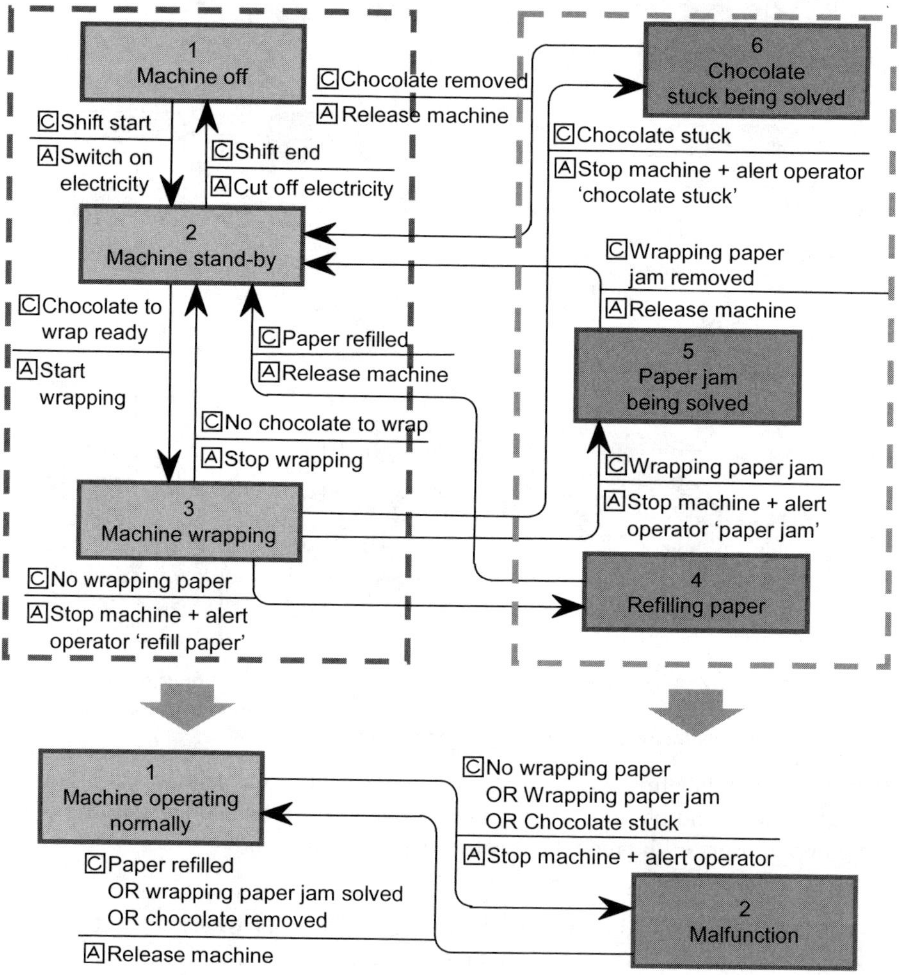

Diagram D1-22: Condensing states into parent states.

Diagram D1-22 shows the bottom-level State Chart of a chocolate wrapping machine and the condensation into the parent states on the next higher level State Chart. In this case, the normal operating states of the machine are grouped into parent state 1 “Machine operating normally” and the malfunctions into parent state 2 “Malfunction.”

When condensing several states into parent states, the following points have to be considered:

- The bottom-level states are reorganized into several bottom-level State Charts as illustrated by the bold dashed gray boxes in *Diagram D1-22*.
- The bottom-level states organized into several State Charts are renumbered according to the hierarchical numbering of the model.
- *Rule SC11* (Only one transition can occur between two states in a given direction.) must not be violated. If several transitions exist between two parent states, they are grouped together according to the following two points.
- Conditions of the different transitions are transformed into conditional statements, connected by the logical operator OR. Example from *Diagram D1-22*:

Conditions on bottom-level State Chart:	Condition on condensed parent State Chart:
No wrapping paper Wrapping paper jam Chocolate stuck	No wrapping paper OR wrapping paper jam OR chocolate stuck

- Actions of the child transition stay when ever possible the same on the parent transition. Example from *Diagram D1-22:* “Release machine.”
- If different actions become one action on the parent State Chart, they are formulated new. Example from *Diagram D1-22:* “Stop machine + alert operator.”

Diagram D1-23 shows the hierarchy of the State Chart model after the condensation of the complex State Chart took place.

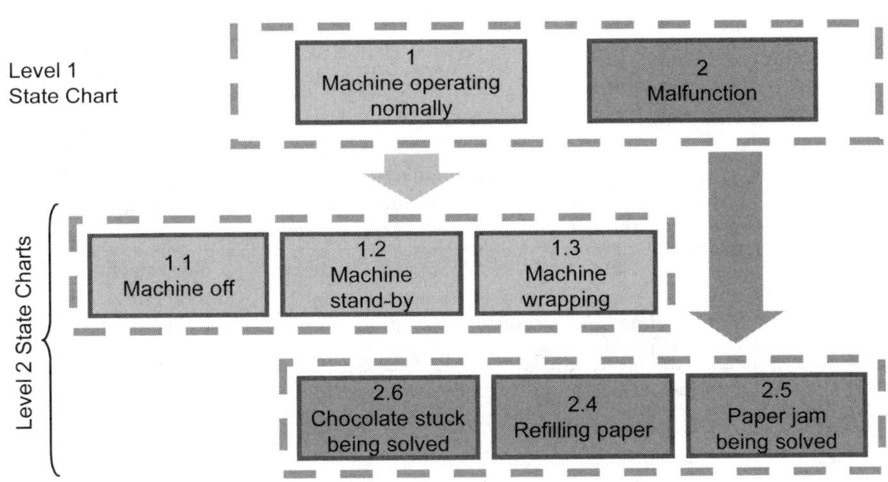

Diagram D1-23: Hierarchy of the State Chart after the condensing.

1.6.3 Components of the Model

The POA model is made for the analysis of a system and, at the same time, serves as documentation for the analyzed system. The model does not only consist of the diagrams, but also incorporates various additional components. *Table D1-7* lists the components that set up the complete model documentation. For the setup of a model in general, please refer to Chapter 1.7 in Section I1.

Table D1-7: The model components

Model component	Explanation
Model name	Like every project, every model has a proper name. It is chosen according to the investigated system and the purpose of the analysis.
Goal of the project	First, the target of the project is stated.
Statement of purpose and viewpoint of the model	It is useful to formulate the purpose of the model with regard to the goal of the project. The viewpoint specifies what the model describes and from which perspective.
Diagrams in hierarchical structure	All diagrams, Flow Diagrams as well as State Charts, from all hierarchical levels, make up the complete diagram set of a model. A list of the diagrams is included in the report as an overview.
Information text	Besides the elements process, flow, and external entity respectively state and transition, the diagrams also contain additional information text. This text includes, at a minimum, the date of the last change and the name of the diagram's designer.
Element specification	A comment and the important parameters for the behavior of the elements are entered in the element specifications. All specifications together form the complete description of the investigated system.
Data dictionary	Every element and its attributes are stored in the DD. Normally, this information is listed automatically by a computer (CASE tool).
Glossary	A standard terminology is used throughout the whole model to ensure precise communication. The fully defined terms and element names are provided in the glossary. It also includes abbreviations, acronyms, and synonyms.
Comments	The model's designer gives comments regarding the entire model. This includes what was done in the model, how to proceed, specific explanations, highlights, etc. The organization of the model might also be mentioned.
Computer program (in Section D2 and D3)	The source code of the programmed simulation model or the real-time control is part of the model description, together with the screen shots of the different program windows.
Evaluation report	A report is written in order to compile the results and recommendations of the study based on the model. It includes the achieved results, problems, discoveries, and proposals.

Purpose and viewpoint of the model: The purpose and the viewpoint for the model should be presented in a short statement. The viewpoint determines what can be "seen" within the model context and from which perspective. Depending on the model's designer and the audience, different viewpoints may be adopted to emphasize different aspects of the subject. Important things in one model may not even appear in another model describing the same system from a different viewpoint.

The statement of purpose expresses the reason why the model is created, what the intension of the investigation is, and the result expected of it. This has an impact on the structure of the diagrams. The highest-level diagram, Flow Diagram or State Chart, determines the main structure of the model. This way, the most important features come first in the hierarchy.

By exploding the top-level states, the system is structured into subsystems. These subsystems are further detailed until all the relevant details of the actual viewpoint of the system and purpose of the investigation are adequately exposed and documented.

Glossary: A standard terminology shall be used throughout the whole model to ensure precise communication. This is not easy in a large model. Normally acronyms, abbreviations, key words, or phrases are used in names. The fully defined terms are provided in the glossary. It helps to standardize the model and to keep order.

Computer program: The source code of the simulation program or the real-time control program is supplemented with comments and explanations, directly in the code, in order to make the source code understandable for other persons. The source code together with screenshots of the user interface is part of the project documentation.

If a commercial simulation tool is used for the simulation model, the created model with its settings and programmed methods and routines must be explained and commented on.

Report and evaluation: The project report summarizes the results of the whole project. The POA model is part of it. It describes what was accomplished with the model and what was found by the As-Is Analysis, what the considerations for the To-Be Model are, as well as what is still to be done, what the advantages are, where the problems lie, and what solutions are suggested for the problems.

The results and conclusions are noted together with the general conditions, and the effects on the system are discussed. The diagrams of the whole model or the most important diagrams are included in the written report to support the conclusions.

1.7 Application Example: Gas Station

Define goal and purpose

The gas station introduced in the application example of Section S1 is further analyzed from the time-dependent aspect. The owner of the gas station would like to investigate the procedure and the behavior of the drivers at the gas station. This is done by describing the system using the State Chart. This may lead to a simulation for optimization.

Supplement Flow Diagram with control flows

The Flow Diagram from Section S1 is depicted here in *Diagram D1-24*. Before drawing the State Chart, the Flow Diagram is supplemented by control flows: "Gasoline.paid," "StartFilling," and "StopFilling."

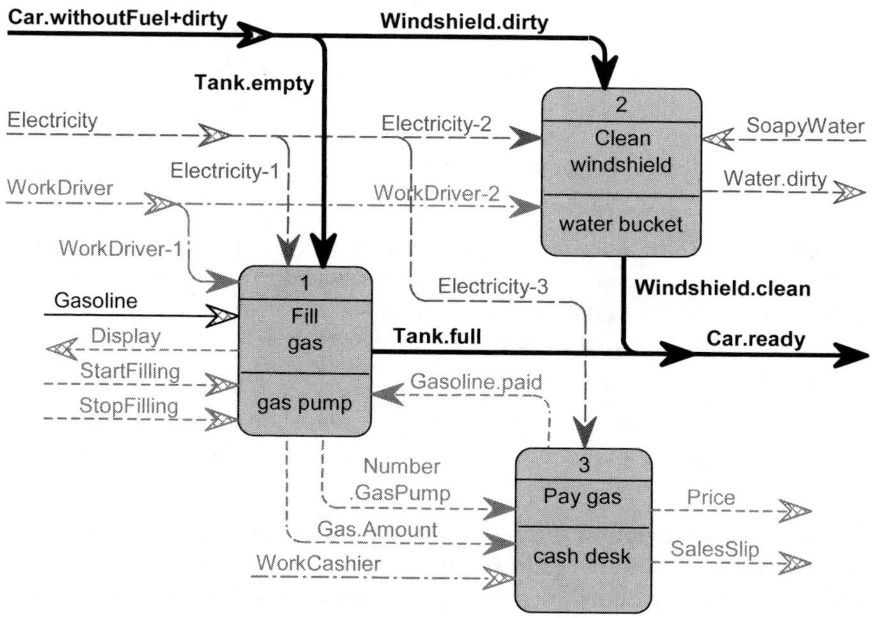

Diagram D1-24: Flow Diagram "Refuel car" at gas station.

Specify process behavior with State Chart

To analyze the behavior of the driver at the gas station, a State Chart is drawn for process 1 "Fill gas" (*Diagram D1-25*). It depicts the states that are typical for the procedure of a driver at a gas station. When arriving at the gas station, a gas pump is chosen. If no gas pump is available, the car waits in line. As soon as a gas pump is available, the car moves to the pump and operates it to fill up the car tank.

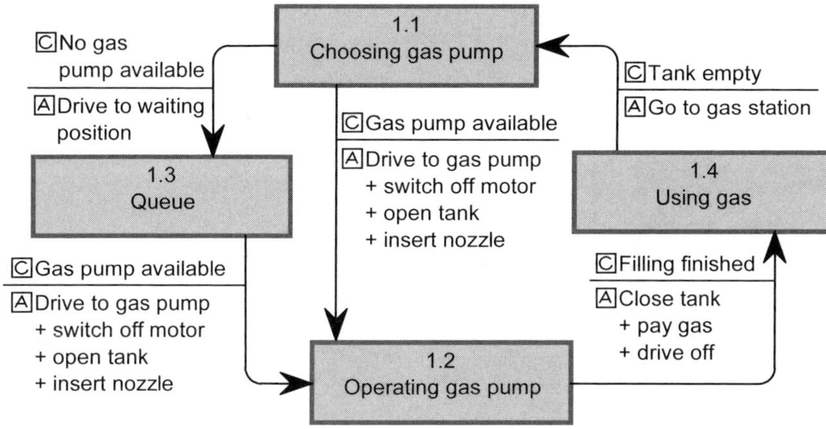

Diagram D1-25: State Chart "Fill gas."

State and transition specification

Now, for a detailed system analysis, a specification is written for every state and every transition of the State Chart. State 1.2 "Operating gas pump" is described in *Specification D1-5*.

Specification D1-5: State "Operating gas pump"

1.3	**Operating gas pump**
Description	The gas pump is operated for the refueling of the car. It has to be checked for the correct type of gasoline for the car. The display on the gas pump has to show 0 liters and 0.00 $ when starting to refuel.

Specification D1-6 shows the transition from state 1.3 to state 1.2. The action of this transition is composed in form of a set of actions.

Specification D1-6: Transition from state 1.3 to state 1.2

Transition	**from 1.3 to 1.2**
Condition	Gas pump available
Set of actions	Drive to gas pump + switch off motor + open tank + insert nozzle

In order to further investigate the behavior of the driver at the gas pump, the state 1.2 "Operating gas pump" is detailed. The child State Chart is shown in *Diagram D1-26* with the states of the car tank while filling. On the child State Chart, off-page transitions appear. These off-page transitions lead to other states on the parent

State Chart and to other child State Charts respectively, if the concerning states are detailed.

The transitions on *Diagram D1-26* show detailed sets of actions, which describe the behavior of the driver. This is useful for programming later a simulation model or a real-time control. Based on this information, the user interface of a gas pump is designed.

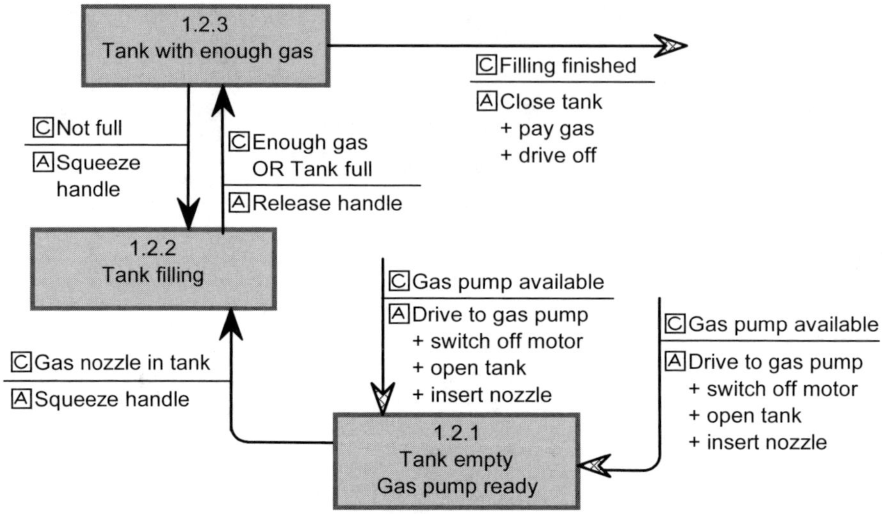

Diagram D1-26: State Chart "Operating gas pump."

Data dictionary

As the model is created, the information on the states and transitions are entered into a data dictionary. The state list of the State Charts "Fill gas" and "Operating gas pump" is given in *Table D1-8.*

Table D1-8: State list of State Chart "Operating gas pump"

Identification number	State name
1.1	Choosing gas pump
1.2	Operating gas pump
1.2.1	Tank empty, gas pump ready
1.2.2	Tank filling
1.2.3	Tank with enough gas
1.3	Queue
1.4	Using gas

The transitions of the State Chart and their conditions and actions are summarized in the transition list in *Table D1-9*.

Table D1-9: Extract from the transition list

Origin State	Destination State	Condition	Action
1.1	1.2	Gas pump available	Drive to gas pump + switch off motor + open tank + insert nozzle
1.2	1.3	No gas pump available	Drive to waiting position
1.3	1.2	Gas pump available	Drive to gas pump + switch off motor + open tank + insert nozzle
1.2.1	1.2.2	Gas nozzle in the tank	Press handle
...			

Check model

The model is checked while drawing and when finished. First, the model's designer verifies the completeness of the states. In addition to the normal working states, exceptional and de-energized states need to be introduced in *Diagram D1-22*. Examples for an exceptional state are the spilling of gas or the running out of gas, among others. The de-energized state for the gas station is a secure state, in which the system can change when a problem occurs. These states are supplemented in the application example of Sections D2 and D3.

For missing connections between the states on the State Chart, transitions must be filled in. However, only one transition between a pair of states in a given direction is allowed. Every transition has to be specified by a condition and, optionally, by an action. A transition takes place momentarily and without delay. If a transition cannot be regarded as timeless, intermediate states must be filled in.

Table D1-10 provides a checklist for the verification of completeness and correctness of the State Chart model. This general checklist can be used for any kind of State Chart.

Table D1-10: Checklist State Chart

Question	Checked
A State Chart describes each process for which the behavior needs to be analyzed for the investigation.	✓
The State Chart has at least two states and two transitions.	✓

The state names reflect the meaning of the represented states. The form of the name indicates that the state lasts some time.	✓
The system has an intrinsically safe or de-energized state, into which it can be brought anytime.	not existing
The states are sufficiently detailed in the hierarchy.	✓
All transitions are labeled by a condition and an action.	✓
There are no unconnected states or transitions.	✓
There is only one transition from one state to another in the same direction.	✓
Every state has at least one entering and one leaving transition.	✓
The top-level State Chart or the child State Chart of a process has no off-page transitions.	✓
Each transition of a specific state has a unique, exclusive set of conditions.	✓
A transition takes place momentarily, which means it is timeless.	✓
Every element, state and transition, in the model has a specification.	✓
Every element has an entry in the data dictionary.	✓

Program coding

The behavior of a driver at a gas station was analyzed by the State Chart. This dynamic model can be used for a later simulation model or the programming of a real-time control, as listed in *Table D1-11*.

Table D1-11: Next steps with POA

Aspect of investigation and optimization	**POA method**
Assessment of sustainability	Based on the State Chart, a system can be inspected concerning its sustainability state. Benchmark parameters are introduced to assess the performance of the system.
Analysis of labor and work place	Using the State Chart, the operating states of production lines can be analyzed and checked concerning the - possibility for automation.
Optimization simulation	The State Chart can be directly transferred into program code and serves as a starting point for the simulation model of a production system in order to do an optimization. This procedure is described in Section D2.
Real-time control	The State Chart is also the starting point for the programming of a real-time control. This can be for a single machine or a whole production plant. The procedure to design a real-time control is explained in Section D3.

1.8 Apply Your Knowledge

1) Elements of Flow Diagram and State Chart

What are the elements of the Flow Diagram and the ones of the State Chart? Give the rules of the naming of those elements.

2) Naming of states

The State Chart in *Diagram D1-27* describes a cornflakes wrapping machine. However, the states are not cleverly labeled. Find correct names for these states. The purpose of this State Chart is to program a real-time control based on it. Therefore, the states are given from the point of view of the machine.

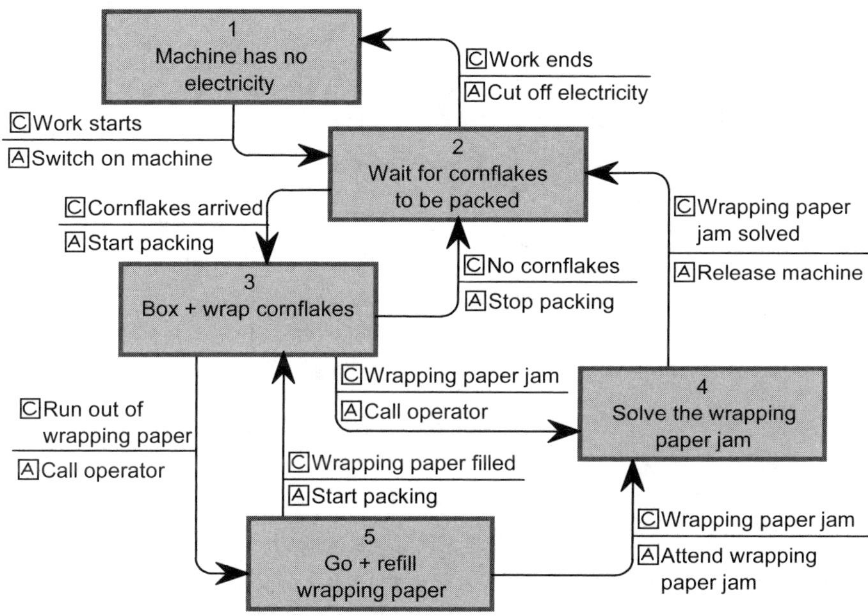

Diagram D1-27: State Chart of a wrapping machine for cornflakes.

3) Basic states of a system

There are five types of basic states for a technical system. List them and give one or more examples for each type of state for: a motorboat, an elevator, a cable car, and a lawn mower.

4) Draw State Chart

Draw the State Chart of a power drill, as it is used by everybody for do-it-yourself jobs. It is helpful to draw first a Flow Diagram to set the boundary of the investigated system. This Flow Diagram probably consists only of one process.

5) Mistakes in State Chart

List the mistakes in the following State Chart (*Diagram D1-28*) of a coffee maker. It is the highest level State Chart in the hierarchy. Explain, which rules (*Rules SC1 to SC27*) are violated, and give the correct solution.

The coffee maker here is a filter machine. You fill a filter by hand with coffee powder, then put the filter into the machine and turn on the brewing. The machine heats the water and lets it run through the filter into the coffee pot. The brewed coffee is kept hot on a heating plate.

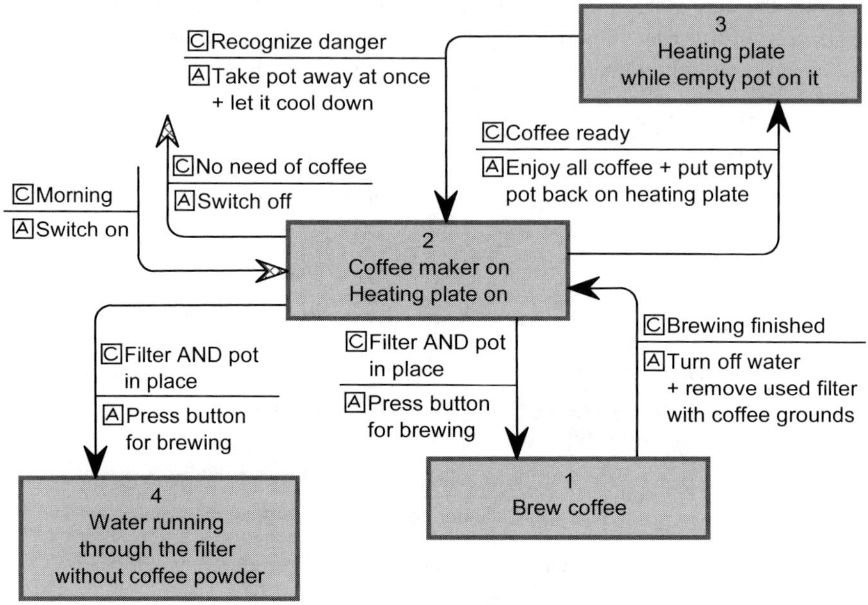

Diagram D1-28: State Chart of a coffee maker.

While the coffee maker is on (stand-by state), the heating plate is on too. As long as there is no coffee pot on it, it does not matter. But the heating plate does not automatically turn off if the pot is empty. It is not a highly sophisticated machine. A sensor for this situation would be convenient.

6) Transitions

Complete the State Chart in *Diagram D1-29*. So far only the states are drawn. The State Chart describes the behavior of a revolving entrance door of a bank building. Supplement the missing transition. Label the transitions with actions and conditions. One transition is given as an example.

As example for an exceptional state, the state "Fire alarm, Building open" was built into the State Chart. What happens to the door if a bank robbery takes place? Try to imagine and put in the according transition with its condition and action.

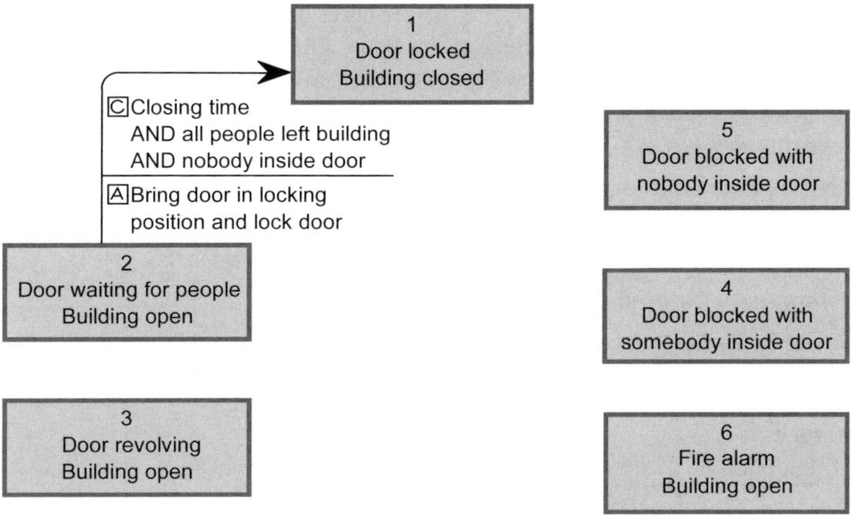

Diagram D1-29: State Chart of a revolving door.

7) Child State Chart

The State Chart in *Diagram D1-30* describes the states for attending a football match in a stadium. The conditional statements going from state 2 to state 1 or from state 3 to state 4 are composed of conditional expressions combined by OR. This indicates that the State Chart should be further detailed.

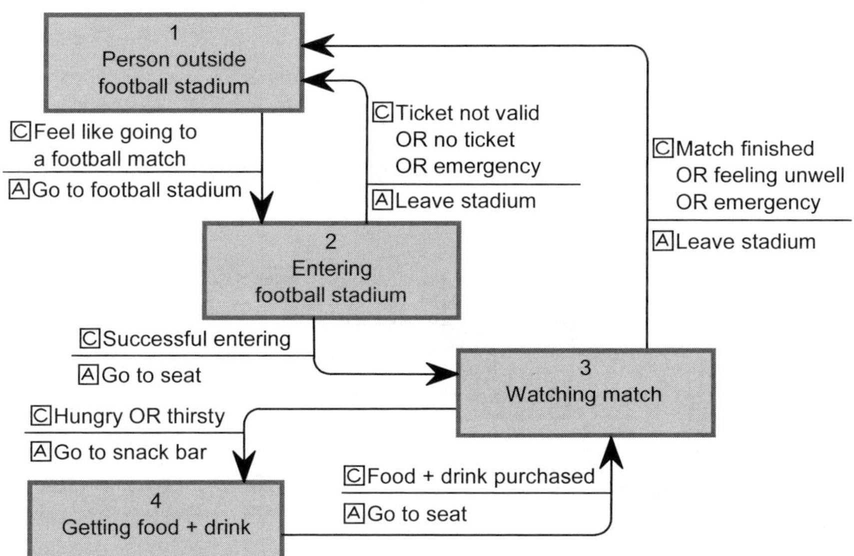

Diagram D1-30: State Chart of the entering of a football stadium.

Draw the child State Charts of state 2 "Entering football stadium" or state 4 "Getting food + drink." For the detailing of the state "Entering football stadium," think about what security check you have to go through. What happens if the ticket is not valid? Or you could also include the aspect of buying a ticket. And there are a lot of people in a stadium who want to enter or get something to drink or eat.

What happens to the conditional statements? How do the conditions of the off-page transitions look on the child State Charts?

8) From Flow Diagram to State Chart

In *Diagram D1-31,* process "Control iron" is detailed into a State Chart. Complete the State Chart by the actions and conditions of the transitions. Which flows from the Flow Diagram are transferred in what transitions?

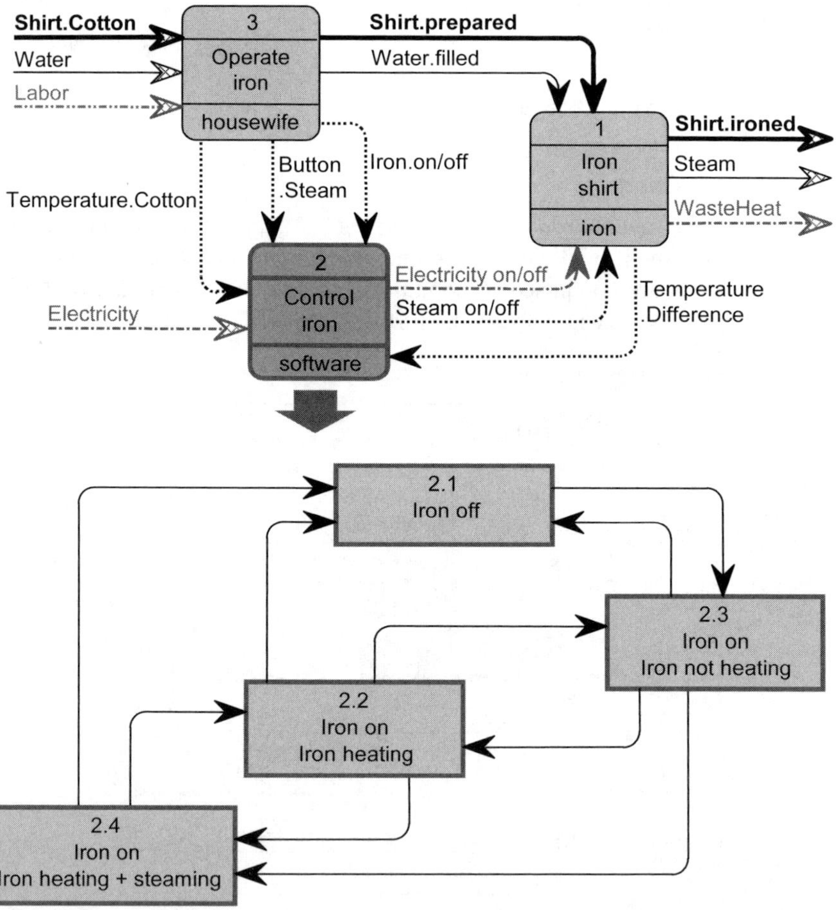

Diagram D1-31: Flow Diagram and child State Chart of the control of an iron.

The housewife chooses the temperature according to the type of fabric. In this case, the shirt is made of cotton. The iron heats up until it reaches the set temperature for cotton. Then the heating is automatically turned off. The temperature drops. When it reaches a certain minimum threshold, as programmed for cotton, it starts heating again. That means an iron does not hold an exact temperature. It cycles around the temperature as set.

The iron has also the option of steaming. To get steam, the housewife presses the steam button. As soon as she releases the button, the steaming ends.

When writing the actions and conditions, remember that several input flows may be combined to a set of actions and several output flows to a conditional statement. A flow can also be used more than once as an action or a condition. For example, the iron can be turned off from any state "Iron on." Therefore, the switch off electricity is an action from any of these states to the state "Iron off," possibly combined with other actions to a set of actions.

9) Conditional statement

In the State Chart in *Diagram D1-32,* the transition from state 2 to state 1 is labeled with the short expression "Parking position" for the condition. What does the conditional statement look like in detail? Write it down as you would write it in the transition specification.

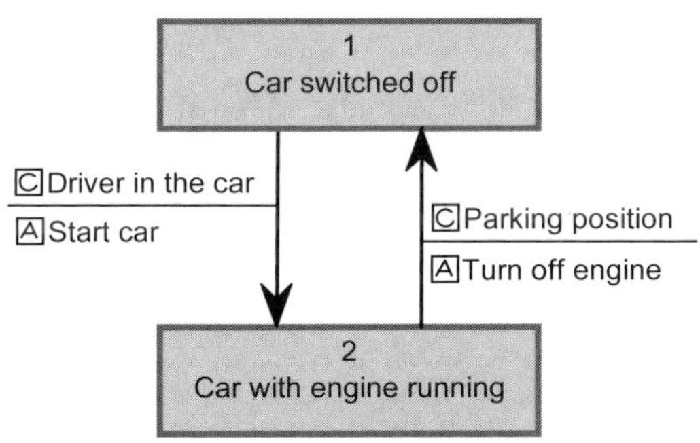

Diagram D1-32: Conditional statement and set of actions.

The transition from state 1 to state 2 is labeled with the action "Start car" that is a set of actions. Write the detailed set of action in the transition specification. Consider what the driver has to do after he entered the car, until he has the motor running and can drive off.

10) Petri Net versus State Chart

For those who are familiar with Petri Nets: Turn the Petri Net of a guest ordering and eating a meal in a restaurant from *Diagram D1-33* into a State Chart. The elements of the Petri Net are:

- Circle: place.
- Square: transition or event.
- Arrow: Arc, connection between place and transition.
- Black little circle: Token or marker.

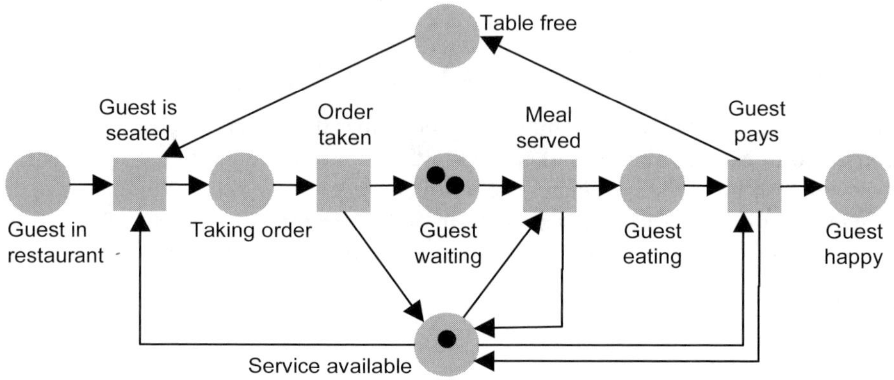

Diagram D1-33: Petri Net of a guest ordering a meal in a restaurant.

Can you see some regularity for translating a Petri Net into a State Chart? Which element of the Petri Net becomes which element of the State Chart?

D2 SIMULATION MODEL

Goals of the Chapter

- Make the connection between the static Flow Diagram and the dynamic State Chart.
- See the principle of design and coding of a simulation model.
- Get familiar with the program modules of a simulation program.
- Code a simulation model based on the State Chart.

Figure D2-1: Industrial textile storage system (Photo: Meiko Meier AG).

2.1 Introduction

Simulations were originally used to test machine controllers before exposing them to the challenge of real-world operation. Now, simulation is applied in a wide range of tasks in conception, design, and specification of manufacturing systems including order handling and management of supply chains. Ultimately, simulation is used for concurrent engineering and design of future manufacturing processes and equipment, to reduce the risk of development before full scale implementation.

In fact, the decision to simulate a system in the phase of its conception and specification is governed by cost and revenue, both in time and money. Most applications of simulation are related to troubleshooting and optimization of existing systems in manufacturing and logistics productions. Here, it is easier to establish the cost efficiency of an analysis tool, since the expected benefit is given by the situation.

A key purpose of POA is to make simulation a more common engineering tool by cutting back the time required to set it up and to lower the threshold of specific knowledge necessary to implement and use it.

An automated manufacturing plant involves more than a hundred active elements, such as motors, valves, pistons, heaters, and a similar number of sensor devices, in addition to the operators. Systems of this size are often too large and too complex for the well-known simulation packages, or they contain functional elements that are not provided therein. For such complex systems, experience shows that it is more efficient to set up a customized POA simulation model than using one of the special simulation programs. These are mainly suited for modeling material flow and jobs of limited complexity.

The basic two methods of POA, the static and dynamic view of the system, are connected with each other. This is indispensable for the setup of a simulation model. POA enforces the consistent use of the Flow Diagram as a basis for any static or dynamic evaluation. In the Flow Diagram the processes and flows are introduced and arranged. The State Charts pertaining to the processes are the graphical basis of the simulation model. This facilitates the communication between the modeler and programmer. Following this procedure of POA, the modules of the source code are always connected to the real system modules. This simplifies error handling and maintenance of the program.

Section D2 introduces the step-by-step procedure from Flow Diagram to State Chart to program code. The design and setup of a simulation model is shown. The application example of a gas station explains the procedure from Flow Diagram to the finished code, and, at the same time, gives the design of a simulation model code. A further step from simulation model to real-time programming is easy, as shown in Section D3. Case Study C5 treats the simulation of the workflow in an industrial production.

2.2 Simulation Model: Why?

2.2.1 Purpose

Check manufacturing processes: The purpose of using a simulation model for a manufacturing plant is two-fold. First, a simulation model is a powerful method to check new manufacturing processes in their conceptual phase and to optimize and debug existing manufacturing systems. Second, a simulation model serves to establish the performance of investments in upgrading manufacturing plants. Typical targets are cutting back manpower or waste, streamlining and debottlenecking production, or evaluation new processes and machinery.

Consistency with environment: By programming a simulation model with POA, based on the Flow Diagram and the State Chart, the consistency of the program with the simulated environment is enforced. The structured drawing of the Flow Diagrams of the process or system and the setting-up of a State Chart allows programming the code in a fast and comprehensive way. Because Flow Diagram and State Chart serve at the same time as the documentation of the model, it makes it easier for other persons than the programmer to overview, maintain, adjust and update the code.

Optimization simulation and test simulation: There are two types of simulations treated in this book: the optimization simulation and the test simulation as a development tool of the real-time program. The procedure from State Chart to code is both applicable for simulation models (optimization simulation) as well as for real-time programs.

In the case of a real-time program for a machine or production plant controller, the real-time program itself is made along by a test simulation. This model serves to check the code based on a simulated environment. After testing, the simulated environment is replaced by sensors and actuators, and the core of the simulation model develops into a real-time control program.

2.2.2 Application

Machine or process simulation

A POA simulation model is used for different purposes. For example, a simulation model of a machine or a production plant may be used to check the behavior and interfaces of a new machine, or the consequences of the replacement of old machines with new machines. Such a simulation model project is presented in the Case Study C5. This simulation model is created for the analysis and optimization of the workflow of an industrial production based on a much faster machine that is still in development. In order to decide on options to include in the concept of the new machine, especially if some functions should be automated or not, the simulation was set up.

Optimization of a plant

The POA simulation model is used to optimize production plants, e.g. improve the behavior of a system with a big number of independent elements. *Figure D2-1* on the first page of Section D2 shows an apparel distribution and warehouse shop. The logistics and setup of such a system are typically simulated first, before designing and building the warehouse. Computer simulation helps in specifying the tracks and switches, deciding on the type of conveyor, conceiving the software for the operation of the warehouse, and purchasing the whole equipment.

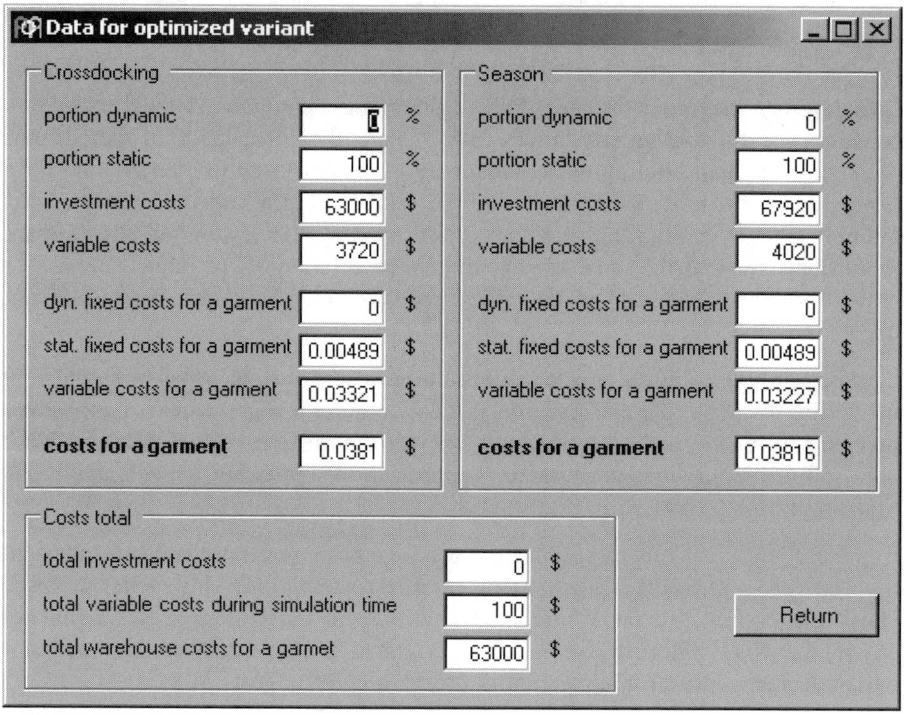

Figure D2-2: User interface of apparel warehouse.

The parameters, such as lot size, cost, stock level, and conveyor speed, of the warehouse, are optimized. The user interface in *Figure D2-2* shows the simulation result, focused on costs and efficiency. In this case, the State Chart prepares and supports programming of the simulation model so that it corresponds best to the real-world processes and therefore allows an accurate logistical setup.

The simulation model will determine the optimal way to store the maximum amount of clothes in the minimum amount of space, the fastest way to recollect these clothes again in order to compile the shipment to retailers, the number of inlet and outlet stations, the length of the tracks of the automatic conveyors, and the number of operators necessary to run the warehouse.

2.2.3 Delimitation

Type of simulation

Continuous, discrete, and static system behavior are treated by different methods. *Figure D2-3* shows several analysis methods, whereas the dark shaded sections are covered by POA. These are mainly discrete and static system descriptions. The Flow Diagram is applied for static system behavior, the State Chart for the dynamic system behavior.

POA is mainly used for the elaboration of discrete simulations as shown in *Figure D2-3*. The application on continuous systems is also possible if the concerning functions are suitably discretized. This issue is not treated in this book. The diagrams, which constitute the POA simulation model, pertain only to static and dynamic discrete systems.

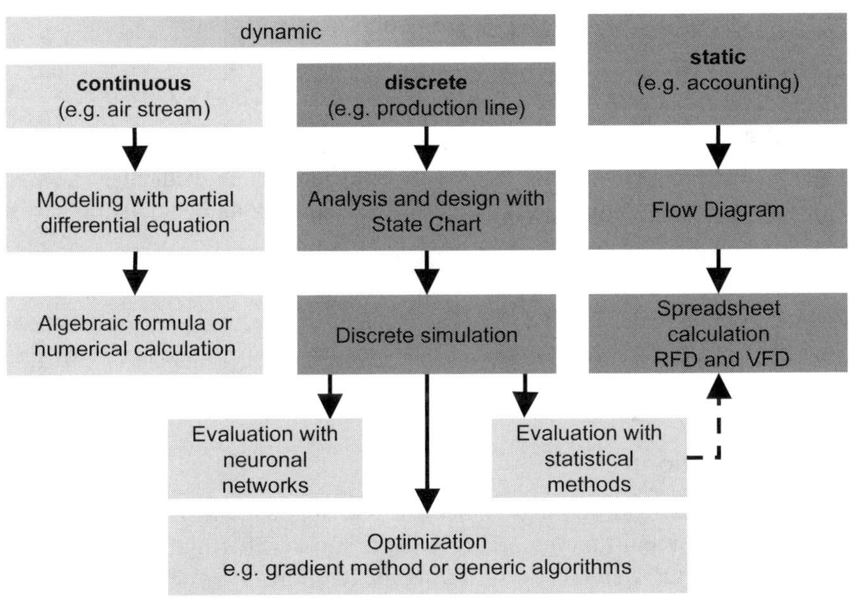

Figure D2-3: Tools for continuous, discrete, and static system behavior.

Programming with support of POA versus simulation package

POA is a method for analysis and design. The resulting simulation model may be coded by any common programming language, the code being derived from the POA diagrams. POA provides the structure and the variables, as well as the flow of statements of the code. Alternatively, a current simulation package may be used to set up the simulation. In this case, POA is used first to check the feasibility of the proposed simulation package for the job, second to provide the system conception by defining the basis of structure, terms and specifications needed for setting up the simulation model with such a package, and third to document the whole project.

Various of such simulation tools are available that offer predefined modules for convenience of programming simulations. These modules are represented by icons as shown in *Figure D2-4*. The icons are assembled to build up the simulation model. Examples of simulation packages commercially available are SimFactory, ProModel, Arena, eM-Plant. These programs are useful for systems with few parameters and a limited number of independent processes. They are typically applied for problems in logistics, as e.g. waiting queues. On larger and more complex systems, these programs hit their limits regarding the number of manageable processes and functions. Real-world manufacturing systems often consist of hundreds of processes and flow, so that programming by setting and connecting icons is simply no longer feasible.

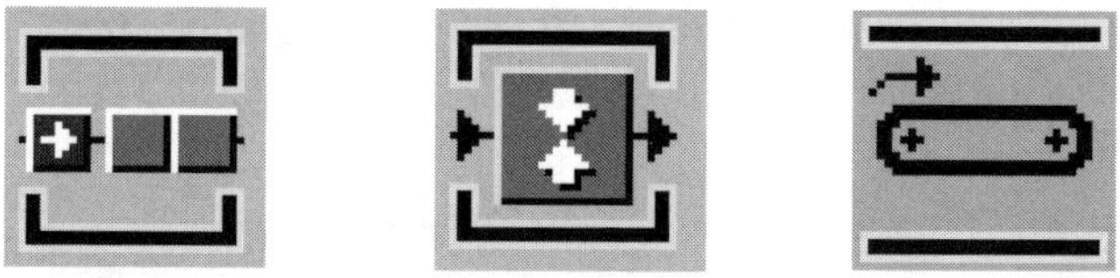

Figure D2-4: Icons for buffer, assembly station, and conveyor belt from eM-Plant.

POA offers a method to simulate the behavior of manufacturing lines, machinery, and machine components with a large number of interacting elements. The size of the system to treat is limited only by the storage capacity and calculating performance of the computer.

Following to the simulation job, parts of the program may be further used to control a machine or plant in real time. This way, analysis, design, and parts of the code are again useful for the real-world system. This is a distinct advantage in cost of system development and design.

2.2.4 Definitions

A simulation model can be programmed for two applications, either for a simulation model or for a real-time program. In the first case, an optimization simulation is set up, in the latter, a simulator for testing the real-time program.

The **optimization simulation** is done by a program that serves to plan, calculate, streamline, debottleneck, improve, and optimize an entire production line or plant.

The optimization simulation is treated in this Section D2. The test simulation is the subject of Section D3.

The **test simulation** aims to test the program and to check the real-time control in all its functions before using it in reality on the machine. In addition, such a simulation can be used to train the operator of the machine.

In *Figure D2-5*, the difference between the optimization and test simulation is given. The starting point of the simulation setup is the same for both types: the context diagram of the investigated system. In the next step, the system and the environment are simulated.

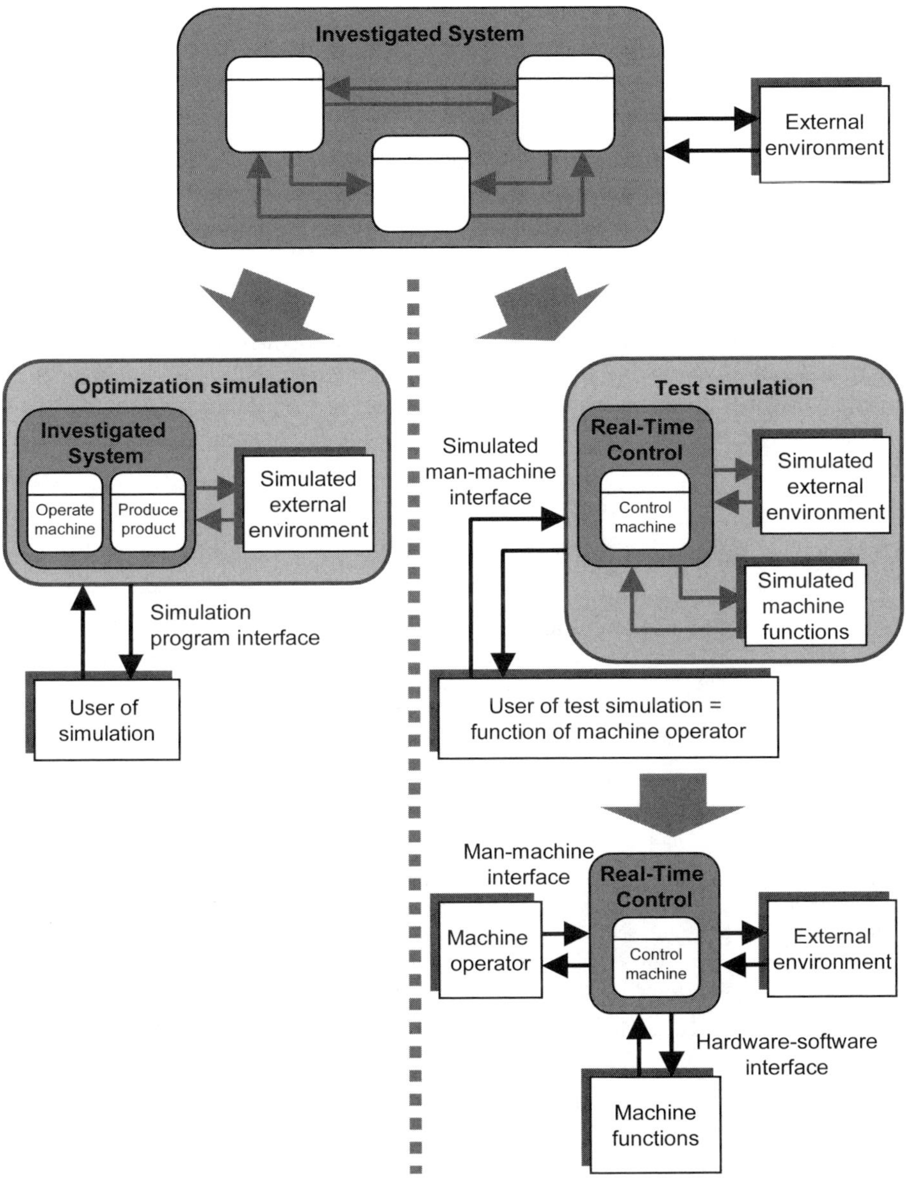

Figure D2-5: Procedure for optimization and test simulation.

In the case of the optimization simulation, machine and operator are part of the whole investigated system. The optimization simulation is used by a simulation user. The simulation model is a complete system. It runs, apart from the start and stop action of the user, on its own. The user enters the operation variables before the simulation starts. During the simulation run, there is no interaction from the outside.

In contrast, the test simulation of real-time control is in continuous interaction with the user of the simulated real-time control. The user acts like an operator and tells the simulated machine what to do. In the case of the test simulation, the operator of the machine becomes the environment of the simulated system, being at the same time the user of the test simulation and later the operator of the real-time control. The test simulation only serves during the test phase in development of the real-time control.

The differences between the optimization simulation and the test simulation for the real-time control are summarized in *Table D-1* from the aspect of the operators in the investigated system and of the user of the simulation. The point to consider is, whether the operator or user is outside the system and therefore represented by an external entity, or if the analysis of his or her behavior is essential and needs to be investigated. In this case, the operator is included as a process within the system boundaries.

Table D2-1: User of optimization and test simulation

	Optimization simulation	**Test simulation for real-time control**
Analysis of machine operator	The operator is analyzed as part of the simulated system. His or her behavior is simulated within the simulation model.	The behavior of the operator is analyzed.
User of the simulation program	User ≠ operator The user is an external entity of the simulation model.	User = operator The user is an external entity of the test simulation, behaving as the machine operator.
Interactions with the simulation program	The user starts and ends the simulation program, sets the values for the variables, and evaluates the results at the end of the program run. There is no interactions during the simulation run.	The same interactions as on the machine.

2.3 From Flow Diagram to Code

2.3.1 Simulation Theory

The steps of developing a simulation model according to the simulation theory and the subsequent steps are shown in *Figure D2-6*. The first four steps are covered by POA as indicated. Further information about the other steps are found in simulation literature. [Law, 1991; Page, 1991; Shannon, 1975]

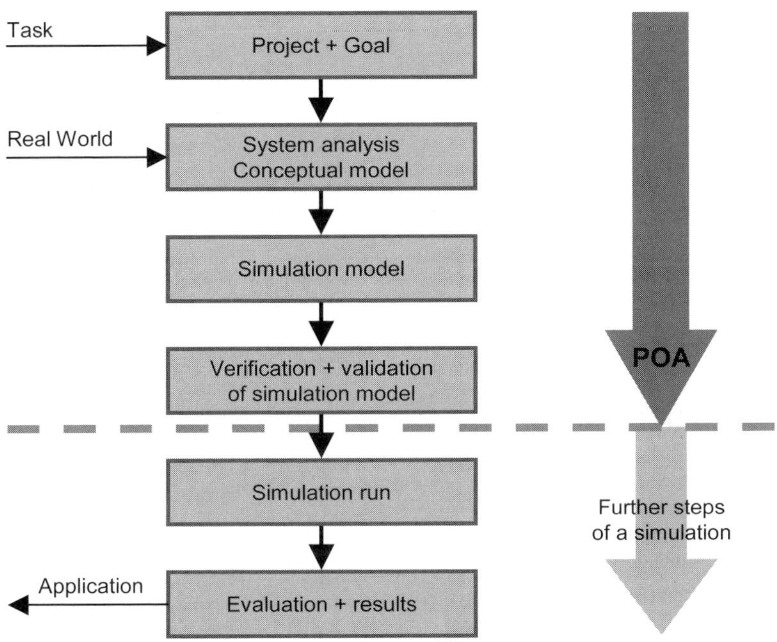

Figure D2-6: Simulation steps for optimizing a system.

In the simulation theory, the model to be simulated described by diagrams or other means is called "conceptual model." Within POA, the conceptual model is formed by the Flow Diagram and State Chart. The simulation model is either programmed by any common programming language or set up by a simulation package.

During the verification, the logistic of the simulation model is checked. It is tested if the model is working and if it is semantically and syntactically correct. The State Chart supports the check of the functions and their connections. The validation checks if the simulation model represents the investigated real-time world and corresponds to the requirements of the user insofar, as it covers the desired aspects and supplies the expected results.

Comprehensive insight into the system and its behavior and optimization possibilities are not only discovered by the simulation itself, but also during its preparation and setup while applying POA. A good part of it becomes obvious already during the conceptual modeling phase with POA (As-Is Analysis), especially while drawing the Flow Diagrams and the State Charts.

2.3.2 Step-By-Step Procedure

The procedure using POA to set up the simulation model, which works as a dynamic model of the plant on the computer, consists of seven steps that are introduced in the upcoming chapters. The steps are listed in *Table D2-2*. In the third column, the corresponding step of the simulation theory, as shown in *Figure D2-6,* is given.

Table D2-2: Procedure steps

Step POA	Explanation of step	Step in simulation theory	Example
1	Specify purpose and goal of the system. Context diagram: Set the system boundaries.	Project and goal	Analyze a chocolate wrapping line for improved throughput
2	Flow Diagram: Specify the system and its structure.	System analysis and conceptual model	Make POA of chocolate wrapping line.
3	State Chart: Specify the behavior of the processes in time.		
4	Specify the requirements of the program and design the user interface.	Simulation model	Transfer POA into code, benchmark operation of wrapping line.
5	Write each module in program code. Declare variables.		
6	Code and setup the simulation model.		
7	Check and evaluate the system behavior by running the program.	Verification and validation	Check wrapping line on model and in real time.

This step-by-step procedure corresponds to the POA toolbox. It leads from the specification of the project to the Flow Diagram, State Chart, and, ultimately, to the simulation model or real-time control program. The steps are done iteratively, insights gained during a later step are integrated in the model by redoing earlier steps within the step-by-step procedure. It is important to update the Flow Diagrams and State Charts anytime the program is enlarged or modified, going back to steps 2 and 3 of the above workflow.

While system analysis is mostly done top-down, the subsequent building up of a simulation program should go the opposite way, bottom-up. By using the POA method, each process corresponds to a program module. The final simulation program is an assembly of such modules, its structure identical to the real-world system. This makes it possible to go through each of the process individually and finally assemble, debug, and check the system as a whole. This corresponds exactly the real-world procedure in setting up a manufacturing plant.

2.3.3 Step 1: Purpose and Goal of System and System Boundaries

The first step in a project is to define the purpose and goal of the study. This goes along with the drawing of the context diagram, where the investigated system and its boundary to the environment are specified.

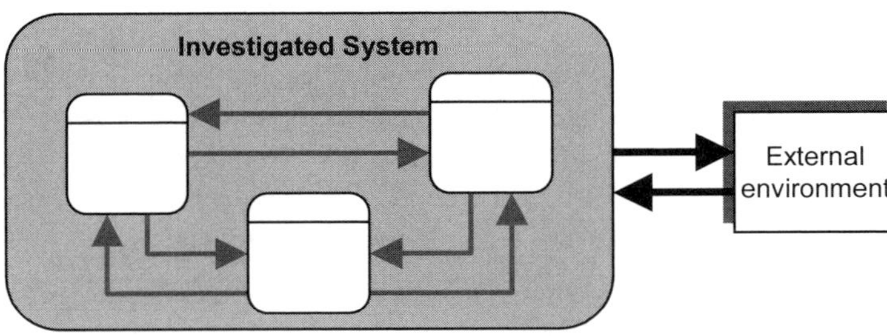

Figure D2-7: Boundaries of the investigated system.

Figure D2-7 shows the context diagram of a manufacturing system, typically containing operating, control and producing processes. The interfaces to the environment specify the interaction with the external environment. The goal of the system is expressed by the process on the context diagram.

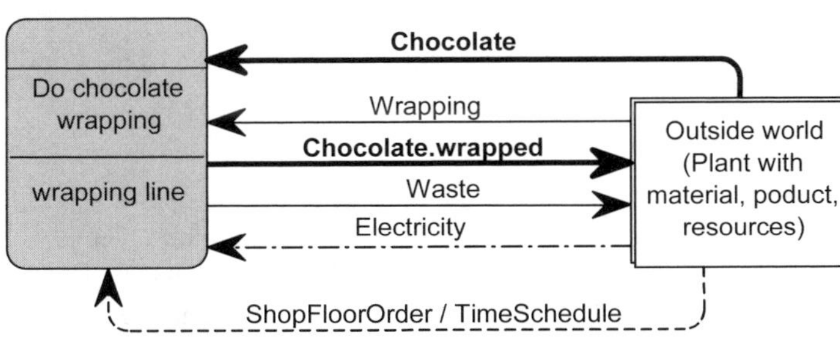

Diagram D2-1: Context diagram of chocolate wrapping line.

Diagram D2-1 shows the context diagram of a chocolate wrapping line. The investigated system is the process "Do chocolate wrapping." The context diagram shows the interfaces to the environment, consisting of resources, electricity, and information. This example of the chocolate wrapping is further elaborated in the next chapters by the step-by-step procedure, as given in *Table D2-2*.

2.3.4 Step 2: Specify System by the Flow Diagram

The next step is to detail the context diagram into a Flow Diagram model, as introduced in Section S1, in order to make sure that the structure of the investigated system is reflected and all interfaces connecting the process to the external environment are included. The detailing of the system into single processes defines at the same time the structure of the program code, since every process is described by a program module in the code.

For the program structure, the degree of hierarchical decomposition of a system is essential. If the system is decomposed in too much detail, and too many processes are programmed in individual program modules with the corresponding interfaces, the program code might end up unnecessarily large.

The context diagram, given in *Diagram D2-1* is further detailed into *Diagram D2-2* with the two processes "Wrap chocolate" and "Operate machine," and the interaction between these two processes. The diagram shows the processes necessary to program a simulation module. These are the producing and operating processes.

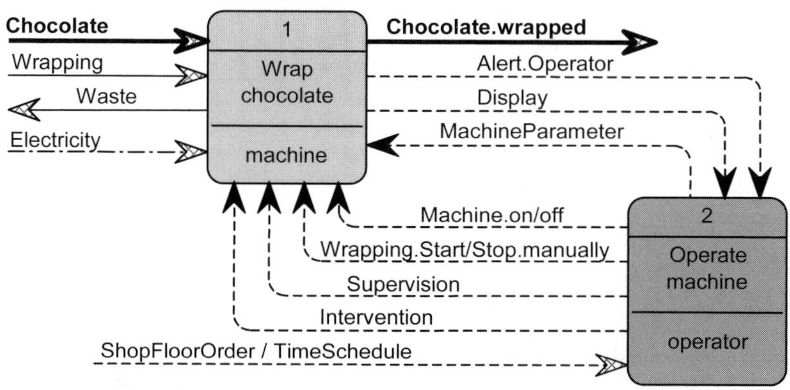

Diagram D2-2: Flow Diagram of chocolate wrapping.

Most manufacturing systems require the interaction of an operator with a machine, an associate with another associate, or a user with a computer. The person involved has to be considered as process within the system. It is important to take into account, how the human beings in the system behave. This ensures an integral model of a manufacturing system.

The information, control, signal, and resource flows are specified in the flow specification. The flows "Machine.on/off" and "Wrapping.Start/Stop.manually" are commands given by the operator. "Alert.Operator" and "Display" provide information to the operator. The flow "Intervention" stands for all the interventions of the operator with the machine. For example, it indicates the removal of a paper jam in the wrapping machine.

By the "MachineParameter," the operator sets up the machine for a specific product. The parameters for the wrapping process are indicated in the *Specification D2-1*. Some of the parameters are entered by the control panel. Other parameters are set by the hardware of the machine, for example to accommodate a different type and size of chocolate bars.

Specification D2-1: Flow "MachineParameter"

MachineParameter			
Description	Wrapping movement depends on the type and size of the chocolate bar.		
Flow type	Information		
Parameters	Batch size	optional 24 or 36	bars/batch
	Production rate	1000–5000	bars/hour
	Type + size of chocolate bar	optional 100 or 200	g

2.3.5 Step 3: Specify Behavior of Processes in Time

A computer model is composed of program modules, which divide the system to be simulated into single parts according to the structure of the real world. Each module matches a specific part of a machine and represents the behavior of the pertaining process.

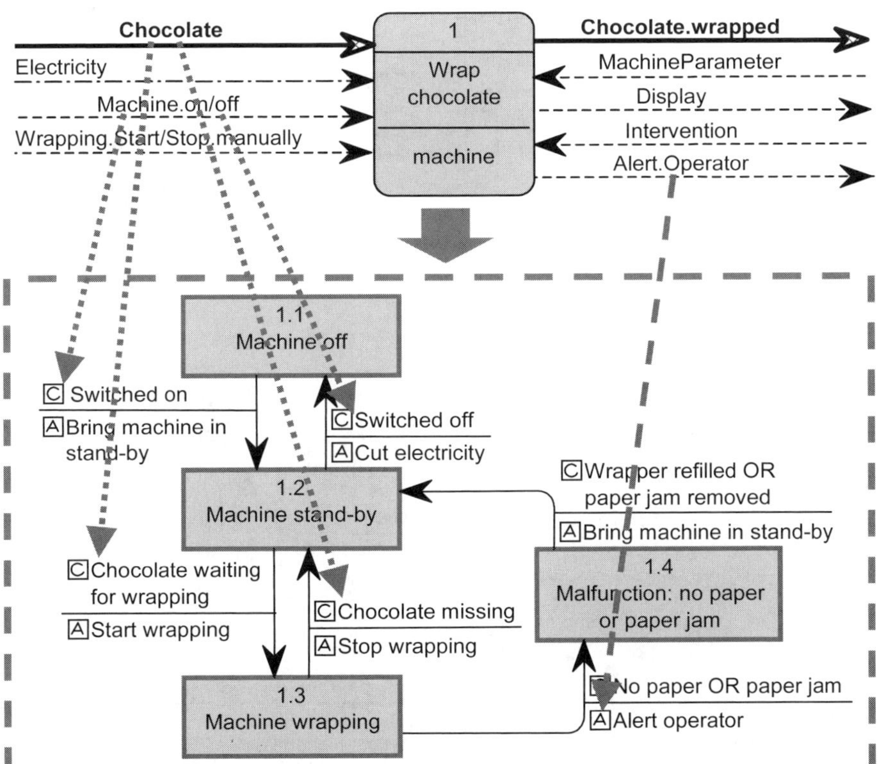

Diagram D2-3: State Chart for "Wrap chocolate."

Processes are specified in time by their states. That means, for each process, for which the behavior needs to be described, a State Chart is drawn. The behavior of every state is determined by the input and output flows of its parent process and the details given in the flow and process specifications. When taking the step from State Chart to program code, each process is eventually programmed in a program module.

Diagram D2-3 illustrates the transition from Flow Diagram to State Chart. From the Flow Diagram, only the parent process and its attached flows are shown. The process “Wrap chocolate” has four states that are connected with transitions, which in turn are defined by conditions and actions. The State Chart given here is simplified, the state “Malfunction: no paper or paper jam” represents all the risk and exceptional states of the wrapping line.

An input data flow may be used to formulate a condition in the State Chart, and thereby triggers a transition to another state of the process. An output data flow may define an action of a state impacting another process. “Alert.Operator” is an output flow of process “Wrap chocolate” and therefore becomes an action in the child State Chart of this process. “Alert.Operator” is an input flow of process “Operate machine” and becomes a condition in its child State Chart.

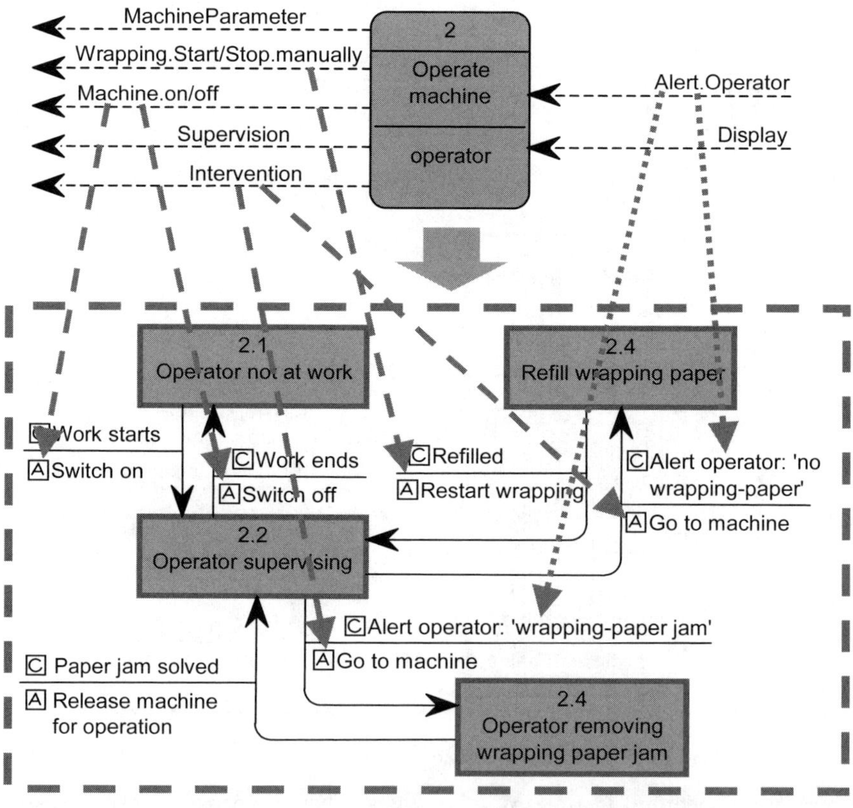

Diagram D2-4: State Chart of process “Operate machine.”

Diagram D2-4 shows the State Chart for process "Operate machine." The State Charts of the processes of the employees, machine operators, and program users are necessary to specify their behavior and interactions with machines and computers. For the optimization simulation, this is important because the operator is part of the simulated environment and is also included among the program modules.

The State Chart describing the work of the operator is essential for the coding of the real-time control in order to include all important, possible interactions on the machines (e.g. switches, keyboard, display).

The description of a process helps to draw a State Chart. The description in *Specification D2-2* of process "Wrap chocolate" calls for a more detailed State Chart than given above in *Diagram D2-3*. The change between the different chocolate bar types and sizes must also be considered.

Specification D2-2: Process "Wrap chocolate"

1	**Wrap chocolate**	
Description	Plant consists of 4 wrapping lines. The product to be wrapped is supplied by the chocolate production lines. The production switches from the white chocolate to the brown and then to the black chocolate, from the pure kind of chocolate to the mixed one. Before changing back to the white and homogenous chocolate, the machine must be cleaned.	
Parameter	Start up time Cleaning time	4.50 h 2.30 h

A State Chart describing a process can also be detailed further into child State Charts (see Section D1). The concept of developing a structure with different levels of detail is represented in the code by the nesting of the program modules. Every child State Chart is represented in the code by its own program module.

Rules D2-1: From process to State Chart

SM1	Each process necessary for the simulation model is specified time-dependently by a State Chart.
SM2	The states of each process are available to all other processes in the system.
SM3	The conditions in a State Chart are defined by the flows and parameters of the relevant process and by time.
SM4	The actions on one State Chart may serve to define conditions used in the State Chart of another process. On the pertaining Flow Diagram, they are indicated by a flow leading from the origin process to the destination process.

Note! While drawing the child State Charts of the processes, new insight in the model is gained. Consistency between Flow Diagrams and State Charts must be checked whenever conditions or actions are added. The flows on the Flow Diagram should be updated accordingly. Newly introduced actions and conditions may also affect other State Charts.

2.3.6 Step 4: Program Requirements and User Interface

Requirements of simulation model

The program's requirements are set up after clearly defining its purpose. The accessible parameters and interactions that can be changed by the user must be clearly distinguished from the functions that are carried out by the program modules. The interactions between the computer program and the outside world represent the communication between user and program. From the programmer's view, the interface of the process with the external entity represents the computer console, which is where the user has to give orders to the machine and receives data.

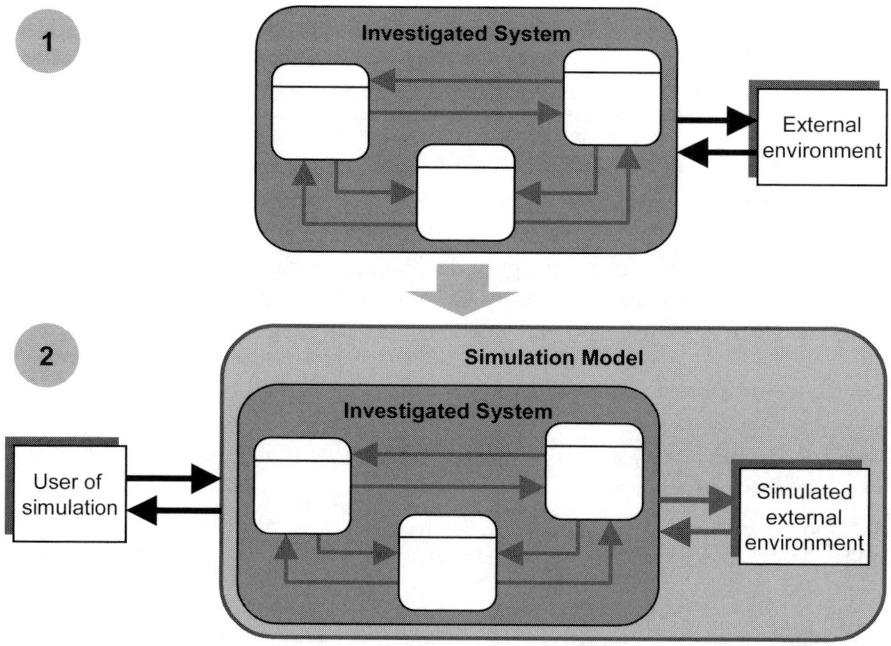

Figure D2-8: Context diagram of investigated system and simulation model.

In order to specify the requirements of the simulation model, the context diagram of the program is drawn. This context diagram is different from the context diagram of the investigated system, as shown in *Figure D2-8*.

Context diagram (1) in *Figure D2-8* specifies the system to be investigated and shows how the system interacts with the environment. This context diagram was drawn in Step 1 of the step-by-step procedure. Within the system, there are processes involving operators and machinery, and flows representing interactions and interfaces between them. This means, the human beings, workers, employees, and operators, including their behavior, are taken into account and are part of the investigated system.

Context diagram (2) of *Figure D2-8* sets the boundaries of the simulation program. Machine and operators are both part of the simulated system and integrated into the simulation model. The plant is simulated as a whole, including the behavior of the employees.

The external environment, treated as external entity in context diagram 1, becomes part of the simulated system in context diagram 2. The interfaces to the external environment are analyzed because they are simulated, too. The only external entity of the simulation model is the user of the simulation program. The interfaces between the user and the program specify the requirements of the simulation model.

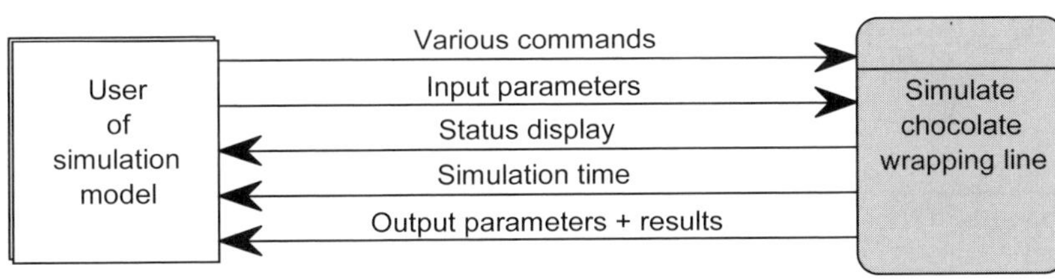

Diagram D2-5: Context diagram "Simulate chocolate wrapping line."

As an example, the context diagram in *Diagram D2-5* shows the simulated system of the chocolate wrapping system. The user of the simulation model communicates with the program via commands and input variables. The program supplies the user with the simulation time, output variables, and results.

Design user interface

The user manipulates the program through windows that represent the interfaces of the program modules. A minimal number of windows is needed for the program preparation, the simulation run, and the evaluation of results. Typically, the user interface for preparing the simulation run requires parameter inputs from the user, enables the user to check the status of the program run, and allows the operator to interfere or to stop the simulation. The results of a simulation run are calculated values (e.g. efficiency of production) or changes of monitored parameters, which are stored in the simulation run history. The results show the values obtained for the parameters representing the system performance.

The design of a program begins by setting-up the user interface. The program designer compiles the parameters that must be entered by the user, and the results that are to be calculated and displayed by the program.

Table D2-3 shows five basic modules of a user interface, their design corresponding to the goal and complexity of the machine or process. The third column gives examples taken from the user interface in *Figure D2-9* on the next page.

Table D2-3: User interface modules of a program

User interface modules	**Description**	**Examples from User Interface in Figure D2-9**
Input fields	Fields where the user assigns values to variables. A definite parameter input field gives a choice or range of values for the input.	Menu parameter
Output fields	Display of results of the simulation, which are calculated at run time.	Menu results
Status window	Display of the actual states of the system and parameter values at runtime.	Status of machine Status of operator
Control buttons	Start, stop, and resume the program run. Allow the user to control the system behavior.	Initialization Set simulation time Start simulation Stop Simulation Reset
Time field	The elapsed simulated time.	Elapsed simulation time
Animation	A graphical representation in motion or in changing color of the simulated object or process, for on-line check of system behavior by plausibility.	Colored fields changing according to actual states

Connection State Chart and user interface

Some Flows from the Flow Diagram and some actions from the State Chart will become elements on the user interface, such as a control button or an input field. For example, the action “Start machine” from the State Chart becomes a “Start machine” button on the user interface. Actions and conditions appear as elements on the user interface if the user interacts during the simulation run with the program.

User interface derived from context diagram

The context diagram of the simulation program as well as the Flow Diagrams and State Charts describing the investigated system help to define the user interface.

Figure D2-9 shows, how elements of the user interface are derived from the context diagram that specifies the simulation program. The user interface in *Figure D2-9* displays the status of machines and operator. It is combined with the animation, which changes the color of the machines and operators according to the actual state.

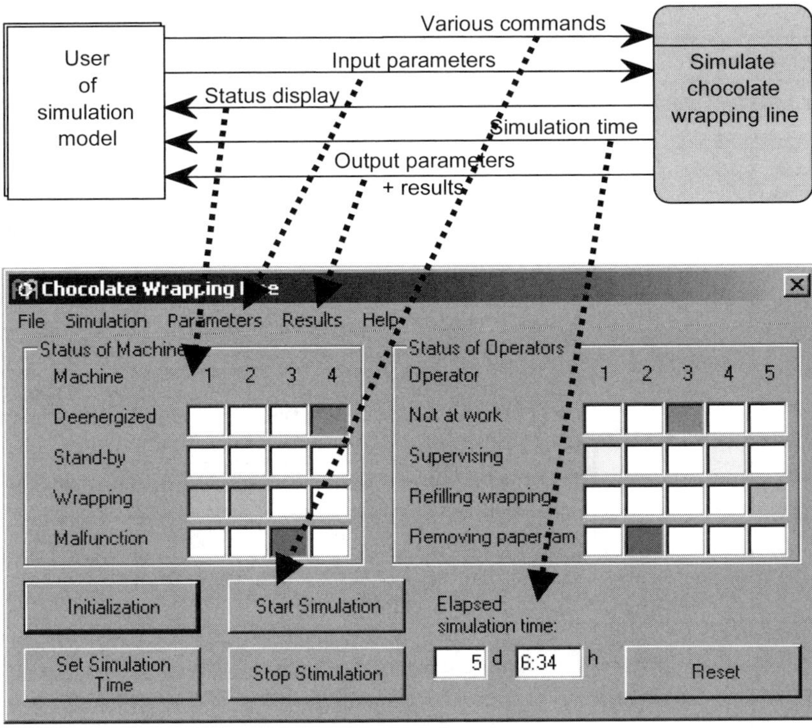

Figure D2-9: User interface derived from context diagram.

The interface flows of the context diagram are represented on the user interface as fields for parameter inputs and outputs, command buttons and display fields. For example the simulation time is displayed in the field: elapsed simulation time.

Rules D2-2: Requirements of the program

SM5	The requirements of the program are defined by its interfaces to the user on the context diagram.

2.3.7 Step 5: Write Each Program Module in Code

Program modules

The simulation program is set up by different modules: process modules, user interface, and auxiliary modules. The *process modules* represent the various processes. The *user interface module* contains the code for the graphical definition of the display components (*Table D2-3*). The *auxiliary modules* represent code parts of general program functions, as end of simulation, store result, and reset. *Table D2-4* lists the modules of the simulation model with their descriptions.

Table D2-4: Program modules

Program module		**Description**
User interface module		It graphically describes and codes the individual components of the user interface (*Table D2-3*).
Process modules	System State	The *states*, according to the State Charts represented by groups of lines of code, are chosen by a `Select-Case` construct. Within the active case (state) conditions for branching to other states are checked, and transitions are triggered.
		The *conditions* of the State Chart are programmed within each of these cases; one or more conditional statements change the state of the relevant process by triggering a transition.
		The *transitions* of the State Chart are programmed to switch the active state to the target state, and to fulfill other calculation tasks.
		The System State i.e. the state of every system module is updated after every transition and may be displayed at runtime in the status window on the user interface.
	Calculation of values	The calculation of values is based on equations. The results are displayed in the output fields.
Auxiliary modules	Variable declaration	Declaration of all the variables is essential for efficient programming and debugging. Modern program languages require strictly the declaration of all variables.
	Initialization	Setting of the initial values of the variables.
	System clock	Choice of time increment for the simulated time (resolution in time). Display of the elapsed simulation time.
	Store	Store setup, initial values, and results during and after simulation run.

State Charts are each coded within a process module. They are composed of the code controlling the states and transitions of the process; therefore, they are programmed based on the State Chart. Inside such a process module, the different sets of equations that represent the behavior of the process are chosen by a `Select-Case` construct containing the individual states of the process.

Definition: A process module is the part of code pertaining to the behavior of a single process, as given by the relating State Chart.

An example of code for the State Chart "Wrap Chocolate" is displayed along with the Diagram in *Figure D2-10*. The structure of the `Select-Case` construct is derived from the State Chart. The shades of the background underscore the correspondence of states in the State Chart with the code.

In the code, a state becomes a group of statements. The change from one state to another state may be coded by an `If-Then` construct that represents the condition and action of the transition from the State Chart. This can be seen in *Figure D2-10*. The states and pertaining parts of the code are set side by side. Within the `If-Then` construct, it can be seen how the system is set to a new state.

```
Public Sub WrapChocolate()
 Select Case StateMachine
   Case 11 'Machine off
     StateMach_11.Color = orange
     ...
     If Machine_on = True Then
       StateMachine = 12
     End If
   Case 12 'Machine stand-by
     StateMach_12.Color = yellow
     ...
     If Chocolate_waiting = True Then
       StateMachine = 13
       StateOperator = 22
     End If
   Case 13 'Machine wrapping
     StateMach_13.Color = green
     ...
     If PaperJam = True Then
       StateMachine = 14
       StateOperator = 24
       Alert_Operator_on = True
     End If
   Case 14 'Malfunction
     StateMach_14.Color = red
     ...
     If Time_PaperJam > 720 Then
       Alert_Operator_on = False
       PaperJam = False
       StateMachine = 12
       StateOperator = 22
     End If
 End Select
End Sub
```

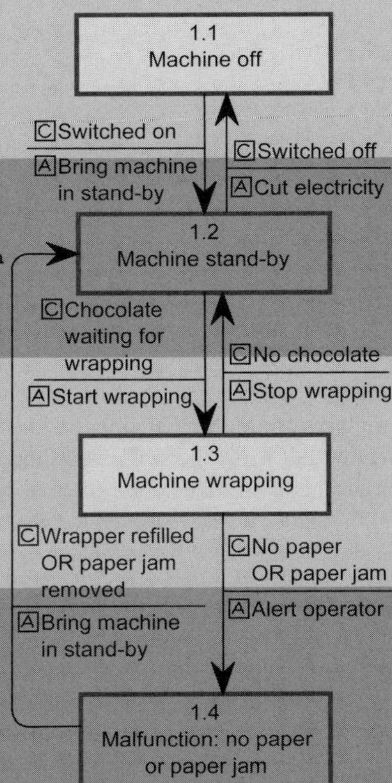

Figure D2-10: Code structure by Select-Case construct.

It is good practice to prepare the following before starting to code:

- variables and their type to declare
- the initial default values of system parameters
- the way to support the assignment of values to user-specified parameters
- rules and algorithms for the calculation
- the initial and default values of variables used to control the program at run time

Variable declaration in the auxiliary module

The parameters respectively the variables are declared in the auxiliary module "variable declaration." The following code example shows variable declarations of the type Decimal, Boolean, Integer, and String. It is advisable to include also the physical unit of the variable as comment in the declaration, for example second, liter, hour.

```
'Declaration of various variables:
Public SimTime As Decimal           'given in seconds [s]
Public Velocity As Decimal          'given in [bars/hour]
Public Wrapping As Boolean          'Boolean [on/off]
Public PaperJam As Boolean          'Boolean [True/False]
Public Operator_Display As String 'Message on Display screen

'Declaration of the states as variable for select case:
Public StateMachine As Integer     'States of machine
Public StateOperator As Integer    'States of operator
```

Variables in the code

Basically, each parameter from a flow or a process specification in the Flow Diagram is a variable in the code. In addition to this, the state of a process is a variable. Last, but not least, the simulated time and the time increment chosen for the simulation are also variables. The transitions (condition and action) into new states may contain values and thresholds for variables.

Variables from element specifications

The element specifications of the Flow Diagrams contain various parameters. These are translated into variables for the code. The *Specification D2-3* of the flow "MachineParameter" contains the variables "Batch size," "Production rate," and "Size of chocolate bar." The "Batch size" will become an integer variable in the program code.

Specification D2-3: Flow "MachineParameter"

MachineParameter			
Description	Wrapping movement depends on the type and size of the chocolate bar.		
Flow type	Information		
Parameters	Batch size	optional 24 or 36	bars/batch
	Production rate	1000–5000	bars/hour
	Type + size of chocolate bar	optional 100 or 200	g

The parameters pertaining to a certain process have to be considered as variables. These may comprise the elapsed time for start up, set up, and shut down of the production line or machine, possible are also downtime, repair and cleaning time.

Specification D2-4 of process "Wrap chocolate" gives the "Start up time" and the "Cleaning time."

Specification D2-4: Process "Wrap chocolate"

1	**Wrap chocolate**		
Description	Plant consists of 4 wrapping lines/machines. The product to be wrapped is supplied by the chocolate production lines. The production switches from the white chocolate to the brown and then to the black chocolate, from the pure kind of chocolate to the mixed one. Before changing back to the white and homogenous chocolate, the machine must be cleaned.		
Parameter	Start up time	4.50	h
	Cleaning time	2.30	h

Figure D2-11 shows how the parameters defined in the flow and process specification become elements on the user interface of the simulation program. By this, the user sets the parameters respectively the variables to a specific value.

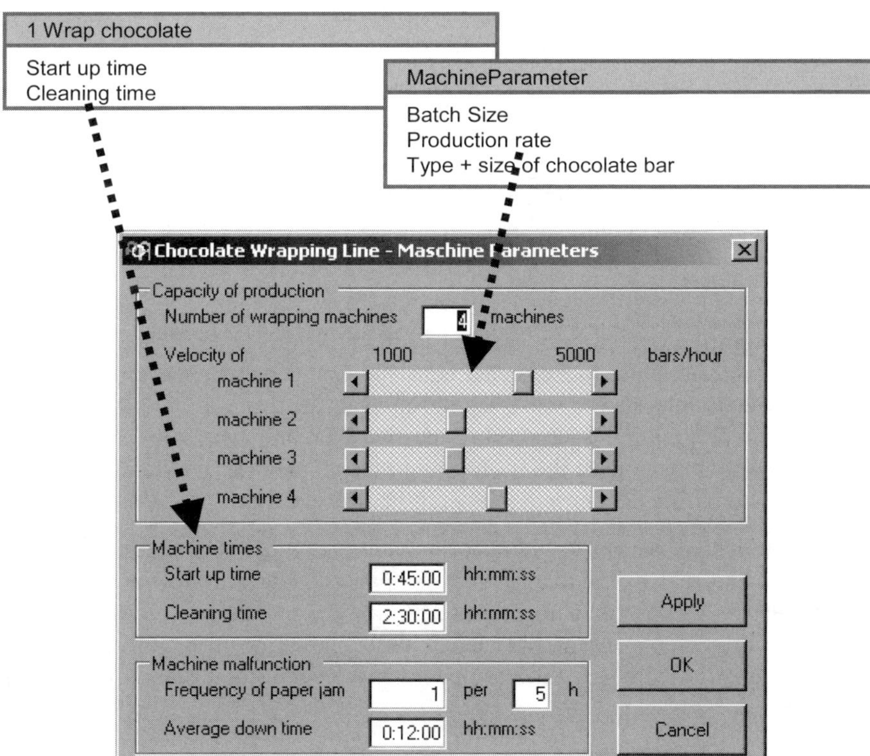

Figure D2-11: Input variables on user interface.

In this example, the flow “MachineParameters” contains in its specification the machine variables, and the process “WrapChocolate” describes in its specification the various durations in time.

States as variable

The state (in the general sense) of a process is handled as a variable. The individual states of a State Chart are the values of this variable. For example: The State Chart “Wrap chocolate” (*Diagram D2-3*) calls the variable “StateMachine,” and the State Chart “Operate machine” (*Diagram D2-4*) provides the variable named “State Operator.”

The variable derived from the state assumes discrete values, which may be represented by numbers to make a clean code. The code example of *Figure D2-10* takes the state numbers from the State Chart. Alternatively, plain language designators may be used as string variables. By this, the names of the states can be used directly in the code, which is easier to read, but affords a longer code.

Conditions as variables with Boolean expressions

Variables are also derived from the condition of the transitions, or from elements of the Boolean expressions within the conditions. Actions may impact all kinds of variables and assign new values, including the state of the own process module.

Conditions are Boolean expressions, meaning they are true when fulfilled, false when not fulfilled. They are set up in the form of an equation, using operators of the type “equal,” “not equal,” “greater than,” “greater equal,” “less than,” or “less equal.” This equation is put into an `If-Then` conditional statement to trigger a transition.

Equations using the operators mentioned in the paragraph above may comprise a wide choice of arithmetic expressions, consisting of variables and operators. The choice is limited only by the variables declared in the program and the operators offered by the programming language. Since all data, and especially all states in a system, are available to describe its behavior, all of these variables should be declared common within the borders of the physical system.

The object-oriented structure within a simulation program has to be strictly set according to the real world that is to be simulated. The introduction of objects, classes, and methods is recommended for multiple identical processes. Examples are the axis drives of a machining center, the spinning positions of an open-end spinning machine, or the servo drives operating the valves of a large air conditioning system.

Examples of conditions, how they are stated on the State Chart, and the equivalent structure of code are given in *Table D2-5*. (This table is an enlargement of *Table D1-3* given in Sections D1.)

Table D2-5: Examples for conditions

General expression	Example of condition on the State Chart	Structured code
free text for object and attribute, event taking place: *k: Boolen variable*	C key pressed C part arrived at assembly machine C part arrived C car stopped	`If` Key_pressed = `True` `Then` do action
parameter value: e = d *e: variable* *d: numerical value*	C count of bottles = 12	`If` Count_bottle = 12 `Then` do action
limit value of parameter: b < a b > a b ≤ a b ≥ a *a, b: any numerical variable*	C temperature < 30 °C	`If` Temperature < 30 `Then` do action
composed expression: e = w AND v e = w OR u *e, u, v, w: any Boolean variable*	C Water = warm AND clean C Weather = rain OR snow	`If` Water_warm = `True` AND Water_clean = `True` `Then` do action
conditional statement: *The combination of any of the above conditions.*	C Participants ≥ 12 AND (Weather nice OR Holiday)	`If` Participant ≥ 12 AND (Weather_nice = `True` OR Holiday = `True`) `Then` do action

Assignment of value to variable

The following part of code shows the change of the value of a variable, by the example of assigning new states of two processes. This two processes describe the machine and the operator and are detailed into child State Charts. The values of the variable in this example are the identification numbers of the states in the State Chart. In this case, it is done within an `If-Then` construct. If the variable "Paper-Jam" is true, then the variables "StateMachine" and "StateOperator" are set to a new state.

```
If PaperJam = True Then
      StateMachine = 14
      StateOperator = 24
      Alert_Operator_on = True
End If
```

Summary: From Flow Diagram and State Chart to code

Table D2-6 gives an overview of the guidelines how structure and code can be derived from the diagrams with POA. It summarizes this chapter. The table lists the elements of the Flow Diagram and State Chart and explains how they are converted into code and especially into the variables and the assignments of their values. The clearer the system is specified by the Flow Diagram and State Chart, the easier is the translation from them to code.

Table D2-6: From Flow Diagram and State Chart to program code

Flow Diagram and State Chart and elements	**Part of program code**	**Variables used**
Context diagram of simulation model	User interface modules	
Data flow (to external entity)	Input and output fields on user interface, status display. Command button on user interface.	
Flow Diagram	Auxiliary/process modules	
Process	Process module (subroutine)	Specification provides: name, type, range, and possible default/initial values.
Flow		Flow name may be variable name. Specification provides: name, type, range, and possible default/initial values.
State Chart	Process module (subroutine)	Variable "state"
State	Group of statements	Number (or name) of state becomes value of variable. Assignment of value within the program module (e.g. by `Select-Case` construct).
Condition	Condition for `If-Then` construct.	Values for comparison, threshold
Action	Executable statement of `If-Then` construct	Value assignment for variable representing state. Further calculations and reassignment of other variables.

Naming

The names of the components of the system may be chosen at will; giving names to things makes them easier to recognize and to distinguish. Structures and general rules for assigning names, as stipulated by the Flow Diagram (Section S1) and the

State Chart (Section D1), provide an easy to understand model. Therefore, it makes sense to set up and adhere to such rules even in a quick-and-dirty approach.

Rules D2-3: Program coding

SM6	Each State Chart is represented in code by a process module.
SM7	Variables valid in the whole program carry a unique name and are declared as global.

2.3.8 Step 6: Code and Setup of the Simulation Model

The elements of source code are variables, their values, and a step-by-step procedure for calculating and changing these values. The specific programming language imposes rules of how to name the variables and how to change their values, along with many services associated with the setup and use of the program.

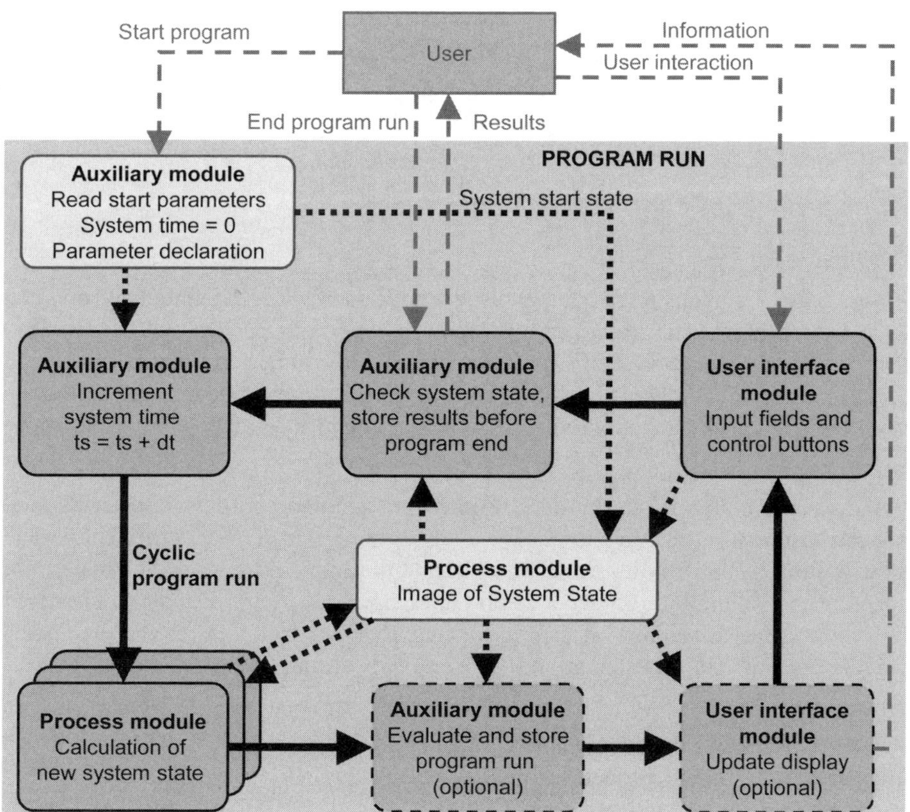

Figure D2-12: Setup simulation program.

A program is composed of individual logical and arithmetic operations put into a specific sequence. *Figure D2-12* demonstrates the interactions of the different program components between each other and the user. The user may interfere with the program at different points in the program run (indicated with dashed arrows).

The loop of the bold arrows shows the cyclic program run through the program modules. The dotted black arrows stand for optional actions happening on demand in the program.

Depicted in the center of the cyclic program run is the System State, which comprises the actual states of the process modules. The System State is continuously updated by each process module during the cyclic program run and is available to all the other program modules.

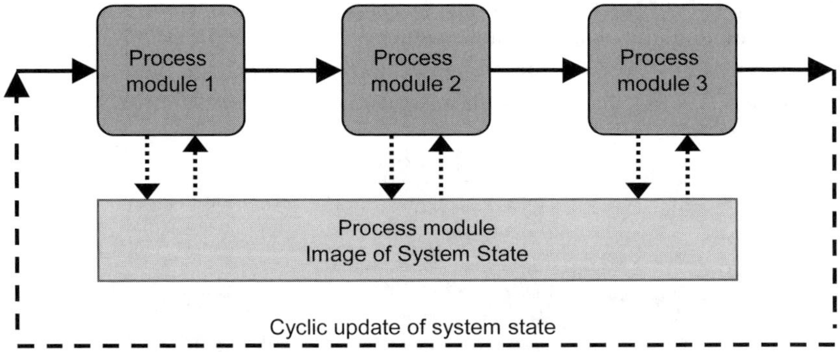

Figure D2-13: Cyclic program run.

Figure D2-13 demonstrates again how the System State is continuously updated during the program run. Every process module reports its current state to the System State and vice versa. In the process modules, the states and transitions of the processes are coded. The transitions are programmed with a `Select-Case` construct, which checks the conditions periodically and changes the actual state whenever required.

In the simulation model, the simulated time for a simulation run may be accelerated or slowed down. After the simulation start, the program runs independently. The slow motion is made by inserting wait states into the program loop. The maximum speed of the simulation run depends on the computer capacity. It may be increased by omitting the animation or adapting the intervals to the states of the processes, leaving out superfluous calculation loops, and only storing needed results.

Auxiliary module system clock

The system clock may be implemented in two ways. For maximum speed of the simulation run, the simulation time is incremented instantly whenever this module of the system clock is called. In case this is not practical, e.g. to allow following up the system behavior online on the computer display, this increment is programmed to wait for a keystroke (single-step operation) or for a predetermined time for fast or slow motion display.

The code part below shows the program module for the system clock. From this program module, other modules are called for each increment in time.

```
Private Sub Clock_Tick
        SimulationTime = SimmulationTime + Clock.Interval
        [Call process module 1]
        [Call process module 2]
        ...
End Sub
```

Rules D2-4: Program coding

SM8	Every process module is cyclically called up by the program.
SM9	In each process module and for every increment in time, all conditions are checked.
SM10	The increment in time is set shorter than the quickest change in state that actually occurs in real time.

2.3.9 Step 7: Check and Evaluate System Behavior

In the verification phase, the simulation model is tested concerning function, plausible behavior, and agreement with the real world. By the State Chart, the functions and interactions are debugged. The validation has to confirm that the simulation model corresponds to the reality. This means that the investigated aspect has to be simulated realistically, and the results have to be comparable to the real values.

Benchmarks

To run a simulation model, a specific benchmark is defined with a given set of realistic start variables. Within these fixed boundaries, single program variables may be changed, while the system behavior is compared to other models or to the reality. The target and the variables available for the system optimization are part of this benchmark.

Check simulation and system behavior

A manufacturing system is developed and commissioned by the same basic steps in real life as when simulated on the computer. The phase of testing and debugging serves to recognize and eliminate errors in the program itself (assignment of values, equations, system representation), in the program run (run-time errors, branches, loops), in the user interface (operator input, console display, real-time interfaces), and last but not least, in faults that have their origin in the original real-world system. *Table D2-7* lists and describes the four phases of system check for a program.

Table D2-7: Phases of program check

Phase	Description
First phase	Check the operator console in its static appearance. Go through the basic steps in operation while observing windows, forms, and dialog.
Second phase	Check the system clock by starting and stopping the program from the operator console.
Third phase	Load the default input parameters, start the program run, and terminate immediately at the shortest delay convenient. Check the System State and compare to the start up state. Make sure that start parameters are introduced and processed correctly by the system, and that the initial behavior of the system is plausible.
Fourth phase	The individual processes are put into operation, step-by-step. The program is started with specific start conditions in order to test the individual process modules. The behavior of the process and the system has to be deterministic and predictable based on the chosen start conditions and start parameters.

Since many processes may only be checked while running concurrently with other processes, the whole system is put into service in a certain order. Processes are added either one-by-one along the path of material in production, or from the peripheral processes of the system towards its core.

It is important to first test control loops in an open loop configuration to make sure that each element behaves properly. This means that the controller is switched off (patched), so that it does not try to move the actual value according to the set value. Thereby the function of the actuating element is tested. The speed of the running program has to be reduced as much as required to observe the proper sequence of states and transitions. Finally, set up a benchmark run that makes use of all system functions, to verify anytime later the integrity of the system.

Simulation run and evaluation

With the finished simulation model, several simulation runs are done with different parameter settings. Settings and results are documented. The results are evaluated and compared to the real-world system or to a benchmark, and conclusions are taken concerning the goal of the project. If random events are introduced for trueness of the simulation, multiple runs provide a statistical base for further investigation, as it is the case in real world. The simulation report includes recommendations for the investigated system based on the simulation results.

2.4 Application of Commercial Simulation Packages

2.4.1 Connection POA and Commercial Simulation Packages

This chapter treats the application of commercial simulation packages and their relationships to the tools of POA. Up to now the simulation model of POA was treated on the basis of the static and dynamic model and the self-programmed code with a common programming language, as for example Visual Basic, C, or Delphi. Instead of a standard programming language, a simulation package may be used to elaborate a simulation model. Also in combination with a commercial simulation tool, POA is very useful.

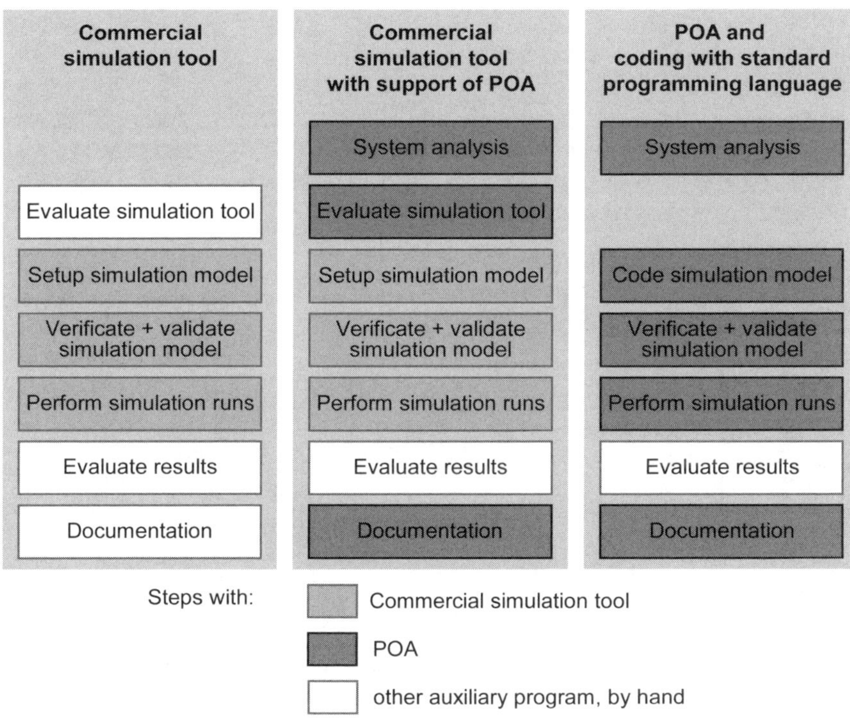

Figure D2-14: Application of commercial simulation packages.

Figure D2-14 shows the application of a commercial simulation package compared to the use of POA and the combination of both. Most of the simulation packages start with the build-up of the simulation model. They do neither support the system analysis nor the conceptual model. To set up a conceptual model, a CASE tool like POA is required. The static and dynamic models of POA include the system specification and documentation. POA helps to understand and get to know the system to be simulated and the critical parameters for the simulation.

2.4.2 Evaluation of Commercial Simulation Packages

Simulation packages work with elements or components that are generally represented by specific icons serving also for the animation. These components are ready to carry out a range of specific functions. To every component belongs a specification to define its behavior. If a function is required for the model that is not integrated, is has to be programmed by hand in the programming mode offered by the package.

For getting to know and evaluate such a simulation package, POA may be used to define the behavior of these components. By this, the user can see quickly, which components of the simulation package correspond to the processes and states of the system to be investigated and to the required level of detail.

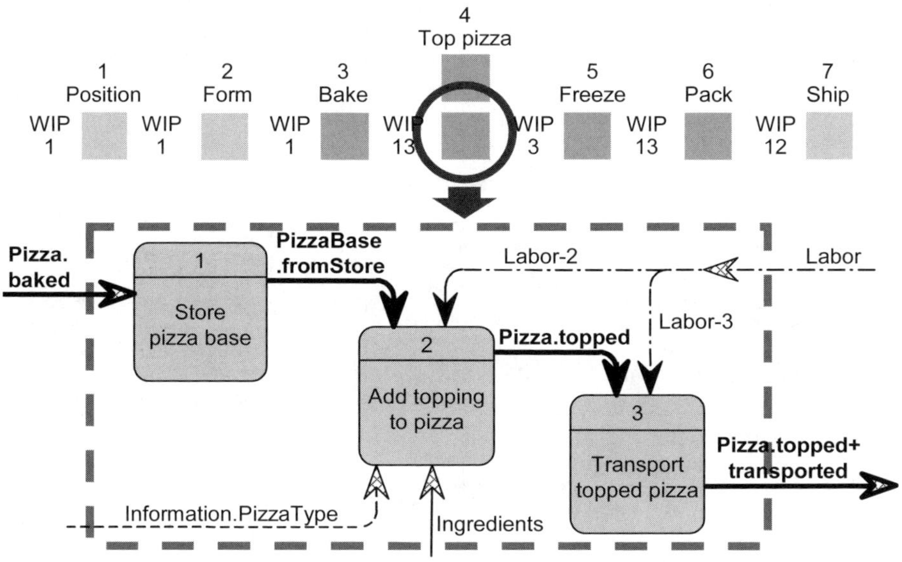

Figure D2-15: Description of a component of the simulation tool MMS.

Figure D2-15 gives an example of one component of the simulation package Modular Manufacturing Simulator (MMS). The model to be simulated is a pizza production line. The component, which is specified by a Flow Diagram of POA, represents the adding of the topping to the pizza. Obviously the component is on a very coarse level compared to the detailed functional analysis with the Flow Diagram. MMS does not show detailed functions with its components. The process of topping the pizza is simulated by a component representing a machine station with an operator.

To the machine station belongs a specification, one for the machine and one for the operator. In the operator specification, the number of operators, their jobs, their assignments, and the rules for their actions are set up. In the machine specification, the time dependent parameters for the manufacturing of a product unit and the number of machines for each process step are fixed. The parameter WIP (work in progress) on the screen shows how many pizzas are in process for the topping.

The program MMS just offers stations as components. The transport of the material is integrated in these stations. Only a production with a single line of sequential processes may be modeled in this way. Networks, recycling loops of material, and parallel production paths with plurality of processes cannot be modeled. However, MMS is quick and easy to use.

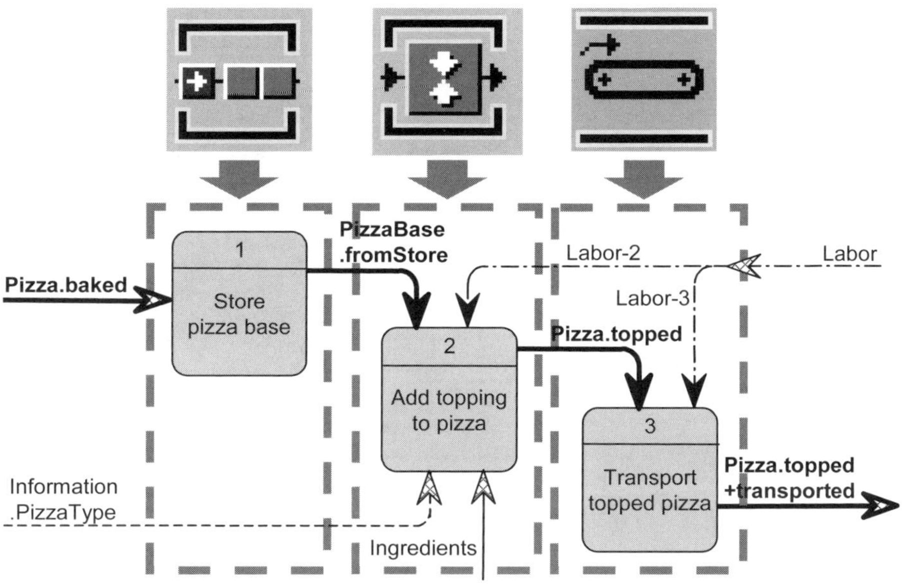

Figure D2-16: Description of a component of the simulation tool eM-Plant.

By modeling with typical simulation packages, a further abstraction is done compared to the Flow Diagram. In general only the material flows are considered. The system information, which is contained in the other flows in POA, has to be added to the specifications of the component, and is not visualized anymore.

Figure D2-16 shows again the pizza production analyzed with another simulation package: eM-Plant. The components of eM-Plant are more detailed than those of MMS. The single processes of the Flow Diagram can be modeled individually with well adapted components of different types. A specific component represents the material buffer. The topping of the pizza is represented by the component for an assembly process. A transport component, specified as conveyor belt, handles the transporting process of the pizza.

In POA, the behavior in time of a process is described by a State Chart. The time dependent states and behavior of a component for a simulation package is defined in its specification. Therefore, the behavior in time of such a component can be described by a State Chart, which is very useful for the evaluation of a simulation package. *Figure D2-17* shows the specification of a "Work station" and the specification of it by a State Chart.

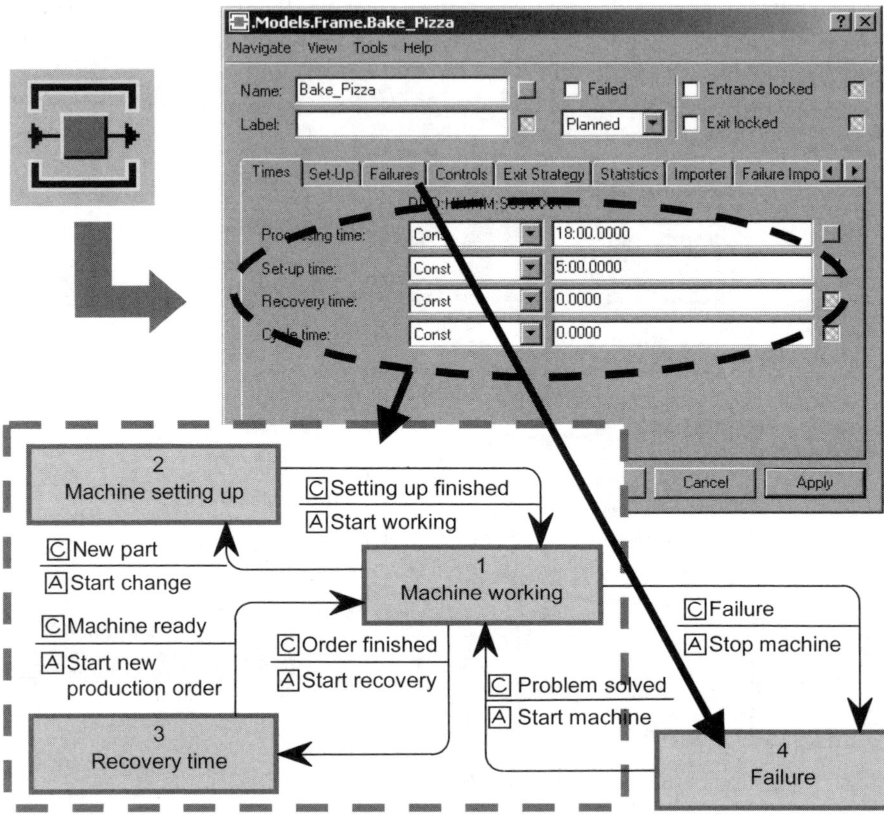

Figure D2-17: Description of the component "Working station" of eM-Plant by a State Chart.

By comparing the POA diagrams of the system to be investigated with the POA diagrams describing the elements of a simulation package, it becomes immediately visible, how well a system is being represented by a simulation package.

For the simulation with a simulation package, POA is useful, for the evaluation and selection of the simulation package, for the generation of the conceptual model, and for the detailed specification of the simulation package elements.

2.4.3 Example with Simulation Package: Gas Station

The application of a commercial simulation package is demonstrated with the example of the gas station. *Diagram D2-6* shows the processes of a gas station. It is the same gas station as in the application example of Section D1. Goal of the analysis is the optimization of the efficiency of the gas station. The use of the gas station should be as efficient as possible, and the waiting time for the customers as short as possible. The following questions arise: How many gas pumps are concurrently occupied? Under which conditions are cars waiting, and for how long? What is the impact on the waiting time if a gas pump is blocked by a car using the gas pump preceding in line?

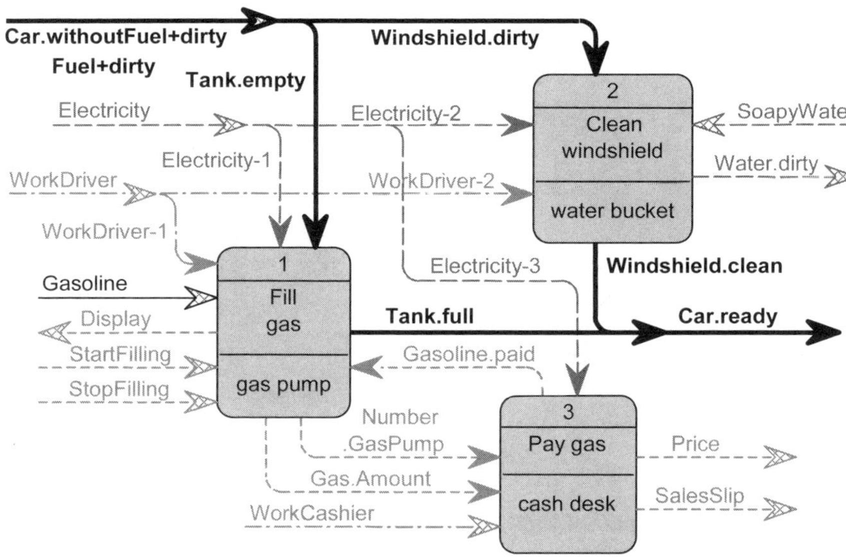

Diagram D2-6: Flow Diagram for gas station.

A simulation model with eM-Plant is built in order to give answers to the above questions. In *Diagram D2-7*, process 1, "Fill gas" is further detailed as a State Chart to show the states of the customers with their cars at the gas station.

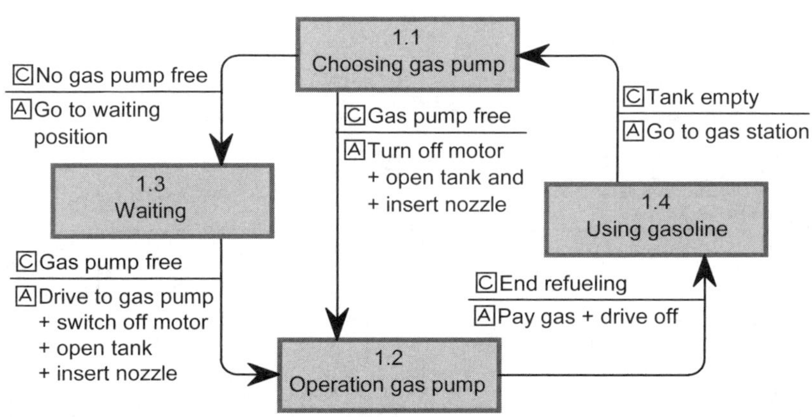

Diagram D2-7: State Chart "Refuel car" for eM-Plant.

Based on the Flow Diagram and State Chart, the components offered by the simulation model are assembled. The gas station consists of eight gas pumps. As shown on the animation of eM-Plant in *Figure D2-18,* there is a waiting area in front of each gas pump group.

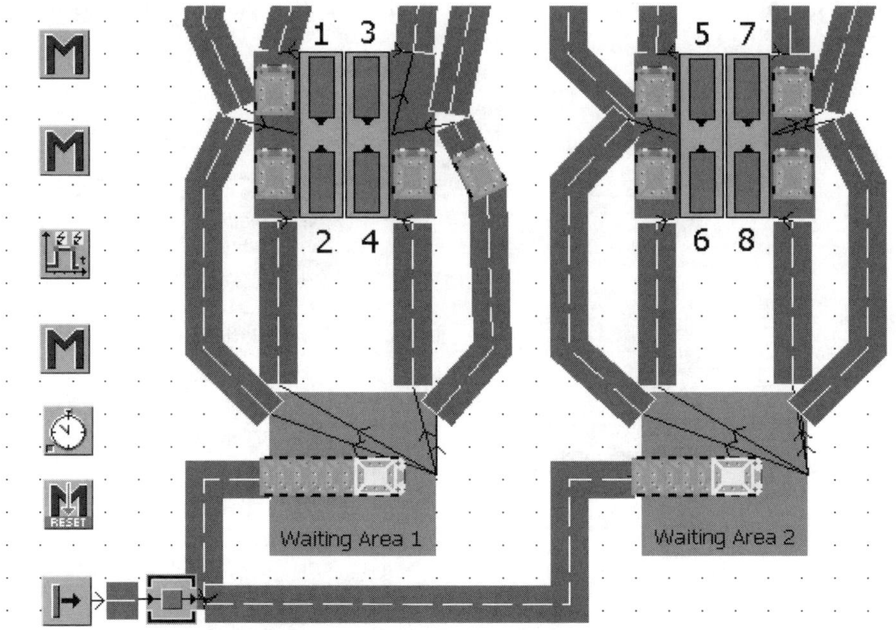

Figure D2-18: Runtime screenshot of eM-Plant animation of a gas station.

A "Work station" component is applied for the gas pump. The path of the cars is built by the component "Way." The cars are adapted from the material units of the simulation package. On the left hand side of the runtime screenshot, the buttons appear that are used for the different methods (procedures for the simulation) used in the simulation, e.g. reset.

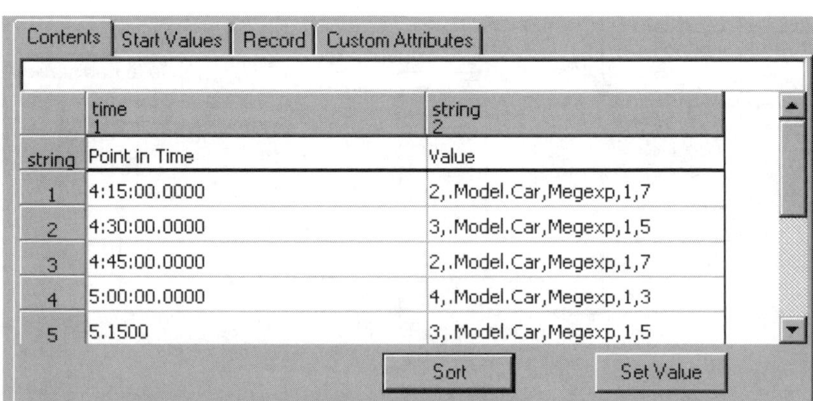

Contents | Start Values | Record | Custom Attributes

	time 1	string 2
string	Point in Time	Value
1	4:15:00.0000	2,.Model.Car,Megexp,1,7
2	4:30:00.0000	3,.Model.Car,Megexp,1,5
3	4:45:00.0000	2,.Model.Car,Megexp,1,7
4	5:00:00.0000	4,.Model.Car,Megexp,1,3
5	5.1500	3,.Model.Car,Megexp,1,5

Sort Set Value

Figure D2-19: Configuration of the component "Source."

The specific data of the simulated system are entered into the specifications of the components. Depending on the type of the component, various options for its configuration are offered, and the user has to choose the ones required. *Figure D2-19* shows the specification of the element "Source." It contains an order list with the releasing time. In this case, the order list represents the number of cars that enter the gas station within a specific period in time. The incoming car listing in *Figure D2-19* for the gas station covers 24 hours a day with different numbers of cars depending on the time of day (e.g. rush hour, night).

This example shows that the common components offered by a simulation package are usually not sufficient for a realistic simulation model. The components have to be extensively specified, modified, and in many cases program code has to be added. The following code is the specification of the element, which distributes the arriving cars on the two waiting areas in front of the gas pump groups. It calculates the number of cars waiting in the two areas and the amount of time the gas pumps are occupied. An incoming car always joins the waiting area with less cars waiting.

The programming language within eM-Plant is called SimTALK.

```
/*--------------------------------------------------------------
Distribution of cars to the two waiting areas
It calculates how many cars are at the gas pumps 1, 2, 3, 4
and on the waiting area 1. The same calculation is done for
the gas pumps 5, 6, 7, 8 and the waiting area 2. The smaller
count decides, on which waiting area the car is going.
--------------------------------------------------------------*/
is
  count_WaitingArea1, count_waitingArea2 : integer;
do
  count_WaitingArea1 := GasPump_l_1.countMOs
    + GasPump_l_2.countMOs + GasPump_r_3.countMOs
    + GasPump_r_4.countMOs + WaitingArea1.countMOs;
  count_WaitingArea2 := GasPump_l_5.countMOs
    + GasPump_l_6.countMOs + GasPump_r_7.countMOs
    + GasPump_r_8.countMOs + WaitingArea2.countMOs;
        if count_WaitingArea1 <= count_WaitingArea2
        then @.transfer(WaitingArea1);
        else @.transfer(WaitingArea2);
        end;
end;
```

This application example and the detailing and description of the elements with the Flow Diagram and the State Chart demonstrates that POA is very useful and highly recommended for the build up of a simulation model, with or without a simulation package. It also demonstrates that a fairly simple analysis problem requires the setup of a quite complicated model, which is difficult to implement by an icon-based simulation package.

2.5 Application Example: Gas Station

2.5.1 Static Model

For the gas station, analyzed in Section S1 and Section D1, a simulation model of the operation of a gas pump is made. The model concentrates on process 1 "Fill gas." The example is kept simple, in order to be able to give here the complete code. For the simulation model, new State Charts are drawn, because the structuring differs from the one given in Section D1.

The computer model comprises the driver and the gas pump, which fills gas into the tank of a car. The program is set up interactively: The user of the simulation program acts as the driver, who operates the gas pump. He or she carries out the typical activities on a gas pump. First, the tank cap of the car is opened. Then the nozzle is inserted into the tank filler and the handle of the nozzle squeezed. When the tank is full, the gas pump turns off automatically. These functions of the gas pump are described by the two processes "Pump gas" and "Operate gas pump" (*Diagram D2-8*). It is a child diagram of *Diagram D1-24* in Section D1.

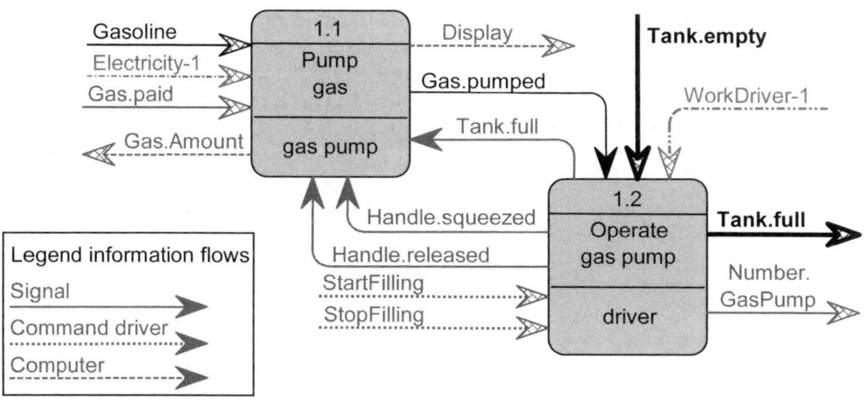

Diagram D2-8: Flow Diagram "Fill gas."

Specification D2-5 of the process 1.2 "Operate gas pump" states that the control of the gas pump is integrated in this process. Furthermore, the content of the tank in liters is given.

Specification D2-5: Process "Operate gas pump"

1.2	Operate gas pump	driver
Description	The control of the refuelling is integrated in this process. The driver can stop the filling of the tank anytime by the handle on the gas nozzle.	
Parameter	Content of tank	60 l

In *Specification D2-6* of the flow "Gas.pumped," the filling rate of the pump is specified as parameter.

Specification D2-6: Flow "Gas.pumped"

Gas.pumped		
Description	Gas is pumped into the tank of the car. The pumped quantity is continually calculated.	
Flow type	Gasoline	
Parameter	Filling rate	1 liter/s

The user's commands are depicted on the diagram by gray dashed arrows. The command "StartFilling" is triggered by squeezing the handle of the gas nozzle, the command "StopFilling" by releasing the handle. The position of the handle on the other hand corresponds to the two control flows "Handle.squeezed" and "Handle. released," depicted as gray arrows. They go from the process 1.2 to process 1.1. Following, *Specification D2-7* of the signal "Handle.released" is shown.

Specification D2-7: Flow "Handle.released"

Handle.released	
Description	The driver can stop the filling anytime by releasing the handle. If the tank of the car is full, the handle is deactivated automatically.
Flow type	Signal (Control flow)

2.5.2 Dynamic Model

Diagram D2-9 shows the State Chart of the process "Pump gas" with its two basic states "Pump off" and "Pump pumping." The information flows "Handle. squeezed," "Handle released" and "Tank.full" going into the process 1.1 correspond to the conditions of the state transitions on the State Chart "Pump gas."

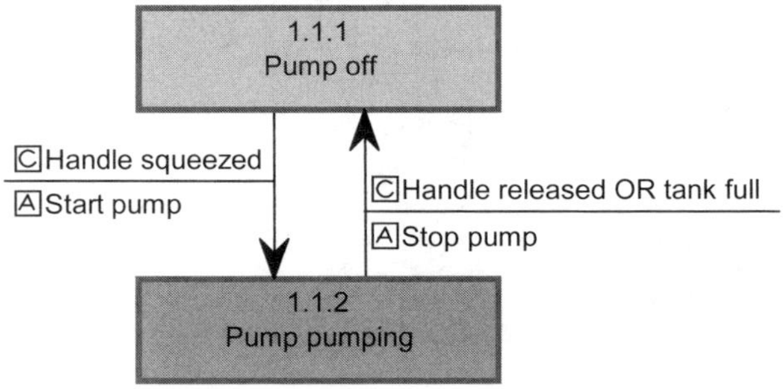

Diagram D2-9: State Chart "Pump gas."

Diagram D2-10 shows the State Chart of the process "Operate gas pump," which has four states, from the view of the driver. The output information flows of process 1.2 "Handle.squeezed" and "Handle.released" become actions in the State Chart "Operate gas pump."

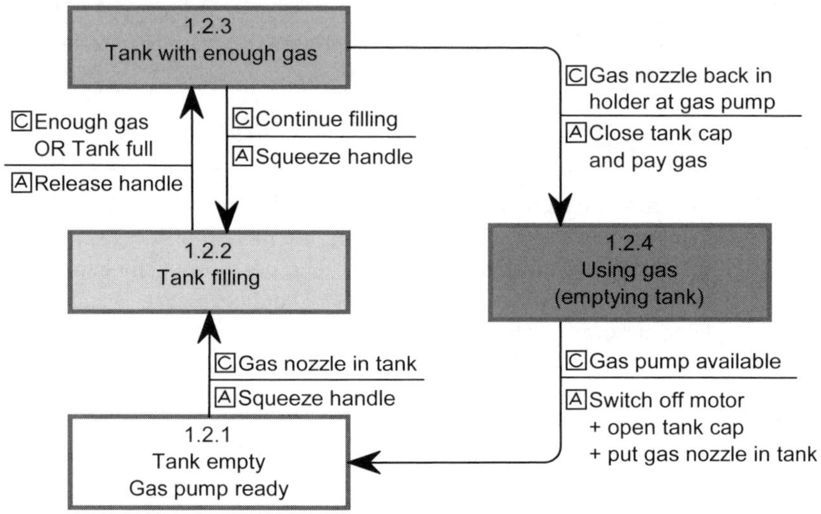

Diagram D2-10: State Chart "Operate gas pump."

There are no alarm and risky states implemented in this simple example. In *Diagram D2-11,* the State Chart "Pump gas" is supplemented as an example with an alarm state. As soon as the pressure in the filling pipe of the gas pump drops, for example gasoline is leaking because the filling pipe is ripped off, the gas pump is turned off and an alarm is triggered. Only if the problem is cleared, the gas pump checked, and the alarm is reset, the pump can be put into operation again.

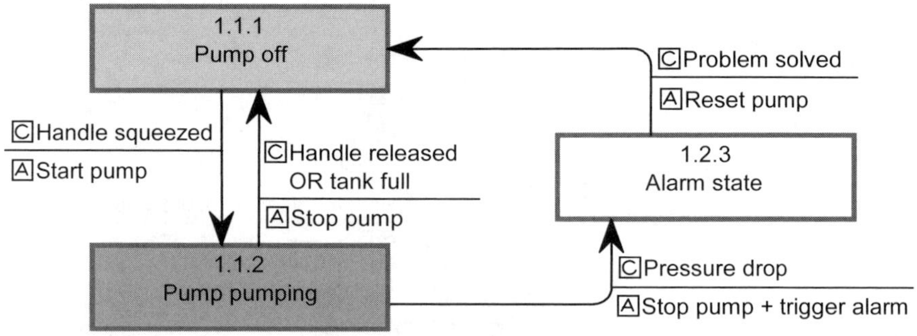

Diagram D2-11: State Chart "Pump gas" with alarm state.

The alarm state for this application example is not treated further and not coded, in favor of a program code short enough to be shown as a whole in the book.

2.5.3 User Interface

This chapter shows and explains the user interface of this small simulation program. The user interface consists of one window, as displayed in *Figure D2-20*. In the output fields, the current states of the pump and of the car tank, the elapsed time, the animation, and the display on the gas pump are indicated. Two different times are measured in this simulation. The “Simulation Time” for the total run time of the simulation begins when the simulation is started by the button “Start Simulation.” The “Filling Time” counts for the duration of the tank filling until the tank is full. The pump does not start together with the simulation. First, the simulation is started. Then, later, as soon as the user is ready and wishes to do so, he or she may start the pump.

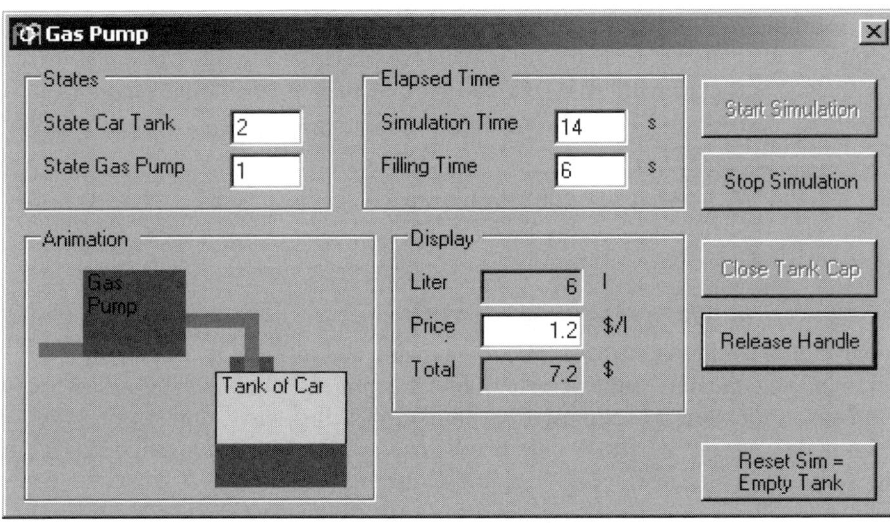

Figure D2-20: User interface of the simulation program.

The user can control the filling of the tank by two buttons: “Squeeze/Release Handle” and “Open/Close Tank Cap.” When the simulation is running, the user can start and stop the gas pump by “Squeeze/Release Handle” as often as desired. The simulation time keeps running, only the filling is interrupted.

Depending upon the state that the simulation occupies, the buttons that allow the user to interact with the system during the simulation may or may not be enabled. For example, the button “Close Tank Cap” is not enabled, as long as the gas nozzle is in the tank and the pump is running.

Animation

The screen shows a simple animation of the system. It acts mainly by changing the colors for the car tank and for the gas pump. The color of the animation corresponds to the current colors of the states. The picture of the pump appears red when the system is in the state "Pump off," and it is green during the state "Pump pumping." The tank cap, for which no State Chart was modeled, is colored red when it is closed and green when it is open. The gasoline level in the tank is violet during the filling and changes to blue when the state "Tank with enough gas" is reached.

In this example, 10 liters in the tank are represented by one pixel on the screen of the animation. The maximum content of the tank is 60 liters, making the tank in the animation 60 pixels high.

2.5.4 Coding of the Simulation Model

Convention in the code

This example was written according to the convention of Visual Basic .NET. In this chapter, the complete code for the simulation model of the gas pump is given. Comments and explanations in the code are written in italic and begin with an apostrophe.

Abbreviations in the code:	btn	button
	Sub	subroutine
	txt	text field

Variable declaration

In general, a parameter is declared with its name and its type of variable. It is helpful to indicate the unit of the parameter in the comment. The time in this simulation is given in seconds. For the sake of simplicity in this small model, the gasoline level in the car tank is calculated in pixels; one pixel equals one liter. The states for the `Select-Case` construct are declared as parameters as well.

```
'Declaration of time, given in seconds [s]:
Public ActualTime As Decimal      'Absolute simulation time
Public FillTime As Decimal

'Declaration of gas level in the tank: 1 pixel = 10 liters:
Dim glevel As Decimal

'Declaration of the states of pump and tank:
Public StatePump As Integer       'States of PUMP
Public StateTank As Integer       'States of TANK
```

User interface

By designing the user interface directly on the screen, the code of the user interface and its components is automatically created by Visual Basic .NET.

Initialization

The simulation program begins with the initialization. In this example, the initialization is coded in two subroutines. The Sub "btnSimStart" starts the simulation.

```
Private Sub btnSimStart_Click
    Clock.Enabled = True
    StatePump = 0
    StateTank = 0
    btnStopSim.Enabled = True
    btnSimStart.Enabled = False
    'Resume pumping where left after simulation stop:
    If FillTime > 1 Then
        StatePump = 1
    End If
End Sub
```

The Sub "btnReset" is the second part of the initialization. If the program is open and a new simulation is started, the parameters are first set to the initial value. When the user clicks on the reset button, the reset function is carried out. In this simulation, the emptying of the car tank is realized by resetting the content of the car tank to zero in order to start a new simulation run with an empty tank. This corresponds in the real world to the release of the gas pump after the paying.

```
Private Sub btnReset_Click
    Clock.Enabled = False
    'Set times to 0:
    SimTime = 0
    FillTime = 0
    'Reset animation:
    GasLevel.Visible = False
    GasLevel.Height = 8
    Pump.BackColor = Color.Red
    TankCap.BackColor = Color.Red
    Neck1.Visible = True
    Neck2.Left = 104
    Neck2.Top = 32
    Neck3.Width = 24
    'Reset text fields:
    txtClock.Text = "0"
    txtFillTime.Text = "0"
    txtStatePump.Text = ""
    txtStateTank.Text = ""
    txtLiter.Text = "0"
    txtPrice.Text = "0"
    txtPriceLiter.Text = "2"
    'Reset buttons on user interface:
    btnSimStart.Enabled = True
    btnStopSim.Enabled = False
    btnTankCap.Enabled = False
    btnHandle.Enabled = False
End Sub
```

System clock

The system clock counts the elapsed time. After having started the clock for the simulation, the public subroutines for the simulation process ("Pumping" and

"Tank"), the "Display," and the "Animation" are activated. For each time increment, the Sub "Clock_Tick" makes a cyclic call of all subroutines.

```
Private Sub Clock_Tick
        SimTime = SimTime + Clock.Interval
        Pumping()
        Tank()
        Display()
        Animation()
End Sub
```

System States

For each State Chart, a public subroutine is written. The shade of gray of the `Select-Case` construct correspond to the shade of gray of the states in the State Charts *Diagram D2-9* and *Diagram D2-10*.

```
Public Sub Pumping
Select Case StatePump                       'States of the PUMP
    Case 0                                  '1.1.1 Pump off
        txtStatePump.Text = 0
        btnTankCap.Enabled = True
    Case 1                                  '1.1.2 Pump pumping
        txtStatePump.Text = 1
        'Filling rate of the car tank:
        FillTime = FillTime + 1
        StateTank = 2
        btnTankCap.Enabled = False
End Select
End Sub
```

In every case of the `Select-Case` construct, the code line txtStatePump.Text or txtStateTank.Text writes the number of the current state into the output textboxes on the display, so that the user can supervise the simulation.

```
Public Sub Tank()
Select Case StateTank               'States of the TANK
    Case 0                      '1.2.1 Tank empty, gas pump ready
        txtStateTank.Text = 1
        glevel = 0
        btnHandle.Enabled = True
        btnHandle.Text = "Squeeze Handle"
    Case 1                      '1.2.2 Tank filling
        txtStateTank.Text = 2
        glevel = FillTime
        btnHandle.Enabled = True
        btnHandle.Text = "Release Handle"
        'If Tank full then:
        If GasLevel.Height = 60 Then
            StateTank = 3
            StatePump = 0
        End If
```

```
        Case 2                          '1.2.3 Tank with enough gas
            txtStateTank.Text = 3
            glevel = FillTime
            If GasLevel.Height = 60 Then
                btnHandle.Enabled = False
            Else
                btnHandle.Enabled = True
                btnHandle.Text = "Squeeze Handle"
            End If
            btnSimStart.Enabled = False
        Case 3                          '1.2.4 Using gas (Emptying tank)
            txtStateTank.Text = 4
            btnHandle.Enabled = False
End Select
End Sub
```

Display of time, quantity, and prices of gasoline

The subroutine “Display” fills the text of the calculated times into the output textboxes of the time. It calculates also the quantity of the gasoline, which was filled in the car tank, and the price for this quantity.

```
Public Sub Display()
    'Show time of filling and emptying in text fields
    'The division by 1000 converts the simulation time
    'into seconds for indication in textbox.
    txtClock.Text = SimTime / 1000
    txtFillTime.Text = FillTime
    'Show liter and total price:
    txtLiter.Text = FillTime * 1
    txtPrice.Text = txtLiter.Text * txtPriceLiter.Text
End Sub
```

Code for animation

The code in the Sub “Animation” handles the rising of the gasoline level and the color changing of the pump symbol and the tank cap. The gasoline level is a simple label that becomes bigger according to the elapsed filling time in the Public Sub “Tank.” The value of the gasoline level is transferred by the variable “glevel” to the Sub “Animation,” where it is calculated into the height of the label.

```
Public Sub Animation()
    'GasLevel rises in steps of 1 pixel:
    GasLevel.Top = 144 - glevel * 4
    GasLevel.Height = glevel * 4
    'Color of pump
    If StatePump = 0 Then           'Pump off
        Pump.BackColor = Color.Red
    ElseIf StatePump = 1 Then       'Pump pumping
        Pump.BackColor = Color.Green
    End If
```

```
    'Color of tank cap:
    If StateTank = 0 Then
        TankCap.BackColor = Color.Red
        btnTankCap.Text = "Open Tank Cap"
    Else
        TankCap.BackColor = Color.Green
        btnTankCap.Text = "Close Tank Cap"
    End If
    'Color of gas level in car tank:
    If StateTank = 0 Then
        GasLevel.Visible = False
    Else : GasLevel.Visible = True
    End If
    'Color Petrol in Tank
    If StateTank = 2 Then
        GasLevel.BackColor = Color.BlueViolet
    ElseIf StateTank = 3 Then
        GasLevel.BackColor = Color.Blue
    End If
    'Position of gas nozzle
    If StateTank = 0 Then
        Neck1.Visible = True
        Neck2.Left = 104
        Neck2.Top = 32
        Neck3.Width = 24
    Else
        Neck1.Visible = False
        Neck2.Left = 120
        Neck2.Top = 56
        Neck3.Width = 40
    End If
End Sub
```

Command by user

Commands from the user by the buttons on the interface bring the system into a specific state. The buttons btnTankCap "Open/Close Cap" and btnHandle "Squeeze /Release Handle" toggle twin functions. They change their purpose depending to the current state of the system. The texts on the buttons change accordingly.

```
Private Sub btnTankCap_Click
    If StateTank = 0 Then
        'Change to: Close Tank Cap:
        StateTank = 1
    ElseIf StateTank = 1 Then
        'Change to: Open Tank Cap:
        StateTank = 0
    ElseIf StateTank = 3 Then
        'Change to: Open Tank Cap:
        StateTank = 0
    End If
End Sub
```

According to the System State, the button btnHandle starts or stops the pumping. At the same time, the text on the button changes. The text appearing on the button is determined in the `Select-Case` construct of the private Sub "Tank."

```
Private Sub btnHandle_Click
  If StateTank = 1 Then
    'Change to Release Handle:
    StatePump = 1
    'Resume pumping after a Stop:
    FillTime = txtFillTime.Text
  ElseIf StateTank = 2 Then
    'Change to: Squeeze Handle
    StatePump = 0
    StateTank = 3
  ElseIf StateTank = 3 Then
    'Change to: Release Handle:
    StatePump = 1
    'Resume pumping after a Stop:
    FillTime = txtFillTime.Text
  End If
End Sub
```

```
Private Sub btnStopSim_Click
    Clock.Enabled = False
    btnTankCap.Enabled = False
    btnHandle.Enabled = False
    btnSimStart.Enabled = True
    btnStopSim.Enabled = False
End Sub
```

Go ahead and try it yourself! It's not difficult.

2.6 Apply Your Knowledge

1) From Flow Diagram to code

What are the steps from Flow Diagram and State Chart to code? List them.

2) Transition from Flow Diagram to State Chart

The Flow Diagram in *Diagram D2-12* describes the use of a common microwave oven.

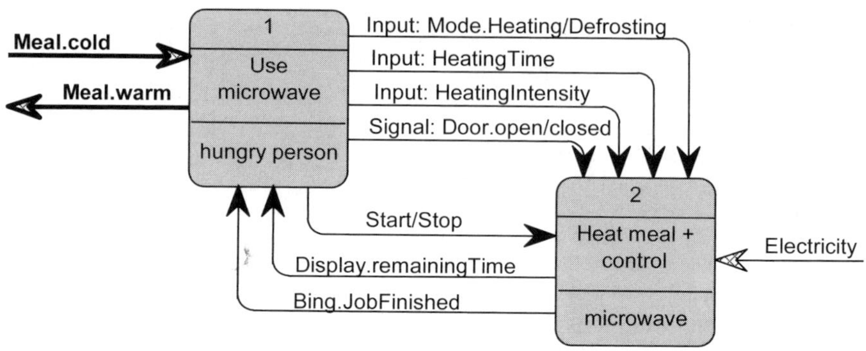

Diagram D2-12: Microwave oven.

Diagram D2-13 depicts the State Chart for process "Heat meal + control." Write the conditions and actions based on the input and output flows of process 2. Some additional transitions are needed to get a complete State Chart.

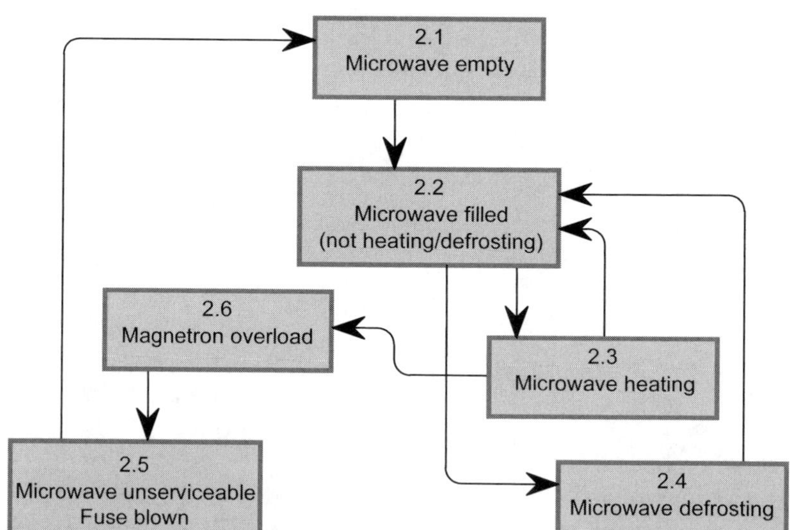

Diagram D2-13: State Chart "Heat meal + control."

3) Flow Diagram or State Chart

Figure D2-21 shows a sketch of a roller coaster with two trains. The station consists of a front section for boarding and a back section for getting off the train. This enables also both trains to be at the same time in the station.

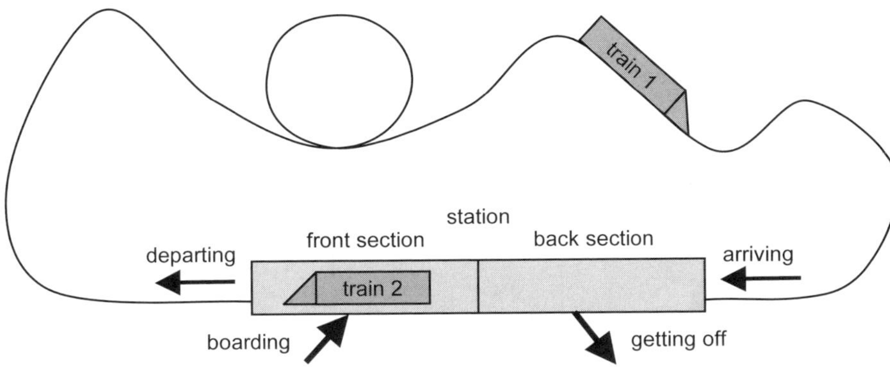

Figure D2-21: Roller coaster.

Diagram D2-14 describes the roller coaster on a first level with two processes.

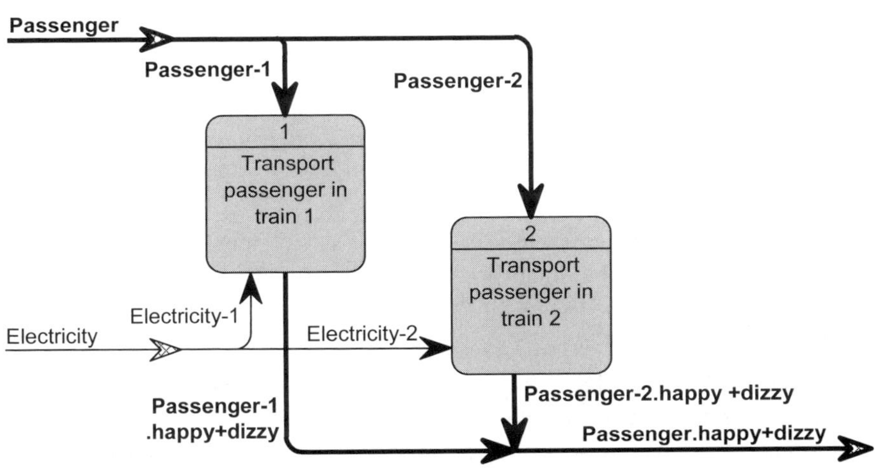

Diagram D2-14: Roller coaster.

Process 1 "Transport passenger in train 1" of this Flow Diagram is detailed into a child Flow Diagram (*Diagram D2-15*).

The passenger flow runs straight through all processes in a sequential manner. Processes and flows form a circuit. There is no network of flows. Somebody started to draw a Flow Diagram and got stuck. In this case, drawing a State Chart is preferable. Replace the Flow Diagram in *Diagram D2-15* with a State Chart. It is the child State Chart of process "Transport passenger in train 1."

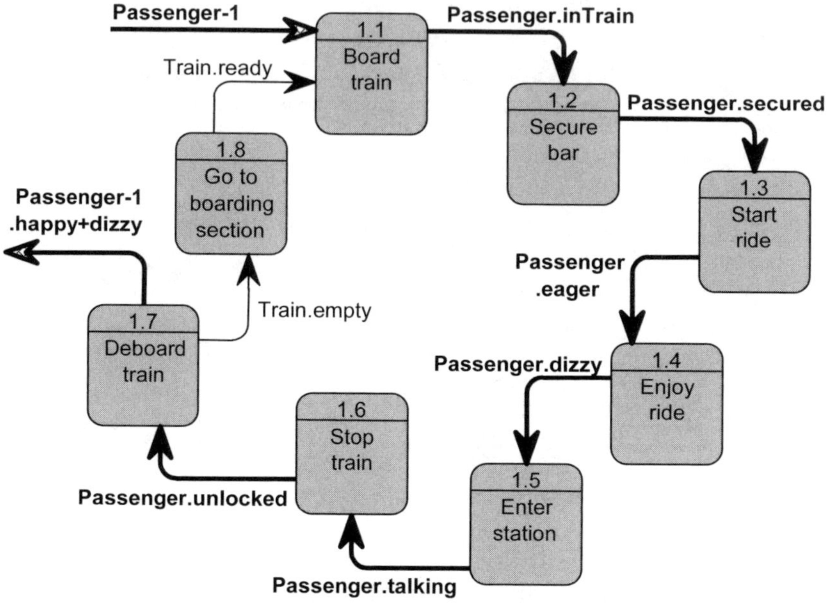

Diagram D2-15: Child Flow Diagram of process 1 "Transport passenger in train 1."

When drawing the State Chart, it is easier to insert states that all describe the process from the same point of view: either the point of view of the passenger or of the train.

4) User interface

Draw the context diagram for a user interface for a simulation of an elevator for the transport of persons in a building. What is the purpose of such a simulation? What are the interactions of the user with the simulation program? What are the inputs and outputs of this simulation?

In a second step, design the user interface with the components known from every computer program. The most common components are listed in *Table D2-3*. In this case, would an animation be useful? How does it look like? An animation may be quite simple. In most cases, it is sufficient to show the different parts of the system as boxes, which change the color or text according to the states they occupy at every point in time.

5) Code based on State Chart

Traffic is an often investigated and simulated problem. *Figure D2-22* illustrates a pedestrian crossing, directed by a traffic light. The pedestrian can trigger the green light for crossing by pressing a button at the traffic light.

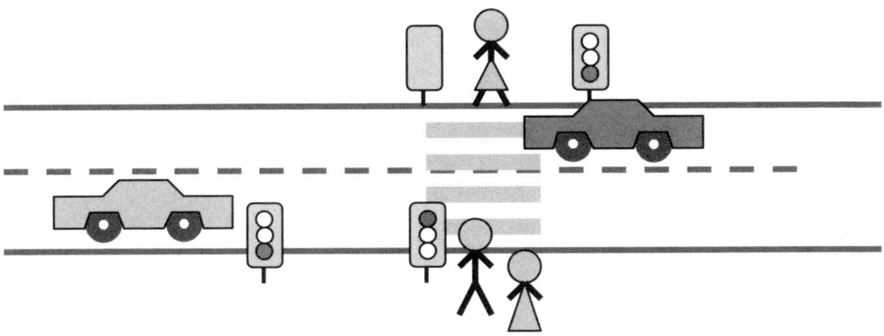

Figure D2-22: Pedestrian crossing.

State Chart in *Diagram D2-16* describes the traffic light for the pedestrian crossing by four states. The condition to change the traffic light in order to stop the cars is triggered by pressing the button at the pedestrian light. The other conditions become true after a certain amount of time elapsed. (TL Traffic light)

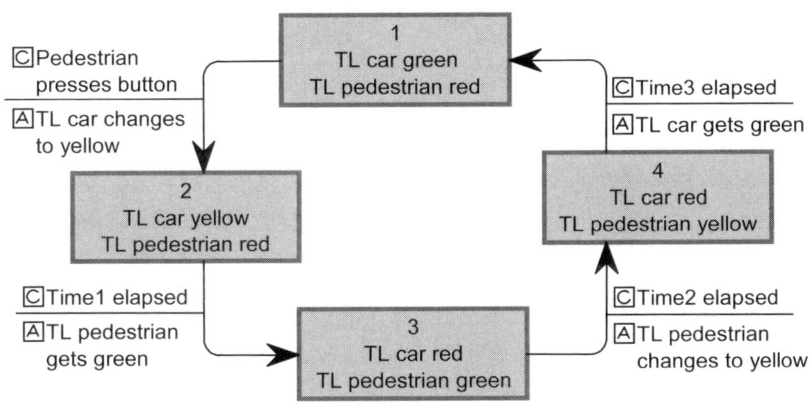

Diagram D2-16: State Chart of traffic light.

Based on the State Chart above, complete the `Select-Case` construct below. How and when does the system change the state? If you are not familiar with coding, just insert into the cases in plain language what the program has to do.

```
Public Subroutine Control_TrafficLight ()
      Select Case TrafficLight
            Case 1    'TL_Car_Green (TL_Pedestrian_Red)



            Case 2    'TL_Car_Yellow (TL_Pedestrian_Red)
```

```
                Case 3    'TL_Pedestrian_Green (TL_Car_Red)


                Case 4    'TL_Pedestrian_Yellow (TL_Car_Red)


        End Select
End Subroutine
```

6) Program modules

List the usual program modules and explain them. Draw the cyclic program run that contains all these modules and indicate how the program runs through these modules. Where is the Start/Stop button for the simulation located?

7) Evaluation of a simulation software

How can you use the Flow Diagram and the State Chart to evaluate a common simulation software package? How do you use Flow Diagram and State Chart in evaluating the performance of a simulation tool? How are the simulation program components represented by Flow Diagram and State Chart?

8) Difference optimization simulation and real-time control

Explain the difference of a simulation for optimization of a plant and the simulation to test a real-time control. How does the user of the test simulation interact with the program? What is the purpose of animation in a simulation program? Is a fast/slow motion function also useful on a real-time control test simulation?

D3 REAL-TIME CONTROL

Goals of the Chapter

- Distinguish operating and exceptional states of a production process and model them on the State Chart.
- Learn to design and code a real-time control by using the State Chart.
- Get to know the specification of a system by its State Domain.
- Code the program of a real-time control based on the State Chart.

Figure D3-1: Setup, coding, and text of a real-time control (Photo: ETH Zurich).

3.1 Introduction

Errors within computer programs make trains collide, space shuttle launch fail, airplanes crash, and a lot of nicely manufactured products ending up as garbage. Most functions of real-time controls can ultimately only be tested and validated in a real-world environment. Regarding source code debugging and maintenance of programs, a systematical, error-free, and complete documentation is crucial, starting with the analysis and leading over the entire life-cycle of the system. The instruments used in this respect are CASE tools. The tool presented here is Process Oriented Analysis.

Before putting a real-time control into operation, a simulation is used to test the control in a simulated environment. The test simulation is developed in the phases of system conception and programming, concurrently to the real-time control. The difference between a real-time control and a test simulation lies in the interfaces to the environments. A test simulation model includes the interfaces of the control to the simulated environment and requires several inputs from the user acting as operator. A real-time control on the other hand is connected to the real outside world, including operators and supervisors.

Design and maintenance of real-time programs use up an ever-increasing part of the development efforts for modern production machinery. The approach to make this more efficient is given by CASE tools. Common characteristics of all CASE tools is the use of graphical representation in addition to text used in classical programming. There are CASE tools on the market that convert a State Chart automatically into source code. This kind of CASE tool is called Lower CASE. POA is an Upper CASE method and supports the system conception and structuring by diagrams (see also Chapter I2.2). POA diagrams specify a complete system and support traditional programming with any of the lines-of-code based universal programming languages. The choice of the programming language is up to the programmer. The POA State Chart may also be converted directly into structured text as PLC (Programmable Logic Controller). The programming interface of most PLC are Lower CASE tools.

POA may be used to prepare such programming by diagrams, but its primary application is analysis and design of systems. POA leaves the method of coding open to the discretion of the engineer.

This Section D3 is based on the State Chart introduced in Section D1. It contains the design and setup of a program for real-time control. The State Domain and the State Map are explained. They visualize the correlation between several State Charts describing the same system and serve for the specification of the normal operation states and the identification of exceptional and risky states of the system. An application example of a gasoline pump explains the design and coding of real-time programs in practice. Case Study C6 shows the concept for a real-time control of a cable car.

3.2 POA for Real-Time Control: Why?

3.2.1 Purpose

Real-time control programs with digital processors are used for the operation of simple applications, of complex machines, or even entire production plants. The programming of these programs is supported by POA, which allows a straightforward translation of graphical system analysis into code. The method is applicable for any kind of system with discrete behavior in time, especially for microprocessors-based controls.

Program inputs from outside: The growing complexity of production machinery calls for powerful methods for design and maintenance of machine control systems. Within a machine control, the computer interacts with the environment in real time, taking care of a multitude of activities in the ongoing process. The system is event-driven by real-world effects. Also the activities carried out by human beings, such as setup, operation, troubleshooting, and maintenance, are included. The behavior of the program is controlled by a multitude of inputs from the outside world by machine interfaces, such as sensors and contacts, and by operator interfaces, such as displays, keyboards, and touch screens.

Program outputs to user or machine: The computer in turn impacts the outside world by actuators, motors, valves, pneumatic and hydraulic cylinders, magnets, clutches, and brakes and updates the operator display. In contrast to a software program, the computer program here has an immediate impact on the physical surroundings. The completeness and correctness of program analysis is essential and a prerequisite for safe operation.

The controller may be a dedicated single chip processor or a board-level computer; it may be a compact PLC or a full-size personal computer. The choice of available hardware, programming systems, and operating systems is extremely wide, and getting even larger by the day.

Real-time control and simulation: The procedure of designing and coding a machine control program or a simulation model is the same up to a certain point. The difference between these two kinds of program lies in the parameter input and output of the program at run time. In the case of a machine control, the interaction occurs by various hardware interfaces; the data inputs, such as actual temperature, position, or velocity, are supplied by sensors on the machine. For safety reasons, all processes are monitored constantly by threshold values. The real-time control program operates in actual real time.

In a simulation, process parameters are delivered partly by the simulation program, reference to the real time is missing, and safety functions can be omitted. Inputs are provided by a simulated machine and a simulated environment, running concurrently within the simulation program. The simulation model operates in accelerated or slow motion time.

Design and integration of the process: The programming of a real-time control by POA is based strictly on the real-world function and hardware of the system. The naming of system components is also based on real-world designations. This ensures that the program is as clear to any user as the functions it represents.

3.2.2 Application

The function of an automatic control of a dishwasher, as an example, becomes clear using a State Chart that shows the states of the process. The dishwasher contains several modules: water inlet and outlet valves, a detergent dispenser, a heating element for the water, the main pump for the rotating arm that sprays water on the dishes, a drain pump, and a fan to dry the dishes.

A state-based view of the system describing the interaction of these system modules is used to develop a machine control. The State Chart allows the user to define the state of the pertaining system module at any point in time (e.g. the dispenser is opened only when water is present). Based on the State Chart shown in *Diagram D3-1,* the control of a dishwasher can be programmed.

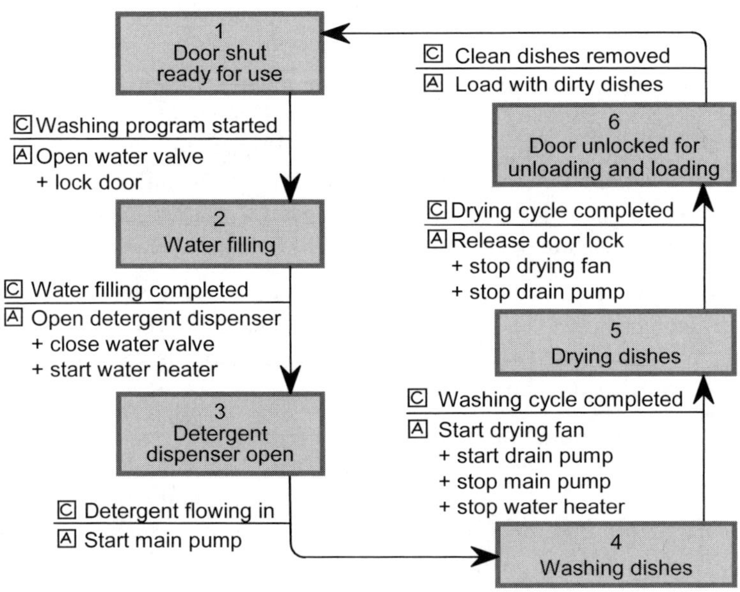

Diagram D3-1: State Chart of a dishwasher.

It is essential for the safety of human beings and the environment, to consider all possible System States for a real-time control: normal operational states as well as problem and risk states. Often, a system is described by several State Charts. The overall System State is composed of states from these various State Charts. The System Domain is a representation that helps in specifying the System State composed in this way.

In Case Study C6, a new concept for the operation of an underground cable car is worked out and implemented in a real-time control program. It is tested with a simulation in order to check its functionality.

3.2.3 Delimitation

In conventional coding, the ways to make a simulation program for optimization of a plant differs from those to make a real-time control, as shown in *Figure D3-2*. A simulation specialist will program the simulation, while a real-time program specialist will program a real-time program including a simulated test environment. A simulation specialist can define his system according to the desired completeness and accuracy. For a simulation, the comfort and performance offered for data evaluation is important, the duration of the simulation runs should to be as short as possible. For the real-time control programmer, the system is defined by the given hardware and its interface to the environment, which also includes the time axis.

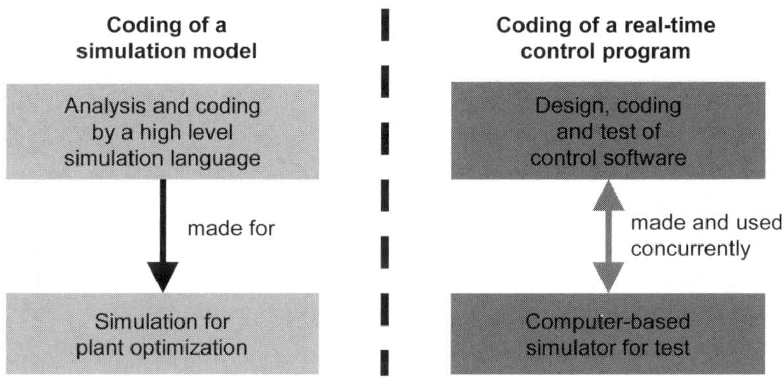

Figure D3-2: Procedure with conventional programming.

POA offers a new way of coding a simulation model or a real-time control for both types of programs. *Figure D3-3* shows that there is a common set of preparatory steps that serve to either purpose. The analysis phase comprises a static analysis by Flow Diagrams and a dynamic analysis by State Charts. The Flow Diagram ensures that the program structure is correct and corresponds to the real-world physical system. The behavior of the system in time is described by the State Chart. On the basis of this analysis, the same procedure serves to code a simulation model or a real-time control program. Either a simulation model for a production plant optimization is created, or a real-time program is coded, along with a simulation program for testing.

For many applications, it is sufficient to describe a system only by its static behavior. A process specifies a transformation of material or data. This transformation is identical to the task of a program module in computing. As long as we disregard its behavior in time, such a module is just a set of equations that are permanently valid for the process. The means to solve such a set of equations is the well-known spreadsheet calculation. Any diagram with processes and flows may be converted into a spreadsheet; flows are represented by variables, and processes are represented by equations that relate the variables to each other. However, the calculation with a spreadsheet does not take into account the behavior of the system in time. To overcome this, POA offers the method of dynamic modeling and coding.

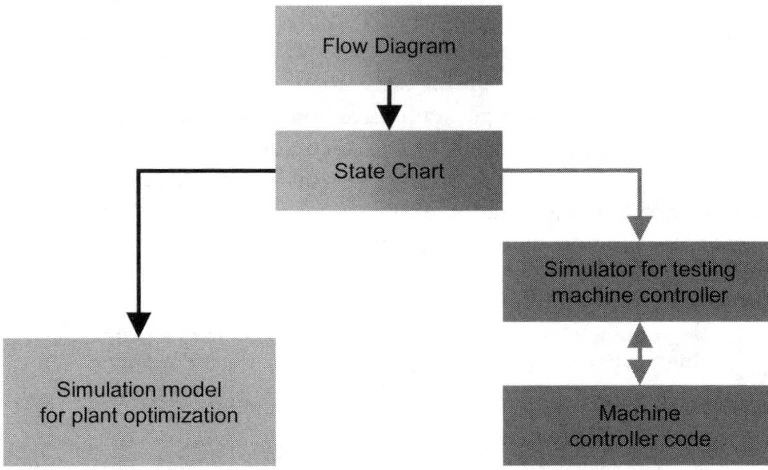

Figure D3-3: POA path to simulation and real-time coding.

Typical instruments to program structured code are Lower CASE tools (see also Section I2). Existing Lower CASE tools are suitable for coding short programs of limited complexity. Examples are programming languages based on Petri Nets for PLC, such as the Sequential Function Chart.

The workload to generate source code however, is not really smaller than with other methods. The use of Lower CASE tools is not popular for programming PC controls. In general, they do not support a structured view of the system with different levels of detail. For this purpose, a higher level of abstraction is required, for which Upper CASE tools are used. POA is an Upper CASE method with some of the Lower CASE tool features.

3.2.4 Definitions

Program module: A computer program consists of program modules. The system under investigation is split up into modules representing individual functions and processes. The program module may be an user interface module, a process module, or an auxiliary module. Interface modules contain the code for the operator surface or console. Process modules correspond to the State Charts and calculate the momentary System State. Auxiliary modules are for example the variable declaration, the initialization, and the system time clock.

Definition: A program module represents a single function of a program. A program module may be a process, interface, or auxiliary module.

System State: A system consists of at least one, but generally a multitude of process modules. At a given point in time, each process module is in one specific state, the active state. The combination of these active states is the System State.

Definition: The System State is a combination of the active states of the processes within the system at a specific point in time.

State Domain and State Map: The State Domain is a multi-dimensional space representing the System States. It is a theoretical construct. In order to be able to graphically represent specific aspects of the System State in an easy way, the State Map is used. The State Map is a two-dimensional graph, and as such an extract of the multi-dimensional State Domain. It depicts two State Charts against each other. The System State, the State Domain, and the State Map are interlinked.

Definition: The State Map is a two-dimensional graphical representation of the State Domain of the system.

3.2.5 History of Manufacturing Automation

The historic development from manual labor to automated production can be explained using Flow Diagrams. It is also the history of the development of the real-time control. In the following, this development of automation is shown on the example of the grinding of wheat to flour.

Figure D3-4: Person using a hand mill (Photo: Römerstadt Augusta Raurica, Ursi Schild).

If the work is done manually, the worker provides the resources, the power, as well as the intelligence for the control. All of this is completely described by one single process. *Diagram D3-2* illustrates the manual grinding of wheat between two stones, as shown in *Figure D3-4*.

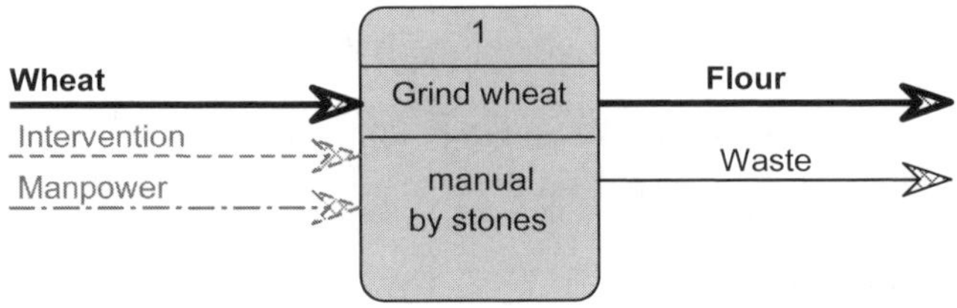

Diagram D3-2: Manual grinding.

Later in history, machines were developed to increase production by using external power. The operator still controlled the device completely by himself and provided all the intelligence for this job. However, power was supplied by an external source, not by the operator, for example by water, steam, wind, or electricity.

Figure D3-5: Water wheel of a mill (Photo: ETH Zurich).

Diagram D3-3 describes the grinding of wheat in a mill driven by a water wheel, as depicted in *Figure D3-5*. Process 2 "Operate mill" includes also the control of the grinding by the operator. He supervises the grinding, adjusts the flow of water, and interacts, when a problem occurs.

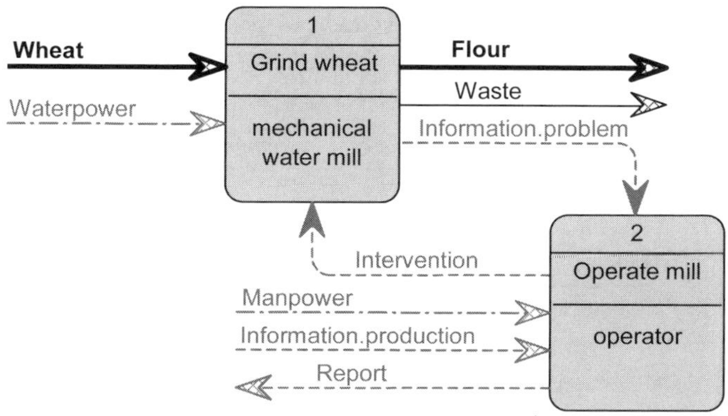

Diagram D3-3: Mechanical grinding.

Computers made a further step possible: automation. Software now takes over the control of the machines. The intelligence, formerly provided entirely by the operator, is now to a great extent supplied by the machine controller. The job of the operator is cut back to interventions on the control and the machine.

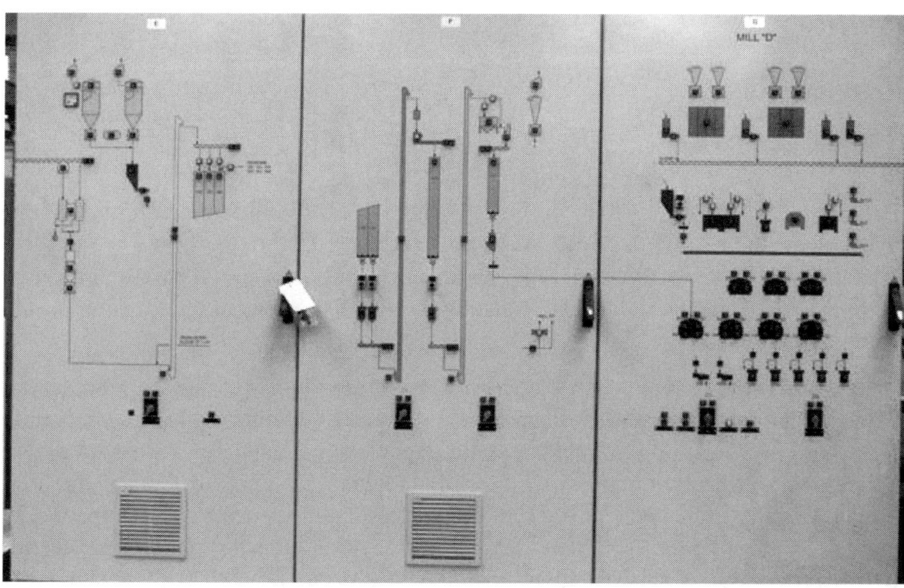

Figure D3-6: Control panel of a modern fully automated wheat mill (Photo: Bühler AG).

Automation means to remove the jobs that are directly involved in handling and processing the product. By this, not only manual operations are taken over by machines, but also the observation of machinery while running and the check of materials in process. It is essential to make comprehensive information on the status of production available to the remaining persons in the plant, so that they can intervene in time in case of failures. The elements used to implement this are plant-wide date networks and centralized process-control systems.

The Flow Diagram serves to specify the data interfaces of the different processes, the structure of the data communication in the plant, and the man-machine interface on the centralized control panel. *Figure D3-6* shows the panel of a modern fully automated mill in the control room.

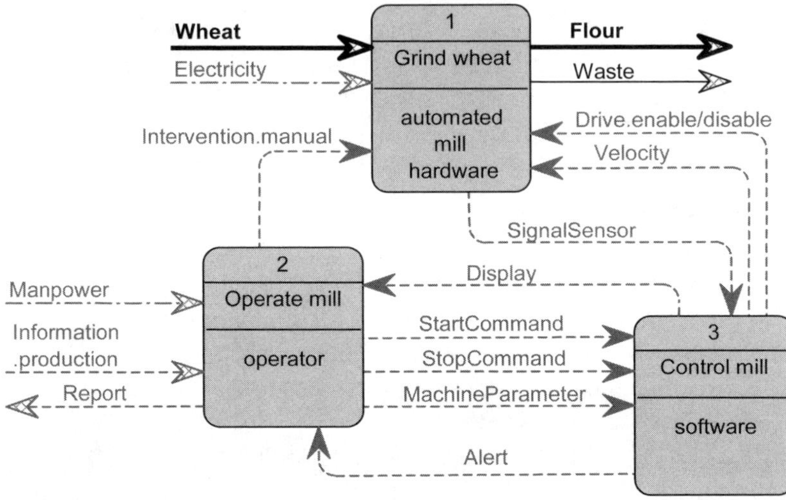

Diagram D3-4: Automated grinding.

Diagram D3-4 describes the grinding of wheat in a fully automated mill. It shows the typical division of the processes of an automated production in three aspects: operator, functions of the machine (hardware), and software. The two processes "Grind wheat" and "Control mill" distinguish the aspects of the hardware and the software.

The information flows (dashed arrows) show that most of the communication between the operator and the machine is handled by the control. The manual interventions remaining for the operator are cleaning and clearing of stops and breakdowns. If a malfunction of a machine occurs, the need for manual interventions is usually indicated to the operator by a display, if necessary accompanied by optical or acoustic alerts. These manual interventions are the only interfaces between process 1 of the machine "Grind wheat" and process 2 "Operate mill," specified as flow "Intervention.manual."

3.3 Machinery States of Manufacturing Processes

3.3.1 Operating and Non-Operating States

The dynamic description of a machine is based on the Flow Diagram. For each process, the behavior in time is defined by a State Chart with several states. All production machines and plants require a minimum set of states, which is determined by their basic functions. Theses basic states are listed in *Table D3-1*. The same list of those states is also found in *Table D1-1* of Section D1.

Table D3-1: Minimum set of states for a production machine

State	Description	Examples
De-energized state	Machine inoperable and safe for the environment. Intrinsically safe state. It precedes a cold start and terminates a shutdown.	External power cut off
Stand-by state	Machine is ready and available any time for operation. Power saving mode, while control circuits stay active.	Waiting for material
Target state	Active operation. Represents the machine working in its normal operating mode. Purpose of the system.	Motor running Machine producing
Exceptional state	Machine not correctly operating, not following its intended functions without risking damage. Demands a human intervention.	Alert Processing interrupted, waiting for intervention
Risk state	Machine posing risk to man, machine, and/or environment. This requires immediate attention and intervention to bring the system into a safe state.	Motor overheating Leakage of liquids Loss of control

The *target state* of production machinery is the state of active operation according to its purpose. This is usually defined first in machine design, since it is what the customer pays for and expects.

Another typical state for automatic machines is *stand-by,* waiting for upcoming operation. This often happens when incoming material is temporarily missing. It is essential to distinguish this state from other non-operating states, since the machine, by being ready anytime for operation, provides a service to the total system, which is in this case the plant.

It is required by law that at any point in time a machine must be able to be stopped and brought into an intrinsically safe, *de-energized state* by the activation of an emergency stop command. The manufacturer of a machine has to confirm this as part of his assessment of safety of operation. It is therefore essential to define this intrinsically safe state of the system early in the design of the machine functions. In this state, the machine is disconnected from all sources needed for operation, such as power and compressed air. Starting up the machine, in the sense of a cold start, begins with this state.

Exceptional states represent limited capability in operation or a malfunction that does not harm the safety of personnel or environment. They may allow operation to continue to a limited degree, while waiting for corrective action by an operator. An example of an exceptional state is with the operation of an injection molding machine: This machine melts plastic pellets and injects the melt into a mold where it solidifies. If the molded part remains stuck in the tool, the machine stops and waits for manual removal of the part, while the heating of the injection barrel is maintained.

A *risk state* poses a direct danger to operators, machine, and environment. It is mandatory to bring the system to a de-energized and therefore safe state by an instant intervention. This begins by triggering an alarm, and ends with the shut-down of the system.

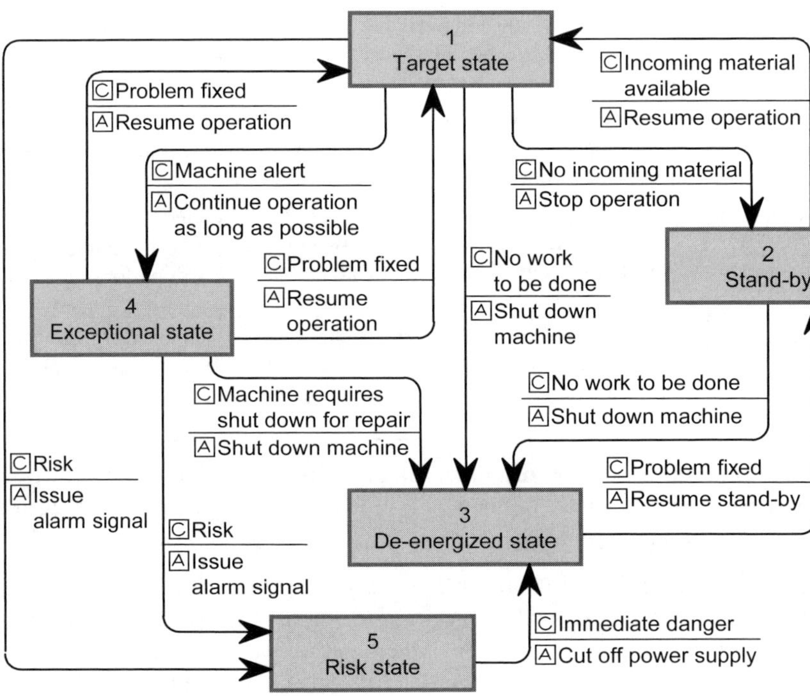

Diagram D3-5: General State Chart of machinery states.

Diagram D3-5 shows the basic states of a manufacturing machine, as explained above and listed in *Table D3-1*.

The following rules apply to the basic states of the State Chart. The rules for the real-time control are marked with the prefix RT.

Rules D3-1: Basic states

RT1	Normal operation of the machine includes the target state, representing the machine fulfilling its intended purpose without restriction, and the stand-by state, representing the machine being available instantly for operation.
RT2	In addition to all intended operational states (target and stand-by states), the State Chart must contain all technically possible states.
RT3	Risky or useless states are not allowed in production and must be avoided by suitable means. Transitions leading into states that are risky or do not serve a useful purpose, are not allowed in the real system.
RT4	Any state of the system not belonging to normal operation will trigger an alert and must have a transition to a recovery action.
RT5	Transitions must be available that will immediately switch the system from any risky state into the designated de-energized state.

3.3.2 Monitoring of System States

To maintain product quality and ensure safety of equipment, it is essential to continuously monitor and check the correct operation of the system. Traditionally, operation is stopped whenever a dangerous situation occurs. This is achieved by a specific action, e.g. by pushing an emergency stop button. This leads the system into a safe state, e.g. by switching off the main power supply. The critical issue here is to ensure that the potentially dangerous situation is detected in time.

In state-of-the-art digitally controlled systems, a different approach is taken. The process controller continuously monitors the system and ensures it is on the correct track of states. Any deviation creates an alarm or triggers a shutdown. In the State Chart, this shutdown of the system is represented by transitions that lead from any of the operating states to a safe, de-energized state.

Safety regulations for machinery make it mandatory that this de-energized state is safe for human beings under all circumstances and is accessible by distinctive transitions from any other state. If a controlled activity for shutdown is necessary to reach the safe state, e.g. applying a brake in order to dissipate the energy stored by a rotating mass, this transient period is again treated as a specific state of the system.

The following example shows the operation of an aerial cable car and the correct function of its drive system. The behavior of the cable car is described in relation to its position on the track. *Figure D3-7* schematically explains its course when approaching the terminal station.

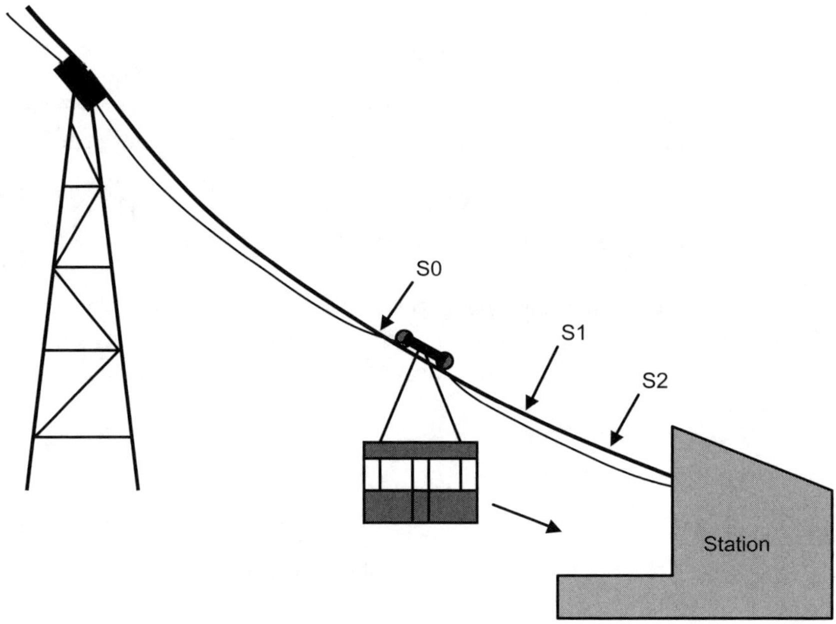

Figure D3-7: Aerial cable car on station approach.

S0, S1, and S2 refer to position initiators along the track of the car. S2 is the final position (gate) within the station. The speed of the cable car versus its position is shown on the graph in *Diagram D3-6*. The cable car cruises at a speed v_0 until passing position S0. After S0, the car decelerates to approach speed v_1 and holds this until passing position S1. At S1, the cable car decelerates to crawling speed v_2 until S2, where the crawling speed is finally reduced to standstill in the station.

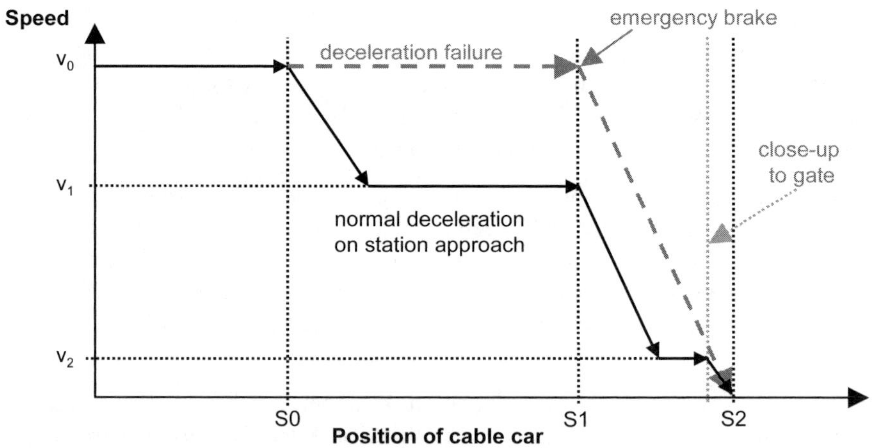

Diagram D3-6: Speed versus position of cable car.

In case of failure of the speed controller, the cable car would continue at cruise speed after passing position S0. This means, it would pass position S1 too early. If this happens, the emergency brake is activated and the cable car is stopped immediately.

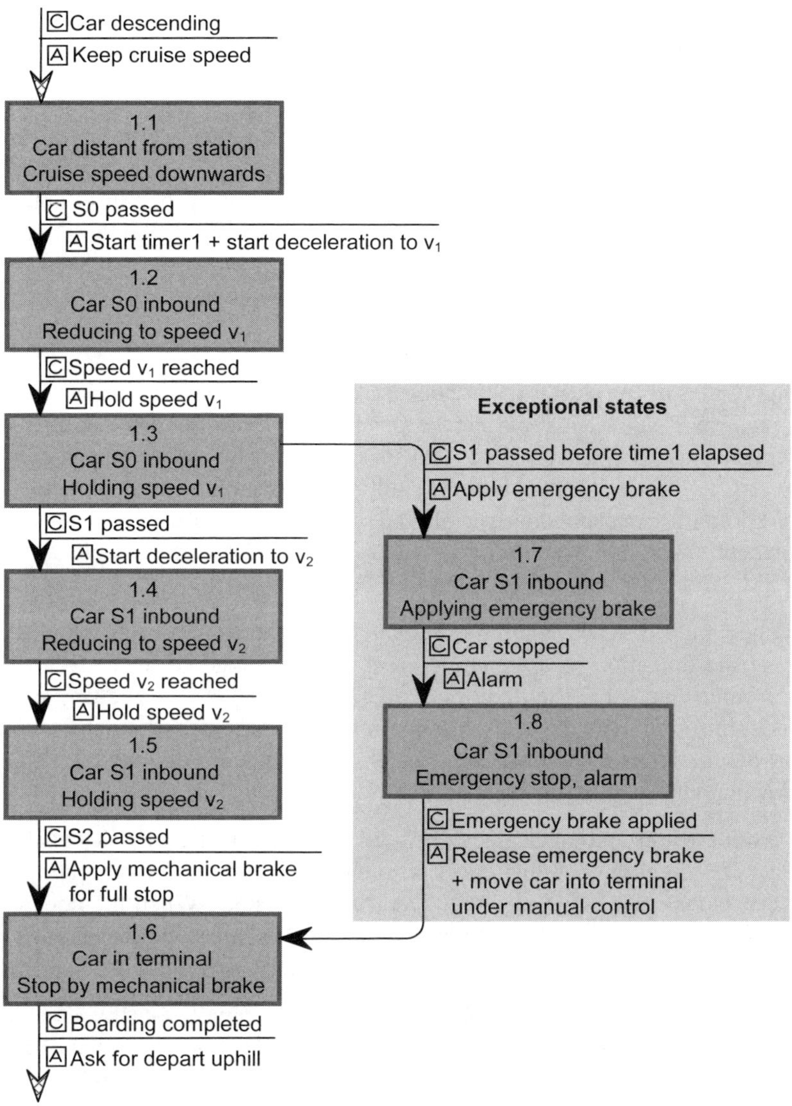

Diagram D3-7: State Chart of an aerial cable car descending to station.

Diagram D3-7 shows the State Chart of the speed controller of the aerial cable car. It enters exceptional states when the position S1 is passed too early in case the cable car did not decelerate sufficiently. This is detected by checking the time delay between passing S0 and S1. Any excess speed shortens this delay. A time-controlled relay triggers an alarm and initiates emergency braking instantly.

The continuous monitoring of the System State is an important feature of automation. It is realized in system design by the following methods:

- Parameters that are held constant within a given state are checked by applying a maximum and minimum limit. For example, the temperature of the oil in a pan used to prepare French fries is kept between 170 °C and 190 °C for frying.
- For parameters that change their value with time, either a critical threshold of this parameter is checked at a specific point in time or the duration of the pertinent state is checked by a time limit. The latter option is the most popular one, since it requires only an elapsed time counter based on the internal computer clock.

3.3.3 Failure Handling

An important side product of this concept of programming is the handling of a system failure. It is a basic requirement for machine control to detect faulty sensors and actuators as early as possible, and to safeguard the safety of personnel and machinery. Theoretically, the case of a faulty system component may be treated by simply adding a corresponding set of states. In reality, the challenge of handling these exceptions in the behavior of a system is first to detect the trouble and then to get an indication of its source and nature.

This is generally done by adding a double-check criterion on each state, as stated above. If the transition to leave the current state is triggered manually or by parameter other than time, the duration of the state is double-checked by minimum and maximum limits for the elapsed time. In the inverse case, where the duration in time of a state is controlled, other physical parameters are double-checked, e.g. a total throughput or a maximum speed.

The example of an automatic door operated by a pneumatic cylinder is shown in *Diagram D3-8*. On the command to shut the door, the system makes the transition from the "Door open" state into the "Door closing" state. When the door is completely shut, as confirmed by a limit switch, the system is in the "Door closed" state.

The controller checks the duration of the "Door closing" state by clocking the elapsed time between the arrival of the command to shut and the signal of the limit switch confirming the "Door closed" position. If this delay in time exceeds the duration of a normal door closing action, the fault is either in the door closing drive or in the limit switch. If this time is shorter than usual, there is either a fault at the "Door closed" switch (it could by mistake be closed all the time), the door was blocked in a intermediate position, or the door was not completely open before. This would point to a defective "Door open" switch.

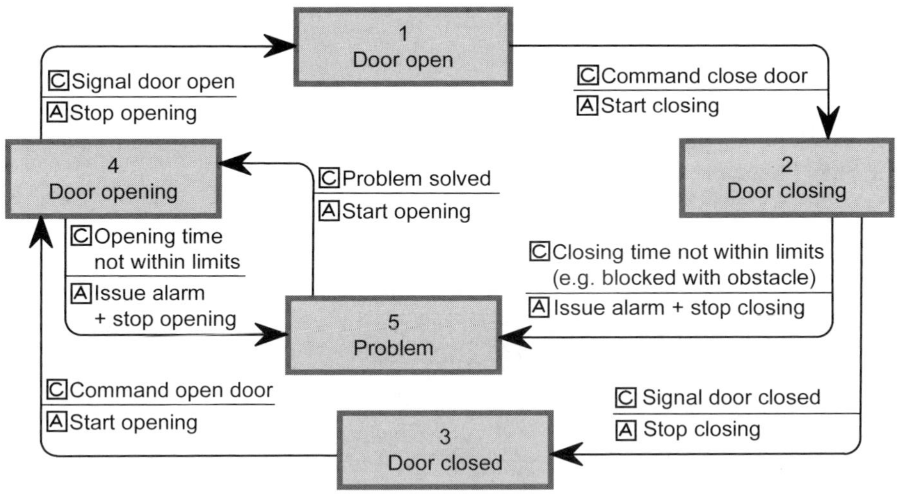

Diagram D3-8: State Chart "Operate door."

This kind of continuous monitoring of system behavior is very popular since it is executed by just a few lines of code that rely on data supplied by the internal computer clock and on sensors that are already used for control. Much of the reliability demonstrated by controllers of complex automated systems is due to this comprehensive and continuous monitoring of the sequence of System States. However, three specific points have to be observed in order to avoid trouble in the real-world shop-floor application:

- The state previous to shutdown and all conditions leading to the failure state must be memorized permanently for debugging. Only by this, the chain of events leading to failure can be reconstructed later.
- The thresholds used on sensor signals should be parameterized and adjustable by maintenance personnel, in order to make them helpful rather than a nuisance in the daily operation of the plant.
- These thresholds are also part of the safety concept for the pertaining machine. Settings for safety relevant functions are only accessible to a specified group of persons that has the necessary system knowledge, is aware of possible risks, and knows the safety concept.

The risk state is further treated by a risk analysis, which is part of the machine documentation and is required in most countries for the commissioning of machinery. The State Chart is an important basis for the FMEA (Failure Mode Effect Analysis), which is an important part of risk assessment. A security concept for the whole system involves further detailing of the risk states and the specification of their effects on the environment.

3.4 System View in the State Domain

3.4.1 Purpose of the State Domain

Figure D3-8 shows the relationship between the State Chart, the System State, and the State Domain.

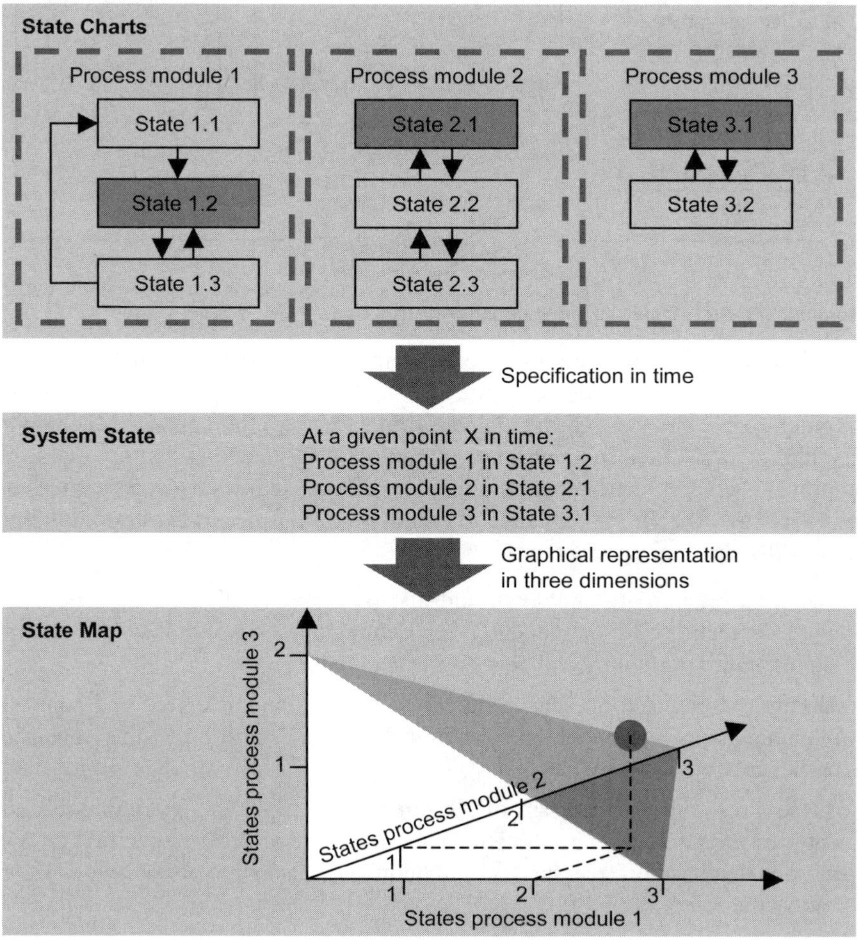

Figure D3-8: Connection between State Chart, System State, and State Map.

The states and transitions of a system are defined by the process modules of the program. At each point in time in a program run, each of these process modules is in a certain state. The combination of the states of the process modules defines the System State. The parameters describing the states of the system may be regarded as the dimensions of the State Domain. If the State Domain is represented in a two-dimensional graph, it is called State Map.

3.4.2 System with Discrete Parameters

The example of a tape recorder demonstrates how a system with two processes is depicted in the State Domain. The analysis starts by modeling of the system of the tape recorder on a Flow Diagram, shown in *Diagram D3-9*.

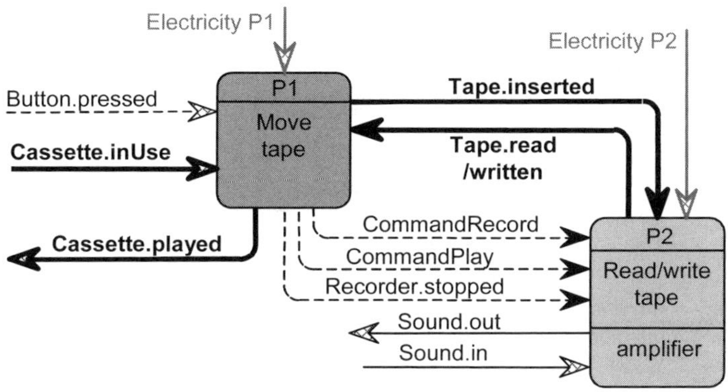

Diagram D3-9: Flow Diagram of a tape recorder.

A tape recorder basically consists of a drive assembly that moves the tape, and an electronic amplifier that handles all the signals involved with recording and play back of the tape. In order to define the two processes of the Flow Diagram in the State Domain, a State Chart is drawn for each process.

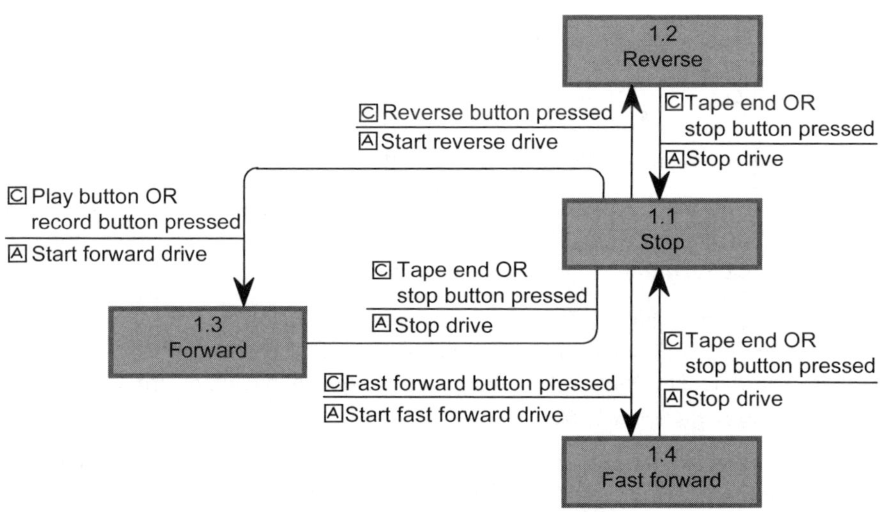

Diagram D3-10: State Chart "Move tape."

Diagram D3-10 shows the State Chart of the process "Move tape." This State Chart has four states that describe the direction and velocity of the tape movement. The Flow "Button.pressed" from *Diagram D3-9* bundles the actuation of the different buttons of the tape recorder. This flow transforms into different actions in the State Chart, depending on which button is pressed.

The State Chart of the process "Read/write tape" in *Diagram D3-11* describes the states of the amplifier.

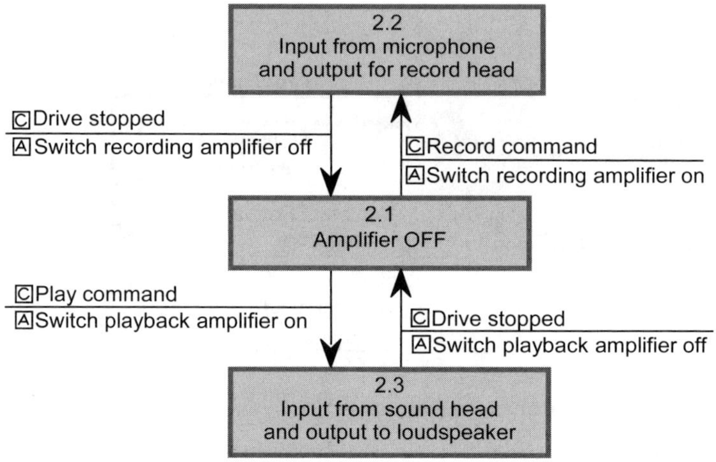

Diagram D3-11: State Chart "Read/Write tape."

These two State Charts may be combined in the State Domain to give an overview on all states of the complete system. The State Domain is used to study the system behavior for all possible combinations of states, including the cases of component failure. Mapping the states of the tape recorder in a State Domain results in a two-dimensional State Map of four states by three states. The State Map given in *Diagram D3-12* has two axes. The x-axis shows the states of the State Chart "Move tape," and the y-axis shows the states of the State Chart "Read/write tape."

Of the twelve possible System States on the State Map, five System States are needed for the standard operations of a tape recorder. These include stop, recording, playing, rewinding, and fast forwarding the tape.

There are seven possible System States that indicate a failure or are destructive. The five states with downward triangle are useless and correspond to a failure of the tape recorder. The two states that are represented by a upward triangle are destructive and will erase or damage the tape.

In the EC directive "Safety of Machinery," the design of production machinery must include the investigation of all possible System States, their consequences on the safety for human beings and equipment, and the implementation of measures to counteract risky situations. The manufacturer of EC compliant machinery has to keep appropriate documentation ready for proof of compliance of his product. A key part of this is the specification of the system in the State Domain by giving an overview of all possible states.

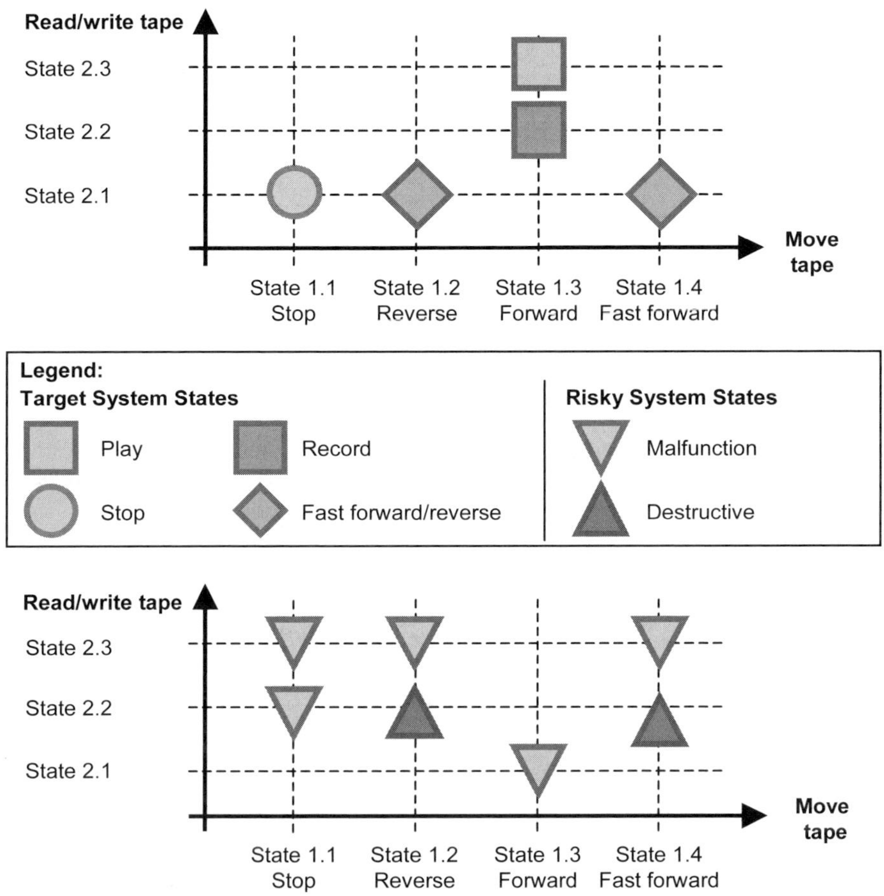

Diagram D3-12: State Map of tape recorder.

3.4.3 System with Continuous and Discrete Parameters

Systems that have both discrete and continuous parameters involved in the specification of parameters require a conversion of continuous parameters to discrete parameters. This is done by comparing continuous parameters with thresholds, or by analog-to-digital conversion. The first method yields a two-state result, such as "above/below." By setting two thresholds, a certain value range can be checked with the result "in-range/out-of-range."

The comparison with a threshold can take place in a sensor (example: thermostat switch) or in the control with analog or digital circuitry. The analog-to-digital conversion creates an integer number that the digital controller uses to define specific states by comparison to thresholds set as numeric values.

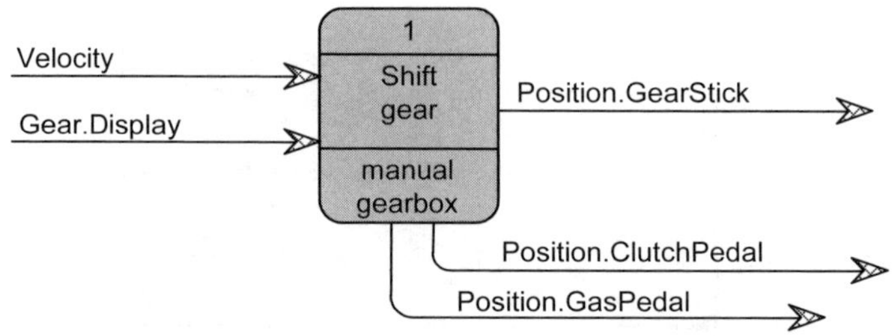

Diagram D3-13: Process "Shift gear."

An example of a system with discrete and continuous states or parameters is a car transmission. The Flow Diagram of the car transmission is shown in *Diagram D3-13*. Here, the operation of a car with manual transmission is treated.

The State Chart in *Diagram D3-14* describes process "Shift gear" and shows the shifting of gears. A gearshift is triggered as soon as the velocity reaches a threshold specified by the driver.

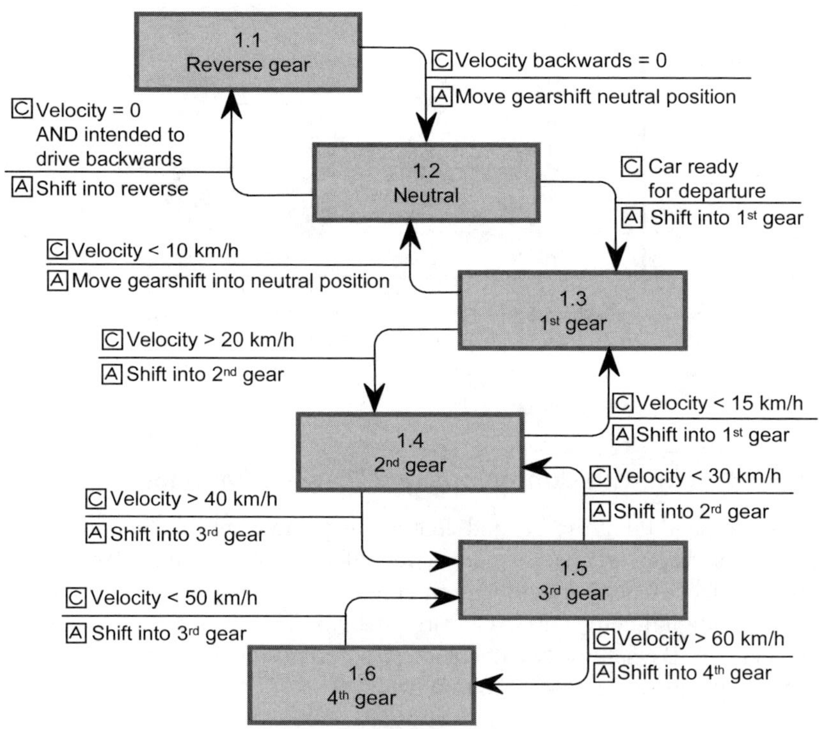

Diagram D3-14: Child State Chart of process "Shift gear."

The gears are discrete parameters, while the speed of the car is a continuous parameter. It is made discrete by introducing thresholds in car speed for every gear. The change of the gears is assumed to take place instantaneously.

The State Map in *Diagram D3-15* is two-dimensional and depicts two different types of parameters against each other. On the x-axis are the gears as discrete parameters, equivalent to the single states from the State Chart in *Diagram D3-14*. On the y-axis, the continuous parameter velocity is depicted. The continuous parameter velocity is discretized by introducing velocity thresholds. For example a minimum velocity of 40 km/h is required to shift into "3rd gear" in order not to kill the motor. Similarly, the maximum possible RPM of the motor limits the upper threshold for a given gear.

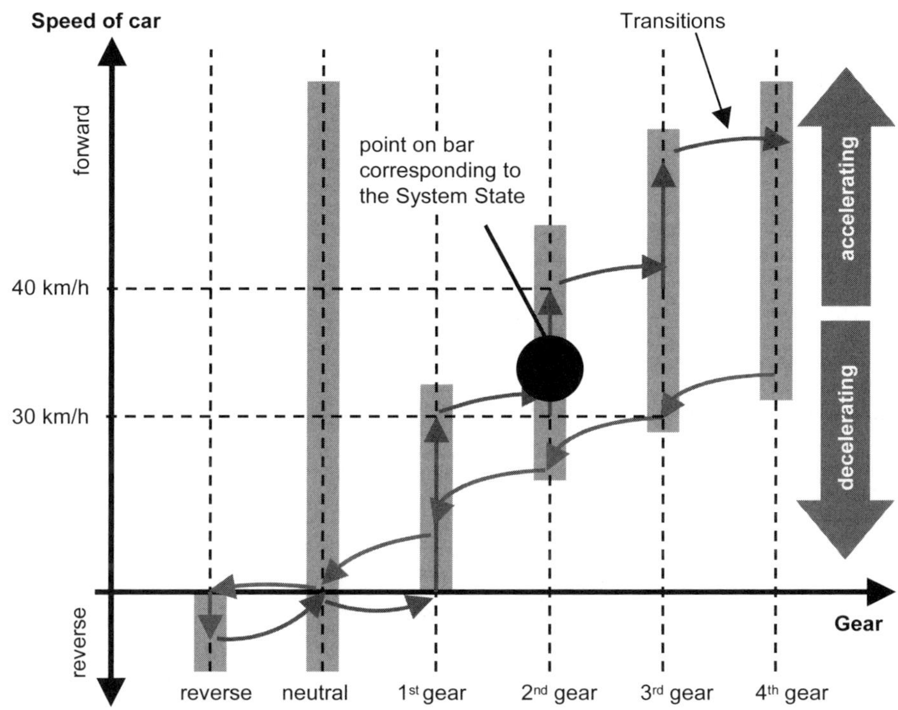

Diagram D3-15: State Map "Change gear."

The State Map shows intended and unintended System States of a car transmission. The permissible speed ranges for a specific gear are represented by gray bars in the State Map, which indicate the gear to select. A System State is a point on one of the gray bars in the State Map. Points outside of the gray bars are unintentional System States.

Starting from the state "neutral," which is depicted by the yellow bar, the transition to "1st gear" takes place and is assumed to be instantaneous. The state of the first gear allows a certain range of speed while accelerating. The transition to the state of the "2nd gear" again is instantaneous. The range of the "2nd gear" is depicted again as a gray bar. The gear change goes in both direction, at different thresholds of speed.

3.4.4 System with Continuous Parameters

Only few processes are really static in time. One example is a fountain running continuously day and night. The same amount of water flows in and flows out, and the water level in the basin is constant.

Consequently, most processes call for an analysis of the dynamic behavior. The example of a bathtub is taken here as a dynamic system with continuous parameters. In *Diagram D3-16* the Flow Diagram of a bathtub is given. The process "Take bath" has one input resource flow "WaterOutput" and two output flows: an output resource flow "WaterOutput" and an information flow "WaterLevel." The water level is an information flow from the bathtub to the environment, in this case to the person taking the bath. Other physical effects on water, such as loss of water due to overflow, leakage, or evaporation, are not considered.

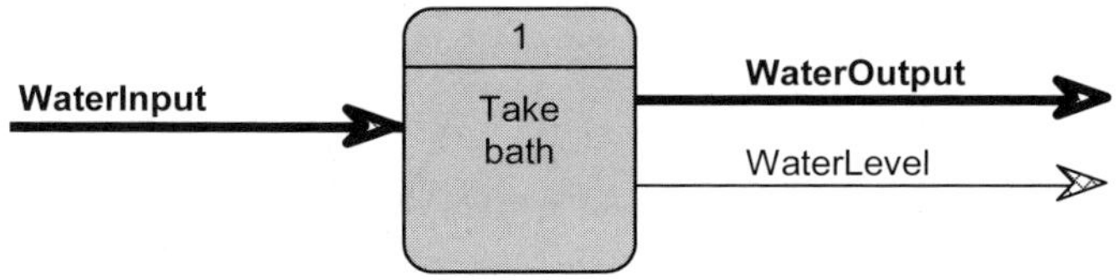

Diagram D3-16: Process "Take bath."

The State Chart of the bathtub is shown in *Diagram D3-17*. Three parameters specify the active System State of the bathtub at a given point in time: the rate of the water flowing in, the rate of the water flowing out, and the current water level. These three parameters correspond to the flows on the Flow Diagram.

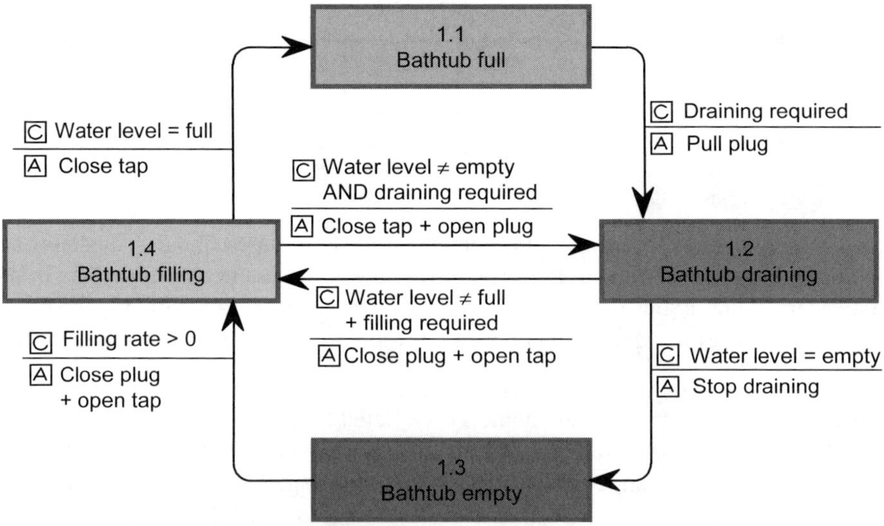

Diagram D3-17: State Chart "Take bath."

At the beginning, the bathtub is empty. The only way to leave this state is to start filling in water. Once the bathtub is in the filling state, the water level is adjusted either by changing the filling rate with the tap or by pulling the drain plug. Upon reaching full level, filling is stopped. Draining is now the only option to get out of the "Bathtub full" state.

The System States for systems with three and more continuous parameters is shown in the State Domain on a three- and more-dimensional graph. The State Domain of the bathtub is given in *Diagram D3-18*. Since the three parameters describing the system are independent of each other, they define a three-dimensional space, each axis representing one parameter. The states correspond to the planes within this three-dimensional space and are colored accordingly.

The System State is represented by a point in the three-dimensional space limited by water level empty/full, filling rate low/high, and draining rate low/high. System States outside of this space are dangerous and useless. For example, a dangerous System State would be a point higher up in the graph than the plane given by the state "Bathtub full." In this state, the bathtub would be overflowing.

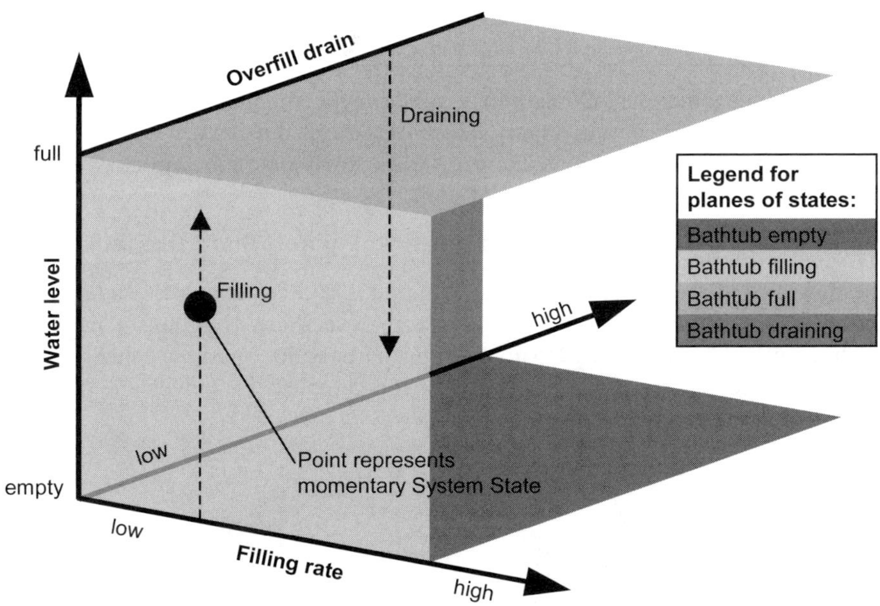

Diagram D3-18: System Domain "Take bath."

Possible System States are located within a space limited by six planes. In this example, four planes are shown. The two horizontal planes stand for the states "empty" and "full." The two vertical planes represent the filling and the draining. In this example, system draining and filling at the same time is regarded as useless. Therefore, useful and intentional System States are located on of the vertical planes of the states "Bathtub filling" or "Bathtub draining." The black point given in

Diagram D3-18 is an example. It corresponds to the state “Bathtub filling,” the “WaterLevel” being half full.

System States within the space limited by the planes but not exactly on the filling or draining plane represent a simultaneous filling and draining. They are not investigated in this system and are not shown in the State Chart.

The algorithm to calculate the water level is defined in the *Specification D3-1* of the information flow “WaterLevel.” The equation is shown in the section “value” of the flow specification.

Specification D3-1: Flow “WaterLevel”

WaterLevel	
Description	The water level indicates how full the bathtub is.
Flow type	Information
Parameter	FillingRate DrainingRate Time change Δt
Value	WaterLevel = Old_WaterLevel + (FillingRate * Δt) – (DrainingRate * Δt)

As we have seen, the State Domain for a system with continuous parameters is used to show and evaluate the maximum values of these parameters, which specify the allowed System States. The State Domain shows the limits of the system.

3.4.5 Considerations for Model Hierarchy and State Domain

The detailing in the hierarchy of Flow Diagrams and State Charts leading to the System States can be different. If the hierarchy of the Flow Diagrams goes into too much details, the drawing of feasible State Charts is not possible. If the detailing level is too coarse, the State Charts become too complex. How does one decide on which level to begin to describe the processes by State Charts?

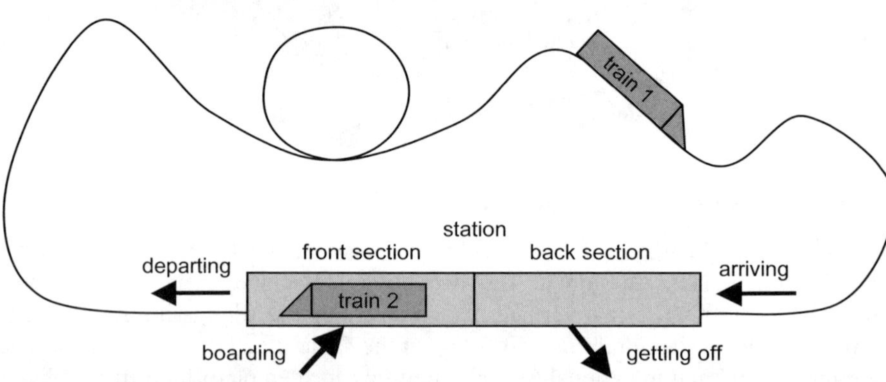

Figure D3-9: Roller coaster.

In the following, a system is detailed in two different ways. The example is a roller coaster. *Figure D3-9* shows the setup. Two trains transport the passengers. The station for boarding and getting off has room for both trains. In the departure or front section of the station, the passengers board the train. In the arrival or back section of the station, the passengers get off the train.

For the first version of hierarchical detailing, the top process "Control roller coaster system" is detailed into a child Flow Diagram with two processes (*Diagram D3-19*). Then, these two processes are each detailed into a child State Chart. The State Charts have each three states, one for train 1 (T1), the other one for train 2 (T2). A train can either be in the front section of the station, be on the track during the ride, or be in the back section of the station.

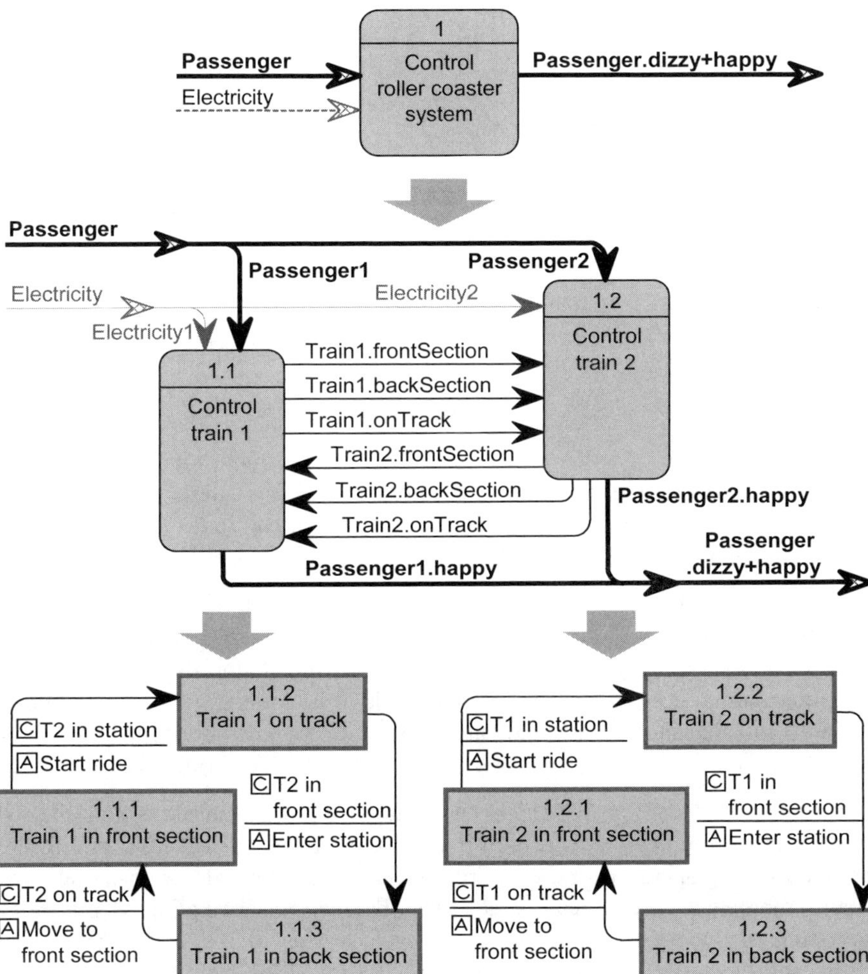

Diagram D3-19: State Charts for functional processes.

From the two State Charts of the trains, a State Map can be combined, as done in *Diagram D3-20*. It shows the System States of the roller coaster that are allowed (operational states) and those that are not allowed (risky and, in the case of the roller coaster, even catastrophic states).

If one train is in the back section of the station and the second train enters the station, we have a crash. This System State, where both trains are in the back section, is indicated by the upward triangle. The same applies when both trains are in the front section. The downward rectangle indicates that both trains are on the track, which is a status not allowed for safety reasons.

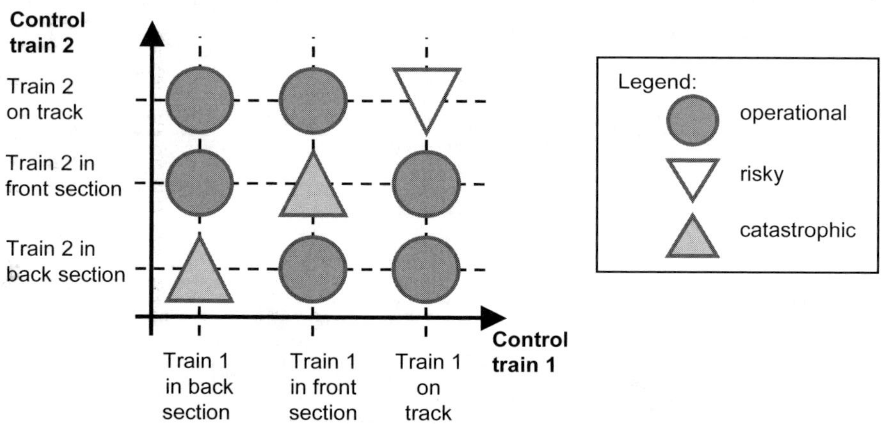

Diagram D3-20: State Charts for functional processes.

Diagram D3-21 shows another way of detailing the process "Control roller coaster system." The process is directly detailed into a child State Chart. This State Chart has nine states that correspond to the nine possible System States in the State Map of the first version (*Diagram D3-20*).

States indicated in light gray on *Diagram D3-21* are the allowed operational states of the system. The states 1.8 and 1.9 indicated in dark gray are the catastrophic states, when both trains are in the same section of the station. State 1.7 is highly risky, representing both trains being on the track at the same time. The dashed gray transitions leading into the risk and catastrophic states must not occur in the roller coaster system

For the moment, *Diagram D3-21* has two states without output transition: the two states 1.8 and 1.9 indicating a crash of trains. To complete the State Chart, a further state for manual operation must be introduced. The details of such a state will contain emergency procedures, as call of ambulance, disentangling of the trains, positioning of the trains in the front and back section of the station by manual control, and the restart of the roller coaster. The transition from this manual state will go to state 1.1 or 1.4, with both trains in the station as a condition for resuming automatic operation.

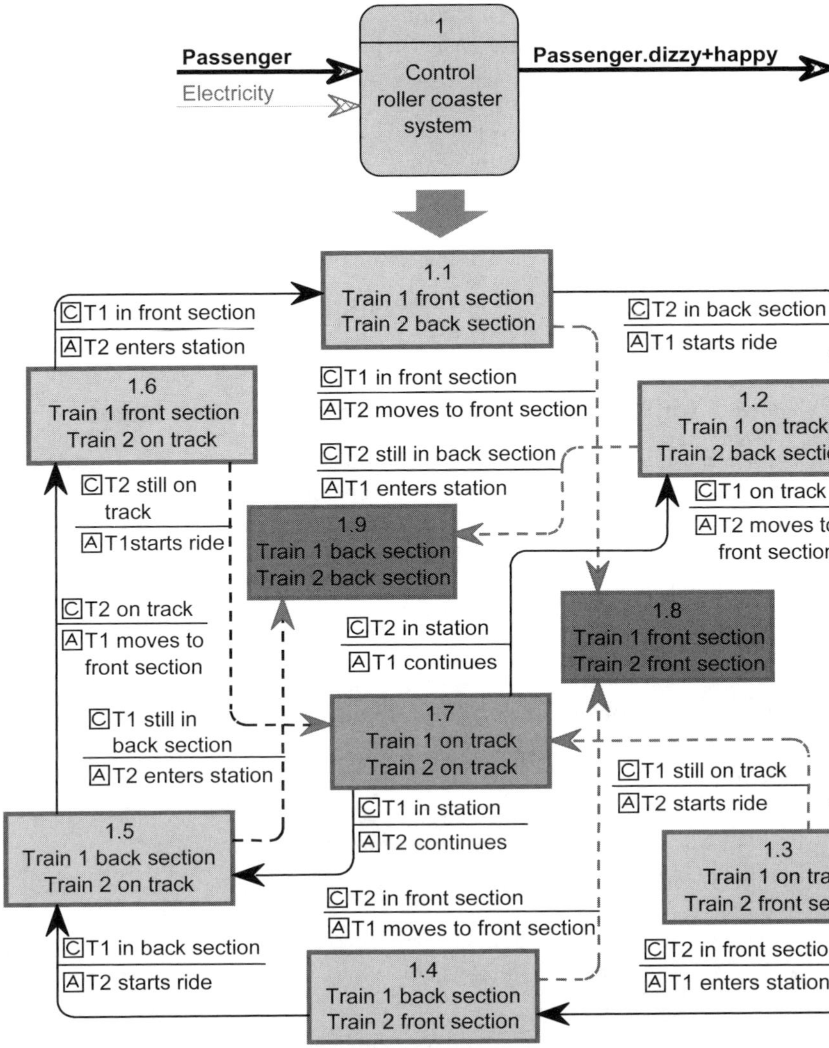

Diagram D3-21: State Chart of a single process.

Conclusion and comparison of the two different versions of detailing

Methodically, both detailing are possible. However, in the case of a real-time control, the first version is the better approach. It is advisable to detail the Flow Diagram down to the level where each function of the machine is described by an elementary process. Each separately controlled function must be described by an own process. In the case of the roller coaster, this means an individual process for the control of each of the trains. These processes are then described by a State Chart.

The check of the System States concerning the target states, risky states, and catastrophic or impossible states are easier established by a State Map than by a State Start.

The second version, whereby a process representing, for example, a whole machine including its different functions is directly described by a State Chart. This becomes quite complex and complicated to understand. The later programming of the real-time control based on such a State Chart will be very difficult and would require some more attention.

Systems with distributed intelligence and low-level modular structure, as in the first version, are easier to grasp.

3.4.6 Rules for State Domain, State Map, and System States

The rules for the State Domain, State Map, and System States are listed in the following table. The State Domain is multi-dimensional, whereas the State Map is the two-dimensional representation of the Systems States and – as such – an extract of the State Domain. As only the State Domain is mentioned in the rules, the State Map is also included.

Rules D3-2: State Domain

RT6	The parameters of the flows and processes of the Flow Diagram specify the parameters of the State Domain or the State Map of the analyzed system.
RT7	Each of the independent parameters of a system opens up a dimension in the State Domain.
RT8	If the State Charts of a system are represented together in a State Domain, each State Chart opens up an axis, and the states are the parameters.
RT9	Every point in the State Domain corresponds unambiguously to an actual System State.
RT10	Points in the State Domain that do not correspond to an intended function or purpose of the system represent a problem or risk situation of the system.
RT11	A discrete state parameter corresponds to one or more points in the State Domain. A continuous state parameter corresponds to a value range or a space in the State Domain.

3.5 Program Design and Coding

3.5.1 Step-by-Step Procedure for Real-Time Coding

This chapter takes up the development of program code based on the analysis by the Flow Diagram and State Chart of Section D2. *Table D3-2* shows the step-by-step procedure as given in *Table D2-2* of Section D2. In principle, this procedure is the same for the coding of a simulation model as for a real-time control. Here, the procedure is supplemented by real-time control specific features, and the focus is set on the analysis and design of the control processes and the man-machine interface. Steps 2, 3 and 6 for real-time control therefore include the job of the operator. Steps 9 and 10 explain the implementation of the real-time control.

Table D3-2: Procedure steps

Phase	Step	Description
1 System analysis	Step 1	State goal and purpose of the project and define the system boundaries.
	Step 2	Specify the system structure by the Flow Diagram including the tasks of human operators.
	Step 3	Specify the behavior of the processes in time by State Charts, including the tasks of operators.
	Step 4	Check the System States and system behavior by the State Map.
2 Program design and test simulation	Step 5	Specify requirements, input and output of program, and design the user interface.
	Step 6	Write each module in program code, including modules representing the outside world.
	Step 7	Assemble and set up the real-time control program.
	Step 8	Check and evaluate system behavior.
3 Implementation of real-time control	Step 9	Reallocate the simulated environment to the real world inputs and outputs.
	Step 10	Check real machine with real-time control and commission machine.

The steps of this procedure are run through iteratively. Whenever a step requires changes that affect previous steps, these should be carefully checked and updated if necessary. Especially Flow Diagrams, State Charts, and their specifications must be kept up-to-date, since they serve as the final documentation of the real-time control program.

Figure D3-10 explains, which parts and aspects of the system are analyzed, which ones are simulated in the test, and what is finally included in the real-time control. It illustrates the three phases from the system analysis to the real-time control by three Flow Diagrams. The external entities and interfaces between the system and the outside world are also depicted to see, how they affect the design of the real-time control in the different phases.

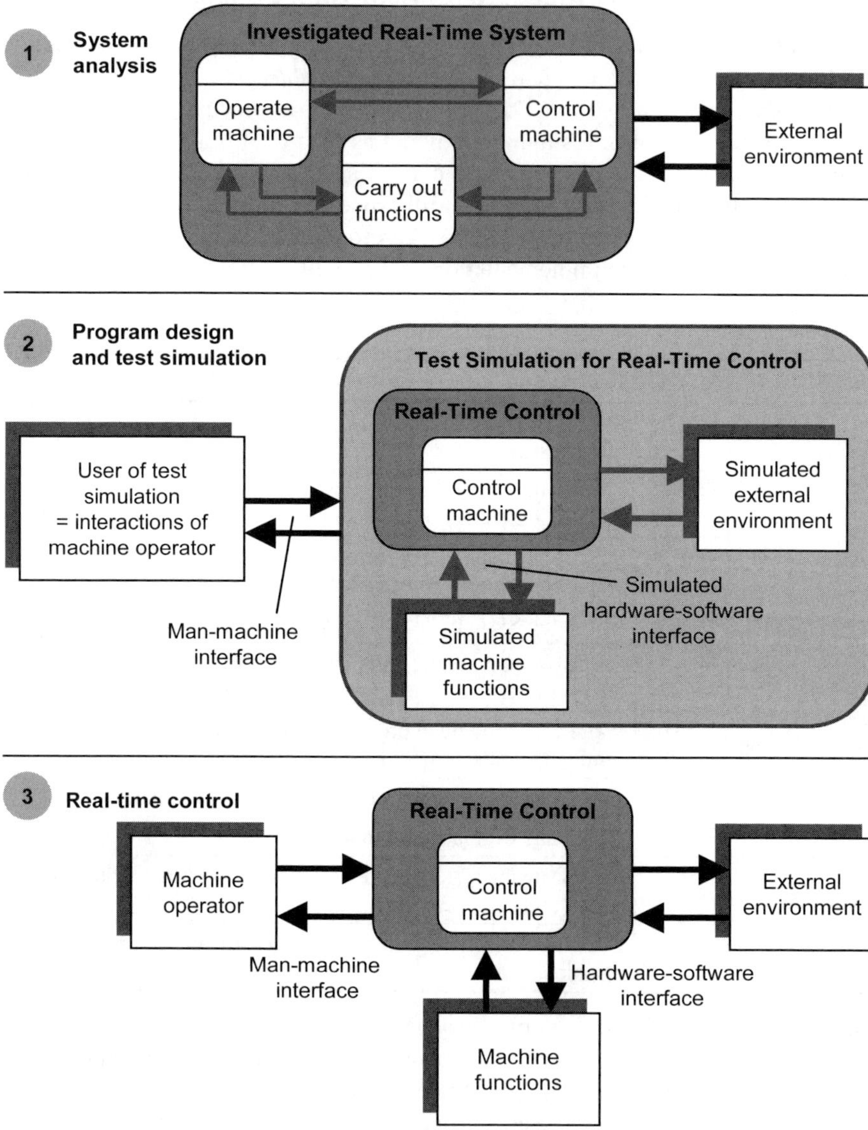

Figure D3-10: Procedure from analysis to real-time control.

Phase 1: System analysis

The first phase is the system analysis, shown by Flow Diagram 1 in *Figure D3-10*. It includes the processes of the operator, the machine function, and the control, describing the three aspects of an automated production. This Flow Diagram also sets the system boundaries by the context diagram. For the real-time control, the interfaces to the outside world as well as between the processes have to be described in detail for later consideration in the code. Next steps of the system analysis are the description of the processes in time by State Charts and the check of the behavior by the State Map.

Phase 2: Program design and test simulation

By simulating real-time control, the processes respectively the behavior of the control are checked (Flow Diagram 2 in *Figure D3-10*). The interfaces of the control process with the "External environment" and the "Machine function" as well as the external entities themselves are simulated too, in order to be able to test all the possible situations that might occur in operation. Therefore, the interfaces and the external entities are shown within the simulated real-time system. In contrast, the process of the operator is outside of the simulated system. It becomes an external entity, as indicated by "User of test simulation = interactions of machine operator." During the simulation and check of the real-time control, the user acts as the operator of the machine. Such a test simulation can also be used for the training of the operators on the machine.

Phase 3: Implementation of real-time control

When taking the step from test simulation to real-time control, the "Simulated external environment" and "Simulated machine functions" are replaced by the real "External environment" and "Machine functions." The real-time control now interacts with the machine and the outside world directly using sensors, actors, displays, keyboard, buttons, switches, motors, valves, magnets, etc. Flow Diagram 3 in *Figure D3-10* shows the system boundaries for the final real-time control.

3.5.2 System Analysis for Real-Time Control

Step 1: State goal of project and define system boundary

The project of designing a real-time control begins as any other project with the statement of its purpose and goal. What is the application for the real-time control? What are the functions of the real-time control for this specific machine? Is it a control for a single machine, a component of it, or an entire production line? Should the control be as far as possible computer controlled, or are there operators who operate the machine, and the machine control only carries out a minimum of safety functions?

The context diagram helps to define the boundaries of the system for which the real-time control will be designed and coded. *Diagram D3-22* shows this on the example of a modern weaving machine with a real-time control.

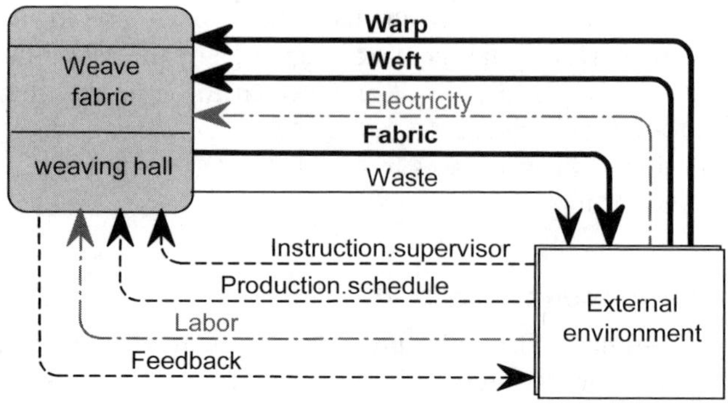

Diagram D3-22: Context diagram of the investigated real-time system.

Step 2: Specification of system structure by Flow Diagram

An automated production with machines is typically specified by three processes defining three different aspects of the machine, as described by the history of automation in Chapter D3.2.5. *Diagram D3-23* shows the three main processes specifying the weaving of a fabric on a loom. It is the child diagram of the context diagram in *Diagram D3-22*.

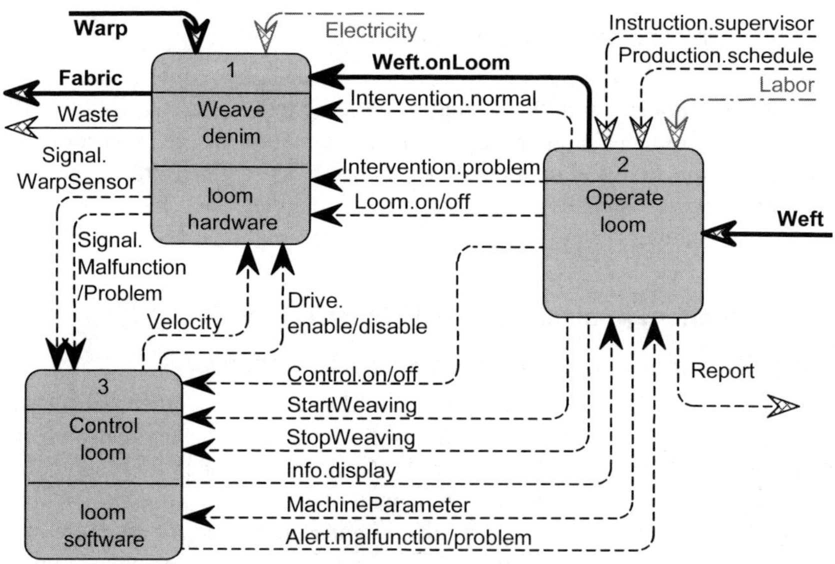

Diagram D3-23: Flow Diagram for the operation of a loom.

The three aspects of automated production processes are:

- Machine hardware (machine functions)
  This includes the different functions of the machine. For the loom in *Diagram D3-23*, these are the let-off of the weft, the insertion of the weft in the warp, the

shed operation, the beat-up of the weft, the rolling up of the fabric, and the supervision of the warp yarn.

- Machine software (control)
  A machine depends on the operator or the real-time control, which have to provide "intelligence" to the machine. The more the machine is automated, the more the control takes over the functions of the operator.
  The loom control, for example, handles the start and stop of the weaving movements, translates operating parameters (for example the velocity) into signals for the loom and issues an alert when the sensors pick up a malfunction.
- Operator
  Supported by the real-time control, the operator has to carry out various tasks. This includes cold start of a machine, complete shutdown, maintenance, failure handling, and supervising quality. In order to code the real-time control, all tasks and interventions of the operator must be defined.

Each one of these three aspects can be described by one or more processes. In the case of the loom here, one process per aspect is used (*Diagram D3-23*). Of course, each of theses processes can be detailed further, if necessary.

The interfaces between these three aspects are also specified:

- Hardware-software interface
  The interface between hardware and software is very extensive. It must be clearly specified. For designing the hardware, it must be known, what signals have to be sent from the hardware to the software in order to install the right sensors. The software needs to fit to the type of sensors, actuators, and the signals they send or receive.
  Examples from *Diagram D3-23:* "Drive.enable/disable" signals start and stop of the machine to the hardware. "Signal.WarpSensor" is sent to the control whenever a warp yarn is broken. The machine is then stopped, and the operator has to fix the broken warp yarn.
- Operator-software interface
  All interactions between the operator and the machine that are part of the program code must be specified by a flow between the processes of the operator and the control.
  Examples from *Diagram D3-23:* The flows "StartWeaving" and "StopWeaving" show the commands of the operator. "Info.display" stands for the display on the screen of the loom that gives the user information about the production process. At the start of a new fabric, the operator has to key in the necessary parameters, indicated by "MachineParameter."
- Operator-hardware interface
  During production, the operator carries out various manual interventions. Only those interactions between operator and machine that do not need coding are represented by flows going directly between the machine hardware and the operator process.

In *Diagram D3-23*, they are given by the flow "Intervention.normal" for normal intervention and "Intervention.problem" for breakdowns and failures of the machine.

As an example from the weaving machine, the operator resolves the problem of a broken warp yarn. He picks up the broken ends himself, as this cannot be done by the machine. After knotting in a short piece of space yarn, he ties up both loose ends and confirms on the user panel that the problem is solved. Now, he restarts the machine. Starting the machine is an interface between operator and software, it is a signal given to the control activating the main drive of the loom.

The rules for the specification of the system structure are given below.

Rules D3-3: Specification of system structure

RT12	The analysis of a real-time control includes the aspects of machine functions, machine control, and operator. Each one is described by processes and the interfaces between them.
RT13	The hardware-software interface is given by sensors and actuators in the machine.
RT14	The operator-software interface transmits the commands of the operator to the machine and the information of the machine to the operator.
RT15	The operator-hardware interface describes the manual interventions performed directly on the machine.

Note! Always consider all three aspects for the analysis and description of the system, for which the control is to be designed. Depending on the degree of detail, the processes representing these aspects are further split up, or two aspects may be incorporated into one process. For example, the processes for the machine and for the machine control may be depicted as one process for a very simple machine.

Step 3: Specify behavior of processes by State Charts

In Step 3, the behavior of the system in time is analyzed by the State Chart. Each process describing the three aspects of the system is detailed into a child State Chart. The following three State Charts detail the processes from *Diagram D3-23*.

The flows between these processes are taken up in the transitions as conditions and actions. The action in one State Chart becomes a condition in the other State Chart if the pertaining processes are connected by a flow on the Flow Diagram. The relationship between Flow Diagram and State Chart is very important for coding later on. See Chapter D1.5.2 and Chapter D2.3.5 for a detailed explanation, of how to do the transition from Flow Diagram to State Chart.

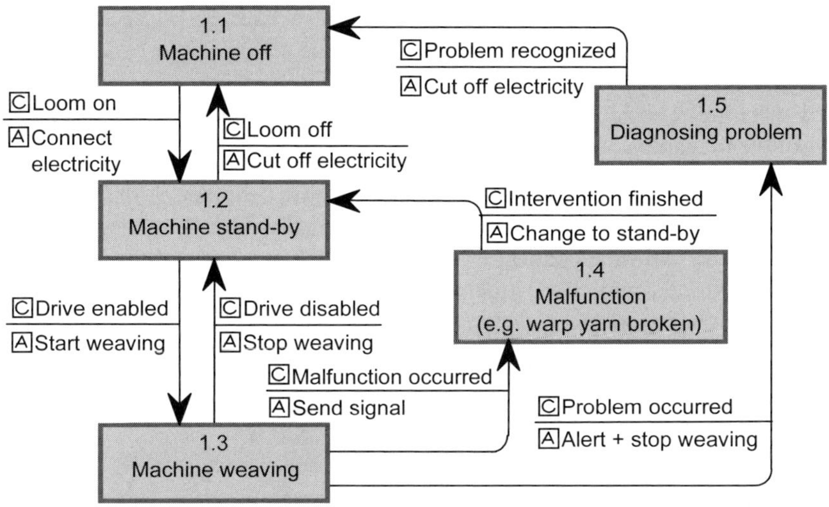

Diagram D3-24: State chart "Weave denim."

Diagram D3-24 gives the State Chart of process "Weave denim" of the hardware of loom, describing the behavior of the machine in time. For a complete description, all five basic states of a machine are used, as listed *Table D3-1*.

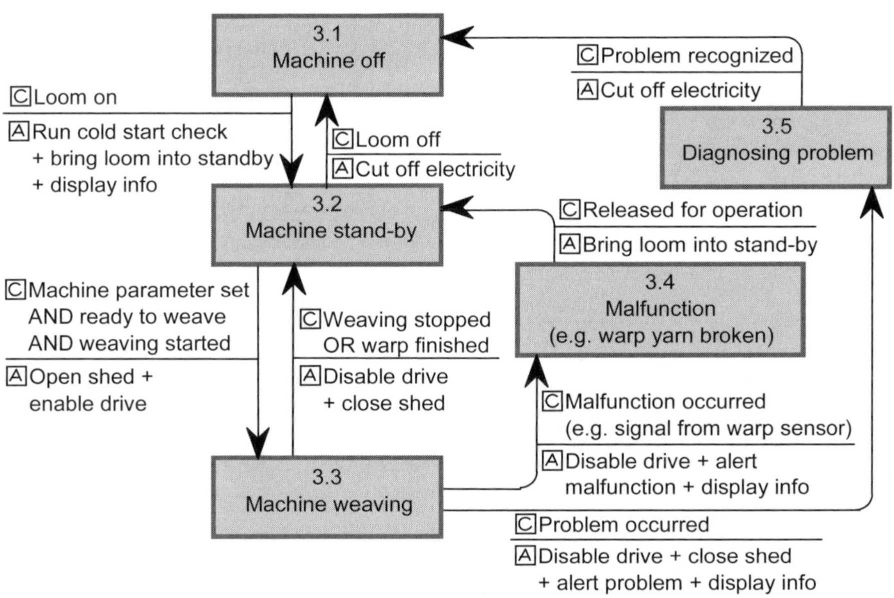

Diagram D3-25: State chart "Control loom."

Diagram D3-25 shows the child State Chart of the process "Control loom." The states on the State Chart of the control are the same as the one for the machine process "Weave denim" with the hardware aspect. The same state names are used.

The conditions and actions however are different, according to the incoming and outgoing flows of the pertaining process. The State Chart of the control has more complex transitions than the State Chart of the loom. This is normal for a automated machine. The higher the degree of automation, the less complex the State Chart of the machine and the more complicated the one of the control.

The State Chart of the operator processes defines the behavior of the operator and the intervention of the operator on the machine. This helps to design the user interface for the real-time control. Even though the State Chart of the operator does not get transformed into code, it helps to understand the behavior of the operator, and therefore the support provided to him by the real-time control.

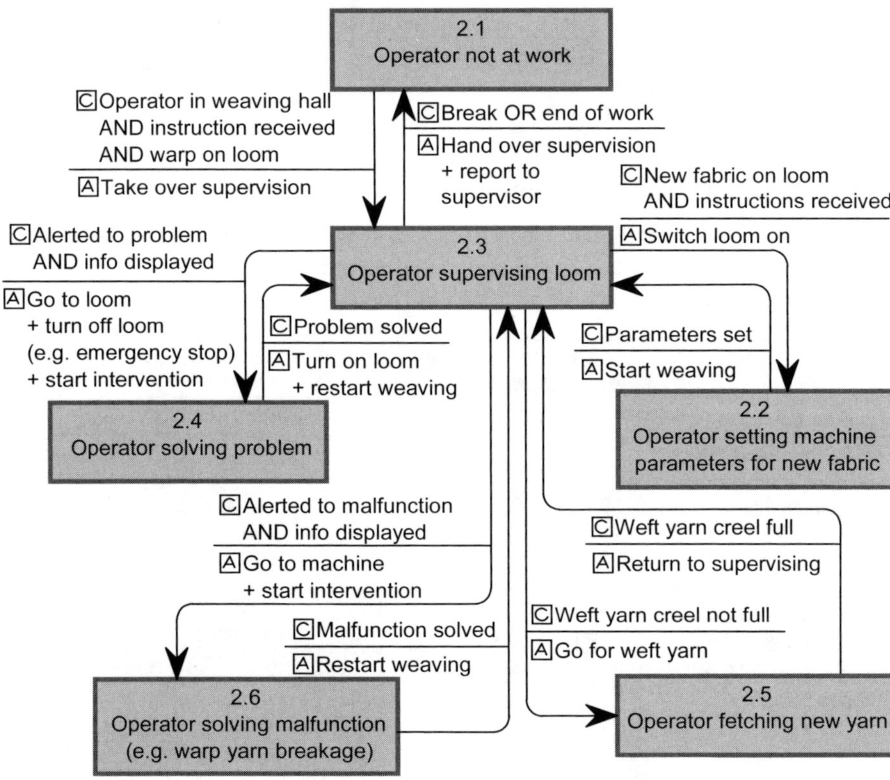

Diagram D3-26: State Chart "Operate loom."

Diagram D3-26 shows the child diagram of the process "Operate loom." The operator supervises the machine. If a problem occurs, as for example the breakage of a warp yarn, he solves the problem and then restarts the machine. When a new fabric is set up, the machine parameters must be changed. The state "Operator not at work" indicates the time when the operator eats lunch, takes a break, or is at home. For the real-time control, it is important to know if the machine can run alone without supervision or if an operator has to be there on place. This may have an impact on the cases that must be covered by the real-time control. For the loom example, the need for constant presence of the operator is assumed.

Based on the five basic states in *Table D3-1,* the states describing the behavior of the operator are drawn. Analog to the basic machine states, the operator has also basic states: Not at work, supervising the machine, normal working states (for example setting machine parameter, fetching new yarn), solving a malfunction, and performing minor repair jobs.

Degree of automation: The degree of automation can be seen by looking at the State Charts of the machine function and the control. A machine requiring a lot of manual work and intervention of the operator has a large and complex State Chart of the machine function. The more the machine is automated, the more the State Chart of the machine function degenerates, while the complexity of the State Chart of the control increases. Normally, when analyzing a system for an automated machine, the State Chart of the machine process is left out. Only the State Chart of the control is depicted.

Step 4: Check of System Behavior by the State Map

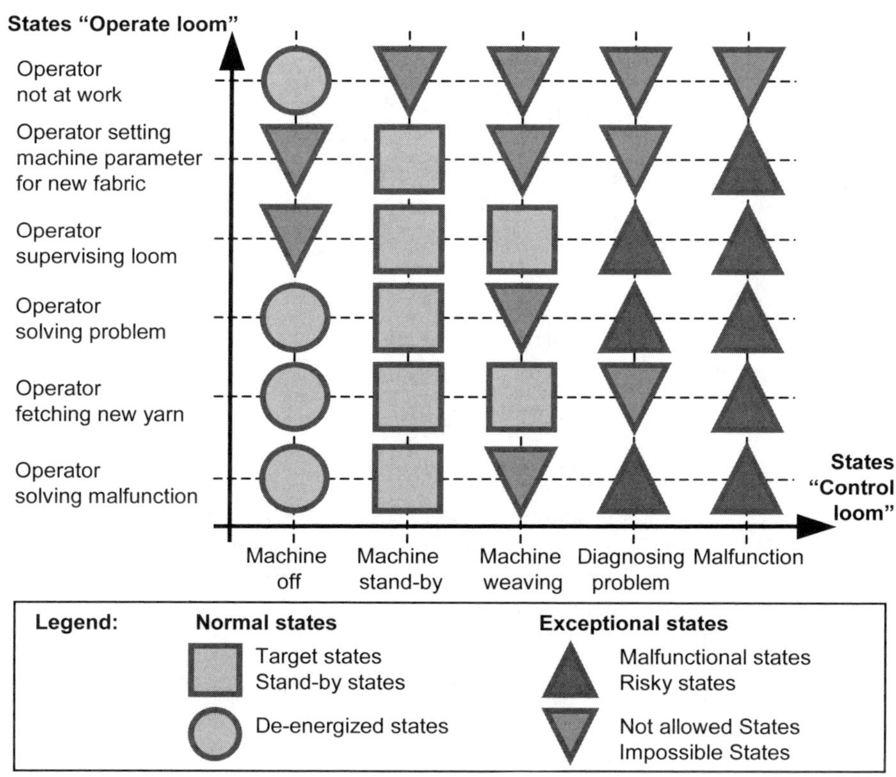

Diagram D3-27: State Map of loom.

The State Map determines, which System States must be considered in coding. *Diagram D3-27* shows the State Map of the loom for the State Charts of the operator and machine control.

3.5.3 Program Design and Test Simulation

Step 5: Requirements, input and output of program, user interface

It is good practice to start by designing the user interface. *Figure D3-11* (an extract of *Figure D3-10*) explains, how the process "Operate machine" becomes the external entity "User of test simulation = machine operator."

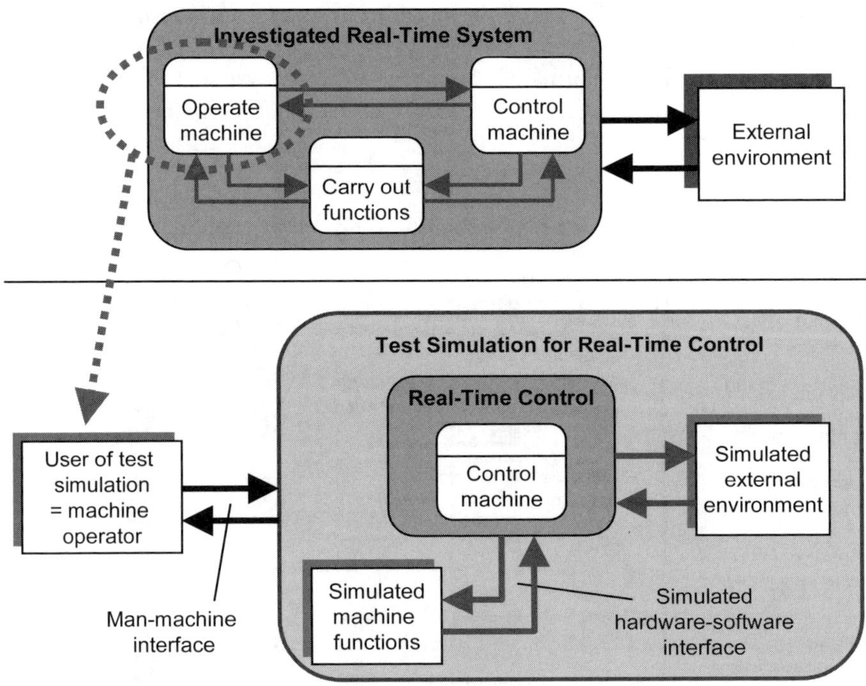

Figure D3-11: Man-machine interface.

The interfaces between the external entity "User of test simulation = machine operator" and the control process become part of the user interface. This way, the input and output parameters and the functions of the program are defined. For detailed explanation for transition from Flow Diagram to user interface see Chapter D2.3.6 in Section D2.

The user interface of the test simulation is turned later during machine design into the operation panel for a machine or a production line, often mounted directly on the machine. *Figure D3-12* shows an embroidery machine with its user interface.

This is a panel with a display at the top and several push buttons. The connected computer serves for designing the embroidery pattern.

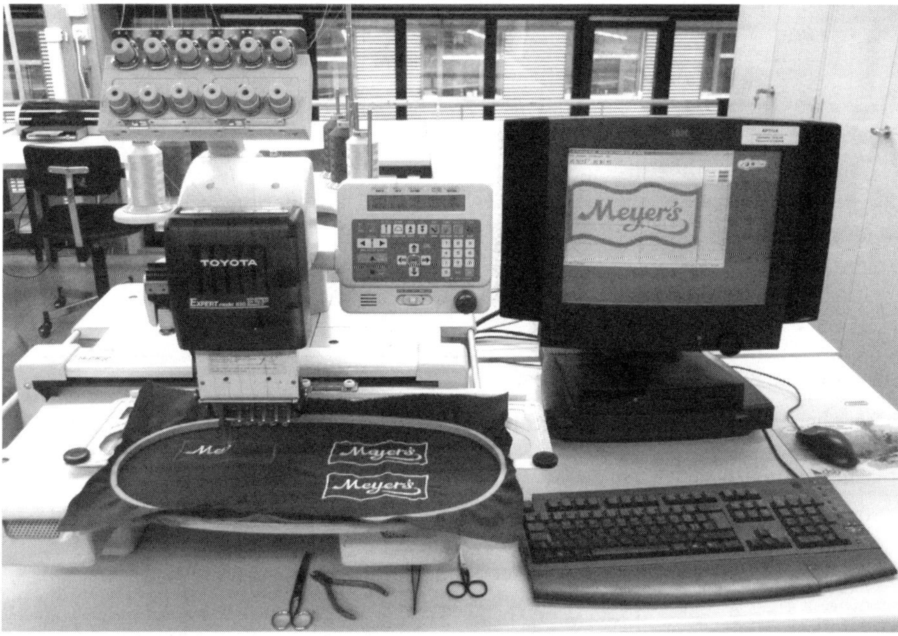

Figure D3-12: Embroidery machine with user interface (Photo: ETH Zurich).

Table D3-3 lists the common components of a user interface module.

Table D3-3: User interface components

User interface component	Description of user interface component
Input interfaces and display	Fields for the input of parameters, or variables respectively, by the operator.
Output interfaces and display	Display of actual values, set values, or other values calculated by the program. They give information about the system functions.
Status window	Display of actual System State at runtime.
Animation	An animation schematically visualizes the System State, sometimes even in motion.
Control buttons	Start, stop, and other buttons that control the behavior of the machine instantaneously.
Time field	The actual real time is displayed.

The rule for the specification of the software requirements is given below.

Rules D3-4: Software requirements

RT16	The software requirements are defined by its interfaces to the external entities on the context diagram.

Step 6: Write each module in program code

A computer program consists of several program modules. The analyzed system is divided into modules, which represent individual production processes or service functions. They are converted into program modules. A list of the different program modules is given in *Table D3-4*.

Table D3-4: Program modules

Program module		**Description**
User interface module		The user interface module includes the code for the single components on the user interface. The components are treated in *Table D3-3*.
Process modules	System state	The *states* of the State Charts of the system are chosen by a `Select-Case` construct. Within the active case (state), conditions for branching to other states are checked, and transitions are triggered.
		The *conditions* of transitions are specified as cases within a `Select-Case` construct and programmed within each of these cases as one or more conditional statements to change the state of the relevant process with an `If-Then` construct.
		The *transitions* of the State Chart command the program to switch to the destination state.
		The System State, i.e. the state of every system module is updated after every transition and may be displayed at runtime in the status window.
	Calculation of values	The calculation of values is based on physical equations. The results are displayed in the output fields on the status window.
Auxiliary modules	Variable declaration	Declaration of all the variables necessary for calculations and operations of the program.
	Initialization	Definition of the values at the start-up.
	System clock	Real-world time (hardware) and watchdog. Call of the actual time.
	Store	Stored data, results, and variables after program run for program evaluation and debugging.
	System check	Test if sensor signals (actual values) and actuator signals (set values) are within specified value ranges. Issuing of alarms and stopping of machine if severe problems occur.

A program module may be a user interface, process, or auxiliary module. A user interface module contains the source code for the components of the graphical user interface. The auxiliary modules contain the variable declaration, the program initiation, and the real-time clock.

The process module controls the processes according to the State Chart and triggers transitions to new states. The source code is based directly on the behavior of the system represented by the State Charts. The transitions are programmed by a `Select-Case` construct. For detailed explanation fro the transfer from State Chart to source code see Chapter D2.3.7 of Section D2.

Rules D3-5: Transition from State Chart to code

RT17	Each State Chart is represented in source code by a process module.
RT18	The actions on one State Chart may serve as references for conditions used in the State Chart of another process. On the pertaining Flow Diagram, they are made available by a flow leading from the origin process to the destination process.
RT19	The system is transparent to the outside. All flow and process parameters within the system are treated as global variables in the program.

Recommendation: For real-time controls, the use of global variables is strongly recommended. This requires and enforces unique names for the variables. For maintenance and spare parts logistics, as well as for security reasons, each component of the system must have anyway a unique identifier. Therefore, names and descriptions are by default globally valid, encapsulation being justified only for physically separated subsystems.

Step 7: Assemble and set up real-time control program

The individual process modules of the program code are run through in a repetitive loop, as shown by the bold arrow in *Figure D3-13*. The System State, which is a combination of the active states of the process modules, is continuously updated from every process module (dotted arrows). The intervention by the user and the communication of the program with the outside world are drawn with gray dashed flows. The program modules with a dashed border are optional.

All objects used for building the software are related to real-world components, their functions, and their properties. To avoid confusion, terms and definitions in software and in the real system must be identical. In many cases, it is in fact necessary to sacrifice convenience of coding and even some proven principles of software development in order to strictly match the source code to the hardware environment. For example, it is a common practice of programming to use recursive functions; however, this is absolutely confusing in a real-world context and should be avoided.

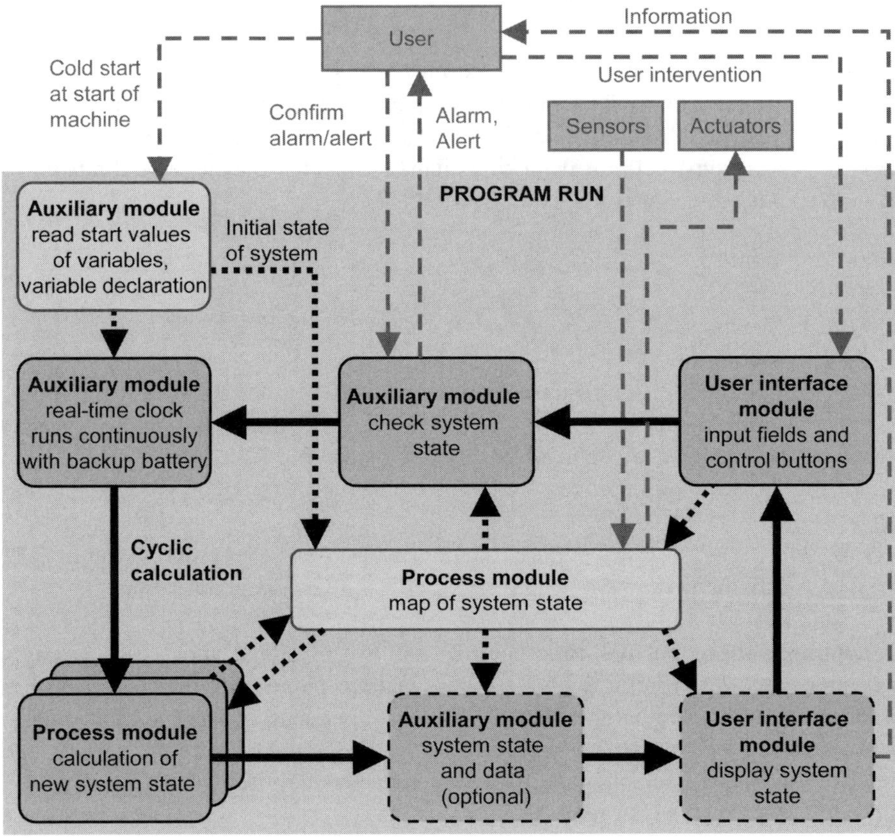

Figure D3-13: Interactions in a control program.

System time: A key global variable is system time. The choice of the time increment is given by the resolution in time of the control system. The size in bits of the system time variable must be sufficiently dimensioned to accommodate the maximum system lifetime. With a time increment of one second and a 32 bit- wide time counter, a maximum lifetime of one million hours is achieved, ample for any manufacturing system. The actual value of time is taken from the internal computer clock.

A real-time control program however operates with real interfaces to a real-world environment, which impacts the system behavior. Time is an absolute parameter, which is independent of the program. The required capacity of the computer is given by the dynamics of the real processes, the resolution in time of the sensors and the velocity of the actuators. A certain computing capacity has to be reserved for the handling of exceptional situations. The loop through the modules takes place within a given interval, to ensure the synchronism of the processes in every operating state. Therefore the timely course of the single program steps and modules has to be treated intensively during programming and test.

Rules D3-6: Naming and time

RT20	Naming and definitions in code and in the real-time system must be identical.
RT21	The time increment of the system time corresponds to the required resolution in time of the control system.
RT22	Every process module is cyclically called up by the program on each increment of the system time.
RT23	In each process module and for every time increment, all conditions are checked.

Step 8: Check and evaluate system behavior

In order to test real-time control systems, program developers use a simulated outside world. This used to be a board full of switches and potentiometers that work on the controller inputs, as well as lights and motors used to show the output of the controller. Nowadays, this kind of hardware simulator is realized by a second computer, or by a second independent part of the control program running on the same computer.

In all these cases, the system, as the outside world for the control, is simulated for a test. This method is prone to errors and very time consuming. To avoid this, it is a good idea to start system design with a simulation of the complete system including control and utilize the controller part of the same program for system control. The sensors and actuators are now interfaces between the two parts of the simulation model simulating on one hand the controller and on the other hand its system outside world.

The phase of testing and debugging serves to recognize and eliminate errors in the program itself (assignment of values, equations, system representation), in the program run (run-time errors, branches, loops), in the user interface (operator input, console display, real-time interfaces), and in faulty behavior of the system itself (stability of control loops). The larger and more complex a system becomes, the more systematically the method of development must be.

Table D3-5 lists and describes the four phases of system check for a control program. Since many processes may only be checked while running concurrently with other processes, the whole system is put into service in a certain order. Processes are added either along the path of material in production or from the peripheral processes of the system towards its core.

Table D3-5: Phases of program check

Phase	Description
1st phase	Check the operator console in its static appearance. Go through the basic steps of operation while observing windows, forms, and dialogs.
2nd phase	Check the system clock by starting and stopping the program from the operator console.
3rd phase	Load the default input parameters, start the program run, and terminate immediately at the shortest convenient delay. Check the System State and compare to the start-up state. Make sure that start parameters are introduced and processed correctly by the system, and that the initial behavior of the system is plausible.
4th phase	Put into operation the individual processes are, step-by-step. Start the program with specific start conditions in order to test the individual process modules. The behavior of the process and the system has to be predictable based on the chosen start conditions and start parameters.

3.5.4 Implementation of Real-Time Control

Step 9: Reallocate simulated environment to real-world inputs and outputs

When the test of the real-time control was successful and all functions work correctly, the simulated environment is reallocated to the real world. The real-time control is connected to the machine and its different parts, as actuators, sensors, etc., and put step-by-step in operation.

Step 10: Check machine with real-time control and commission machine

This procedure starts with the sensor signal inputs, to be checked in a manual mode of the control program. In a second phase, the actuators are checked similarly, while they are physically disabled by cutting off or limiting the power supplied to them. Once theses interfaces are debugged, the system is commissioned process by process, yet without material in work, taking appropriate precautions for safety of the persons involved. This phase ends with checking the supervisory functions and safety devices.

In a next step, material is fed into the machine, and the real world processes are adjusted, tested, and debugged. As soon as the safe operation of the whole system is established, performance tests are run to confirm its ability to fulfill the customer's expectations.

3.6 Programmable Logic Control of a Fan Heater

3.6.1 Structure of the System

The fan heater is a well know household appliance for creating a comfortable surrounding, by either a cooling breeze or quickly heating a room. It consists of a propeller or a tangential blower wheel and an electrical heating element, both mounted with the controller in a housing satisfying contemporary standards of style. Electricity is taken from a wall socket. The user selects the functions "fan" or "heater." In the heater mode, a thermostat controls the heating power in relation to the room temperature.

On the example of this fan-heater controller, the purpose and interaction of the Flow Diagram, the State Chart, and the programming of different kinds of controllers are briefly demonstrated.

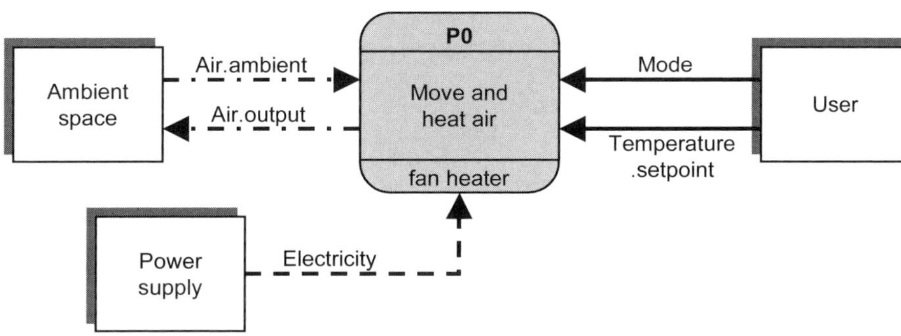

Diagram D3-28: Context diagram of fan heater.

The context diagram in *Diagram D3-28* gives the top level of the hierarchical model. The fan heater exchanges resources with the "Ambient space" and the "Power supply." The "User" controls the fan heater by data flows.

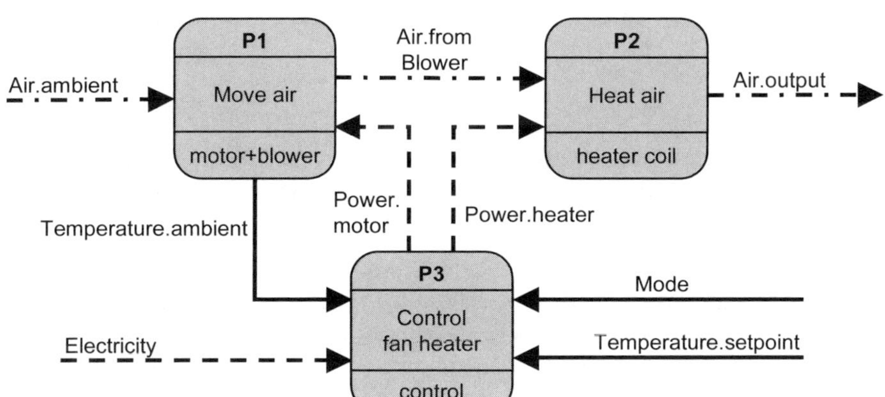

Diagram D3-29: Flow Diagram "Move and heat air."

Diagram D3-29 shows the second-highest level with the processes “Move air,” “Heat air” and “Control fan heater.” The fan takes in “Air.ambient” and moves it by the heater back into the “Ambient space.” The sensor for room temperature is integrated into the fan, which in turn precedes the heater in the fan. It supplies a signal corresponding to the actual value of the ambient temperature. The corresponding “Temperature.setpoint” is given by the user. The controller connects the fan motor and the heater to the line voltage.

3.6.2 System Behavior

Drawing individual State Charts for the fan motor and for the heater would be trivial, as both have only the two states “ON” and “OFF.” The State Chart for the controller (process 3 “Control fan heater”), switching on and off the fan motor and the heater, comprises four possible states. Heater operation while the fan is blocked is not allowed. Due to the missing flow of air, the whole device would overheat within a short time, leading to permanent damage. This state is treated as “Malfunction” state and avoided by specific safety measures.

The states given above in plain language are put into the State Chart along with the transitions (*Diagram D3-30*). In a first step, only the states serving a useful purpose are considered, and all pertaining transitions (black arrows) are put in.

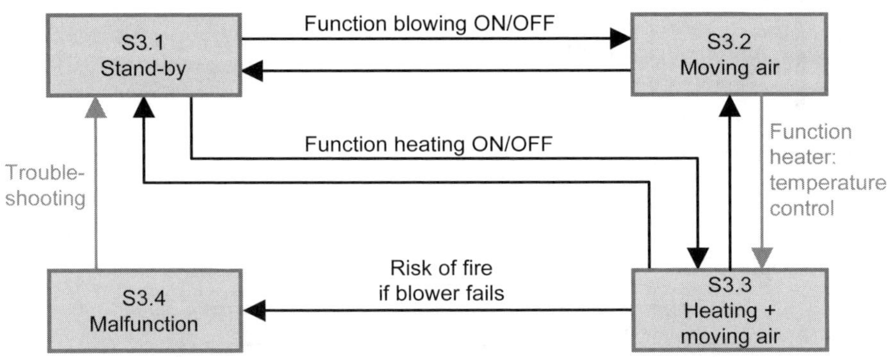

Diagram D3-30: State Chart “Control fan heater.”

In a second step, all states within this chart are checked if each of them may be entered and exited again, in order to make the whole set functional. This may call for additional transitions (gray arrows in a *Diagram D3-30*).

In a third step, the conditions are established and added to trigger the transitions from state to state. As far as possible, these are deducted from information already available in the system, e.g. by states of other processes or by elapsed time. If the required information or criterion for a condition is missing, additional information sources (sensors, timers) must be designed into the system. *Diagram D3-31* shows the result of this step. The following abbreviations are used: OM for operation mode, TR for room temperature, and TS for temperature set point.

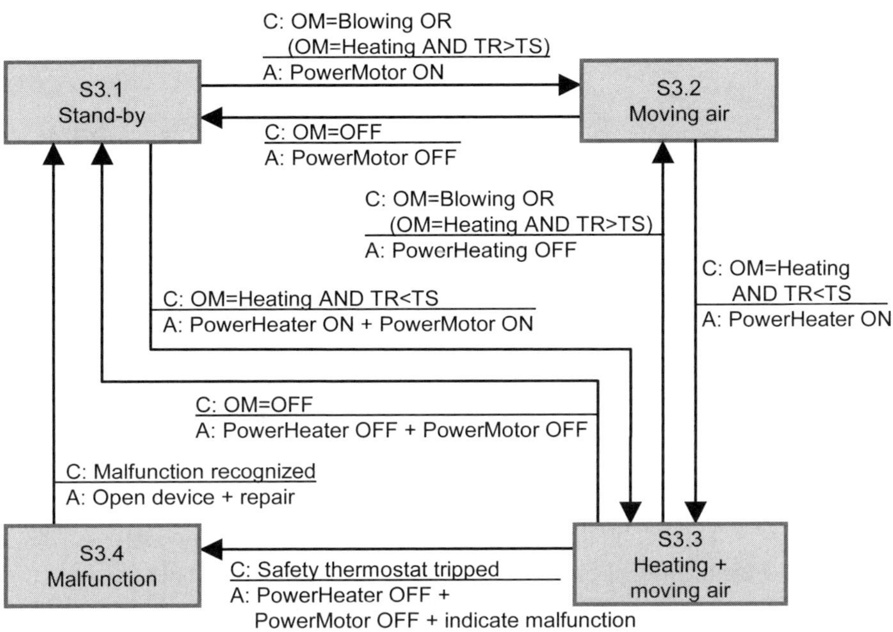

Diagram D3-31: Comprehensive State Chart "Control fan heater."

In a fourth step, the Flow Diagram is updated according to the State Chart, with the additional information and criteria introduced above, in order to keep the model of the total system consistent. In our example, the Data Flow "Overheating: Safety thermostat" is attached to processes P2 and P3. The Flow Diagram completed this way is given in *Diagram D3-32*.

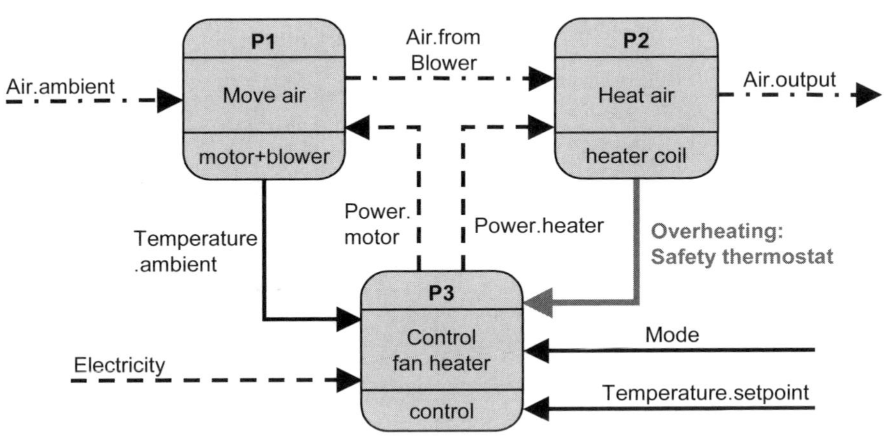

Diagram D3-32: Updated Flow Diagram of fan heater.

3.6.3 Risk Analysis

An additional demand on information, flows, and sensors is often given by a failure mode effect analysis (FMEA). The possibility of failures should be considered concurrently during system conception. While working on the details of system control, handling of failures is updated.

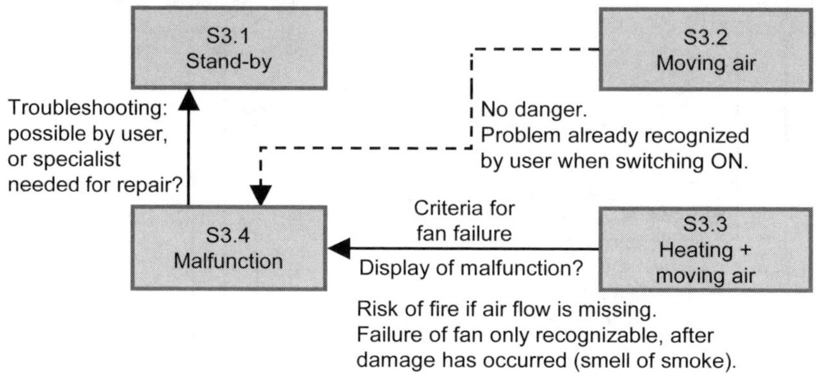

Diagram D3-33: State Chart of fan heater including Failure Analysis.

A worst-case evaluation takes in account that each function may fail at any point in time. The possible damage is assessed and limited by suitable measures. More information on this subject is given by technical safety standards. These call for the earliest possible detection of dangerous states and for measures to reach a safe state immediately. In general, this is achieved by completely cutting off all supply of external energy. *Diagram D3-33* shows the State Chart of the fan heater with an assessment of possible risk and damage, and the appropriate measures.

Table D3-6: Risk Analysis of fan heater

Scenario	**Consequences**	**Risk potential**	**Proba-bility**	**Measures**
Short circuit in controller	Overheating, smoke, fire	High	Moderate	Fuse in power supply
Open circuit in heater coil	Not heating anymore	Low	High	Warning in manual: Take fan heater to repair
Motor, bearing, or winding defect	No air flow, heater glows, smoke, fire	Very high	Moderate	Safety thermostat
Thermostat defect, does not switch off	Continuous heating, overheating environ-ment, smoke, fire	High	Moderate	Limit duration continuous of heating
Thermostat defect, does not switch on	No heating	None	Low	Warning in manual: Take fan heater to repair

Table D3-6 gives an overview on the possible risks of the system and the countermeasures.

3.6.4 Programming Languages for PLC

Starting with the complete State Chart in *Diagram D3-31,* the hardware and software of the controller are now defined, ready to code the control program. Different methods are available for programming, all of them taking advantage of the Flow Diagram and State Chart prepared previously. This chapter gives an overview of programming languages for Programmable Logic Controllers (PLC), as standardized by the International Electrotechnical Committee (IEC).

```
  Contact      Contact         Contact          Output
  (Opener)     (Closer)        (Closer)         (Relay coil)
 Safety                        Control
 thermostat    Heating         thermostat       Heat coil
----] / [-------]    [----------]    [-----------(     )--------
            |
            |                                   Blower motor
             ---]    [-------------------------(     )--------
```

Diagram D3-34: Ladder Diagram (LD).

A simple State Chart is easily converted in code by a Ladder Diagram (LD), as given in *Diagram D3-34*. Coding starts with the transitions between the states. The conditions correspond to the inputs of the PLC, while their logical connection is realized by contacts arranged serially or in parallel. The actions correspond to the outputs of the PLC. The LD corresponds to the circuitry of a conventional hardware controller. Such circuits may also be derived from a State Chart, according to the rules indicated above.

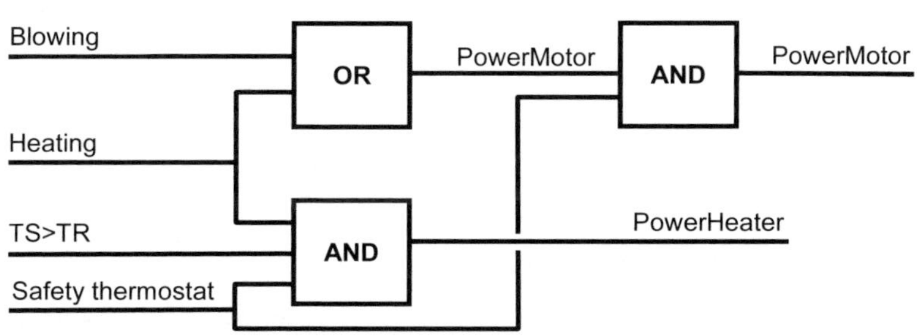

Diagram D3-35: Function Block Diagram (FBD).

The Functional Block Diagram method of graphical programming works in a similar way (*Diagram D3-35*). The contacts and coils of conventional relays are substituted here by the pertaining graphic symbols of Boolean logic.

Programming languages consisting of lines of code are available in the form of Structured Text (ST) or Command List (CL). In this case, the state of each process should be treated as a numerical or a plain language string variable with the name of the pertaining process (*Diagram D3-36*). The transitions between states are put in code in the form of conditional statements that are compiled for each process in a module of the code. By adding comments referring to the State Chart, the source code becomes easy to understand. People preferring classical programming languages along with strict object orientation, as popular in larger computers, define the processes by classes and sub-classes, and treat the pertaining state charts as methods.

```
PowerMotor:= 0
PowerHeater:= 0
IF OM_Blowing OR (OM=Heating AND TR > TS) THEN
     PowerMotor:= 1
END IF
IF OM=Heating AND TR<TS THEN
     PowerHeater:= 1
END IF
```

Diagram D3-36: Structured Text (ST).

More simple, but not necessarily easier to read, are graphical programming languages basing on Petri Nets. The example given in *Diagram D3-37* should be read from top to bottom.

```
INITIAL_STEP Stand-by
      PowerMotor          := 0;
      PowerHeater         := 0;
END_STEP

TRANSITION FROM Stand-by TO Moving-Air
      := OMBLowing;
END_TRANSITION


STEP Moving-Air
      PowerMotor          := 1;
END_STEP
```

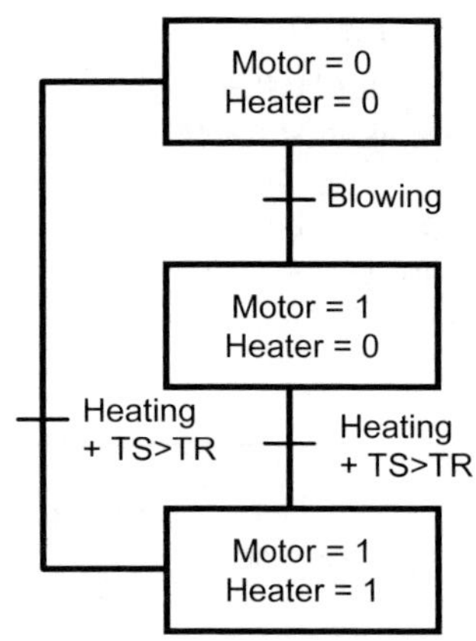

Diagram D3-37: Sequential Function Chart (SFC).

The Sequential Function Chart (SFC) is a single level State Chart with conditions noted on the transitions. The State Chart is directly converted into graphical code by simply adapting the symbols. For the interaction of different processes, flags or semaphores are introduced to transfer information between the different diagrams. In the POA method, these connections are depicted in the Flow Diagram by data flows. For modeling larger systems, it is essential to work with both diagrams, Flow Diagram and State Chart.

3.7 Application Example: Gas Pump

3.7.1 Flow Diagram and Specifications

Purpose of the real-time control

This chapter follows up the case study of a gas pump. In Section D2, a simple simulation model of a gas pump was coded. This chapter develops the real-time control for a gas pump. In order to keep the example reasonably short for this book, the functions of the gas pump are simplified. In reality there are additional complex functions to be taken into account for this control. The functions of the gas pump used in the model are shown in *Diagram D3-38*.

When a car needs gas, the driver drives to a gas station. There, the typical activities on a self service gas pump occur: Open the tank cap, take the filling nozzle from the gas pump, insert it into the car tank, and start the refueling by squeezing the handle on the filling nozzle. By releasing the handle, the driver can stop the filling anytime. As soon as the car tank is full, the pump must turn off automatically. This signal is indicated by the flow "Tank.full." The amount of gas filled into the car and the total price of the gas are displayed on the gas pump. After the driver has paid for the gasoline, the display is set to zero and the gas pump released for the next user (Signal flow "Gas.paid").

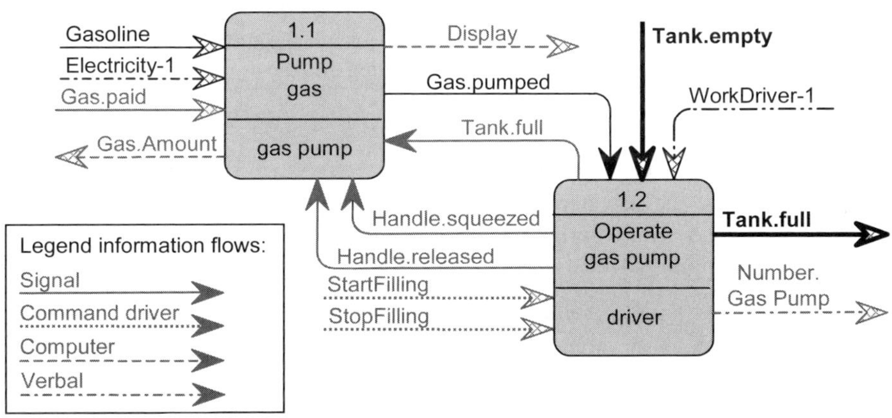

Diagram D3-38: Child Flow Diagram of process "Fill gas."

Diagram D3-38 describes the system "Fill gas" to be analyzed by two processes. The parent process of this diagram can be found on the *Diagram D1-24* "Refuel car" in the application example of Section D1.

Aspects of a machine control

To describe an automated system, the aspect of the machine functions, of the control, and of the operator are described by processes. In this simplified gas pump, the aspects of the functions and of the controller are both integrated into the process "Pump gas." The filling nozzle is part of the process "Operate gas pump."

Element specifications

The element specifications incorporate detailed information that is necessary for the drawing of the State Chart and the following coding of the program for the control of a gas pump.

Specification D3-2: Process "Operate gas pump"

1.2	**Operate gas pump**	**gas pump**
Description	The gas pump transfers gasoline into the car tank as long as the handle of the gas pump is squeezed. If the handle is released, the pump stops.	
	When the tank is full, the pump stops automatically.	
	The pumped amount of gasoline is measured and displayed.	

Specification D3-2 describes the functions of process 1.2 "Operate gas pump." *Specification D3-3* of the flow "Gas.pumped" specifies the filling rate of the gas pump as a parameter.

Specification D3-3: Flow "Gas.pumped"

Gas.pumped			
Description	Pumped gas flowing from the gas pump into the car tank.		
Flow type	Gas		
Parameter	Filling rate	1	liter/second

The commands of the driver are depicted by dotted gray arrows entering process 1.2. The command "StartFilling" is given by squeezing the handle on the filling nozzle, and the command "StopFilling" by releasing the handle.

The signals of the handle position on the filling nozzle are information flows, or signal flows respectively, going from the filling nozzle to the gas pump. They are represented by solid gray arrows going from process 1.2 into process 1.1. The handle of the filling nozzle is deactivated automatically, as soon as the car tank is full (control flow "Tank.full"). *Specification D3-4* describes the signal flow "Handle.released."

Specification D3-4: Flow "Handle.released"

Handle.released	
Description	The driver may stop refueling any time by releasing the handle.
Flow type	Signal (control flow)

3.7.2 State Charts

Diagram D3-39 is the State Chart of process "Operate gas pump" and has four states. It depicts the refueling function from the point of view of the driver, and is not directly included into the program code of the gas pump. However, it is necessary to know the actions of the driver in order to identify his interactions with the gas pump.

When the driver takes the filling nozzle from the gas pump. the pump is activated. By operating the handle of the filling nozzle, the gasoline supply is started and stopped.

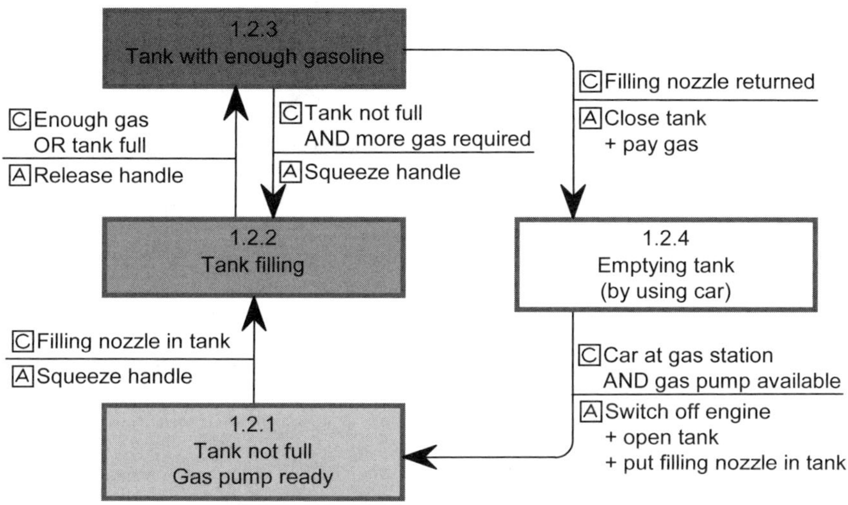

Diagram D3-39: State Chart "Operate gas pump."

The actions "Squeeze handle" and "Release handle" of the State Chart "Operate gas pump" appear as the conditions "Handle squeezed" and "Handle released" on the State Chart "Pump gas." This shows very nicely how the State Charts of a model influence each other, even if no transitions go from one State Chart to the other.

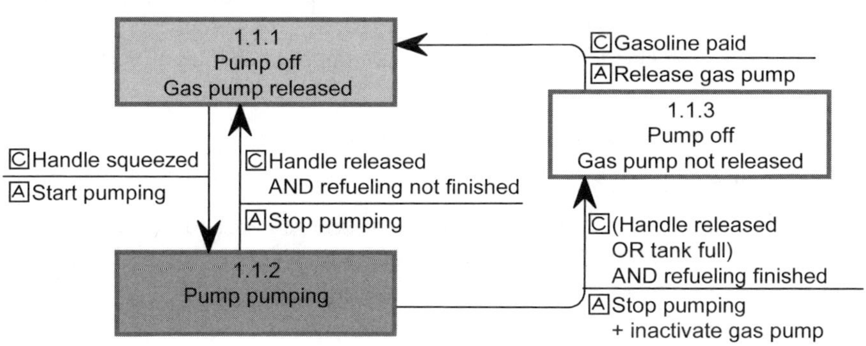

Diagram D3-40: State Chart "Pump gas."

Diagram D3-40 shows the State Chart of the process "Pump gas." The information flows leading into process 1.1 correspond to the conditions of the state transitions in this State Chart.

Diagram D3-41 depicts the State Map of the gas pump. The states of the State Chart "Pump gas" are on the y-axis and the states of the State Chart "Operate gas pump" on the x-axis.

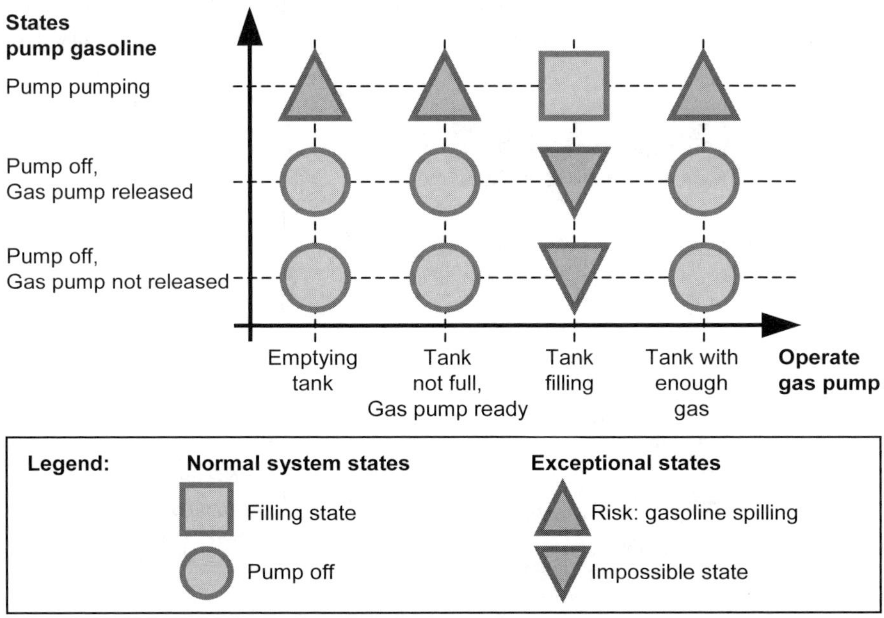

Diagram D3-41: State Map of gas pump.

From the combination of the states, the possible System States emerge. It becomes obvious, which are the intended and which the non-intended or risk states. There are even impossible states. The pump may only supply gasoline if the nozzle is inside the car tank (System State represented by square). Otherwise, gasoline will spill on the ground, shown as an upward triangle. The circular System States represent safe states. The downward triangles are impossible System States.

3.7.3 User Interface and Program Code

Interface to the environment

Figure D3-14 shows the interfaces of the real-time control to the environment. These are the information flows to and from the displays on the gas pump (user interface), the driver actions at the filling nozzle (position of the handle), the signals to the pump motor, and information to and from the cash desk.

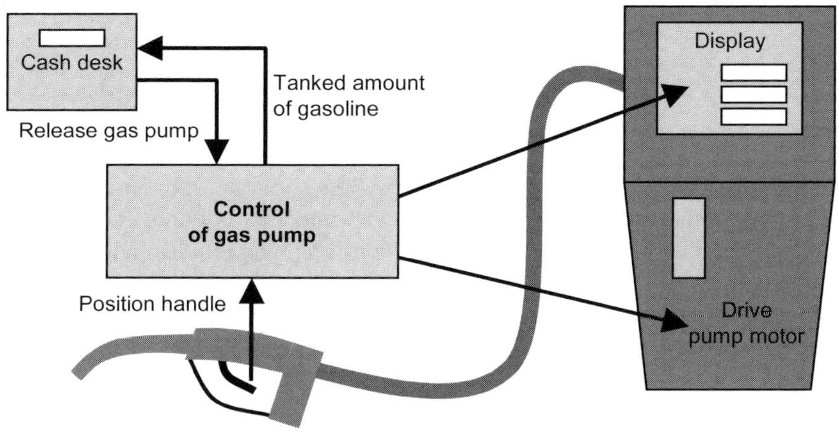

Figure D3-14: Gas pump interfaces.

Program coding

The code example was programmed based on the conventions of Visual Basic .NET. Comments and explications are marked with an apostrophe and written in italics.

Variable declaration: In general, a variable is declared with name and type. It is helpful to indicate the physical dimension and the unit of the variable as a comment directly into the code. The states of the `Select-Case` construct are also declared as variables.

```
'Declaration of variables
Dim GasLevel As Decimal            'Gas level in tank in liters [l]
Dim Liters_pumped As Decimal       'Data for display in liters [l]
Dim LiterPrice As Decimal          'Data for display in Dollars [$]
Public StatePump As Integer        'Declaration of states
Public SystemTime As Decimal       'Time in seconds [s]
```

User interface: The code for the user interface consists of the display of the tanked amount of gasoline and the price to pay.

Initialization, Reset: The start-up of the gas pump control when power supply is connected (cold start) or after a reset (hot start) is not treated here.

The following Private Subroutine (Sub) "ReleaseGaspump" shows the release of the gas pump after a driver used it, and at the same time resets the display to zero. The release of the gas pump happens as soon as the driver paid for the gasoline.

```
Private Sub ReleaseGaspump
    'Release gas pump:
    If Gasoline_paid= True Then
        StatePump = 1
    End If
```

```
    'Reset display on gas pump:
    DisplayLiters.Text = "0"
    DisplayPrice.Text = "0"
End Sub
```

System time: A timing clock gives the pace of update of the gas pump display.

System States: For each State Chart of the machine, a public subroutine is written. In this example the Sub is programmed for the State Chart "Pump gas," since the other State Chart describes the behavior of the driver, who requires no coding. The gray scales of the `Select-Case` construct correspond to those of the State Chart in *Diagram D3-40.*

```
Public Sub PumpGas
  Select Case StatePump
      Case 1            '1.1.1 Pump off, Gas pump released
        Pump_pumps = False
        Pump_released= True
      Case 2            '1.1.2 Pump pumping
        Pump_pumps = True
        If GasLevel = full Then
            Handle_squeezed = False
            State PumpGas = 3
        End If
      Case 3            '1.1.3 Pump off, Gas pump not released
         Pump_pumps = False
         Pump_released = False
  End Select
End Sub
```

Display: The Sub "Display" calculates the amount of pumped gas and the price to be paid and shows this information on the display of the gas pump.

```
Public Sub Display
  'Shows amount and price of gasoline.
  DisplayLiters.Text = Liters_pumped      'Info of counter
  DisplayPrice.Text = Liters_pumped * Price per liter
End Sub
```

User commands: Some activities of the driver at the gas pump put the system into certain states, for example applying the handle of the filling nozzle. Depending on the System State, the operation of the handle starts or stops the pumping of gas.

```
Private Sub Squeeze_handle
    If StatePump = 1 Then
        StatePump = 2
    End If
End Sub
Private Sub Release_handle
    StatePump = 3
End Sub
```

Now, integrate hardware and software and make it work!

3.8 Apply Your Knowledge

1) Processes of a machine

Diagram D3-42 shows a washing machine as you know it from every household. Complete this Flow Diagram with the flows between the three processes already drawn on the diagram.

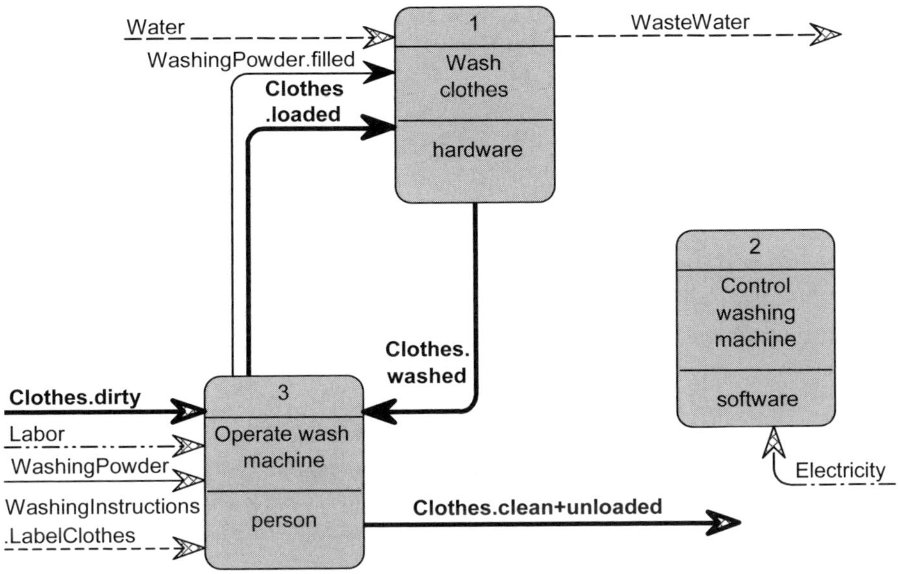

Diagram D3-42: Washing machine.

In addition, draw a Flow Diagram for washing clothes for doing the laundry by hand in the river or a water tub, as in old days.

2) Simulation or real-time control

What is the difference between the simulation for optimization explained in Section D2 and the simulation for testing a real-time control?

3) Step-by-step procedure for real-time coding

On the example of a coffee machine, list the steps of the procedure for setting up a real-time control and explain them.

4) Three aspects

The context diagram in *Diagram D3-43* shows the process "Bake bread." Detail the process "Bake bread" in a child diagram drawing the processes for the three aspects "machine function," "machine control," and "operator." Draw as many interface flows between the processes as you can think of. The bread baking can be assumed alternatively industrial or done in the household.

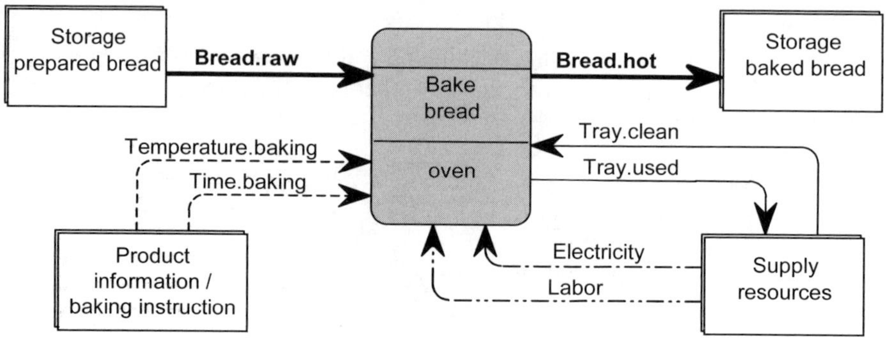

Diagram D3-43: Context diagram for baking bread.

5) Context diagram for a real-time control

The design of a real-time control for a production line or a machine is organized in three phases:
1) system analysis
2) program design and test simulation
3) implementation of real-time control.

Each of these phases has its own context diagram. Draw the context diagram of a injection molding machine for phase 1 and indicate on the diagram how it changes to become the context diagram for the phases 2 and 3.

6) State Map with discrete parameters

The control for a lathe needs to be programmed. *Diagram D3-44* describes the lathe. This machine is not fully automated. The operator has to load the raw piece into the machine and to set the parameters (for example: cutting tool position, spindle speed, feed rate). The machine requires an operator to supervise it.

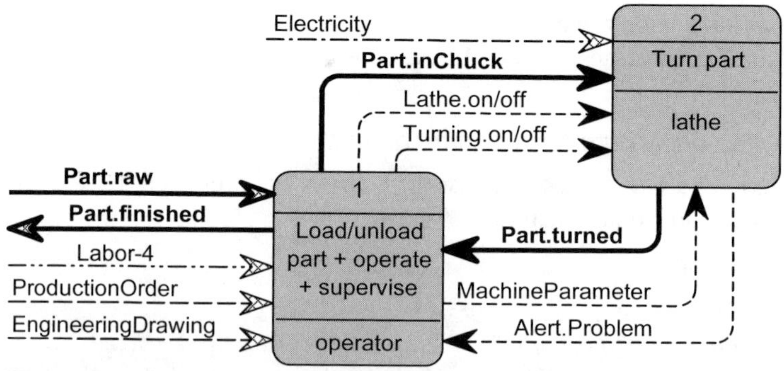

Diagram D3-44: Flow Diagram for working with lathe.

Figure D3-15: Lathe (Photo: Ernst Hedinger AG).

Figure D3-15 shows a machine shop with a lathe. The State Chart describing the lathe may be used for manual or automatic operation.

Diagram D3-45 and *Diagram D3-46* give the State Charts for process 1 "Load/unload part + operate + supervise" and process 2 "Turn part."

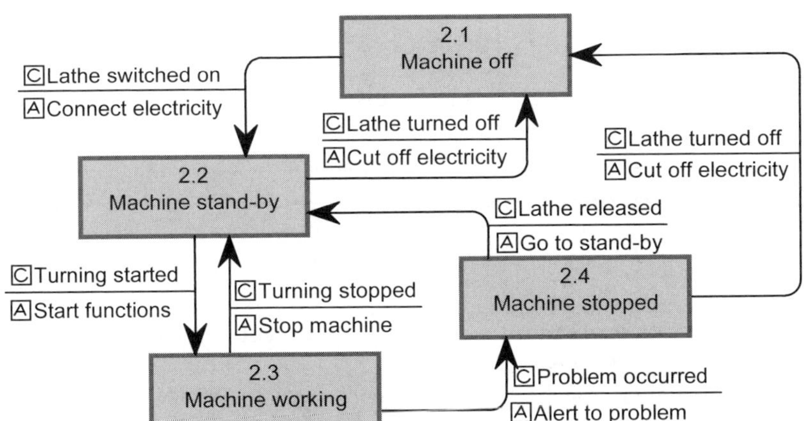

Diagram D3-45: State Chart for process 2 "Turn part."

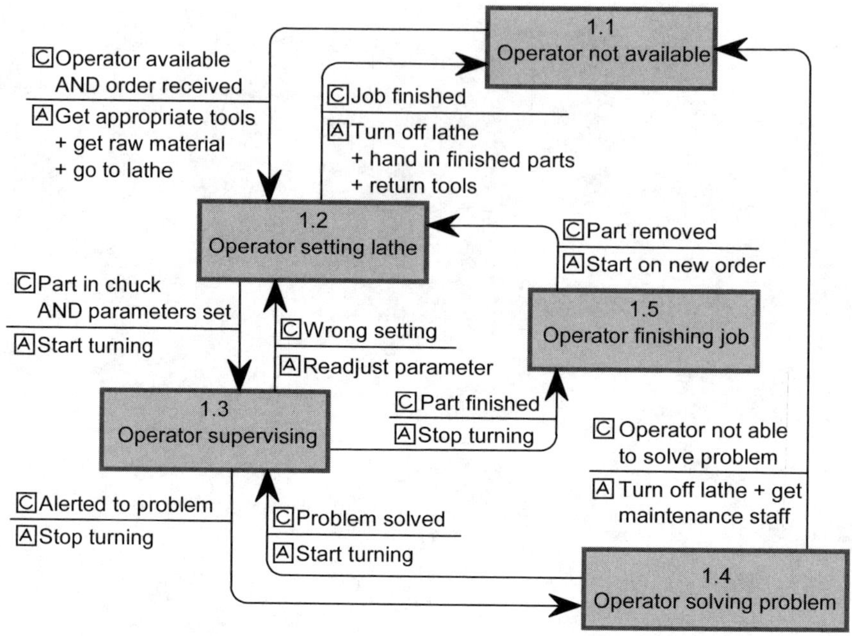

Diagram D3-46: State Chart for process 1 Load/unload part + operate + supervise."

Complete the State Map of a lathe in *Diagram D3-47* based on the given State Charts to check the System States.

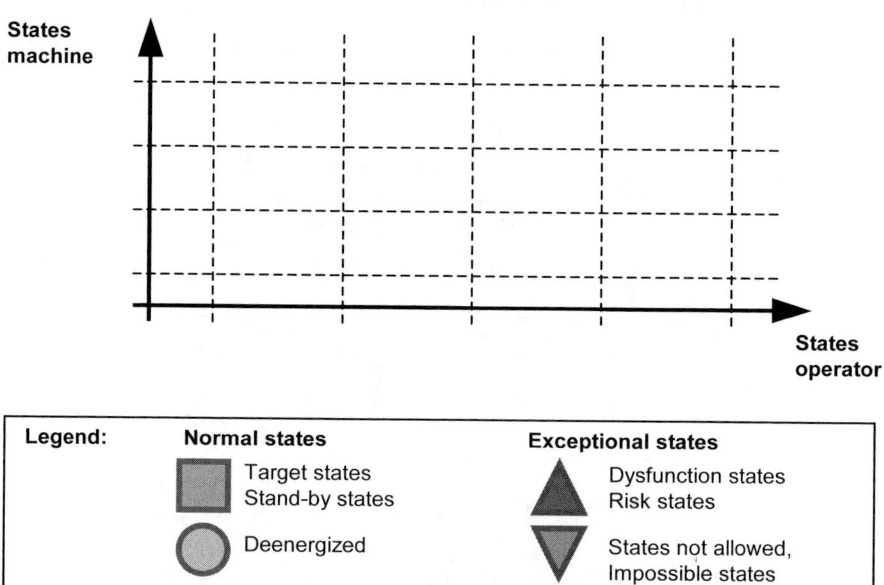

Diagram D3-47: State Map for lathe.

7) State Map of a revolving door

Figure D3-16 shows schematically a revolving door, once in the state when the building is open and people are using the door as indicated, and once when the building is closed and the door is completely closed and inactive.

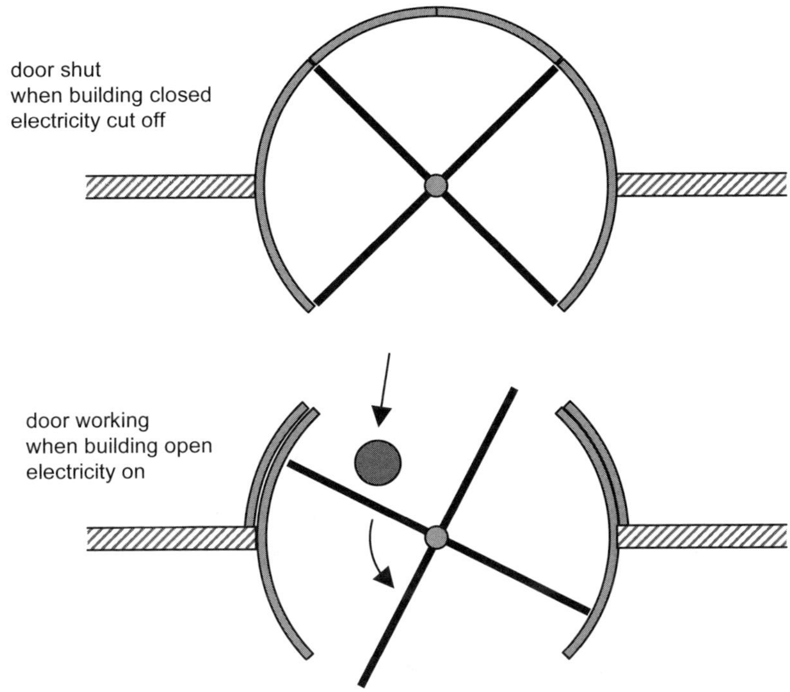

Figure D3-16: Revolving door working and shut.

Diagram D3-48 shows the Flow Diagram with the two processes of the revolving door.

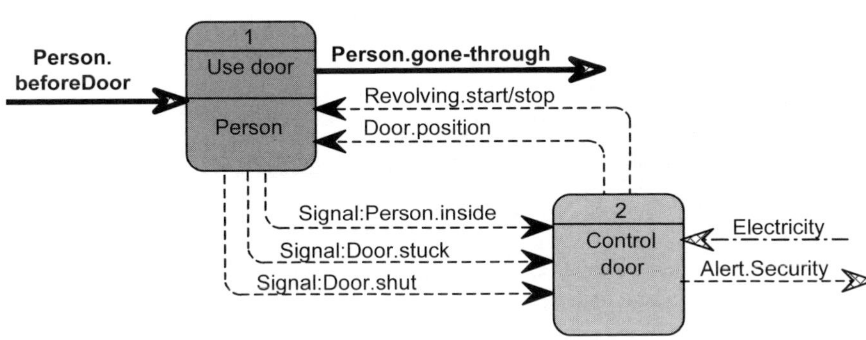

Diagram D3-48: Flow Diagram of a revolving door.

The child State Charts of the two processes of the revolving door are given in *Diagram D3-49*.

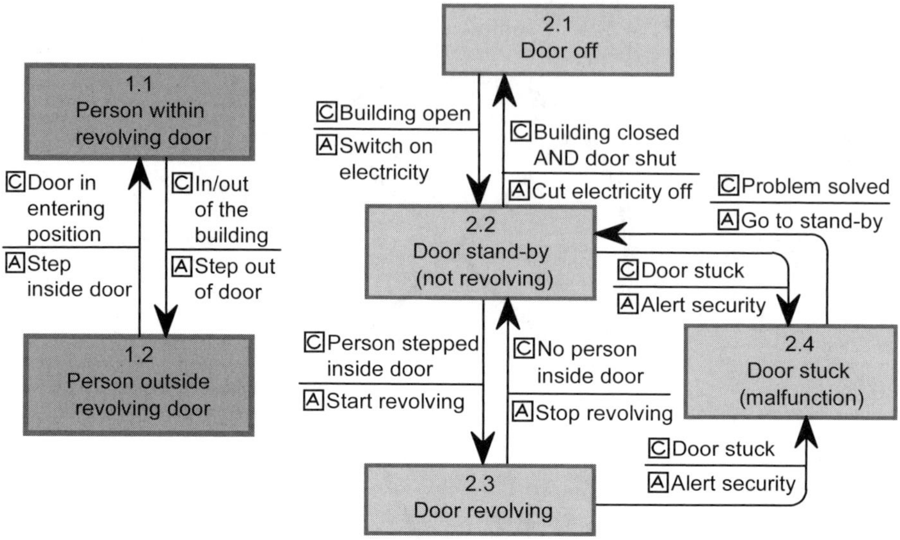

Diagram D3-49: State Charts of processes "Use door" and "Control door."

Draw the State Map based on the given State Charts to check the System States. Be aware of the System States, when a person is trapped in a door. Such states should not occur.

Additional question: Introduce the state "Fire in building" and draw the necessary transitions. Supplement the State Map.

8) State Map for discrete and continuous parameters

Draw a State Map for a system with discrete and continuous parameters. *Diagram D3-50* shows the process of a heating and cooling system for a building. The State Chart depicts the states for heating and cooling (*Diagram D3-51*).

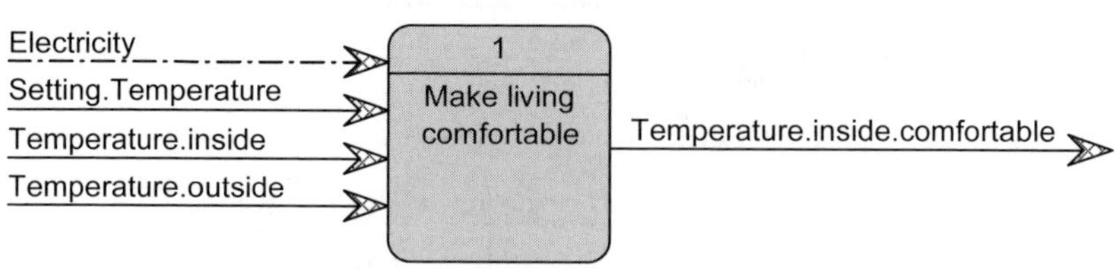

Diagram D3-50: Process for heating and cooling.

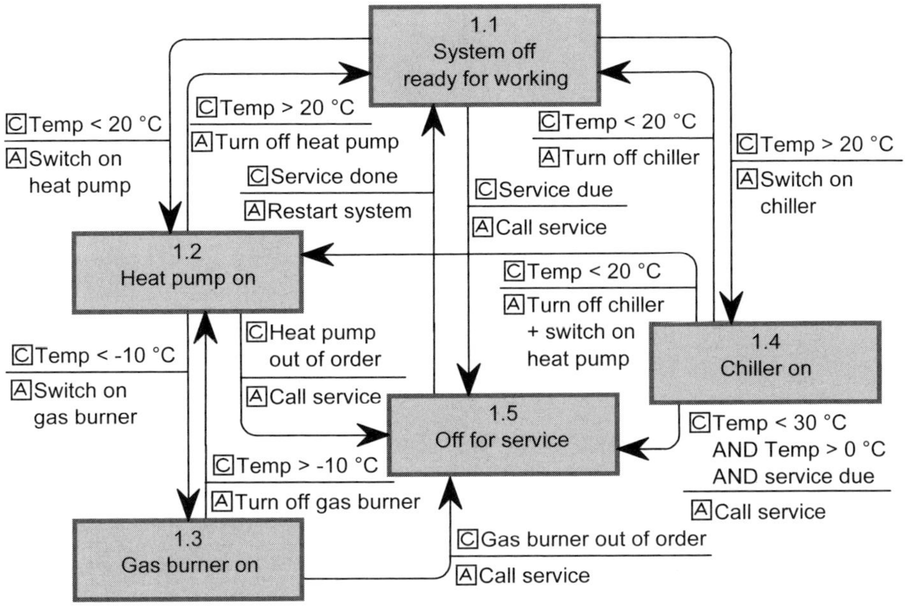

Diagram D3-51: State Chart for heating and cooling of a building.

Complete the State Map in *Diagram D3-52* with the System States and transitions. The states are indicated by gray columns, the transition with arrows. The conditions and actions from the State Chart help complete the State Map.

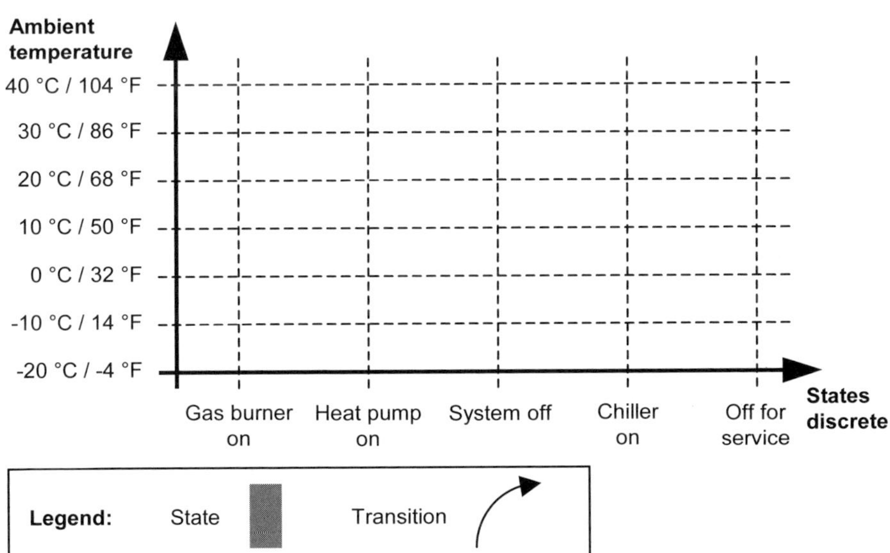

Diagram D3-52: State Map for heating and cooling of a building.

C CASE STUDIES

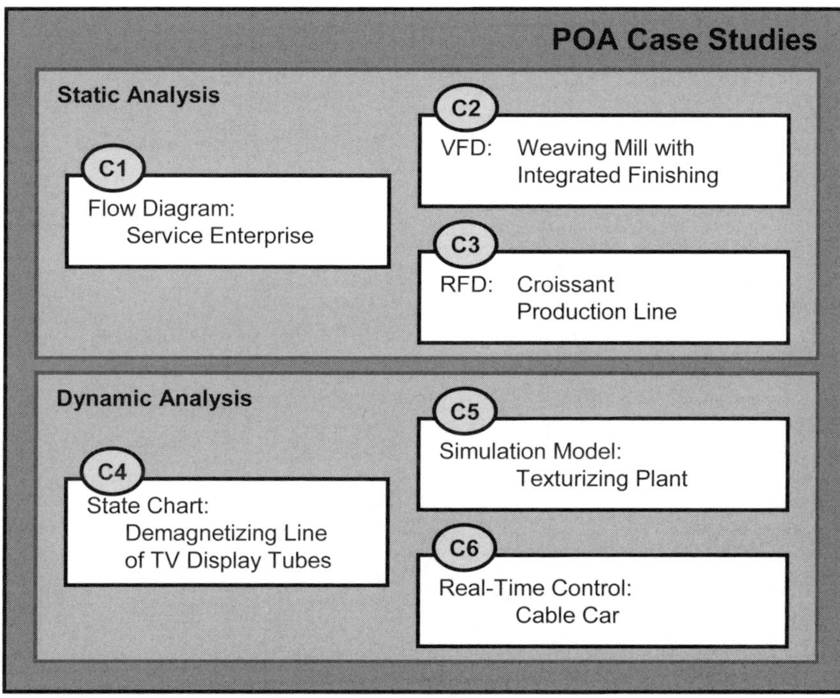

C1 SYSTEM ANALYSIS OF A SERVICE ENTERPRISE

Goals of the Case Study

- Learn to set up a model of a Flow Diagram step-by-step.
- Classify flow types and subtypes.
- Detail processes and set up the hierarchal structure of the model.
- Describe the flows and processes in the element specifications.

Figure C1-1: A bar (Photo: Meyer's Classic Bar AG).

1.1 Getting to Know the Operation of a Bar

This case study presents the analysis of a company in the service industry. The advantage of using Process Oriented Analysis is immediately obvious for large and complex systems. For an enterprise operated by only a few persons and a system being relatively simple, the operation may appear to be straightforward, and the benefits of POA may not be as clear. However, POA can be quite useful and beneficial even when applied to a small-scale operation, as will be shown in this case study that analyses the running of a bar.

Analysis goals

In this case study, the owners of the bar would like to improve the operation of the bar. Therefore, they need to map all the processes and interactions. Thus, it is desired to use POA to fully understand the operation of the bar and to find out where improvements can be made in order to meet the final goal: increase profits. The following should be considered for this bar:

- What furnishing and equipment are already in the bar? What needs to be procured and installed?
- Are renovations necessary? Is there any need for refurbishing?
- Should investments be made?
- How shall the personnel be organized? Who should handle which task and is responsible for what?
- Should certain operations of the bar be computerized?
- What are the costs and the resource consumptions?
- What services should be offered to satisfy the customers?
- Is the bar profitable? Do the prices of food and beverages cover the expenditures?

It is evident already that many things need planning, even in a small enterprise. Here, the Flow Diagram of POA is useful. It gives an overview of the processes, interactions, and workflows.

Setup of the bar

The bar analyzed in this example has the following setup: The staff of the bar consists of two owners and three part time employees. The owners manage the bar and serve the customers; the three employees help with the service, in the kitchen and the cleaning.

1.2 Setting up the Model

1.2.1 Specify the Investigated System

Steps of modeling

This chapter guides you through the application of the method of the Flow Diagram to model a bar. The model is set up according to the following steps:

Step 1: State the purpose of the system.

Step 2: Draw the product flow with the main processes.

Step 3: Define the system boundaries using the context diagram.

Step 4: Classify the flow types and subtypes.

Step 5: Set up the hierarchy of processes and flows.

Step 6: Balance the model.

Step 7: Specify the elements and list the data dictionary.

Step 1: Purpose of the system

First, the model's designer must state the purpose of the system to be investigated. What does the system actually do? In this case study: What is the purpose of a bar? Express it in one sentence.

The owners of the bar want to make profit; this is the goal of every owner of an enterprise. However, from the viewpoint of the investigated system, the purpose of the bar is to make customers happy. Therefore, the purpose of this system is: Make customers happy by serving them good drinks and food.

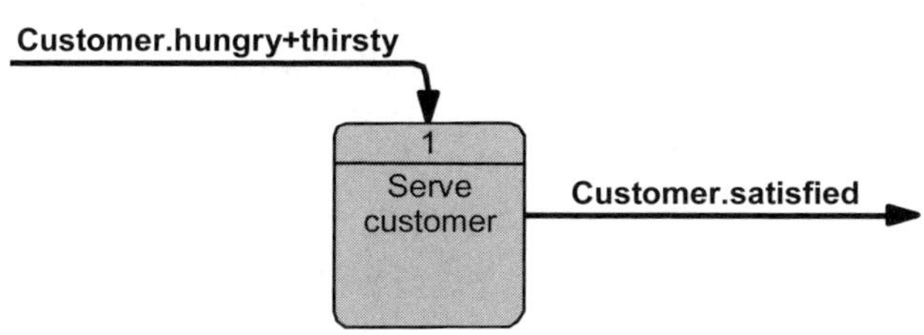

Diagram C1-1: Purpose of the system.

This purpose can be depicted as a simple Flow Diagram with a single process "Serve customer" (*Diagram C1-1*). It illustrates the simplified product flow: A hungry and thirsty customer enters the bar, and a happy customer, satisfied by having enjoyed food, drink, and perhaps even an entertaining chat, leaves the bar. This is the simplest way to begin a model. Sometimes it is a single process, like in this case. For other models, there may be several processes to draw such a simplified diagram in a first step.

Diagram C1-1 is a draft to illustrate the purpose of the bar and to gain an overview of the system that will be analyzed. Later, as more is learned about the system, the diagram will be completed. By then, enough will be known about the system to decide if "Serve customer" is the process, which covers the whole investigated system and can be drawn on the context diagram, or if it is only a part of the system and therefore one of many processes of the level 1 diagram.

Step 2: Product flow with the main processes

In the second step, the process "Serve customer" of *Diagram C1-1* is detailed in the next lower hierarchical level, and the main processes are drawn with the associated product flows. All the processes needed to serve a customer are included in this diagram. We all know these basic activities, even if we never operated a bar or restaurant, because everybody has visited such a facility as a customer.

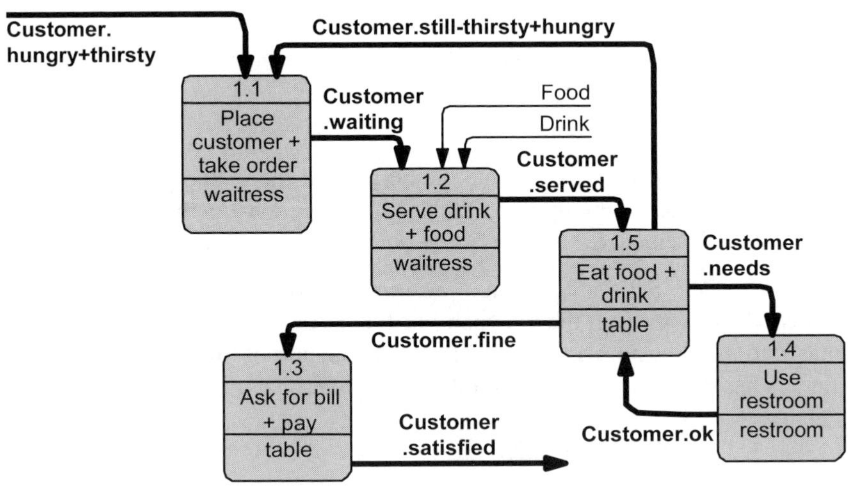

Diagram C1-2: Product flow on level 2.

Follow the product flow. This approach of analysis is called "walk through." *Diagram C1-2* shows the detailed product flow. The customer is directed to and seated at a table, and the waitress takes the order. Then, the customer is left waiting until the drink and/or the food is served. These two material flows are already depicted in the diagram. The customer then enjoys food and drink. Eventually, the customer may need to go to the restroom. Before leaving the bar, the bill for food and drink must be paid.

Looking at *Diagram C1-2*, one can see that only a part of the bar is depicted: The activities related directly to the handling of the customer. For this analysis however, we would like to look at the kitchen and the management, too. Therefore, we go back up to the draft with the single process "Serve customer" on *Diagram C1-1* and add further processes and flows. (See *Diagram C1-3*.)

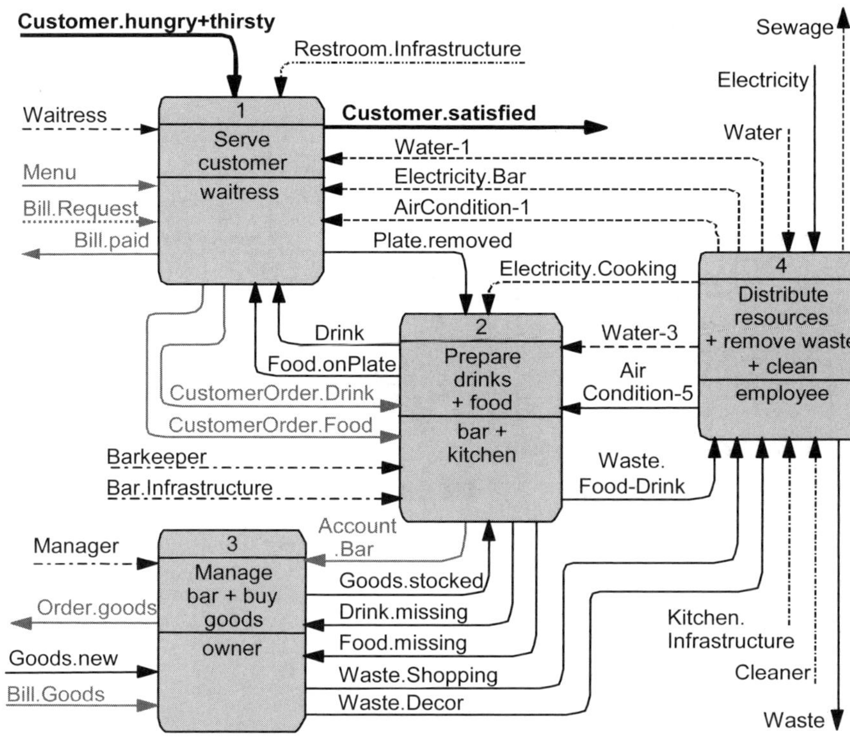

Diagram C1-3: Level 1, all processes.

The bar does not only serve customers. There are many processes carried out in the background that the customer is unaware of, such as the distribution of resources, the management of goods, and the preparation of food and drinks. Only by these processes, the operating of the bar is possible. After drawing additional processes and flows, it becomes clear what we want to investigate and which processes belong to the system. Process "Serve customer" in *Diagram C1-1* is not the entire system of a bar, but only one of the main processes on the level 1 diagram.

Step 3: System boundary with the context diagram

The next step is to define the system boundaries and the interactions with the environment. For this, we draw the top level, called the context diagram, one level above *Diagram C1-3*. The context diagram (*Diagram C1-4*) contains a single process that describes the whole system to be investigated. This process has to incorporate all the processes of level 1. It exchanges resources, information, workforce, and material with the external entities.

In the context diagram, all the flows that interact with the outside world of the model are drawn. At the beginning, it is difficult to catch all of these flows. After the diagrams of lower levels have been depicted, you usually go back to the context diagram, as it is done in this Case Study, redraw the context diagram and balance it with the child diagrams. This means that you add the flows, which were recognized and drawn on the lower levels, to the context diagram.

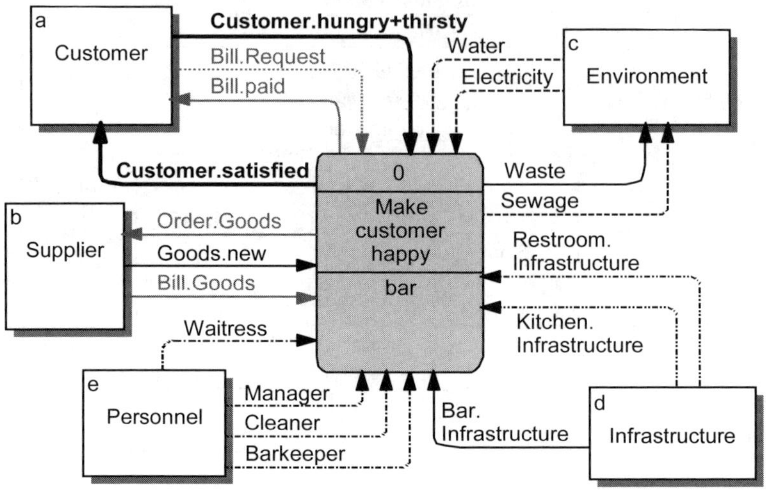

Diagram C1-4: Context diagram.

1.2.2 Detailing of the Diagrams

Step 4: Flow types and subtypes

The diagrams depicted so far show that different flow types are used in the model. As the diagrams get crowded and more and more flows are added, it is especially important to organize the flows in groups and flow types.

Table C1-1: Classification of flows

Flow group	Flow type	Line style
product	product	black solid bold
various material	material	black solid
	waste	black solid
information	paper	gray solid
	verbal	gray dotted
resources	water	black dashed
	electricity	black dashed
	labor	black dot-dashed
	infrastructure	black dot-dashed-dot

Table C1-1 lists the classification of the flows and the line style that are used for the flow types in the diagrams. For example, the product flow is depicted with a bold solid black line labeled in bold.

Step 5: Hierarchy of processes and flows

In the next step, we look at the hierarchy of the Flow Diagram model. So far, the model of the bar consists of the processes displayed in gray and framed with a black line in *Diagram C1-5*. The level numbering in the hierarchy is as follows: The context diagram is the level 0 diagram with the process given in dark gray. The processes displayed in light gray and numbered from 1 to 4 make up the level 1 diagram.

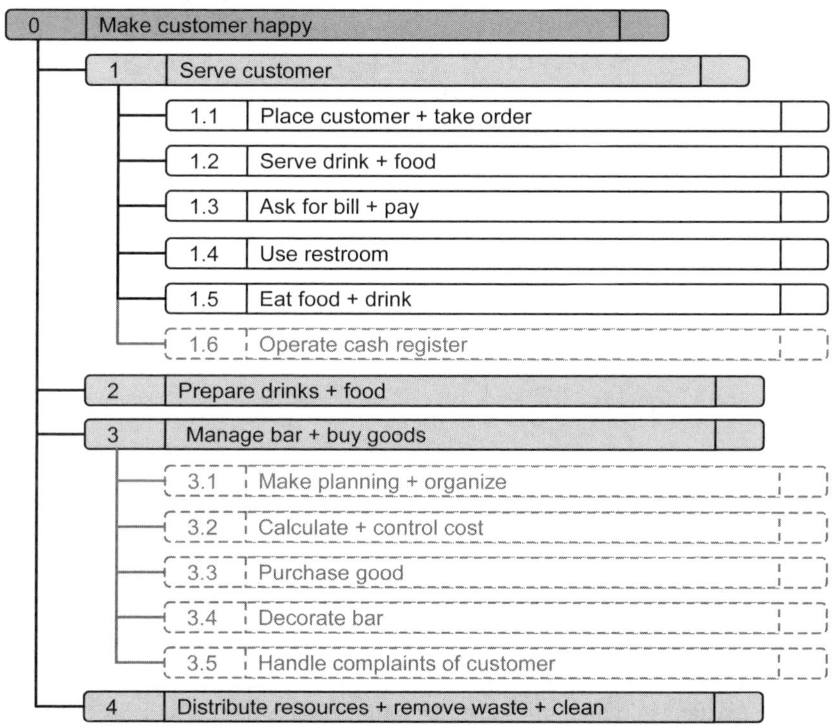

Diagram C1-5: Hierarchy of the processes.

On level 2, we have the child diagram of process 1 with the processes 1.1 to 1.6. The processes represented by white boxes and gray dashed border are built into the model later, on the child diagram of process 2.

Diagram C1-2 included so far only the product flow. It is completed with all the other flows occurring in a bar, as shown in *Diagram C1-6*. This diagram is the second level of the system, the child diagram of the process "Serve customer."

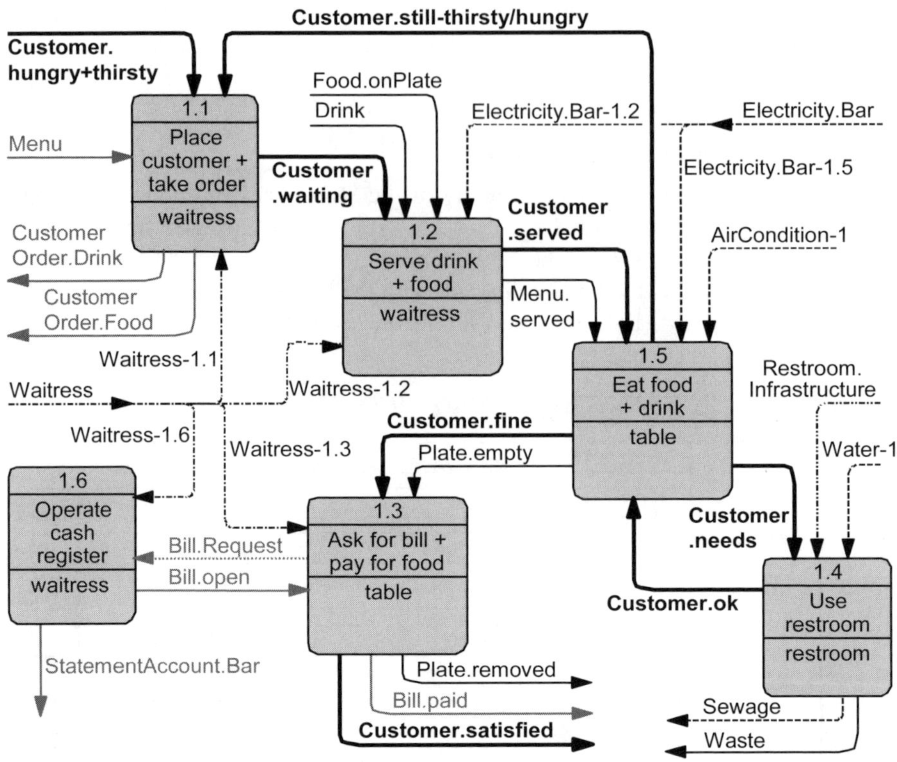

Diagram C1-6: "Serve customer."

Diagram C1-7 on the next page reveals the child diagram of process 3 "Manage bar + buy goods." In the bar, one person does all the planning, organizing, purchasing, and handling of orders, demands, and complaints. If one person does all of these things, it is not always easy to differentiate these processes; nevertheless, it is a worthwhile endeavor. Some possibilities for an efficient organization that were not previously considered may become obvious. In this diagram, there is a large number of information flows, which means that this person must handle a lot of information and data. In this situation, a computer may be helpful to generate a centralized database where all the information can be stored and organized. If the bar and the number of employees grow in the future, the need for a computer will also increase.

Step 6: Balance model

As seen in the system hierarchy on *Diagram C1-5,* the model consists of several diagrams on different levels. While detailing processes into child diagrams, one learns more about the model and draws more flows going in and out of the child diagram than there are flows going in and out of the parent process. This is normal within the procedure of designing a model. Therefore, from time to time you need to check the level balance. Check the flows of the parent processes against the flows of the child diagrams.

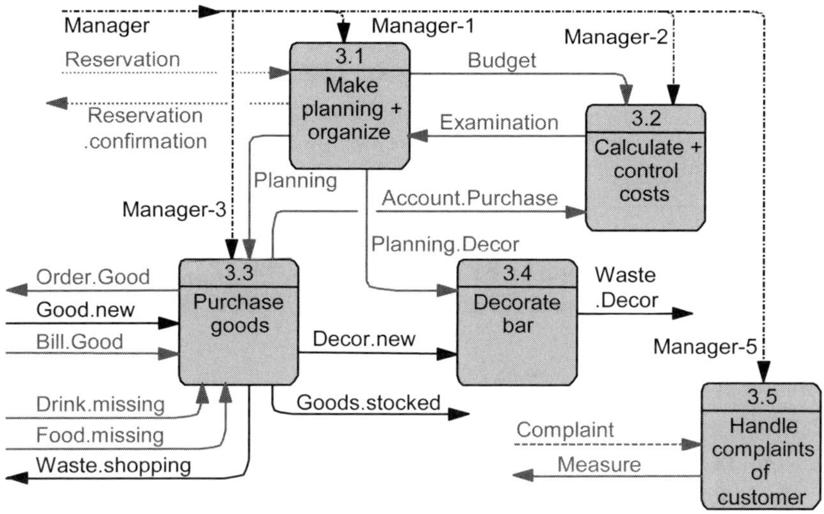

Diagram C1-7: "Manage bar + buy goods."

Doing the check for the diagram of the bar, you will find that several flows are missing. They were not added in the hierarchy during modeling. Difficult in this case are the so-called dangling flows. These are flows that are not connected to a process at both ends. This happens especially to off-page flows because the origin and destination of these flows are on two different diagrams. For a dangling flow, usually the flow on the parent diagram is also missing. Flows that are not balanced in the model hierarchy between child diagram and parent process (error list) are listed below:

- Input of process 3.5: Information "Complaint" and "Measure" are not on the parent diagram, where they should go from process 1 to process 3 (*Diagram C1-3*).
- Input of process 3.1: Information "Reservation" is missing on the parent diagram (*Diagram C1-3*) as input into process 3 and as well as on the context diagram going from external entity "Customer" into the system (*Diagram C1-4*).
- Output of process 3.1: Information "Reservation.confirmation" is missing on the parent diagram (*Diagram C1-3*) coming from process 3 and is missing on the context diagram leaving the system for the external entity "Customer" (*Diagram C1-4*).
- Input of process 1.1: Information "Menu" is a dangling flow on *Diagram C1-6*. Where does the "Menu" come from? From process 2 "Prepare drinks + goods" or process 2 "Manage bar + buy goods"?

The flows “Complaint,” “Measure,” “Reservation,” and “Reservation.confirmation” that are missing on the parent diagram as discussed before clearly illustrate that we had new insights and ideas while detailing the model of the bar.

Step 7: Specifications of elements and data dictionary

Every flow and process in the Flow Diagram model has an element specification. The details in this specification are important; they especially help persons who study and use the diagrams to get to know the system, but did not draw the diagrams. The description clarifies the content of the flow or the exact function of a process. A model should last for a long time, and an element name often does not provide sufficient information to be fully understandable to people other than the model’s designer. The original designer may no longer be able to give any further information.

The section “Flow type” in the flow specification indicates the flow type defined in *Table C1-1.*

Specification C1-1 gives the flow specification of “Customer.hungry+thirsty.” This flow represents a person who wants something to drink and to eat. The flow is part of the flow type “product.” This is all of the information available about this flow at this moment.

Specification C1-1: Flow “Customer.hungry+thirsty”

Customer.hungry+thirsty	
Description	Person who wants something to drink and/or eat.
Flow type	Product

In *Specification C1-2,* the flow “Customer.satisfied” is described. This flow is the end product flow leaving the system. In the description of its specification, the main purpose of the bar is stated, as described at the beginning of the Case Study.

Specification C1-2: Flow “Customer.satisfied”

Customer.satisfied	
Description	Make customers happy by serving them drinks and food. The customer leaving the bar is fine, and no longer hungry or thirsty. He/she had some entertaining chats with other customers and the waitress. He/she liked the food, and he/she is likely to come back.
Flow type	Product

Specification C1-3 defines the flow “Kitchen.Infrastructure.” The description indicates what is included in the flow and how several parameters may be later used for cost calculation.

Specification C1-3: Flow "Kitchen.Infrastructure"

Kitchen.Infrastructure	
Description	Includes all the infrastructure of the kitchen. This includes the space with the pertaining rent, utilities, depreciation, and interests. These serve to calculate costs for preparing meals.
Flow type	Infrastructure

Specification C1-4 defines what activities are part of the process "Manage bar + buy goods" on *Diagram C1-3*. We see these subprocesses detailed in the child diagram of the process in *Diagram C1-7*.

Specification C1-4: Process "Manage bar + buy goods"

3	**Manage bar + buy goods**	**owner**
Description	The management of the bar includes the task of organizing, scheduling, human resources, planning, purchasing of goods, and organization of maintenance and repairs.	

The model is detailed as far as it is necessary to answer the questions asked at the beginning of this project, as described in Chapter C1.1 of this case study.

1.3 Evaluation Report and Benefits of the Method

Several findings and advantages of the Flow Diagram in practice and in this specific case study have been demonstrated.

Service enterprise

This case study clearly shows the benefit of the Flow Diagram in a service enterprise where the product is not a tangible good but rather is a not-so-easy-to-quantify service. The Flow Diagram helps in planning the organization of a bar. The model was drawn before the new owners took over the bar in order to get an overview of the activities in the bar. This assists in deciding on the number and kind of employees needed to hire.

Processes not directly involved with the product

This model illustrates that many processes in a company are not directly involved with the product, which is in this case the "Customer." Such a process is for example "Manage bar + buy goods." These processes are necessary for the functioning of every company. Unfortunately, they are often underestimated, trivialized, and left out of an analysis. It is often assumed that they only create costs because they are not directly involved in the production process.

The flows in this model clearly show that each process creates, collects, and contributes important information and resources to the overall production processes. With the Flow Diagram, these jobs and activities get equal weight in an investigation as the production processes. This is important for any kind of optimization.

Information flow

In this bar, a great amount of information flows in the child diagram of the process "Manage bar + buy goods" are unveiled. This points to activities of the company that could clearly benefit from a computer based information system.

The large amount and variety of information flows in a modern company make it difficult to follow and to document the flows. Often, the information flows seem to appear out of nowhere or are passed on without arriving at their destination. Unlike resource flows, data flows can be created and deleted anywhere, anytime. This makes it necessary to regularly check whether information flows are still necessary and are correctly stored or if and when they may be deleted.

Dangling flow

Loose ends of flows, which occur whenever the designer did not connect both ends of a flow to a process, become immediately obvious when the Flow Diagram is created. Even in this small model, there were some dangling flows without origin or destination process. In reality, dangling flows are usually information flows.

In this case study, we identified a dangling flow on level 1. The information flow "Menu" of process 1.1 on level 1 was forgotten as input of process 1. Otherwise, it would have been discovered that this flow has no origin. Where does the menu come from: process 2 "Prepare drinks + goods" or process 3 "Manage bar + buy goods"? Does the owner or the cook put the menu together? In this bar, this is the same person. Where do we have to place the putting together of the menu? In the kitchen activities or as a part of the organization and planning? According to the answer to this question, the flow "Menu" can be attached as input flow to process 2 or 3.

Consistent naming

In general, the rules of POA force the user to create a clear description and consistent names of processes and flows throughout the whole enterprise. This is very helpful and is greatly appreciated by anybody involved. A consistent naming is often not present in an enterprise; however, the naming standard of the Flow Diagram offers good opportunities to implement order and clearly define what is meant by the flow or process. In practice, different terms for the same activity, area, or flow exist side by side and may create confusion. The Flow Diagram helps to create concise definitions, the basis for every successful engineering job.

C2 ECONOMIC ANALYSIS OF A WEAVING MILL WITH INTEGRATED FINISHING

Goals of the Case Study

- Set up a model of a Value Flow Diagram step-by-step.
- Classify flow types and subtypes.
- Describe the flows and processes in the specifications, define the parameters, and set up the formulas for the value calculation.
- Do the value calculation for a product.

Figure C2-1: Weaving mill (Photo: Lauffenmühle GmbH & Co.).

2.1 Model of a Production Plant

The case study in this chapter presents an example of a Value Flow Diagram (VFD). It is a model of an industrial company, a weaving mill with integrated finishing. The product in this case is a fabric, which is followed throughout the whole production by the VFD. Its value is calculated at each step in production.

This analysis was carried out in an existing company. The modeled plant is called "WeaveFine." This case study is a summary of the original investigation. On the one hand it gives an overview, and on the other hand it highlights some of the interesting details. The model WeaveFine is too big for all its diagrams and specifications to be discussed here completely. Therefore, only a part is displayed here.

The company WeaveFine expanded and changed over the years. Additional plants were acquired and integrated. A weaving company and a finishing plant were merged to become one company that could carry out all processes, from weaving to the finished fabric. Several years after the merger, the processes and flows were still not adjusted. The two former companies are still visible in the structure and naming of processes and flows; this became obvious while drawing the diagrams.

The management did not know the true value and costs of the products within the new merged company. They wanted to verify the product prices and check the cost accounting. In recent years, the taxes on wastewater increased drastically. The influence of the wastewater costs on the product value was not known.

At that point, it was decided to use the VFD to do an analysis of processes and flows to find out where adjustments should be made for optimization. The value of the fabrics would be calculated, beginning with a standard product, and in a second step with a complicated product. In addition, the management wished to investigate the investment of a wastewater pre-treating unit and its influence on the value and the costs of the final product.

2.2 Company and Product

Company

The model company "WeaveFine" is a two-step process company. In the first step, the raw fabric is woven in the weaving mill. In the second step, the raw fabric is dyed and finished.

Investigated product

The product to be investigated here is the final product of the company WeaveFine: It is a dyed and finished fabric. This means that this fabric runs through both weaving and finishing processes. Each fabric in the company is given an article code. In this VFD model, "Fabric02" is analyzed.

Weaving

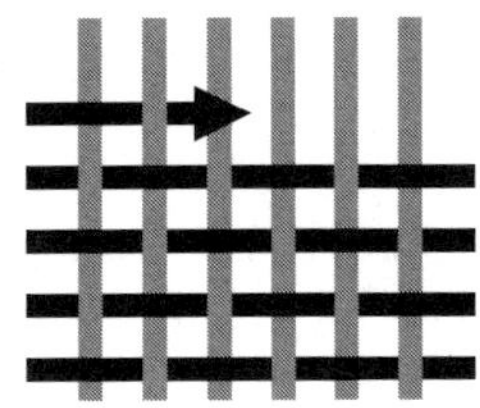

This chapter gives some explanations of the production steps used to create a fabric. The fabric is woven on a loom. During the weaving process, the weft ends are interlaced with the warp ends. In the weaving preparation, the lengthwise yarns in the fabric, the warp ends, are taken up all at once in parallel on the warp beam. The quantity of the yarn used to weave the fabric is calculated from the data "Warp length" and "Weft on loom" given in *Table C2-1*.

The parameter "Weft on loom" is necessary for the calculation of the weaving time for a warp and is given in ends per centimeter. Through the weaving, the warp ends and weft ends in the fabric are not straight, they go around each other in a wavy line. This means that the woven fabric is shorter than the original length of the warp ends. This warp weaving-in can be calculated from the two weft densities before and after the finishing, "Weft on loom" and "Weft density finished." In this case, the shrinkage is quite significant, since the investigated product is an elastic fabric. Such a fabric draws together when taken from the loom, where it is in a stretched state.

As an example, *Table C2-1* shows an extract of the original production papers of the "Fabric02" that is investigated in this case study. This production paper accompanies the order for the weaving into the weaving mill. Not all the data are given here, only the data necessary for the value calculation and explanation in this case study are listed.

Table C2-1: Production paper of the weaving department

Weaving data			**Explanation**
Weft on loom	18	ends/cm	Number of weft ends/cm of the woven raw fabric before finishing.
Weft density finished	26.6	ends/cm	Number of weft ends/cm of the finished fabric.
Warp length	3094	m	Meters on warp beam.
Weave piece raw	90	m	*Raw meter* per piece after the weaving process, before finishing.
Piece length	54	m	Piece length of the finished fabric in *finished meters*. It is a standard piece length, and therefore the same for all articles.
Number of pieces	34		Number of pieces per warp.

Raw fabric

The fabric taken from the loom is the raw fabric; its quantity is given as "raw meter." It is cut in "Weave pieces raw" of 90 m length.

Finishing department and finished fabric

During the dyeing and finishing process the fabric shrinks. The fabric ends are damaged because they are repeatedly sewn together and again separated for different processes, which causes an additional loss. As a consequence, the fabric becomes shorter during its production. The final length of the finished fabric is given in "finished meters." The piece length of a finished fabric is standardized to 54 m. It is the same piece of fabric as the raw fabric that had originally a length of 90 m.

Customer order data

The specific data of a fabric for the value calculation are taken from the production papers (*Table C2-1*) and the customer order (*Table C2-2*).

Table C2-2: Customer order data

Order data			**Explanation**
Fabric code	02		Company internal code.
Order length	14,700	m	The meters of finished fabric ordered by the customer.
Number of warps	8	wp	The customer's order length is converted into number of warps (wp): 14700m / 54m/piece = 272.2 pieces per order 272.2 pieces / 34pieces/warp = 8.0 warps

The data from the customer order and production paper, as well as the resulting conversions, are inserted into the process and flow specifications of the VFD to be used for the value calculation.

2.3 Procedure for Setting up a Model

This chapter describes the step-by-step procedure for creating a VFD (according to Section S2) on the example of this weaving mill with integrated finishing.

1. Start with the product flow on level 1 of the model.
2. Classify the flows into flow types and subtypes.
3. Structure the model hierarchically.
4. Specify the elements process and flow.
5. Define the parameters in the specifications and combine them into formulas.
6. Calculate the values and put the results in the VFD.

Step 1: Product flow on level 1

The simplest way to begin a model is to start with the product flow and the main processes on a high level diagram. In this model, it is the level 1. The main processes with the simplified product flow are drawn, as shown in *Diagram C2-1*.

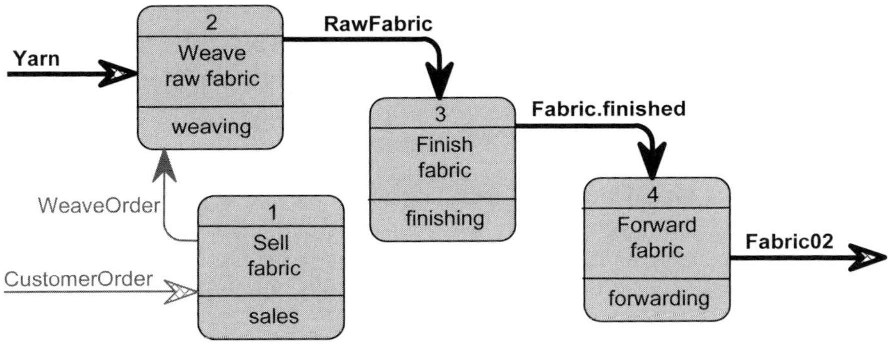

Diagram C2-1: Simplified product flow.

The model describes the system from the sales department to the shipment, from the "CustomerOrder" and the "Yarn" entering the weaving department until the finished fabric "Fabric02."

Recommendation: For the first model of a company, start with a simple product that is typical for the production, is assumed to have no exceptions, and does not need special treatment. Later models will treat more complicated production sequences and networks.

Step 2: Flow classification

In the next step of modeling, various material and resource flows are added to the product flow. The flows are classified into flow types. By the information flows, the administrative and planning processes are added to the model. WeaveFine is a complex and large model. Special measures are therefore taken to make the diagrams easy to read and understand, using the classification into flow groups. On the context diagram, all the flow groups are drawn. All other diagrams in this case study show only the product flow and one of the following flow groups:

- Materials
- Information
- Resources
- Monetary resources

The different flow types within a flow group are distinguished by line styles. For a better readability, the product flow is depicted by a fat line and labeled in bold. The name and value of fictitious value flows are written in italics.

As just explained, only one flow group at a time is depicted on a diagram. However, all processes are shown in every diagram. On single diagrams, this may result in processes that do not seem to have any connected flow, or have no input or output. The reason for this that this process is not involved with any flow from the particular flow group shown on the diagram. Such a process is depicted in white with a dashed border. If only one flow group is visible on a diagram, keep in mind that the rules apply for all the flows, not only for one flow group. This has to be considered especially for the process balance and the calculation rules.

Step 3: Hierarchical structure

The processes are hierarchically refined. The hierarchical structure in this model is functional, based on the structure of the departments in the company. *Diagram C2-2* shows the hierarchy of the processes and diagrams in this model. The context diagram, with only one process, is the level 0 diagram. Processes 1 to 7 are on the level 1 diagram.

Step 4: Element specifications

Every element must be specified with a name, description, set of parameters, set of equations, and value. The element name alone is not sufficient. Take care to comment the elements in the rubric "Description" of the element specification. The element specification gives the true benefit to the model. By fully describing each element in an organized format, the model is still clear at a later point.

Step 5: Parameters

The parameters for the value calculation are defined in the specification and combined into formulas. This work is laborious. First, the parameters must be collected from different sources in the company. Then, the VFD calculates the value for one single product. To do this, the user has to consider how the values and costs of a plant or a department are split up among the different products.

If values, parameters, and allocations are unknown, estimates may be used for an initial value calculation. In the course of detailing the model, the estimates are replaced one after another by the actual values and quantities.

Step 6: Value calculation

The user completes the calculations step-by-step, according to the method of VFD.

The Standard Unit (SU) for the model WeaveFine is 100 m finished fabric. Each value shown on the diagrams refers to the SU of 100 m finished fabric.

Remarks on the model WeaveFine

The diagrams of the model WeaveFine in this case study are all shown from the beginning with the calculated values. The setup was done according to the step-by-step procedure. Typical features of the VFD and special cases in the diagrams or in the specifications are also further discussed in the chapter. In this model, only costs and outputs are given. Money flows of expenses and receipts, as for example the payment to a supplier or from a customer, are not looked at.

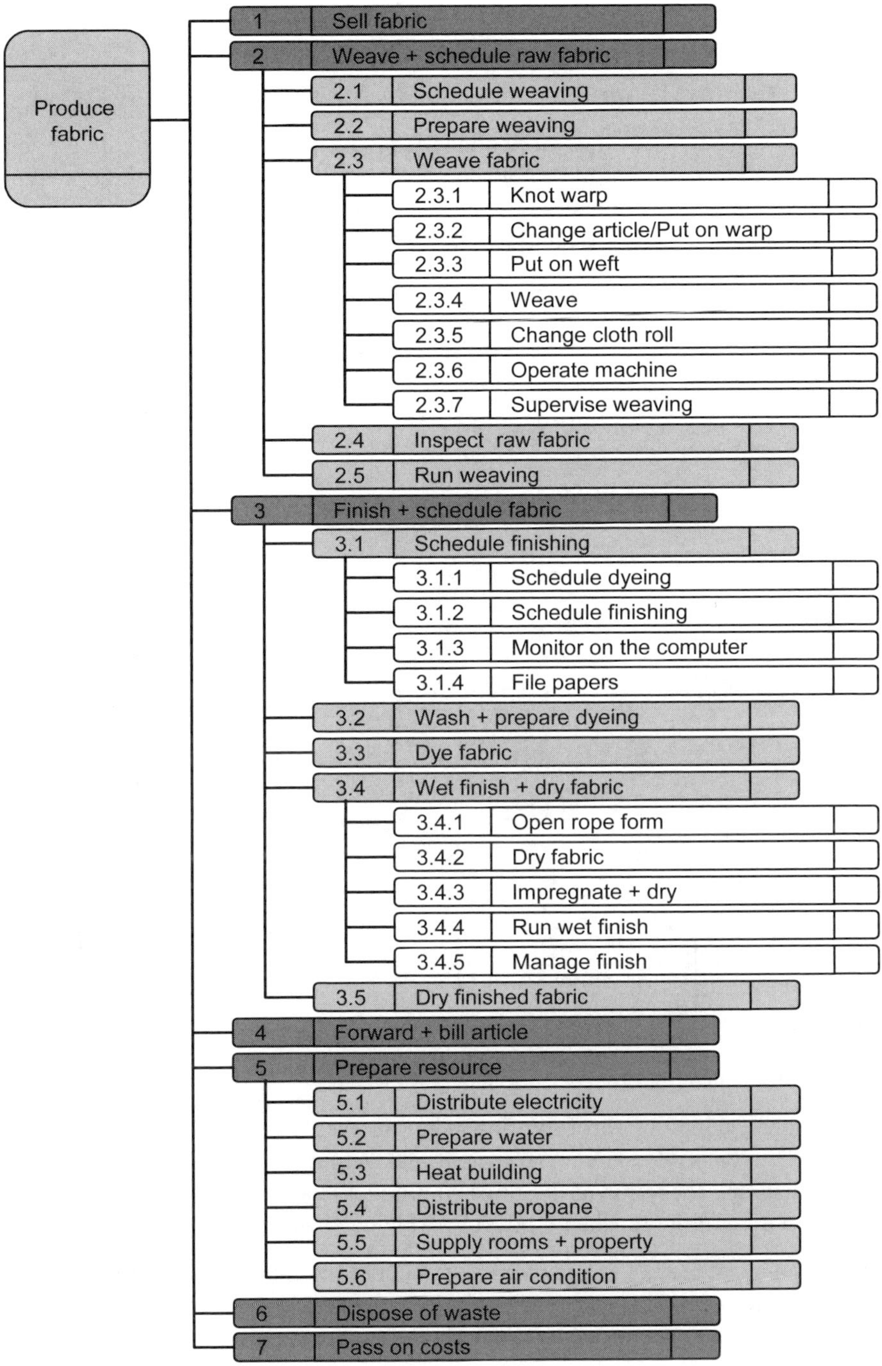

Diagram C2-2: Hierarchy of the model WeaveFine.

2.4 Value Flow Diagrams of WeaveFine

2.4.1 Context Diagram

The context diagram defines the boundary of the investigated system. A model does not need to represent the whole company. In the case of WeaveFine, only the departments directly involved with the production of the fabric and the fulfillment of the customer order are depicted within the system (*Diagram C2-3*).

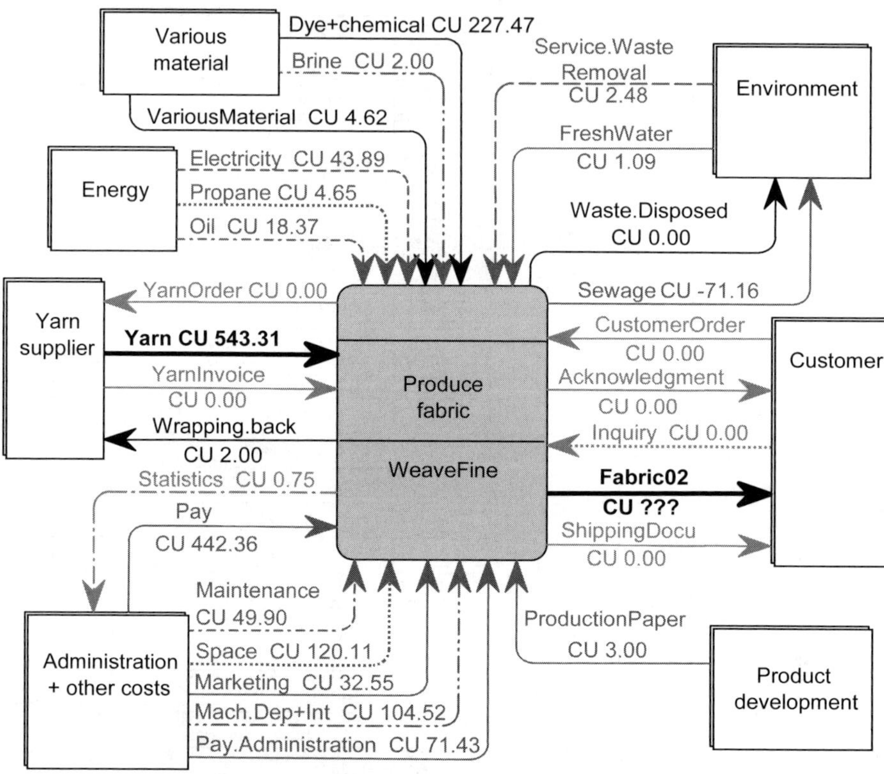

Diagram C2-3: Context diagram of the company WeaveFine.

In the context diagram, all of the flows that interact with the world outside of the model are given while showing the input and output values of the system. In this context diagram, the value of the finished product "Fabric02" is not yet calculated. The calculation of the finished product is explained later in this chapter.

Departments that are not included in the system but exchange value with the model, such as "Administration" or "Product development," are shown as external entities, which bring their costs through various flows into the system.

The external entity "Administration + other costs" provides several services. The cost of the human labor goes into the system through the flow "Pay." The cost of the administration is passed onto the product by "Pay.Administration." This external entity supplies also the "Maintenance" costs, the depreciation and interests ("Mach.Dep+Int") for the machines, capital and maintenance costs for the buildings ("Room"), and the costs from the "Marketing."

The external entity "Environment" includes nature, society, community, and government. The interfaces are the local taxes (i.e. waste taxes by "Waste.Disposed"), as well as "FreshWater" and "Sewage."

The "Product development," in this case the fabric design, is not investigated in this model. It appears as external entity. The costs incurred in this department are passed onto the product through the information flow "ProductionPaper." *Specification C2-1* shows the specification of this flow. All of the data necessary for the production processes and for the mass calculation are recorded on this "ProductionPaper." *Table C2-1* at the beginning of this case study provides an extract of the "ProductionPaper."

Specification C2-1: Flow "ProductionPaper"

ProductionPaper			
Description	Paper Through the production paper, the development costs are passed onto the product. The value of this paper is given by the product and design department; no calculation is needed.		
Flow type	Production paper		
Parameter	Production paper	3.00	CU/100m
Calculation			
Flow value		**3.00**	CU/100m

2.4.2 VFD Level 1: "Produce Fabric"

The VFD "Produce fabric" is the level 1 diagram in the model. Versions of this diagram are shown throughout this paragraph, each time with another flow group displayed.

Flow group "material"

Diagram C2-4 shows the product flow and the flow group "Material." Processes 2, 3, and 4 are the production processes. Processes 1 and 5 are not involved in any activity with a product or material.

The product output "Fabric.withoutPassedOnCosts" of process 4 does not correspond to the product output "Fabric02" on the context diagram. Later, when we know more about the model, this diagram will be completed by adding process 7 "Pass on costs," which has the final product output "Fabric02."

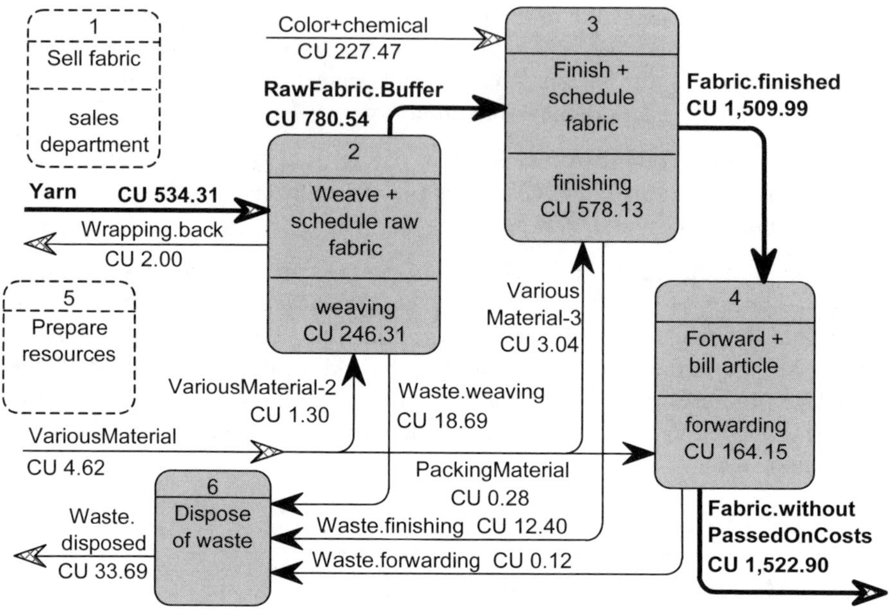

Diagram C2-4: "Produce fabric" – material flows.

Flow group "information"

The information flows in this case study are classified into the flow types paper, verbal, and computer to investigate the actual use of the computer system by the employees, and to see where savings in printouts can be made. Another reason for classifying the information is to see where informal information (verbal) is used, such as the various "Agreements" on *Diagram C2-5*. These "Agreements" exchanges turned out to be a vital production information between the people scheduling the customer orders within the different departments. This information is never written down. If it is lost or if the scheduling person is absent, it creates problems. *Table C2-3* lists the different flow types of the flow group "information" and shows the line styles used in the diagrams.

Table C2-3: Flow group "information"

Line style	Flow types of information	Subtypes
solid	Information on paper.	- Production paper - Order
dotted	Verbal information.	
dot-dash-dot	Data inserted into the computer system. Accessible from each terminal within the company.	

On this high-level diagram, you can see how the information goes through the entire company from department to department (*Diagram C2-5*). Some simplifications were made and not all the information flows included, so as not to overcrowding the model.

The information flows show clearly that the weaving mill and the finishing department work like two independent companies. There is minimal information exchange between weaving and finishing: no paper, only a verbal arrangement between the persons doing the scheduling of the two departments and the identification of the raw material "ID-Coupon" written directly onto the fabric. No documentation is present.

By this different amount of flows between the processes, the diagram clearly reveals that one employee in the finishing department (process 3), who is responsible for the scheduling, also unofficially runs the whole production. The number of information flows between processes 1 and 3 and processes 3 and 4 confirm these findings. When process 3 was further detailed, if became even more obvious. If this employee is absent, scheduling and management, especially for new customer orders, do not run efficiently. The rates of mistakes and miscalculations immediately increase.

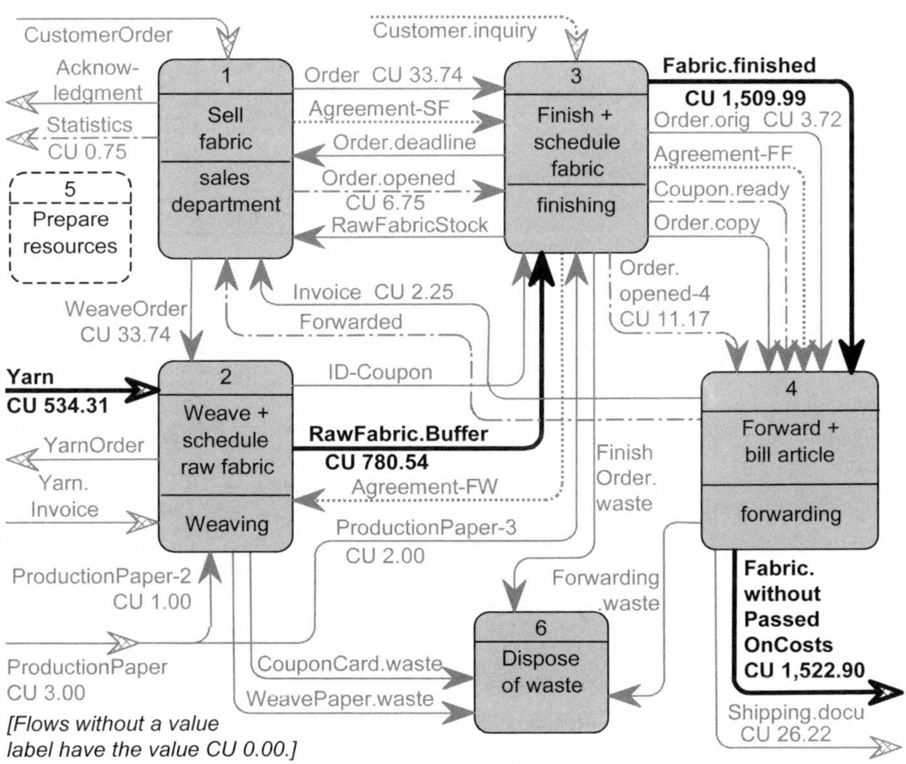

Diagram C2-5: "Produce fabric" – information flows.

The process "Sell fabric" includes the whole internal sales department. First, the customer orders a fabric. Then, there is a meeting where the finishing is scheduled and a verbal agreement between the two departments is made. The weaving department is not included in this agreement. The order is entered into the computer system, so that everyone can check the order status from every computer. A "WeaveOrder" is printed on paper for the weaving department.

Flow group "resources"

On level 1 diagram of the hierarchy (*Diagram C2-6*), the total resource consumption of the processes can be readout. The preparation and distribution of these resources take place in process 5 "Prepare resources."

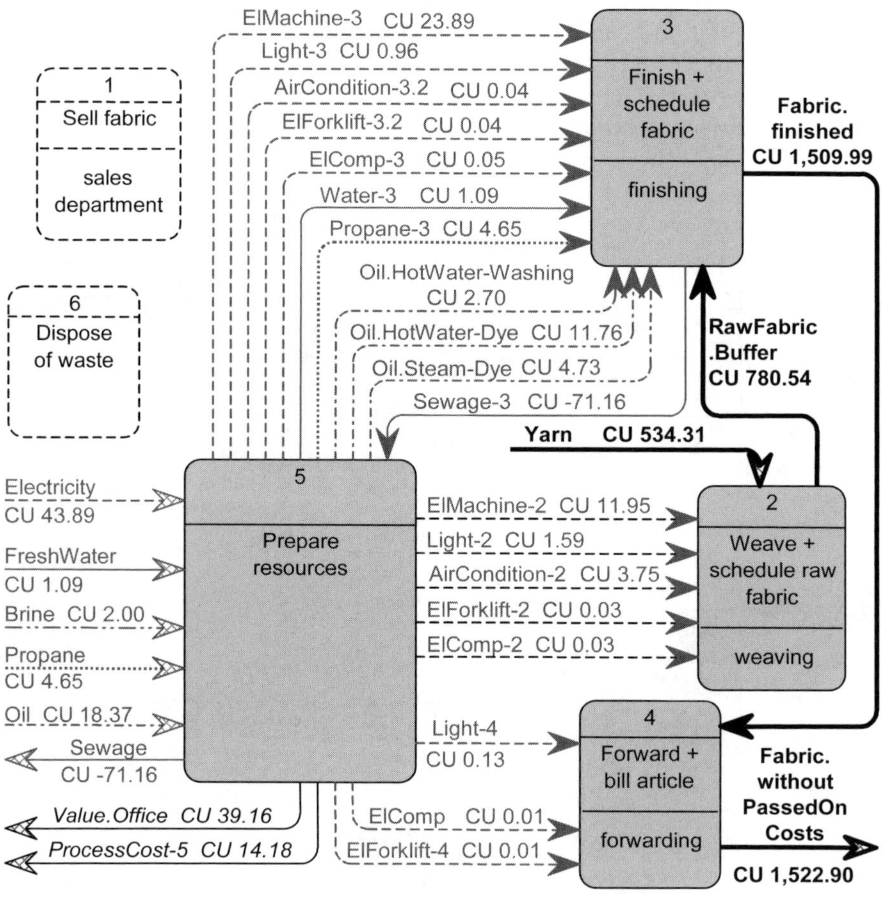

Diagram C2-6: "Produce fabric" – resource flows.

Resources for the offices of the sales department and the administration are not calculated separately, but instead are added to the fictitious flow "Value.Office."

"ProcessCost-5" is also a fictitious value flow that passes on costs to another process. This concept is explained in Chapter C2.4.5.

Sewage costs

A lot of water is used for dyeing fabric in the finishing department. This results in "Sewage" leaving the investigated system through the process "Prepare resources." The "Sewage" is depicted as a negative value flow. The "Sewage" itself goes in the direction indicated by the arrow, eventually leaving the system and going into the external entity "Environment." The company has to pay the community for the disposal of industrial sewage. This service of cleaning the wastewater received from the community goes into the system; therefore it flows opposite to the direction of the arrow of the "Sewage," resulting in a negative flow.

The process "Prepare resources" is further detailed into the processes "Distribute electricity," "Prepare water," "Heat building," "Distribute and store propane," "Supply rooms + property," and "Supply air condition." This diagram is not shown in this case study. The value flow "ProcessCost-5" on *Diagram C2-6* combines maintenance, service, room, and pay from the child diagram of process "Prepare resource."

The different flow types and subtypes of the group "resources flows" are listed in *Table C2-4*.

Table C2-4: Flow group "resources"

Line style	Flow types of resources	Subtypes
solid	water	- "FreshWater," "Water" - "Sewage"
dashed	electricity	- "ElMach" for machines - "Light" - "ElForklift" - "AirCondition" - "ElComp" for computer
dotted	propane	
dot-dash	oil	
dot-dash-dot	brine	

Flow group "monetary resources"

The product value must include all the costs of the buildings, machines, pay for the labor (salaries, wages), as well as other costs that contribute to the end cost of the product. These costs can also be shown as value flows. In this model, they are bundled in the flow group "monetary resources." This flow group consists of the flow types given in *Table C2-5*.

Table C2-5: Flow group "monetary resources"

Line style	Flow types of monetary resources	Explanation
solid	Pay	Salary, wages, etc. for the labor.
dashed	Waste	Local taxes and fees for different kinds of waste.
dotted	Space	Includes maintenance, depreciation, and interests for the buildings.
dot-dash	Maintenance	Maintenance, repair, and spare parts of machines and equipment.
dot-dash-dot	Mach.Dep+Int	Depreciation and interest of machines and equipment.

The cost of each machine consists of "Maintenance," depreciation and interest ("Mach.Dep+Int") on the capital. Their values are normally established for a period of one year in the bookkeeping. They must be allocated to the different products and converted to the SU of 100 m finished fabric. This annual value applies for most of the monetary resource flows. For the building costs ("Room"), the floor space needed (square meter) for the production of one product, "Fabric02," has to be calculated.

Specification C2-2: Flow "Pay.Sales"

Pay.Sales			
Description	Salary in the sales department. Calculation of the salary costs for a normal order without exception handling and customer complaints.		
Flow type	Pay		
Parameter	NumberEmployees PaySales per employee WorkDays per year NumberOrders per day	10 80,000.00 220 50	persons CU/year day/year order/day
Calculation	(PaySales * NumberEmployees) / (WorkDays * NumberOrder) = (80000.00 CU/year * 10 person) / (220 day/year * 50 order/day) =		
Flow value		**72.72**	CU/100m

Specification C2-2 shows the value calculation of the salary flow for the sales department "Pay.Sales." There are 50 customer orders per day to be handled. The size of the order does not matter for the process time. It takes the same time to process an order of 5, 100, or 2000 meters of fabric. Therefore, the annual salaries for the 10 sales employees can be broken down into the salary per order. We assume for this model that the quantity ordered by one customer is 100 m finished fabric.

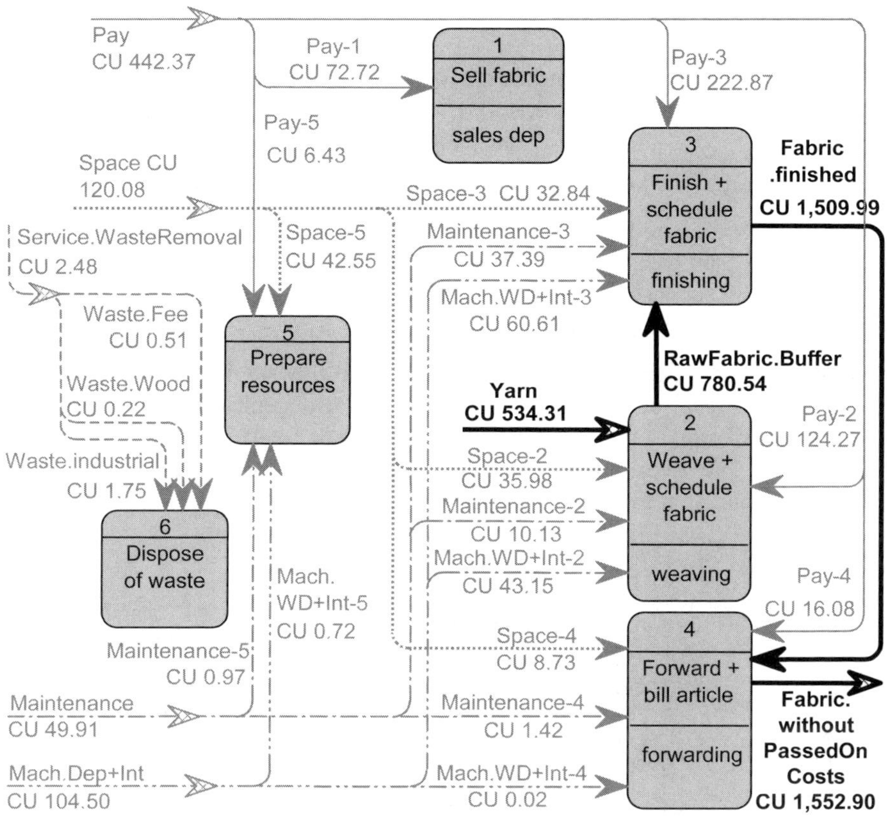

Diagram C2-7: "Produce fabric" — monetary resources.

The various wastes, such as trash, wood, and industrial waste, are disposed of separately. For these, waste taxes and fees for the community must be paid. All of these waste costs are shown in the diagram of the monetary resources (*Diagram C2-7*); the sum is shown as "Service.WasteRemoval." These waste flows go as positive flows into the company because they represent the service of the waste disposal received by the community. This value is first added to the waste, which cannot be seen on this diagram since it does not show the material flows.

The waste collection and removal within the company is carried out by the auxiliary staff of the different departments. The costs of these employees are calculated separately, but together with other activities. It is difficult to distribute these onto the single processes because the waste removal is done in between other activities. Therefore, no additional wages are introduced in the process "Dispose of waste."

The monetary resources provide a good example for the hierarchical addition of flow values. First, the flow values are calculated with the parameters from the consuming processes on a lower level, and then they are summed up in the hierarchy, as shown by the value of the child and parent flows.

2.4.3 VFD Lower Levels

VFD "Weave + schedule raw fabric"

Diagram C2-8 shows the VFD of the weaving department. The process chain in this department ends with the inspected fabric rolled up in the buffer before the washing as is suggested by the flow name "RawFabric.WashBuffer."

The yarn enters the production through the process "Prepare weaving." Some yarn suppliers take back packages, pallets, and tubes. This is shown by the flow "Wrapping.back." The supervisor of the preparation shop is responsible for planning and scheduling the weaving preparation.

In addition to the product flow and various material flows going into the processes, the produced waste, as it leaves the process, is also depicted. One goal of this model is to investigate the amount and cost of waste created by the production of 100 m fabric. The investigation revealed that a substantial amount of waste occurs, which is an indication for potential savings.

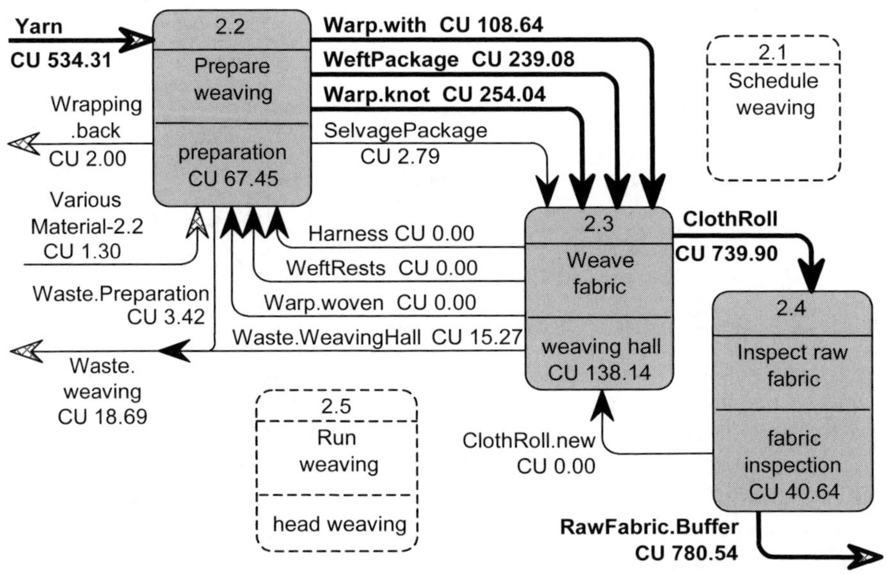

Diagram C2-8: "Weave + schedule raw fabric."

The main task of the weaving preparation is to produce the warp for the weaving process. The product flow "Warp" leaves the preparation process by two flows: either as warp drawn already into the harness ("Warp.with"), or as warp without harness ready to be knotted on a warp already running on a loom ("Warp.knot"). Normally, warps are knotted on the already running warp on the loom. To allow cleaning of the loom and the harness (or if another type of fabric is started on the loom), every forth warp is set up completely new on the loom, the so-called "Put on warp." For this, the warp must be supplied complete with the harness ("Warp.

with") by the weaving preparation. This task involves a lot of manual work und is quite expensive.

Specification C2-3 shows the process specification "Prepare weaving" with the value added. The value added is calculated as the value difference of the total product output minus the total product input. In this case, the value of the total product output is the sum of the product flows "Warp.with," "WeftPackages," and "Warp.knot."

Specification C2-3: Process "Prepare weaving"

2.2	**Prepare weaving**		**Preparation**
Description	From the yarn cleaning until the warp beam or the drawn in warp. Integrated activities: - Transport weft packages to the loom with fork-lift. - Transport yarn to the sectional warping. The preparation operates with 7 shifts per week. Air Conditioning is not further detailed.		
Parameter	Number of warps Air condition Fork lift Light Time prepare weaving	800.00 980.00 55.00 20.00 60.00	warp/year kWh/week kWh/week kWh/week hour
Key	Number of warps	800.00	warp/year
Value added	Σ product output - Σ product input = (Warp.with + Warp.knot + WeftPackages) – Yarn = (108.64 + 254.04 + 239.08) – 543.31	**67.45**	CU/100m

The Reference Unit (RU) in the weaving preparation and the weaving hall is the number of the warps per year. For WeaveFine, this is 800 warps per year. This number is used as a parameter and a key for various value calculations of processes and flows. The parameter weave time, consumption of resources, and the organization of the employees are given with reference to the RU. In some specifications the value for a warp is given as an intermediate result, before converting it to the value per SU of 100 m fabric. *Table C2-1* at the beginning of this case study includes the production data and the necessary information for this conversion.

VFD "Weave fabric"

Diagram C2-9 shows the child diagram of the process "Weave fabric" with the information flows. It illustrates how the value from the process 2.3.7 "Supervise weaving" is taken to the other processes by the information flows "Instruction." Instructions are verbal information given to the operators by the supervisor. The labor and knowledge of the supervisor go into the product through these flows. Thus, these flows carry the value from process 2.3.7 into the product fabric.

In a production, the product is always accompanied by some kind of information flow. The minimum information is the identification of the product. In the process "Change cloth roll," the flow "WeavePaper" leaves the product. From then on, the "CouponCard" goes with every roll of fabric (product flow "FabricRoll") from the loom to the fabric inspection.

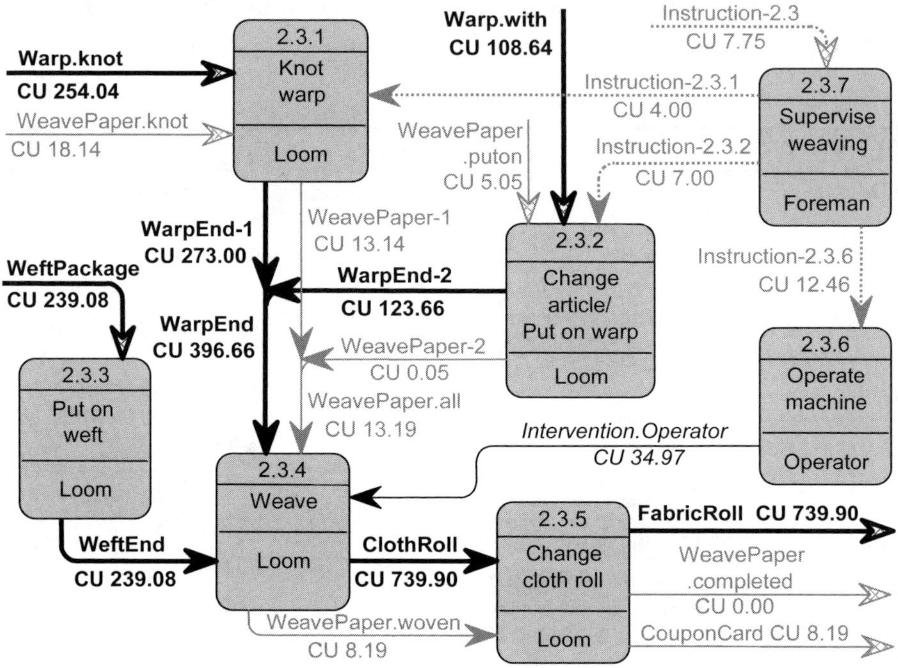

Diagram C2-9: "Weave fabric."

Allocation of information costs

Based on the general production paper from the product development and design, the so-called "WeavePaper" with the specific information for the weaving is derived. It accompanies the product through the entire weaving mill. The value of this paper has also to be passed on the product. Each process that uses this paper is charged with CU 5.00. This value goes into the product. CU 5.00 is an arbitrary value chosen by the model's designer. The implementation of this can be seen through the subtraction of CU 5.00 in the calculation formula in *Specification C2-4*.

Almost the entire value of "WeavePaper" is transferred to the product by the last process "Change cloth roll." After the process "Change cloth roll," "WeavePaper" is completed and is no longer used in the production. The value of "WeavePaper.completed" is set to CU 0.00. The flow "WeavePaper.completed" is returned to the scheduling department for review and as a confirmation that the warp is finished and that the weave order is completed.

Specification C2-4: Flow "WeavePaper.woven"

WeavePaper.woven			
Description	The WeavePaper allocates CU 5.00 to be passed onto the product to every process through which it runs in the weaving hall.		
Flow type	WeavePaper		
Parameter	Allocation to process	5.00	CU/100m
Calculation	WeavePaper.all – Allocation to process = 13.19 – 5.00		
Flow value		**8.19**	CU/100m

Fictitious flow "Value.Operator"

The process "Operate machine" includes all the tasks of an operator carried out in the weaving hall, especially on the loom. This process has no obvious output. No material or information is leaving it. Therefore, a fictitious value flow "Intervention.Operator" is chosen to transfer the value from this process to the process "Weave." Such interventions by the operator with the loom are, for example, repairing stops and yarn breaks or monitoring the machines.

2.4.4 VFD "Finish + Schedule Article"

In the finishing department, the product flow of "Fabric02" does not run straight through the process chain any longer. It switches between several processes. Some of them may be passed more than once by the product.

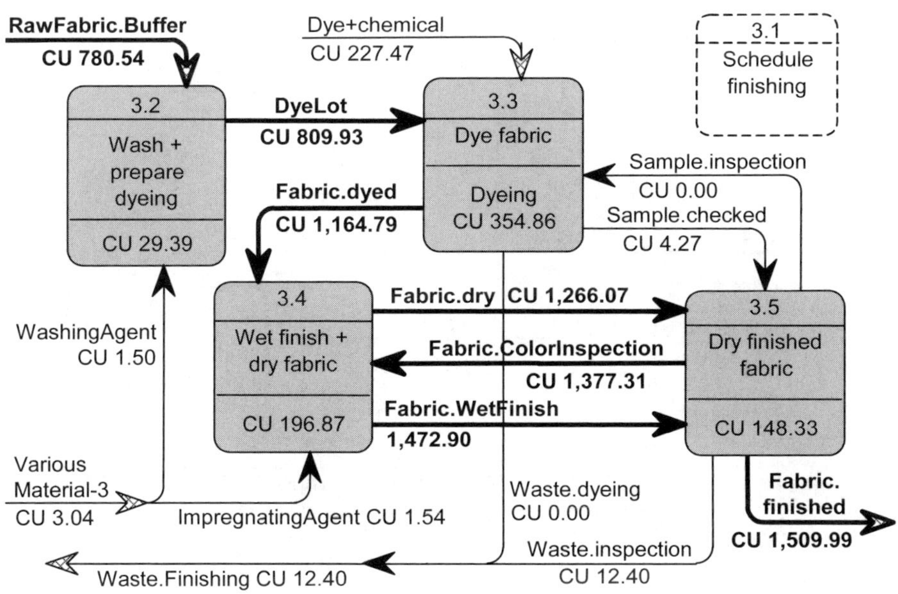

Diagram C2-10: "Finish + schedule article."

After the dyeing, the fabric is taken in containers to the wet-to-wet finishing where it is disentangled, rolled up flat on the roll, and dried on the tenter frame. Then, the fabric moves to the department of dry finishing for the color inspection and thereafter returns to the wet-to-wet finishing for further treatment. Finally, back in the dry finishing department, it runs through the decating (steaming) process and the finished goods inspection.

It is not easy to keep a clear process sequence in the finishing department, since every fabric needs a specific treatment and switches several times between wet-to-wet and dry finishing, as seen in *Diagram C2-10*. In the finishing department, the to-and-fro between the processes, which are structured according to the acquisition of the machines over the years, involves many transport movements. These are done by the employees by hand. An in depth investigation with a further detailing of these processes on the next level may be worthwhile to improve the layout of the shop.

2.4.5 Fictitious Value Flow to Pass on Costs

Before the product leaves the company, various costs must be passed on to the product. Looking back to the diagram of level 1 (*Diagram C2-4*), it can be seen that an additional process is introduced for the passing on of different values and costs in *Diagram C2-11*: Process 7 "Pass on costs." This corresponds to the real world, since the passing on and apportion of costs is a process, even if it is not obvious, since it takes place in accounting only. The product flow runs also through process 7, where the various costs are debited to it.

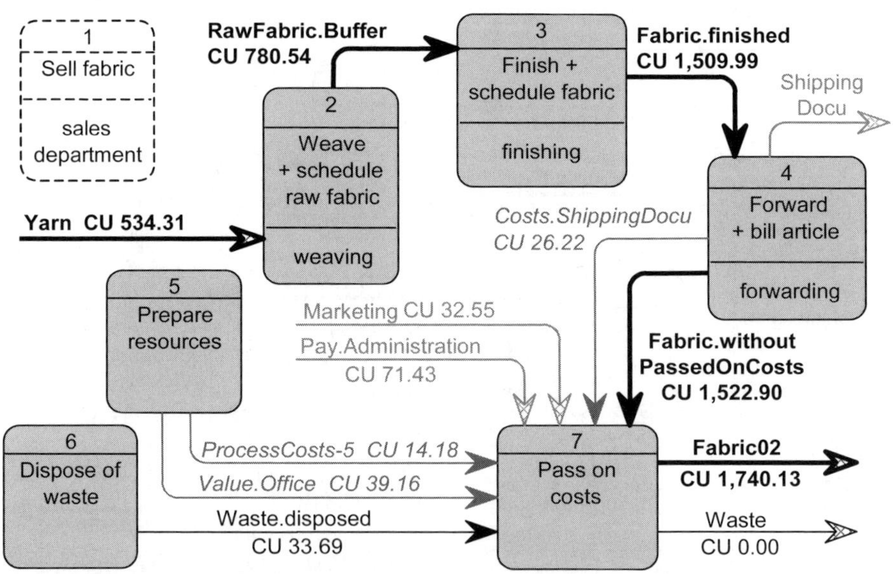

Diagram C2-11: Pass on of costs by fictitious value flows.

The product flow “Fabric02” shows the final product value. This value is a reference for the “rock-bottom” sales price to cover all costs of the production should. The flow “Fabric.withoutPassedOnCosts” gives the product value as it leaves production, before any further costs are passed on.

The apportion of costs is represented with the help of fictitious value flows. The fictitious value flow “ProcessCosts-5” carries the costs for space, maintenance, and machine depreciation and interests of the preparation of resources in process 5. The fictitious value flow "Value.Office" represents the space costs of the offices.

Some of the flows that enter the system from the external entities on the context diagram (*Diagram C2-3*) were not used in lower-level processes. These flows are “Marketing” and “Pay.Administration” (including the salaries of the directors). Their costs are debited to the product by process 7, too.

Other flows, such as “ShippingDocu” and “Waste.disposed,” are only allowed to leave the system with the value CU 0.00, since nobody pays for them and the model would not be in balance anymore. To set the value to CU 0.00, their value must be passed onto the product. On *Diagram C2-11,* two way to do so are demonstrated by the flows “ShippingDocu” and “Waste.

“Waste.disposed” runs through process 7, debits the product, and then leaves the process as “Waste” with the value CU 0.00 the system. The other possibility is shown by the flow “ShippingDocu.” This flow leaves the system with the value CU 0.00 directly from process 4. Its value is passed onto the product by the fictitious value flow “Costs.ShippingDocu” through process 7.

Specification C2-5: Flow “Fabric02”

Fabric02	
Description	Standard article of WeaveFine. Finishing: Dyeing and impregnating.
Flow type	Product
Parameter	
Calculation	(Calculation with process balance.)
Flow value	**1,588.84** CU/100m

Specification C2-5 of the flow “Fabric02” shows the value of the finished product.

As mentioned in Chapter C2-3 Step 3, the diagrams in this case study show only the product flow and one other flow group of the four flow groups. This has to be taken into consideration when doing the process balance. For the process balance, the value of all flows of all flow groups must be summed up.

2.5 Evaluation Report and Benefits of the Method

Using the VFD in practice and in this specific case study provided some important indications for accounting and sales. Take into account that the original case study was much more extensive and that not all diagrams with all the details, specifications, calculations, and comparisons could be given here.

Value of standard product

The VFD in this case study calculated the value of a standard product, "Fabric02," in order to verify the product price. The new cost allocation proved that "Fabric02" was sold below its actual value for decades. The price did not cover the production costs. The actual sewage costs for this specific product were much higher than assumed.

Decision for investment

Over the last few years, the dyeing process became very expensive because the taxes on sewage were raised drastically. The dyeing process produces a lot of "colored water." For years, the company considered the acquisition of a conditioning facility for the sewage. However, they were never quite sure if it was worth the investment. By the child diagram of process 3.3 "Dye fabric," the VFD model showed immediately the potential profit of such an investment, when the originally assumed sewage costs and the real one analyzed with POA were compared. Consequently, the company ordered the conditioning facility immediately.

Fictitious value flows

In this case study, the fictitious value flow was of great help. Especially at the beginning of the model, fictitious value flows simplified the handling of the exchange of the value flows between the two very independent departments: weaving and finishing. Most of the fictitious flows were replaced during the design of the model. The more the model was detailed and the more data were integrated, the more fictitious value flows could be replaced by resource and information flows.

The fictitious value flows also illustrated clearly to the management how often apportion and passing on of costs occur in a company. This process of the cost appointment in costing is not always easily understood by the people involved. Graphically displayed by the fictitious value flow, it becomes clear to everybody which costs are passed on and where the costs originate.

Value of information flow

Nowadays in the companies, the handling of information becomes increasingly important as part of the daily job. The costs of information and their handling are rarely taken into consideration. It is common knowledge that computer systems cost money. However, the impact of computer systems on profitability remains obscure. With the VFD, a tool is available to establish the costs and values of information flows.

In this case study, the information flows were classified to establish the actual use of computer tools on the shop floor. It was distinguished between information on paper, in the computer system, or verbal. It proved that the employees in the finishing department really worked with the computer and did not make print-outs. Other departments were not yet on this level of computer use. The management was pleased to see that the investment in a computer program specially designed for this company was worth while.

Merger of two plants

The paths of the flows, in addition to their naming, still showed that the company WeaveFine was a merger of two companies. Seven years after the merger, the flows were still not running smoothly between the different divisions. The flow names were not consistent throughout the company. A flow leaving a process in the weaving department did not have the same name when it arrived in the finishing department. By drawing these diagrams, the problems were made obvious so that they could be fixed. The diagrams here show the standardized naming.

Orderly production flow

In the finishing department, different fabrics are treated differently, as they go in different paths through various machines. Because of this, the product flow is sometimes very confusing. *Diagram C2-10* shows an example of the product flow in the finishing department. Using the VFD, it is easily understood where the production runs smoothly and orderly and where the interfaces between two departments or processes are not clearly defined. In the case of poorly organized flows, the number of flows between processes increases significantly. The diagrams clearly show this.

C3 EXERGY ANALYSIS OF AN INDUSTRIAL BAKERY

Goals of the Case Study

- Assess a production line that has a high degree of automation.
- Evaluate the energy consumption using the RFD.
- Identify technical optimization possibilities on the production line.

Figure C3-1: Conveyor belt with croissants (Photo: ETH Zurich).

3.1 Energy Analysis of the Croissant Line

Purpose and viewpoint of the model

The system analyzed in this case study is the production line of ready-to-bake croissants. The company, which produces these croissants along with a lot of other products, is a major industrial bakery in Switzerland. *Figure C3-1* and *Figure C3-2* show parts of the ready-to-bake croissant line.

The goal of this analysis is to evaluate the energy consumption and losses in the production. Of particular interest is the thermal energy consumption caused by the baking process. Possibilities of waste energy reuse are assessed by the RFD with an exergy analysis.

Figure C3-2: Triangle performs for croissants (Photo: ETH Zurich).

The production lines of bakery products are semi-automated. Human workforce is mainly needed for supervising the production, for quality control, setting up the line for different products, for cleaning, and for packing up the product. In order to quantify the human workforce the time-dependent unit "workforce minutes per kilogram product (WMK)" is introduced. The human workforce is quantified in order to give an idea of the human workload in this production in relation to the machine work load.

3.2 Resource Flow Diagrams of the Croissant Line

3.2.1 Context Diagram

The context diagram of the croissant production is shown in *Diagram C3-1*. The company receives the raw products, as well as "Energy" in the form of electricity and gas from different suppliers and delivers "Ready-to-bake Croissants" filled-up in bags to the retail market. The "Product waste" is used as cattle food; and the "Broken bags" go to waste treatment plants. "Waste energy," "Waste water," and "Steam" go to the environment.

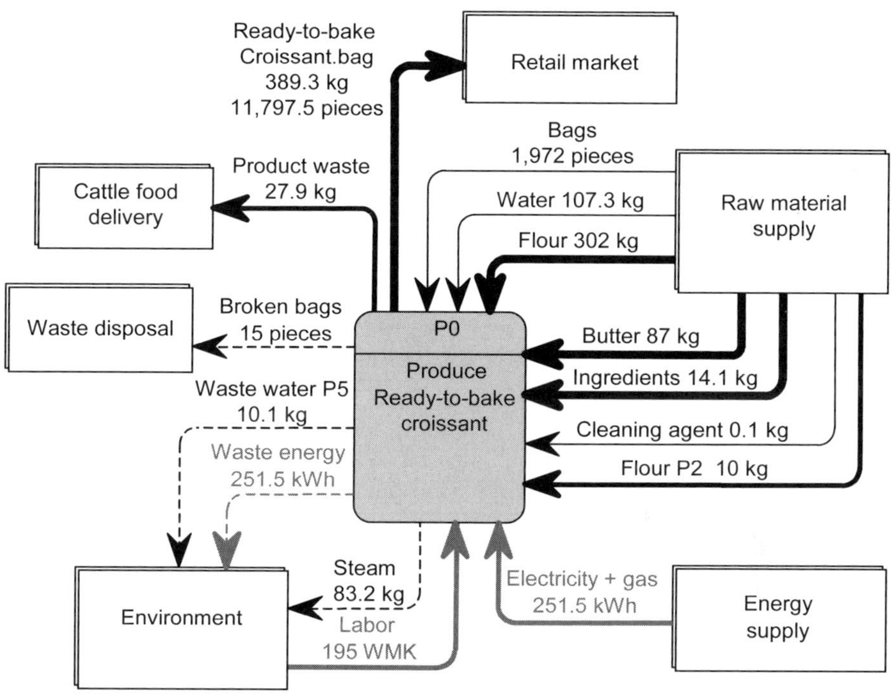

Diagram C3-1: Context diagram of the system croissant production.

The system is further detailed into three levels in order to explore the details and the data needed for the calculation. The model designer is free to determine how to split up the single processes on a child diagram; this depends on the purpose of the analysis. Usually, every level highlights a specific purpose of the production system. On one level, the main production processes may be shown in detail, on another level the auxiliary processes, for example, transport or storage, may be highlighted. The level convention may be done in a geographic, functional, or other manner. The first level of detail in this study (*Diagram C3-2*) shows the plant level with a rough view of the production processes of the ready-to-bake croissants and a detailed view of the auxiliary processes. The second level shows the production processes in detail.

The ready-to-bake croissants production line is used to produce several products. The ready-to-bake croissants are produced within 20 hours each week. During the rest of the production time, the line is modified and other products produced. *Specification C3-1* of process P0 characterizes the product and the settings of the production line.

Specification C3-1: Process "Produce ready-to-bake croissant"

P0	**Produce ready-to-bake croissant**		
Description	Production processes for ready-to-bake croissants. Purpose: Evaluation of energy consumption and losses in production. Assessment of working load of man and machine.		
Parameter	Croissants produced per hour (goal)	12,000	pieces/h
	Mass of dough produced and input per hour	480	kg/h
	Mass of raw croissants	40	g
	Mass of baked croissants	33.3	g

3.2.2 RFD Production Level

The flow classification in *Table C3-1* lists the classification into flow categories, flow types, and subtypes and shows the drawing conventions and flow types used in the model of this case study.

Table C3-1: Flow classification for the case study

Flow category	**Flow type**	**Flow subtype**	**Line style**
Product flow	Product	Product, flour	Black bold line
		Product waste	Black line
Material flow	Water, material	Water, detergent	Black thin line
		Waste water Broken bags	Black dashed thin line
Energy flow	Electricity	Electricity	Gray line
	Thermal energy	Thermal energy	Gray line
		Waste energy	Orange dotted line
Information flow	Labor		Gray line

First, the RFD for level 1 of the croissant production is set up in *Diagram C3-2*. The main production processes and auxiliary processes are identified and all of the input and output flows assessed. The mass of the material flows and the total energy of energy flows are assessed, measured, and calculated. The processes "Prepare dough," "Prepare raw croissant," "Pre-bake croissant," and "Pack croissant" follow the path of the product flow. The auxiliary processes to the production line are "Clean tray" and "Burn gas." The values indicated on the diagram refer to the Standard Unit.

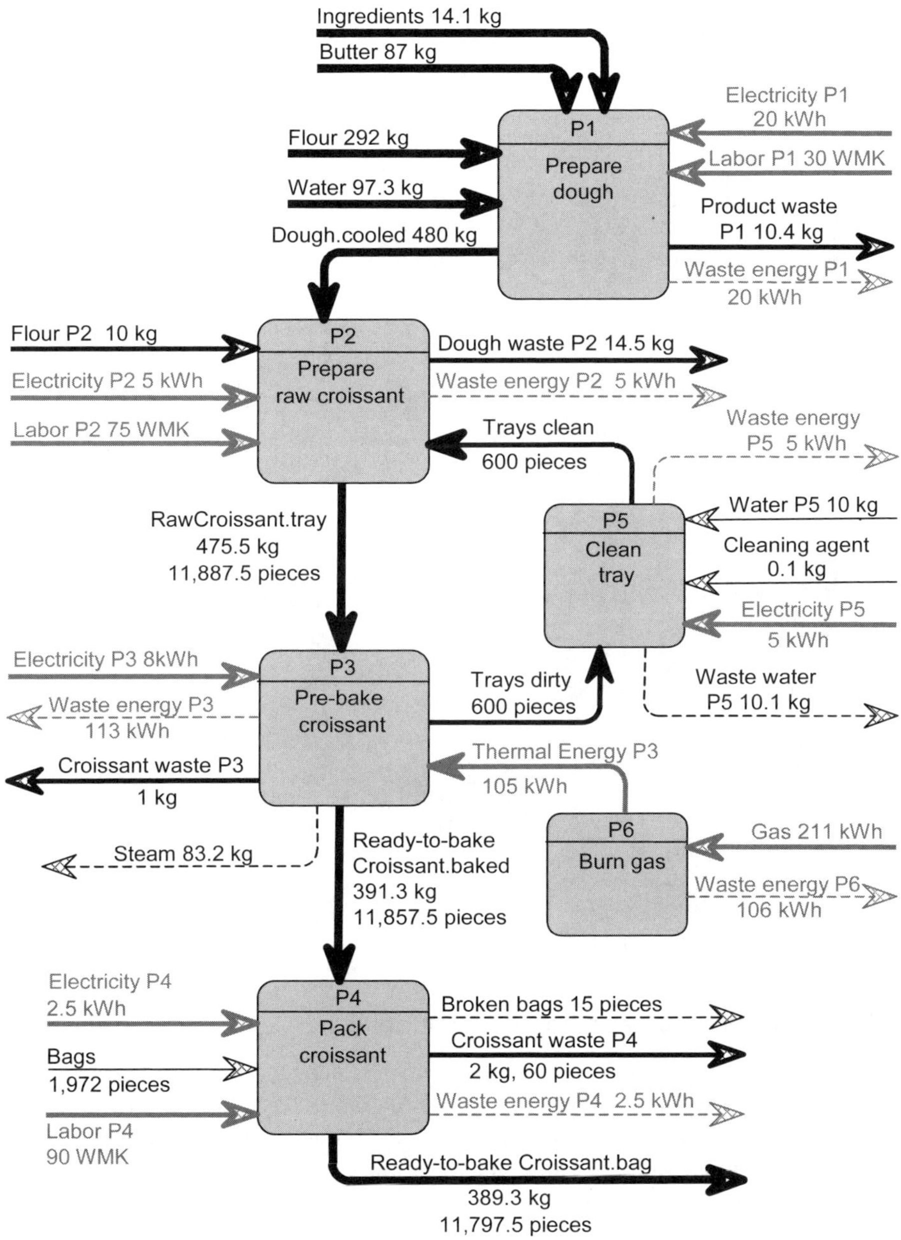

Diagram C3-2: "Produce ready-to-bake Croissant."

The preparation of dough is not detailed further on another level. This is due to the fact that the dough is prepared in an alternating batch process by two kneaders without particular impact on the production line. The dough kneading machines are shown in *Figure C3-3*.

Figure C3-3: Kneading machines (Photo: ETH Zurich).

3.2.3 Mass Calculation of Product Flows

The mass calculation of the product flow, which is depicted bold in *Diagram C3-2*, is explained in the following. The mass calculation is based on the Standard Unit (SU). The SU can be set as the mass of final product, the mass of raw material required or an intermediate product mass. In this case study the mass of the intermediate product "Dough.cooled" was specified as SU. The reason is that the amount of dough used for the production was the secure and measured value of this production line. The mass of raw material needed was calculated by adding the losses of P1 to the mass of "Dough.cooled." To calculate the mass and subsequently the number of produced croissants of the production line, the losses of the processes P2 to P4 are withdrawn from the mass of "Dough.cooled."

Table C3-2: Mass calculation of product flow

P1 Prepare dough	P2 Prepare raw croissant	P3 Pre-bake croissant	P4 Pack croissant

Input P1	**Output P1 = Input P2**	**Output P2 = Input P3**	**Output P3 = Input P4**	**Output P4**
	Standard Unit: 480 kg dough			
Ingredients, flour, water, butter	Dough.cooled	Raw Croissants.tray	Ready-to-bake Croissants.baked	Ready-to-bake Croissants.bag
490.4 kg		475.5 kg	391.3 kg	389.3 kg
	Reference Unit: 12,000 pieces	11,887.5 pieces	11,857.5 pieces	11,797.5 pieces
	Reference Unit: 2,400 bags			1,967 bags

Table C3-2 shows the connection of the single mass values of the product flow as well as the Standard Unit and the Reference Units. The RU are the number of croissants and the number of bags with filled-up croissants. Process P1 needs a surplus of 10.4 kilograms of ingredients and raw products to deliver the SU of 480 kg dough. Therefore, process 1 is supplied with 490.4 kg mass. In the preparation process P2, flour for rolling-out is added and product losses occur.

The baking process P3 reduces the mass per croissant by evaporation of the liquid part in the dough considerably (although not the number of croissants). P4 then delivers a total of 11,797.5 pieces (instead of the target of 12,000), which is a negative deviation of 202.5 croissants from the target production. The remaining croissants have a total weight of 398.3 kg.

Already at this point some conclusions can be drawn from the RFD:

- The target production of 12,000 croissants per hour can not be met with the supplied dough.
- The baking process (P4) causes a considerable loss of mass per croissant.

The mass loss in P4 is due to the evaporation of liquid while pre-baking the croissants. The other process losses are small in comparison. While pre-baking, 8.25% of the raw croissant weight is lost. Before pre-baking, each croissant weighs 40 grams, while after baking it weighs 33.3 grams. Surprisingly, the baking process is only a minor generator of reusable energy. This will be an interesting issue for the exergy analysis.

3.2.4 Energy Calculation of Resource Flows

Based on SU and RU, the mass and total energy of the other resource flows are calculated. The data for the energy consumption per week of the production line is available in the company and can be broken down into individual hours of production for this specific product. The gas input to the oven is measured on site. The energy consumption and heat production of the oven is a major point of interest and is therefore analyzed with an energy/exergy balance. Also of interest is the fact that the croissants themselves carry a significant amount of heat after baking.

The flow values for the labor are estimated and expressed in a unit WMK, a measure, which means working minutes per kilogram of final product.

3.2.5 RFD Second Level of Detail and Production Layout

For better understanding of the production line, the diagrams in this chapter show the RFD of the next detail level together with the layout of the concerning part of the production line.

Diagram C3-3 shows the child diagram of process "Prepare raw croissant." The first production section consists of a conveyor belt, which transports the dough from the left side of the layout plan to the right side. The dough is first rolled into a thin strip (P2.1) and is cut into triangular pads that are rolled and bent into the final shape and controlled manually (P2.2). After the quality control, the raw croissants are transferred to the trays. The raw croissants are carried on trays to be transferred to the curing and baking process.

The number of employees working in the line is symbolized in the production layout by sketched symbols. One employee is required to continuously put the cooled dough on the conveyor belt. A second person, the machine supervisor, retrieves the cut edges of the rolled-out dough, recycles them, and refills the flour spreading units that help prevent the dough from sticking to the belt. If the flour is missing, the line has to be stopped for cleaning of sticking dough. A quality controller checks the form and the appearance of the formed croissants and picks out rejects.

As shown in the RFD of P2, the flows "Electricity P2" and "Labor P2" are split into the child flows. Thereby, the mass and total energy of the parent flow is allocated to the child flows.

Similarly, the waste energy flows and dough waste flows are merged to the parent flows. The flow value of the parent flow is the sum of the flow values of the child flows. The value of the parent flow is transferred to the parent diagram (level 1 diagram).

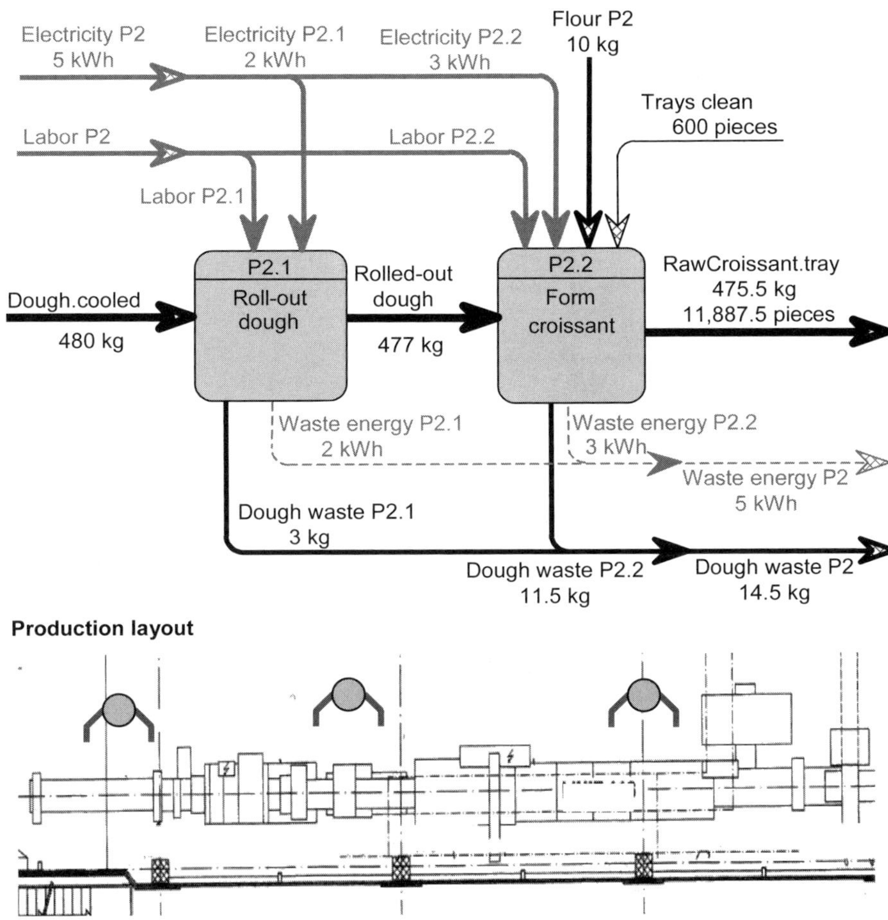

Diagram C3-3: RFD and layout of "Prepare raw croissants."

In *Diagram C3-4,* the child diagram of the process P3 "Pre-bake croissant" is shown. The pre-curing process P3.1, the baking process P3.2, and the cooling process P3.3 are depicted in both the RFD and the schematic view. The pre-curing function is carried out by a paternoster system (buffer transport system), meaning that the croissants circulate on their trays for a certain time in the conditioned containment with a specific temperature and moisture. Then the croissants are pre-baked in the oven.

After the oven, the croissants are passed from the trays to a conveyor belt. The croissants travel along a conveyor belt of 100 meters of length along the ceiling of the production hall for an approximate one-hour cooling period. The croissants then transfer their heat to the surrounding air. Processes P3.1 and P3.2, pre-curing and baking, are fully automated, and do not require human interaction with normal production. Therefore, no operators are indicated in the production layout.

As you see on the diagram, the trays are separated from the product flow in process P3.2 and leave the process to go to the cleaning process. This is not to be mistaken for a split and merge of flows because the trays and the product flow are parts of different flow types. A split and merge is only allowed with flows from the same flow type.

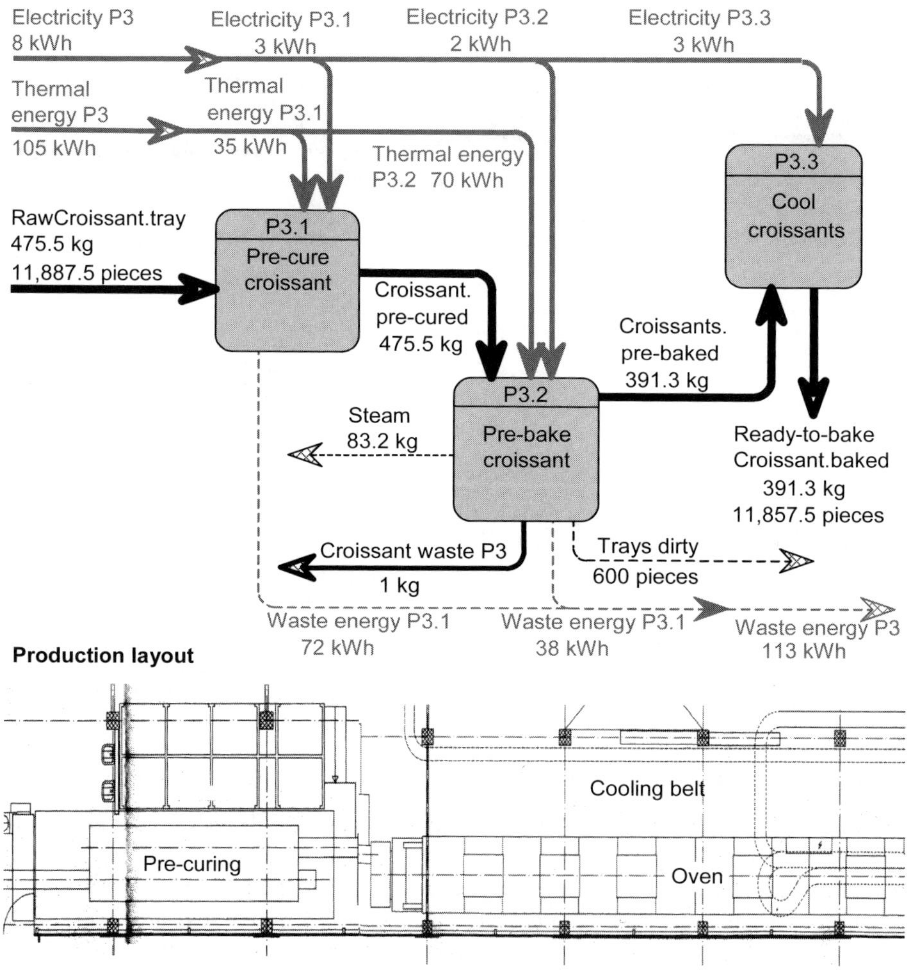

Diagram C3-4: RFD and layout of “Pre-bake croissants.”

The total energy of the electricity and thermal energy for the oven and the precuring are displayed on the diagram. It becomes obvious that these processes consume a significant amount of energy and produce a considerable amount of waste energy. These will be examined later on with an exergy analysis.

Diagram C3-5 is the child diagram of process P4 “Pack croissant” shows the final part of the conveyor belt used for cooling, and the subsequent packaging process.

The production layout depicts a part of the conveyor belt for cooling and subsequent packing.

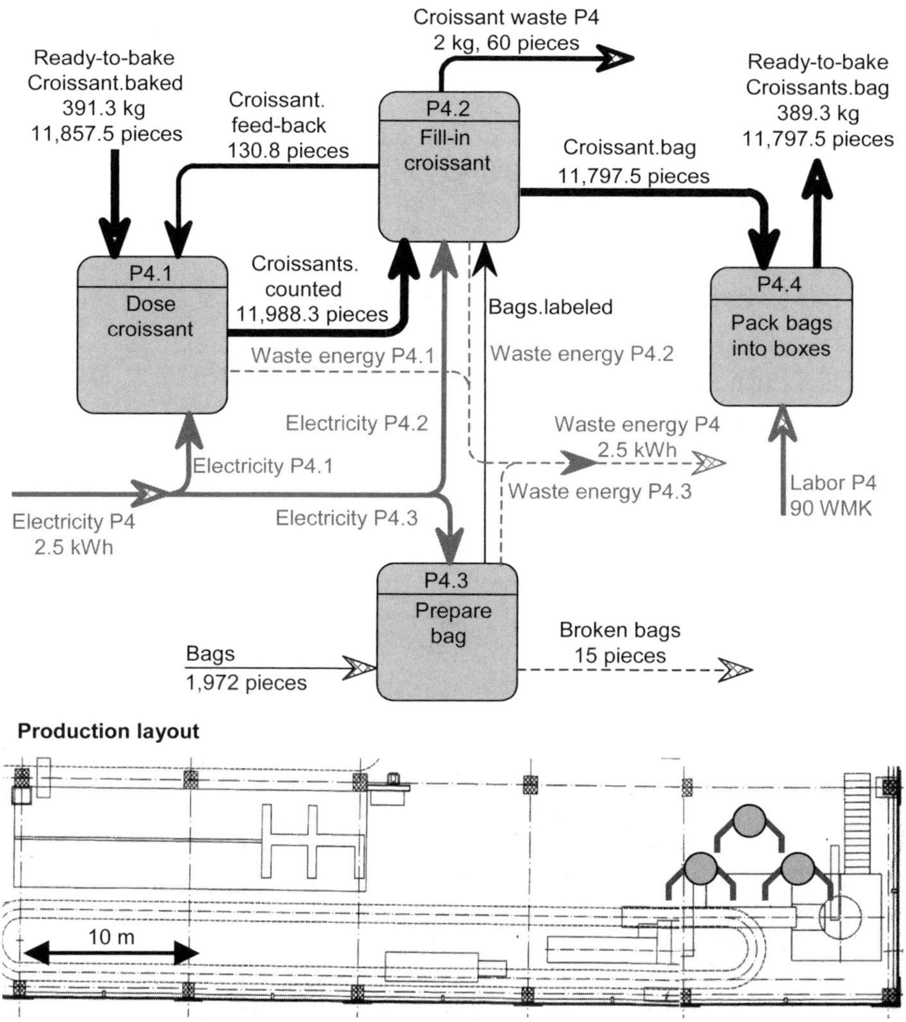

Diagram C3-5: RFD and layout of "Pack croissant."

The packaging is done automatically with a weighing device, which counts the number of croissants for one bag. The croissants fall from the conveyor belt directly inside this weighing device. Before the bags are delivered to the filling-up device where the croissants are weighed, the bags are labeled by another machine. Here some of the bags are torn apart and have to be thrown away. The labeled bags are filled, each with six croissants, sealed and delivered to the employees, who pack them into containers to be transported by truck to the retail market.

As seen on the diagram, there is a feed-back flow "Croissants.feed-back" that we have not seen on the parent diagram. This feed-back flow indicates that a part of the croissants fall out of the bags during the filling process. These croissants are collected and re-enter the weighing process. With this process, numerous croissants fall out of the device and have to be picked up by one of the employees working in the packaging section. The croissants are fed back to the weighing machine.

Process P4 needs more human workforce than all the other processes. One person supervises the filling-up device and the feed-back of croissants. Two or three other persons are busy filling the croissants bags into boxes and removing the boxes.

3.3 Exergy Balance of the Baking Process

3.3.1 Purpose of the Exergy Balance

An exergy analysis is important for Process P3 because there is a high consumption of "Thermal energy" and production of "Waste energy" (heat), as seen from the energy flows in *Diagram C3-2*.

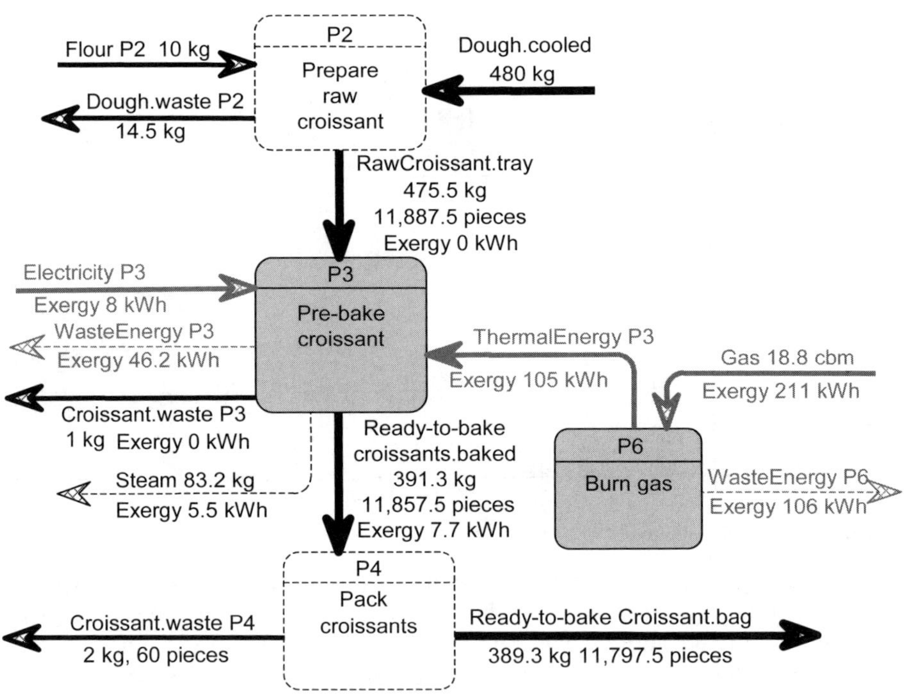

Diagram C 3-6: RFD with exergy values.

The croissants contain a considerable amount of heat after baking that is gradually lost by cooling exposed to the ambient air. The usefulness of this energy and its potential to be reused is of interest. With this in mind, an exergy analysis, based on the energy data calculated is performed. *Diagram C3-6* shows a part of the RFD in *Diagram C3-2*, depicting the product and energy flows for the baking process.

The processes P3 and P6 are displayed with all their resource flows. The resource values given in this diagram are the exergy of the flows. The exergy calculation is performed step-by-step as follows. The exergy of the electricity flows equals the energy amount. The exergy of the material flows and the exergy of the energy flows have to be calculated.

3.3.2 Exergy Calculation of Material Flows

First, the calculation of the exergy of the product flow leaving the oven "Ready-to-bake croissants.baked" is examined. The exergy of material flows is calculated with the *Equation E13* of Section S3. For the exergy calculation of the croissants, the caloric value of the croissant is not included, the focus lies on the thermal energy stored in the croissant after baking.

E13	$e_x = (h - h_0) - T_0(s - s_0) + \frac{1}{2}v^2 + g \cdot z$	e_x flow exergy of material flows h specific enthalpy s specific entropy v velocity g gravity constant z elevation T temperature [K] X_0 subscript 0: at ambient conditions

The kinetic and potential energy, the last two terms of *Equation E13*, can be ignored in this case because they are negligible compared to the thermal energy. This means that the croissants do not carry significant potential energy and energy of movement. The equations to calculate the enthalpy and entropy difference are given here.

$h - h_0 = c_p \cdot dT$	cp specific heat capacity at constant pressure
$s - s_0 = c_p \cdot \ln\left(\frac{T}{T_0}\right)$	

For the calculation of the exergy of a croissant, the thermal energy is a key element. The caloric value of the croissant is not included into the calculation because it is not relevant for the heat emitted by the croissants after baking. The calculation of the exergy of the baked croissants is based on the assumption that the water and fat carry the main energy of the croissant. The heat capacity of water and fat is indicated in tables of basic material properties. The content of the water in the baked croissant is estimated to be 20%, and the content of fat is 18% of the total baked croissant mass. The calculated exergy of the flows "Ready-to-bake Croissants.baked" and "Steam" is shown in *Specification C3-2* and *Specification C3-3*.

Specification C3-2: Flow "Ready-to-bake Croissants.baked"

Ready-to-bake Croissants.baked		
Description	Ready-to-bake croissants after leaving the oven	
Flow type	Product flow	
Parameter	Mass Heat capacity (water) per gram and Kelvin Heat capacity (fat) per gram and Kelvin Temperature (baking) Temperature ambient Mass water Mass fat	387.3 kg 4.18 J/gK 2.15 J/gK 523.15 K 298.15 K 79,200.00 g 71,280.00 g
Calculation	enthalpy difference (water)	$4.18\ \text{J/gK} * (523.15\ \text{K} - 298.15\ \text{K}) = 940.38\ \text{J/g}$
	entropy difference (water)	$4.18\ \text{J/gK} * \ln\left(\frac{523.15\,\text{K}}{298.15\,\text{K}}\right) = 2.35\ \text{J/gK}$
	exergy (water)	$940.38\ \text{J/g} - 298.15\ \text{K} * 2.35\ \text{J/gK} = 239.85\ \text{J/g}$
	enthalpy difference (fat)	$2.15\ \text{J/gK} * (523.15\ \text{K} - 298.15\ \text{K}) = 483.75\ \text{J/g}$
	entropy difference (fat)	$2.15\ \text{J/gK} * \ln\left(\frac{523.15\,\text{K}}{298.15\,\text{K}}\right) = 1.21\ \text{J/gK}$
	exergy (fat)	$483.75\ \text{J/g} - 298.15\ \text{K} * 1.21\ \text{J/gK} = 123.0\ \text{J/g}$
	assumption exergy (for 1 hour)	$239.85\ \text{J/g} * 79{,}200\ \text{g} + 123.0\ \text{J/g} * 71{,}280\ \text{g} = 27{,}763{,}560\ \text{J}$ $27763560\ \text{J} / 1000000\ \text{MJ/J} / 3.6\ \text{kWh/MJ} = 7.71\ \text{kWh}$
Exergy		7.71 kWh

The croissants leaving the oven have an exergy of 7.71 kWh. This part of exergy is lost to the environment by cooling the croissants and exhausting the steam. It is a considerable amount of exergy that might be useful for other applications. This is the first interesting finding from the exergy analysis.

Specification C3-3: Flow "Steam"

Steam		
Description	Evaporated water from croissants while baking	
Flow Type	Energy resource	
Mass		82.2 kg
Calculation	exergy	$239.85\ \text{J/g} * 82{,}200\ \text{g} = 19{,}715{,}670\ \text{J}$ $19715670\ \text{J} / 1000000\ \text{MJ/J} / 3.6\ \text{kWh/MJ} = 5.48\ \text{kWh}$
Exergy		5.48 kWh

The second considerable exergy loss to the environment is the exergy of the steam, which is 5.48 kWh. Also this exergy would be ready for reuse, especially because it is stored in water, which is a good heat storage medium.

3.3.3 Exergy Calculation of Energy Flows

The next step is to calculate the exergy of heat streams. The exergy of the electricity flows corresponds to the total energy, and therefore does not need to be calculated. The equation is found as *Equation E12* in Section S3.

E12	$e_{x,energyflows} = Q \cdot \left(1 - \frac{T_0}{T_1}\right)$	$e_{x,energyflows}$ Q T_0 T_1	exergy thermal energy flows total energy of energy flows ambient temperature [K] temperature of energy flow [K]

The waste energy produced in the baking process is separated into waste heat of mechanical processes and waste heat of the oven because of the different temperature levels of the heat flows. The two parts are summed up in the flow "Waste energy P3." The temperature of the cooling air released from the motors is assumed to be 60 °C; the temperature of the exhaust of the oven is 250 °C. This is a considerable difference that will be noticeable in the amount of exergy available of these waste energy flows. In *Specification C3-4* of the parent flow "Waste energy P3," the exergy of the two child flows is added up.

Specification C3-4: Flow "Waste energy P3"

WasteEnergyP3			
Description	Waste heat produced by the oven and the machinery		
Parameters	Waste energy oven (measured)		105 kWh
	Waste energy machines (estimated)		8 kWh
Calculation	Exergy waste energy oven	$105\,\text{kWh} \cdot \left(1 - \frac{298.15\,\text{K}}{523.15\,\text{K}}\right)$	
	Exergy waste energy machines	$8\,\text{kWh} \cdot \left(1 - \frac{298.15\,\text{K}}{343.15\,\text{K}}\right)$	
Exergy	calculation	45.16 kWh + 1.05 kWh	
	value		46.21 kWh

The results of the exergy of the two child flows show the direct dependency of the exergy value of the temperature of the waste energy. The waste energy of the oven has a much higher temperature than the one of the motors. Therefore the exergy value of the waste energy of the oven is considerably higher.

In process P6, a large part of the exergy contained in the gas is lost by burning it. The exergetic efficiency of a steam generation process is around 50%. This finding is not specific to this production site but rather a general remark for this kind of thermal energy production. The exergy losses are due to the low efficiency of the oven, but also due to the fact that the heat used for baking is on a much lower temperature level than the heat produced by the burning process.

3.3.4 Exergetic Efficiency Calculation

The next step in the energetic process evaluation is the calculation of the exergetic efficiency of process P3. The exergetic efficiency is the quotient between the target exergy and the consumed exergy. The produced or target exergy is the amount of exergy introduced into the product or produced for other processes. For process P3, two exergetic efficiencies are calculated in *Specification C3-5.*

The exergetic efficiency 1 defines the intended exergy as the exergy content of the croissant after baking. The resulting efficiency is rather low. Exergetic efficiency 2 includes the exergy of the steam flow as intended exergy for further reuse in the company. In this case, the efficiency of the process has nearly doubled. This demonstrates that the steam generated in the process could be useful and possibilities for heat recovery should be investigated.

Specification C3-5: Process "bake croissants"

P3	**Bake croissants**	
Description	Production of pre-baked croissants out of raw materials	
Calculation	exergetic efficiency 1 without steam reuse	$\frac{7.71 \text{ kWh}}{105 \text{ kWh}}$
	exergetic efficiency 2 with steam reuse	$\frac{13.19 \text{ kWh}}{105 \text{ kWh}}$
Exergetic efficiency 1		7.34 %
Exergetic efficiency 2		12.56 %

3.4 Benefits of the Method

RFD

The RFD shows the processes and interfaces of a production in an easy way. The values are immediately visible on the diagram. Flows show up that are relevant in the production and for the environment because they carry a lot of energy, e.g. waste energy flows, or mass. Recycling flows can also be traced with the RFD, in order to implement closed-loop flows.

Exergy analysis

Based on the exergy analysis in the previous example, the following can be found: The exergy of the waste heat flow of the oven is very high, the oven should be insulated in a better way, especially at the entry and exit of the croissants. The waste energy of the oven however should be reused in the oven itself, else too much exergy will be lost. This is largely because air is an inefficient heat carrier medium.

The exergy of the steam is considerable and readily usable as hot steam in the heating of the oven or of other facilities. Technical measures for collecting the steam and its reuse in the restaurant kitchen or laundry should be investigated.

The exergy of the "Ready-to-bake Croissants" is low and not easily reusable; major technical changes of the cooling circuit would be necessary. The exergy of the waste heat of the other mechanical processes is low and not easily reusable because it is dispersed all over the place. A high impact would have the reuse of the waste heat of the gas burner by using the heat in the exhaust airflow for the preheating of clean air for baking. For this purpose a heat-exchanger is easily installable and poses no risks.

By calculating the exergetic efficiency, it is shown that the reuse of waste energy may be useful in order to increase the efficiency of the process. In this case, technical recommendations can be worked out for such improvements.

C4 SYSTEM CONTROL FOR THE DEMAGNETIZING OF TV DISPLAY TUBES

Goals of the Case Study

- Create the concept of a new production line with POA.
- Learn how to structure a complex system to generate a basis for a process control.
- Do the step from Flow Diagram to State Chart.

Figure C4-1: Demagnetizing line for TV display tubes (Photo: ETH Zurich).

4.1 Demagnetizing of TV Display Tubes

Television display tubes are a mature product, made in large quantities by a few plants scattered around the world. Color display tubes for TV sets are made in typical mass production plants. The latest and possibly last generation of TV display tubes features large, flat screens with rectangular pixels. To convert energy into visible light, a sheet metal mask (shadow mask) with rectangular holes is placed inside of the tube behind the screen to align an electron beam with dots of phosphorous. The schematic view of such a display tube is shown in *Figure C4-2*. This is also called CRT (cathode-ray-tube).

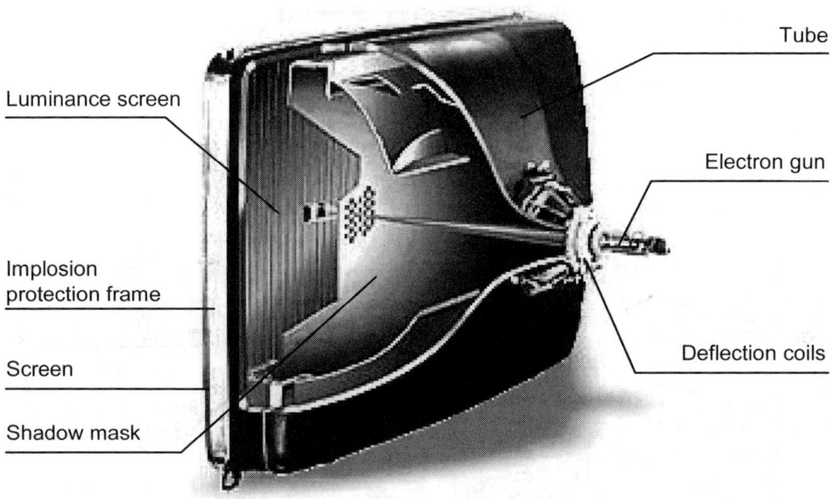

Figure C4-2: Components of display tube.

Material change

The shadow mask inside is held in place by flat springs within the glass body of the tube. To achieve equal color over the whole screen, the dimensions and the position of the mask must maintain narrow tolerances. In addition, any magnetic field other than the one created by the deflection coils must be strictly kept off the screen or suppressed, because they would disturb the electron beam.

Originally made of an expensive nickel alloy, this mask is now made of plain steel in order to cut costs. While steel is easier to shape and therefore much cheaper to process, it picks up traces of residual magnetism, especially by electric welding, like arc or spot welding.

4.2 New Conception of a Demagnetizing Process Line

The use of plain steel instead of nickel alloy requires a sophisticated degaussing process in order to get rid of any magnetism on the finished mask. This part of the production of TV-display tubes had to be developed and planned new, and then implemented in the existing production line.

The degaussing (demagnetizing) process involves applying alternating magnetic fields, with special direction and progress and decreasing amplitudes, targeted in various positions on the mask. This is done by placing the mask in various sets of degaussing coils that are energized by a decaying sinusoidal current. The degaussing line has five sets of coils shown in *Figure C4-3:* Coils A to E.

The principle of demagnetizing is applied in every TV set by an integrated degaussing coil for normal operation when switching on the TV. Since the magnetic field of the earth is sufficient to magnetize ferromagnetic parts in the vicinity of the tube, every time the TV is switched on a demagnetizing puls is created. This is an alternating magnetic field with decaying amplitude. The same principle of demagnetizing is used in tube production; however, much more powerful coils are used, to eliminate residual magnetism coming from welding, machining, and assembly operations.

Some special requirements of the demagnetizing line are the concurrent treatment of different sizes of tubes, the total absence of ferromagnetic parts close to the final degaussing station, and the safety of operation because the coils are working with high voltage and current.

Planning and conception of an automatic demagnetizing line for a complex and sensitive assembly such as a display tube mask is a challenge for an engineering team. POA was applied in order to develop a new production line for the demagnetizing of display tubes and to put the production line into operation.

The whole project had to be completed within only six months. The best available engineering methods were required to avoid time consuming coordination between the engineering team, the customer, and the suppliers of various components and subassemblies and to work as efficiently as possible.

The demagnetizing process was developed and planned in parallel to the conception of the entire line of production of the TV-display tubes. At that point in time the production resources for the production of the new masks where already in place and ready to start production. The customer realized very late that the demagnetizing process he had planned did not meet the requirements of the TV-display tubes. The equipment used during the preliminary tests was not suitable for mass production. Therefore, an offer for a new concept for an automated demagnetizing process had to be worked out within days.

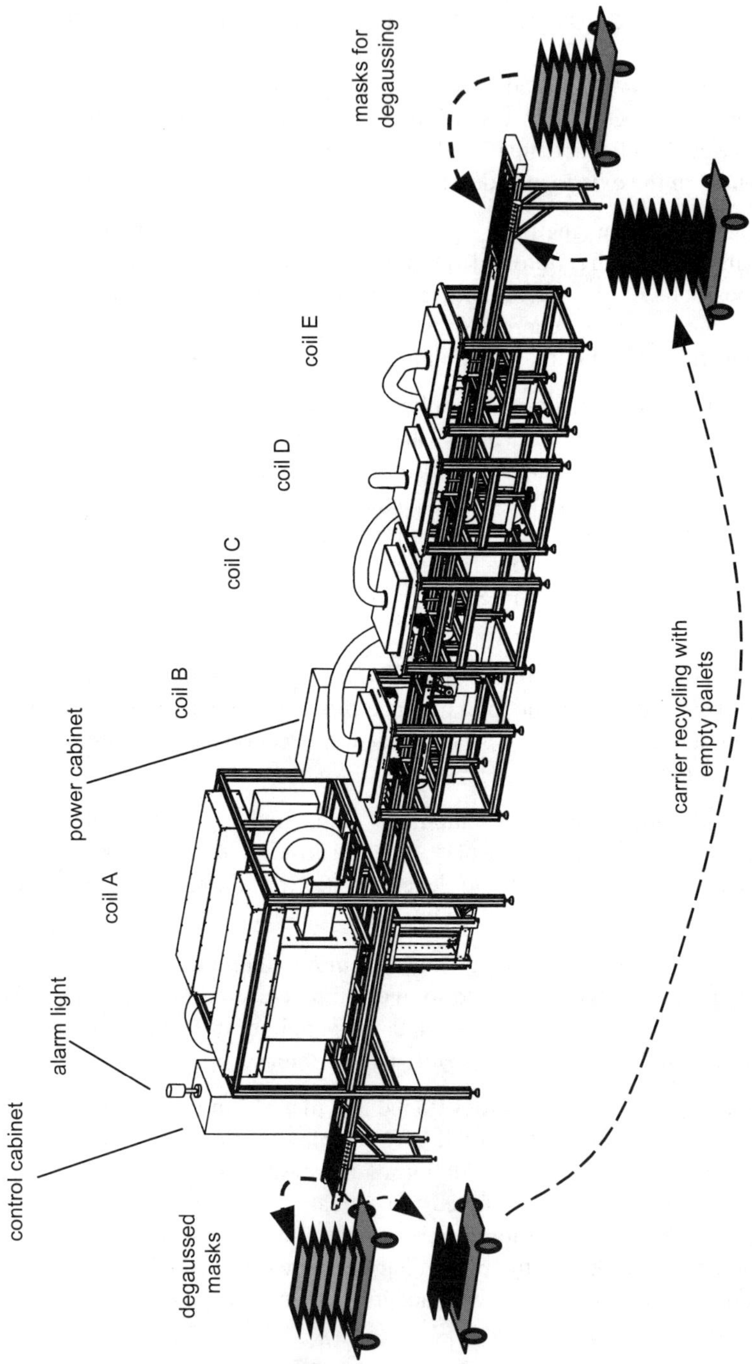

Figure C4-3: Sketch of demagnetizing line.

4.3 System Architecture of the New Production Line

The concept of the newly developed demagnetizing process for the production line of the TV display tubes was started by drawing Flow Diagrams. Thereby, the best possible structure of the demagnetizing line was assessed and the critical interfaces between the various components and environment were determined.

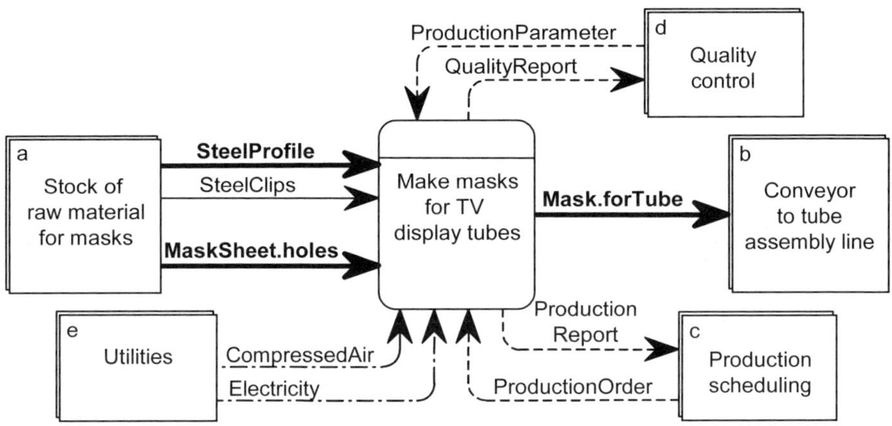

Diagram C4-1: Context diagram "Make masks for TV display tubes."

The context diagram of the complete display tube production line in *Diagram C4-1* was sketched manually during the first meeting with the customer in order to get an overview of the system. Also during this first meeting, the Flow Diagram detailing the process of the context diagram was drawn on paper to establish the interfaces of the display tube production and the demagnetizing line. The necessary information was gathered at a one hour interview with the customer. *Diagram C4-2* shows the first detailing level with the processes of the production line for display tubes.

The path of the product flow was determined first; then the other resource flows were supplemented on the Flow Diagram. The line style of the flows is assigned as follows: solid bold line for product flows, dotted line for information flows, and dot-dashed line for resource flows.

The relevant information for the new concept of the demagnetizing line was caught in the level two diagram "Degauss mask." The customer provided a set of drawings with the dimensions of the different mask types. The following project goals where stated:

- Target production rate is 5 masks per minute with any of the mask types.
- The requirements for spacc and resources have to be specified in the offer.
- The requirements of the European Community directive "Safety of Machinery" have to be strictly followed, although they do not contain any specific recommendations concerning demagnetizing processes [Directive, 1998].

The new line has to cope with the following restrictions:

- The masks are transferred manually to the demagnetizing line and from there again to the TV display tube production line. This transfer will be automated later in future.
- The machine is to operate stand-alone, by unskilled personnel.
- Compressed air is supplied by from a 3/4 inch tube, at 8 bar pressure.
- Electricity is available at 400 V, 50 Hz, 3 phase, and up to 200 A (more if required).

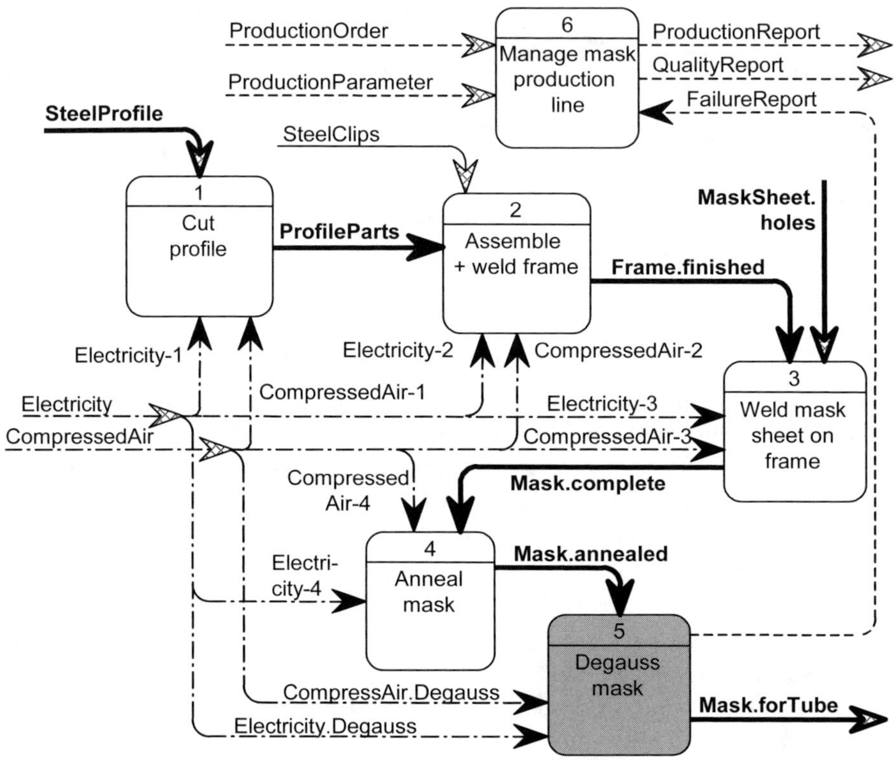

Diagram C4-2: Flow Diagram "Make masks for TV display tubes."

The second detailing level in *Diagram C4-2* was drawn based on the requirements and a sketch of the layouts and dimensions of the line. The Flow Diagram "Make masks for TV display tubes" shows the various activities that are necessary to produce the mask for the new TV display tubes. Process 5 in gray represents the planned demagnetizing line that is newly integrated in the TV display tube production. The line was originally planned and engineered with a single degauss coil and the pertaining energy supply, machine control, transport system, and control for the whole production line.

The engineering partners and subcontractors for these modules were called in for a briefing on the project. The Flow Diagram was used as a common communication platform to explain the different functions and components of the machine. The subcontractors submitted their own proposals and estimates in the same week. The customer received the quotation and the concept within ten days.

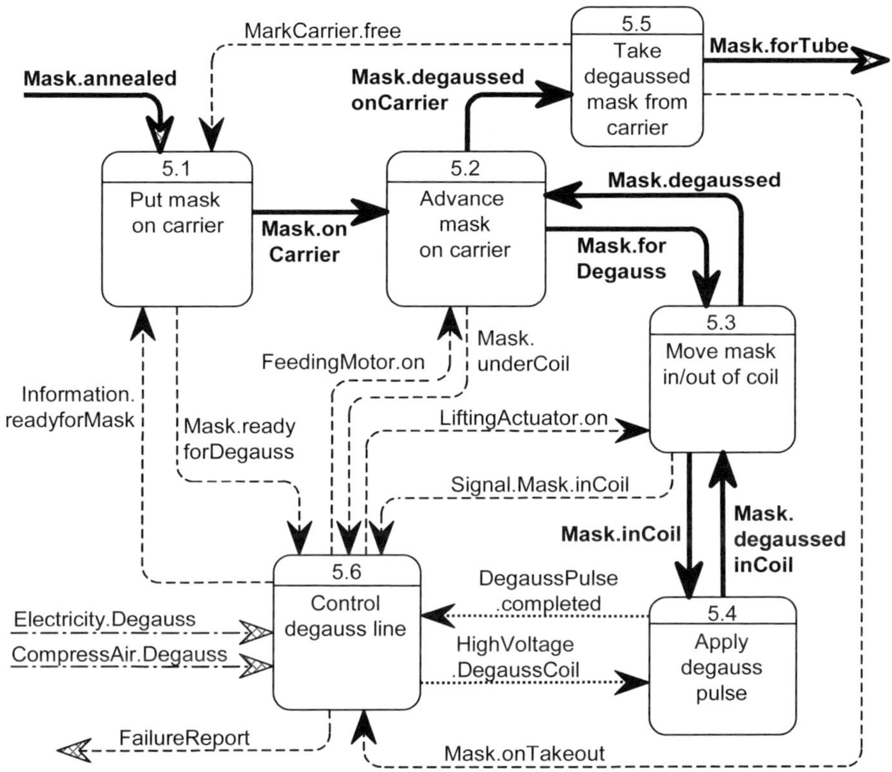

Diagram C4-3: Flow Diagram “Degauss mask”.

The function of the demagnetizing plant from *Figure C4-3* is depicted in *Diagram C4-3*.

As shown in the layout in *Figure C4-4*, the mask is fixed on a carrier, transferred by a belt to a position below the coil, and lifted into the degaussing position within the coil. A magnetic field tailored to the material and shape of the mask is applied. Next, the pallets with the mask are lowered back on the transport belt and transferred to the unloading station. Operators manually load and unload the masks and also return the empty pallets.

The machine would consist of a degaussing coil, a degaussing power controller, a handling and feeding system for the masks, and a controller for the machine.

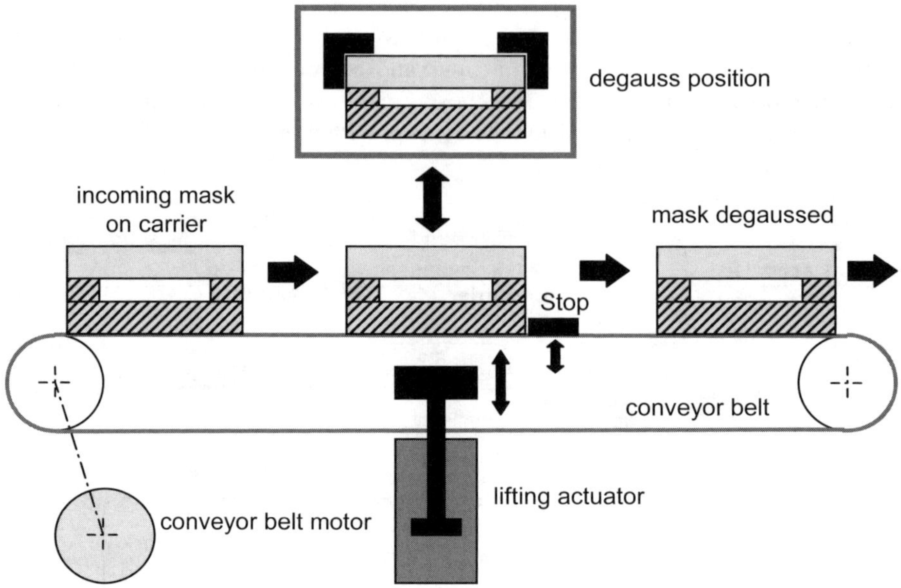

Figure C4-4: Layout demagnetizing process.

4.4 Process Control for Degauss Production Line

For the degauss production line, a certain type of Programmable Logic Controller (PLC) was chosen according to the specific requirements of the customer. The programming was done directly on the machine that was provisionally assembled on the shop floor, as done in the traditional design of machine control. The program was created and put into operation on a notebook computer by pulling icons and typing in parameter values.

At the beginning of the project, the entire line consisted only of one coil, coil A. For this, the programming of the control by PLC was satisfactory. The method was no longer sufficient when coils B to E were added. The addition of more coils was necessary after changes had been made in the mask production process during commissioning of the whole production line. The tests showed that one single coil was not able to eliminate all residual magnetism. So the demagnetizing line had to be enlarged with four more coils (*Figure C4-3*).

This enlargement of the line was technically simple concerning the process by adding more stations, but caused problems with the now enlarged code of the controller with the additional control components for the coils B to E. The enlarged program did not work reliably anymore. A couple of crashes occurred. The program could not be adapted successfully, because the code was completely messed up. The programming engineer was unable to find the reason, gave up, and had to be replaced. The program had to be coded from scratch.

At this point, POA was again put in effect to re-engineer the control system by using State Charts; POA was already used for the initial System Architecture. There was no documentation available to the PLC program, since the original program was made on site during the setup of the line. The whole software had to be re-engineered. If used from the beginning of the project, a State Chart would have allowed the software to be adapted and expanded without problem. In general, it is important to develop the program specification independent from the programming platform, so that it can be easily modified and maintained and ultimately transferred to a different hardware.

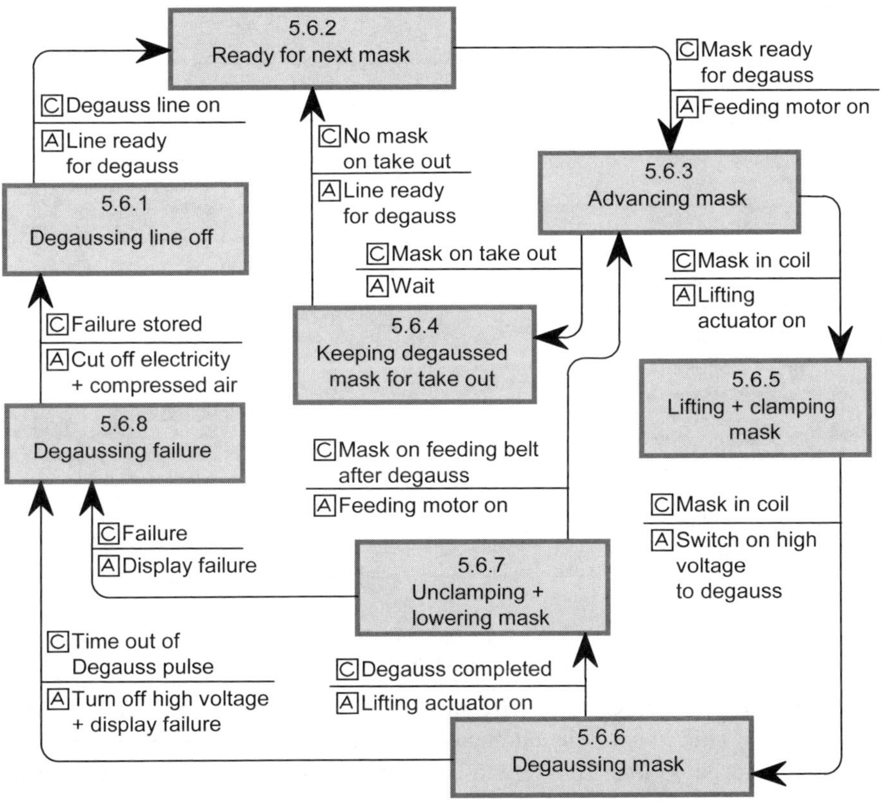

Diagram C4-4: State Chart “Control degauss line.”

Based on the Flow Diagrams, the State Charts were drawn. The State Chart for the degaussing line brought order into the PLC program. By tracking the System States, the real reason for the crashes was finally established. It was a mismatch of the parameters of the inverter within the conveyor belt motor. The inverter settings did not fit the conveyor belt motor, which was not detected originally, but led to occasional stalls later when the conveyor belt was enlarged. After this intervention and completing of the new program, the degauss system operated trouble free around the clock.

The State Chart of process 5.6 "Control degauss line" is depicted in *Diagram C4-4*. Besides the normal operational states, it also shows exceptional and safe states. The safe state is "Degaussing line off," and the exceptional state "Degaussing failure."

In the next step, the behavior of process 5.3 "Move mask in/out of coil" from *Diagram C4-3* is looked at in detail and specified in a State Chart. This process describes the positioning of the mask in a degauss position and the removing of the mask after the demagnetizing. The State Chart of this process is modeled in *Diagram C4-5*.

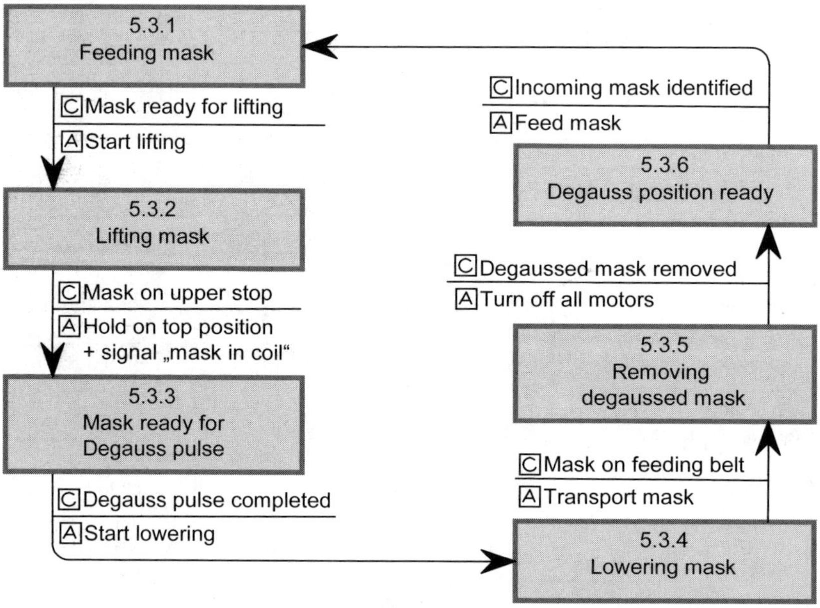

Diagram C4-5: State Chart "Move mask in/out of coil."

In connection with process 5.2 "Advance mask on carrier," process 5.3 "Move mask in/out of coil" the whole automated transport movement of the mask. When enlarging the degauss line from one to five degauss positions, the procedure of the conveyor belt became much more complex, because each mask had to be transported from one coil to the next and lifted into the correct degauss position in an exact time sequence. With regard to the control to be programmed again, the transport process had to be investigated very exactly concerning the distances and positions, since the conveyor belt would now carry not only one, but five masks of any type at any time.

4.5 Benefits of the Method

The application of POA for system conception and graphical system specification was a huge advantage concerning the time required for developing the conception and the communication within the project team. The structuring of the planned demagnetizing line on different levels of detail allowed to keep the overview over the setup, function of the line, and project. Besides, alternative solutions for the process design could be checked quickly during the planning stage. During the realization, the producer of the mask had to do several adjustments to the process, because the mask itself was developed further in the meantime. POA was also useful during these stages, because the concept could be changed easily without much effort and the interfaces to other processes could be updated based on the new requirements.

POA was used in this project for two different reasons: In a first phase for the Flow Diagram served for structuring of a new process line in its conceptual state. In a second phase, the State Charts helped to reengineer the control program of the new entire display tube production line. Process Oriented Analysis contributed to success in the following steps:

Understanding the system setup

- Because flows and processes were used by drawing Flow Diagrams already at the first discussion with the customer, mistakes and time lost by misunderstandings could be minimized.
- Using the Flow Diagram made possible the communication between professionals speaking different languages: English, German, and Japanese. It allowed planning and setup of the time critical process. The diagrams were easy to understand for everybody involved from across the different companies involved.
- Using a Flow Diagram allowed to layout and understand the mask manufacturing process quickly. It was easy to quickly create and check alternatives for the concept. The diagrams allowed for easy visualization of the optimal placement of the degauss process.
- The integration of the degaussing line within the plant, especially regarding resources, such as power supply, compressed air, and manpower, was faultless. The interfaces were clearly defined and the consumption of resources specified. For this purpose, lower-level diagrams were used, and the flows to the outside world were discussed with the customer who had to prepare the site for the installation of the equipment.
- The flows between the lower-level processes and the subassemblies such as coil, power controller, and automatic handling system enabled all components and subassemblies to be ordered “first time right” at the suppliers, since everything was specified in detail.

Coding based on State Chart

Later on, the demagnetize process for the mask was detailed by a State Chart in order to develop the real-time control of the process. The State Chart enables the engineer to code a program with reference to the real world structure and the behavior of the process. With a detailed State Chart, coding is done easily and quickly.

The State Chart showed the interaction between mask feed, mask positioning, and coil power control. It enabled the team to solve the production problems and to reengineer the confusing machine control program. It allowed for the design of a new control program for the expanded production line that is maintainable and independent of the hardware platform.

C5 OPERATIONAL CONCEPT FOR AN AUTOMATED PLANT

Goals of the Case Study

- Model a state of the art production plant.
- Learn the transition from the static model to the simulation model.
- Design a simulation program.
- Run the simulation and evaluate the results.

Figure C5-1: Friction texturizing of yarn for pantyhose (Photo: ETH Zurich).

5.1 New Production Setup

Design of a new machine

In this case study, a production plant for texturizing synthetic yarn is simulated. Texturizing with disc friction units was first established years ago. At the ITMA 2003 (International Textile Machinery Exhibition), the top speed displayed by texturizing machines was 1,600 m/min, while the normal production running speed was 500 to 700 m/min. With a new fluid cooling, developed at the ETH Zurich and shown at this exhibition, the texturizing speed was raised up to 2,500 m/min. The development of this machine had only started one year previously, with a study covering the basic concept of the new machine.

Before starting the final design of the new texturizing machine, two questions arose concerning material flow and operation:

- Should the creel holding the packages that supply yarn on the machine offer space for a second yarn package per position?
- Should the change of packages with the final yarn be done by the operators or automatically by a so-called doffer on the winder?

These two points have a profound impact on the efficiency of the machine and the operators. They can only be answered by simulating the organization of the workflow, including machinery and operators.

Introduction of new machine in the plant

The size and layout of the texturizing machine, with the new cooling system, is different from the traditional design. This, and the much higher speed, implicates the following changes for future texturizing plants:

- Number of machines
- Layout of the plant
- Organisation of the workflow

Specify purpose and goal of the simulation project

For concept, design, and investigation of a plant with the new, much faster texturizing machine, a simulation model is set up. This simulation model should give answers to the questions listed above from the economical point of view and establish the requirements of the new machine in a production context. Since no plant with the new texturizing machine yet exists, the information for the simulation model is taken from existing plants and adjusted to the conditions of the new machine. For the investigation, different options of the machine design are considered, according to the questions above.

5.2 What is Texturizing?

5.2.1 Set System Boundaries

Texturizing is an intermediate step in the production of man-made fibers. Before texturizing, the fibers are extruded on a spinning machine in the form of Partially Oriented Yarn (POY) and wound up on packages (*Specification C5-1*).

Specification C5-1: Flow "POYpackage"

POYpackage		
Description	POY Partially Oriented Yarn Packages come from the spinning machine.	
Flow type	Product flow	
Parameter	Weight, Option 1 + 2 Weight, Option 3	17.5 kg 24.5 kg

The next step in the yarn production is the texturizing process. By texturizing, POY is processed into Draw Texturized Yarn (DTY) and gets the bulk and elasticity required for hosiery and apparel (*Specification C5-2*).

Specification C5-2: Flow "DTYpackage"

DTYpackage		
Description	DTY Draw Texturized Yarn	
Flow type	Product flow	
Parameter	Weight, Option 1 + 2 Weight, Option 3	3.5 kg 2.5 kg

Specification C5-1 and *Specification C5-2* describe the input and output product flows of the texturizing process. They also show how parameters may be varied for different simulation runs. In this case, it is the weight of the packages.

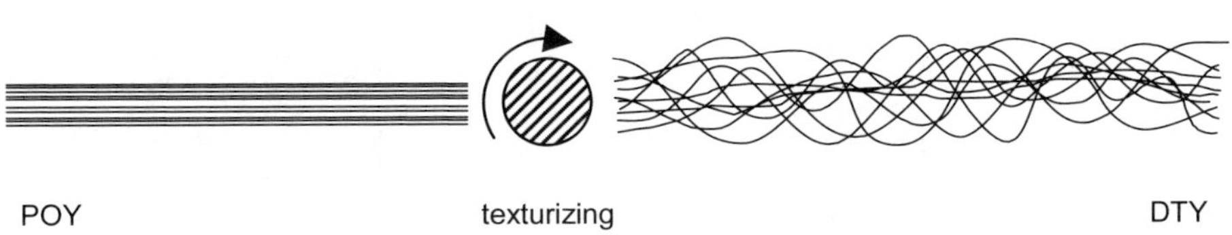

Figure C5-2: Yarn going through texturizing schematically shown.

The reason for texturizing is to draw and curl the fibers. The drawing gives the fiber more strength. By crimping, the yarn becomes more comfortable to wear. In the original state before texturizing, the synthetic POY yarn is flat and smooth, which creates an uncomfortable cold and slippery feeling on the skin. In contrast, natural fibers are already curled and provide yarns with bulk. Such a yarn is more pleasing to the skin, has more volume, and incorporates more air between the fibers building the yarn. To add bulk to the flat yarn, it has to be texturized. During the texturizing process, the fibers are also drawn to improve strength. *Figure C5-2* illustrates how the flat POY turns into crimped DTY.

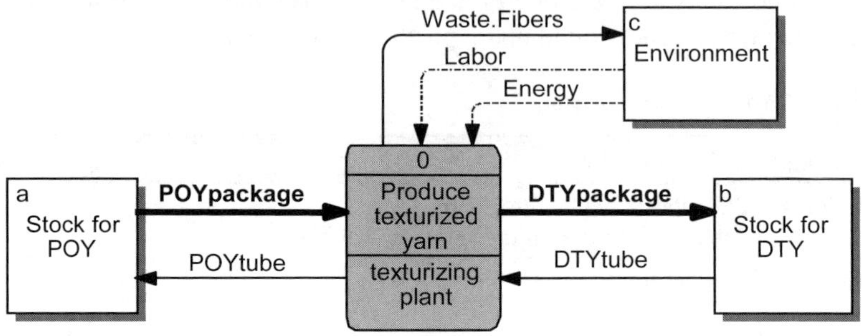

Diagram C5-1: Context diagram for texturizing.

The context diagram sets the boundaries for the system to be investigated by the simulation (*Diagram C5-2*). It is the texturizing process. "POYpackages" go into the process and "DTYpackages" come out.

5.2.2 Specify System and its Structure by Flow Diagrams

Diagram D5-2 gives the first level of the investigated texturizing plant. Process 2 is the actual texturizing process, carried out by the texturizing machine.

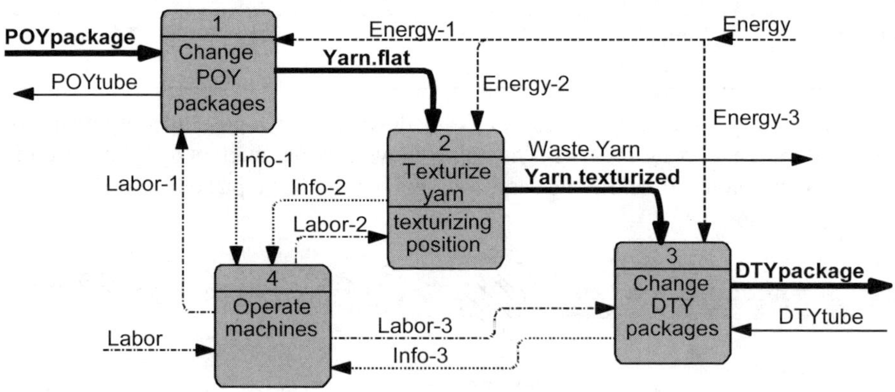

Diagram C5-2: Flow Diagram "Texturize yarn."

The processes also require manual work by the operators of the machine. The jobs, interventions, and supervision by the operators are indicated by the flows "Labor," distributed by process 4 to the different processes. "Labor-1" and "Labor-3" include the handling of the input "POYpackages," output "DTYpackages," and the empty tubes.

Specification C5-2 describes the process "Texturize yarn" carried out on the texturizing position. The texturizing speed is targeted at 2,500 m/min for the new high-speed machine. The figure illustrating schematically the texturizing position is part of the specification. From the POY packages the "Yarn.flat" enters the texturizing position. It goes several times around a first, heated godet and then around a second heated godet. Between the two godets, the yarn passes the texturizing unit. The second godet rotates faster than the first godet, which provides the drawing of the fibers. A machine has multiple texturizing positions, 4 or 8.

Specification C5-3: Process "Texturize yarn"

2	**Texturize yarn**	**texturizing position**
Description	Yarn is drawn and texturized.	
Flow type	Product flow	
Parameter	Texturizing speed	2,500 m/min
Figure	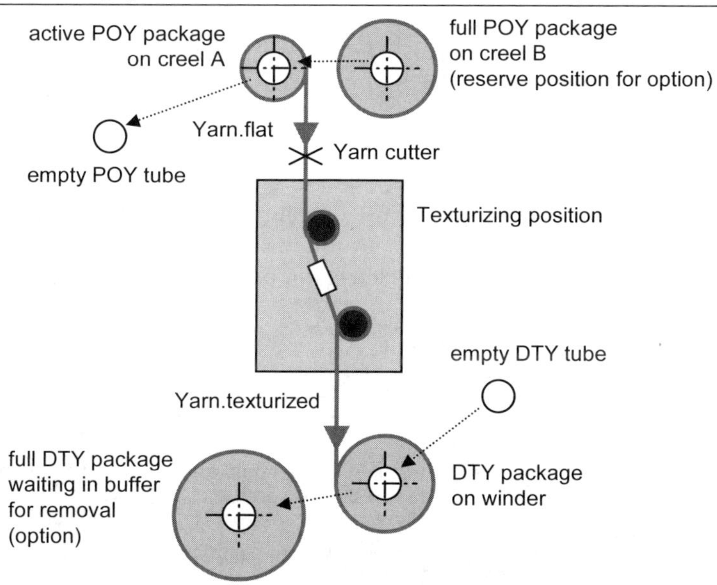	

5.3 Dynamic Model of the Texturizing Plant

5.3.1 Specify Behavior of Processes in Time

The behavior of the machines and the operators in time arc described by the State Charts of the pertaining processes. *Diagram C5-4* describes the view of the machine. The states are depicted for the purpose of the investigation. The simulation shows, how much time a machine is actively producing and how long it stands still.

The standstill is further divided in machine "Waiting for operator" and machine being attended by the operator for "Re-threading OR fixing yarn break."

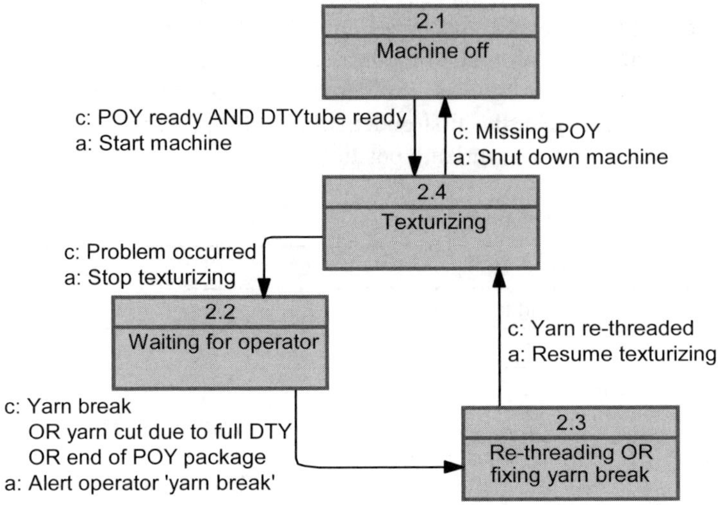

Diagram C5-3: State Chart for process "Texturize yarn."

Diagram C5-4 shows the State Chart describing the process "Operate machines," which is the view and job of the operator. It is of interest to establish with the simulation, if there are too few or too many operators. For this investigation, the states "Operator on patrol" and "Operator busy" are introduced.

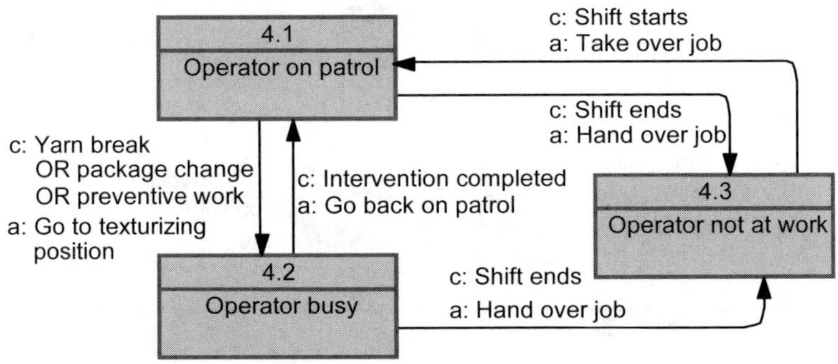

Diagram C5-4: State Chart for process "Operate machine."

5.3.2 State Specification and State List

Where necessary, the states are detailed into child State Charts and/or described by element specifications. *Specification C5-4* of state "Operator busy" lists the jobs of the operator within this state. These jobs are in fact child states of the state "Operator busy." In this case, it was chosen to set up a condensed list of states, giving an overview of the different jobs on the position, instead of drawing a child State

Chart. The child states in the specification are numbered in the hierarchy according to the rules. Details of the operator jobs on a texturizing position and their duration are given in seconds. This list contains all interactions possible.

Specification C5-4: State “Operator busy”

4.2	**Operator busy**			
Description	Operator is busy at a texturizing position.			
Child states	1.3.1	Put the new POY package on the creel, incl. connecting of yarns and taking away the empty tubes.	60	s
	1.3.2	Take away the full DTY packages and fill up new empty yarn tubes.	30	s
	1.3.3	Thread up texturizing position.	150	s
	1.3.4	Walk to other side of the machine	30	s
	1.3.5	Manual change (manual doffing) of POY and DTY packages.	60	s
	1.3.6	Automatic change (automatic doffing) of POY and DTY packages.	30	s
	1.3.7	Remove knot between old and new package into suction.	60	s
	1.3.8	Clear yarn break (end down) on winder.	40	s

5.4 Simulation Program for the Texturizing Plant

5.4.1 Specify Requirement of Program and Design User Interface

The context diagram of the simulation program defines the interface to the user, the input from and the output to the user (*Diagram C5-5*).

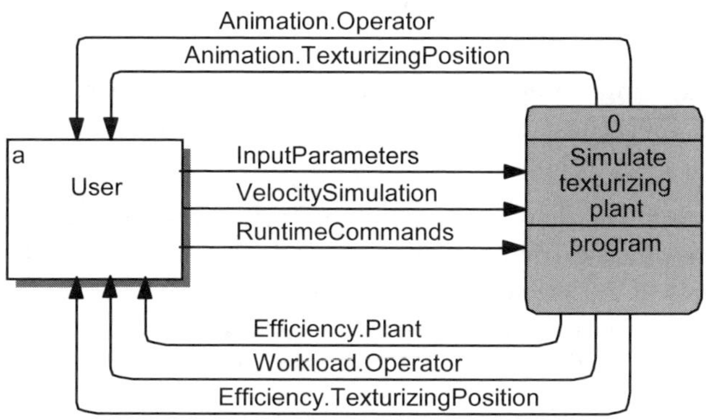

Diagram C5-5: Context diagram for simulation program.

The requirements to specify comprise of the following:

- Range of options of various machine functions and operator modes to be simulated.
- Benchmarks and evaluations to be generated at run time, and final reports.
- Input parameters to match the output parameters required for the evaluations.

5.4.2 Evaluations Required of Simulation

By simulating the plant equipped with the newly developed texturizing machine, benchmark conditions for the plant are set up. Promising configurations of machine and operator jobs are investigated. Machine design options are compared to reach maximum production on a given benchmark.

The simulation should provide the following evaluations and results:

- Indicate state of every single texturizing position for plausibility check and evaluation of efficiency.
- Indicate state for every operator for plausibility check.
- Evaluate efficiency of whole plant. Show workload of machines and operators.
- Test plant for efficiency while changing machine parameters.
- Test plant for time lost on yarn break. Indicate time to clear yarn break for every texturizing position.
- Evaluate efficiency of machinery as a function of workforce available.

5.4.3 Parameters

In order to obtain the desired results and information about certain features of the machine to be designed, the following parameters are defined for the simulation model. The texturizing machine is still in the conception stage, and only a few general specifications exist. This has to be taken into account when defining the parameters and their range of values.

Number of texturizing positions: The number of texturizing positions may vary between 1 and 480. The simulation runs in this case study are done with the benchmark set at 480 positions.

Texturizing speed: The delivery speed is selectable between 1 and 3,000 m/min. The simulation runs are carried out with the benchmark value of 2,500 m/min. This value corresponds to the speed that is the target for a future machine. The state-of-the-art machines reach, in general, a speed of 1,000 m/min, a few machines go up to 1,600 m/min. A practical production runs at a speed of 500 to 700 m/min to make sure the desired quality of the texturized yarn is obtained.

Buffer: The parameter "Buffer" includes two aspects, which are investigated because they have an important impact on the design of the machine:

- Number of creels: A second position for a POY package as a buffer. What is the impact of having one or two packages in the buffer? For two packages, a second creel is required that calls for more room.
- Doffing: Changing the package with the texturized yarn (DTY) to an empty tube. Is this done manually or automatically?

This leads to four buffer options as listed in *Table C5-1*.

Table C5-1: Buffer options

Doffing / Number of creels	manual	automatic
1 creel	Buffer option 10	Buffer option 11
2 creels	Buffer option 20	Buffer option 21

Yarn count: The yarn count can be chosen between 1 and 300 g/km.

Weight and ratio of the packages: Either the weight of the POY packages or the one of the DTY packages can be chosen. Texturizing is most efficient with the biggest POY and DTY packages possible (this means the fewest changes). POY packages, at a weight of more than 20 kg, are handled by a lifting device. DTY packages are generally smaller, since the downstream processes call for packages to be handled manually, which restricts their weight to less than 5 kg.

For avoiding waste, it is essential to have an integer number ratio of POY to DTY package weight, in order that the POY package is finished the same time, as the DTY package is full. This means, for example, 5:1 or 7:1.

Machine stops and yarn breaks: These are defined by the interval of the occurrence and their duration.

Number of operators: The number of operators may be set between 1 and 20. This number has an influence on the efficiency of the whole plant, as well as on the workload of the workforce.

Assignment of operators: The assignment of the operators to the different tasks is rather complex due to the dynamic behavior of the system. The tasks are broken down into four categories of jobs:

- Big change: Both DTY as well as POY packages have to be changed. The DTY package because it is full, and the POY package because it is empty.
- Small change (Doffing): Change of DTY package only. (Since the DTY packages are smaller than the POY packages, this change occurs more often.)
- Preventive work on the buffer: these jobs are not dependent on a specific point in time, but may be carried out during a certain time slot while the machine is producing, to avoid a later stand still of the machine.
- Yarn break (end down): repair break and restart texturizing position.

These categories lead to the lists of states in *Specification C5-4* (Chapter 5.3.2). To form a parameter for the simulation program, which the user can choose, these jobs are combined into two options:

- Assignment option 1: The operators work in groups and attend the *whole plant.* They carry out the task with the highest priority.
- Assignment option 2: The operators work in groups and patrol along *a section of the plant.* At the texturizing position, the task with the highest priority is carried out. The sections of the plant are staggered in time regarding the duration necessary to complete a DTY package.

Mode of packages change: There are two different ways of organizing this job:

- Sequential change: Change at one texturizing position after the other. It is of interest to optimize the time between the change of two texturizing positions.
- Wild change: A texturizing position gives an alarm if a change is due. The changing takes place on demand instead of at a predetermined interval.

The parameters do not all have the same relevance for the investigated problem. In a first step of this investigation, the focus lies on the four most relevant parameters. These are "buffer," "weight and ratio of the packages," "assignment of operators," and "mode of packages change."

5.4.4 Options for Machine Design

The variants and options to be simulated are defined by the parameters within the element specifications. The Flow Diagrams and State Charts are drawn in a general way including all possible combination of options.

5.4.5 User Interface

The user interface contains different windows. The window in *Figure C5-4* provides information about the simulation during the run and includes command buttons for interaction during the simulation.

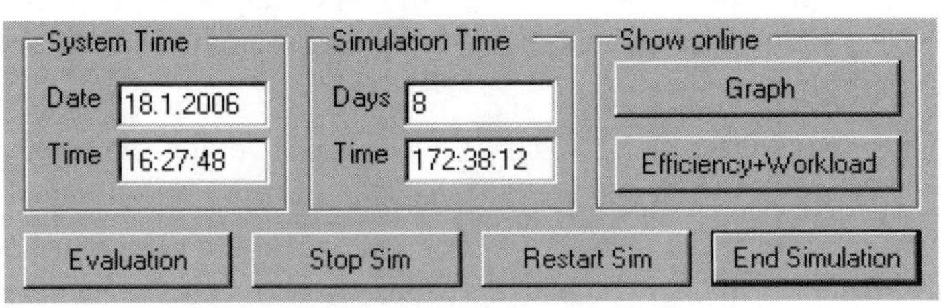

Figure C5-3: Window for user interactions during simulation.

The simulation parameters are specified in the window "Parameter Input" in *Figure C5-4* with several registers for parameters of machine, operator, and simulation. The command button "Parameters OK" confirms the inputs, and the program checks the correctness. If all the simulation parameters are valid, the command button "Start Simulation" for the simulation run appears. In the register "Parameters Simulation," duration, speed, and time increment of the simulation are set.

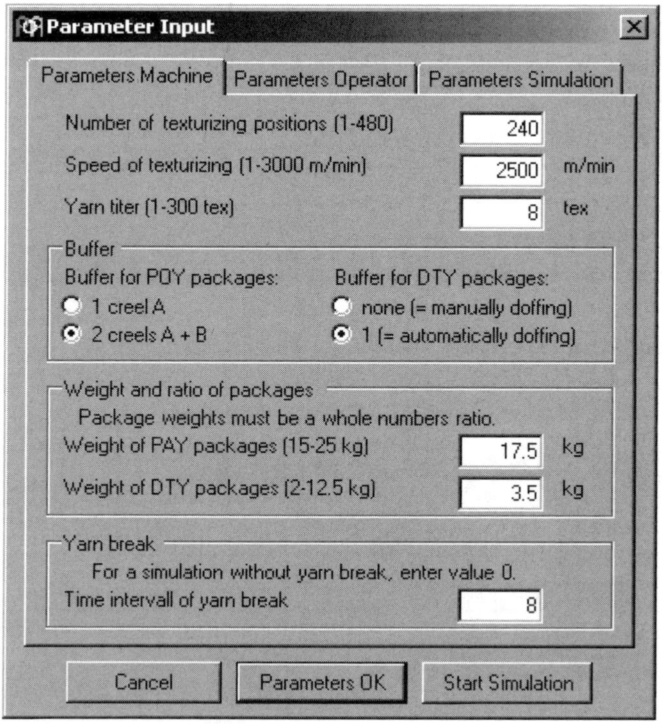

Figure C5-4: Input window for parameters.

The window "SimulationGraphic" in *Figure C5-5* shows the animation of the plant during the simulation. On this window, every single texturizing position of the plant is depicted by several fields indicating various states of the texturizing position. The user of the simulation can define the size of the plant by choosing the number of texturizing positions between 1 and 480.

A texturizing position consists of different sections, the states of which are indicated in six different fields. *Figure C5-5* shows the texturizing position 8 to 11 for explanation of the six fields.

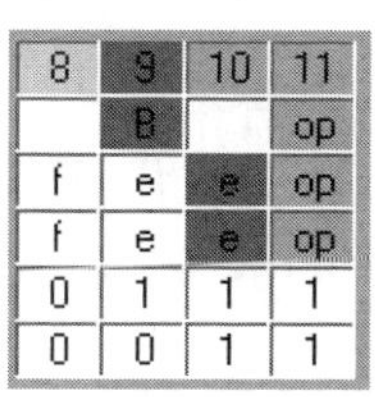

Field 1: number of texturizing position
Field 2: State of yarn break
Field 3: State of POY, creel A
Field 4: State of POY, creel B
Field 5: State of DTY tubes
Field 6: State of DTY packages

Figure C5-5: Fields for indicating state of texturizing position.

- Field 1 with the number of texturizing position indicating state of position by background color: green (light gray) = texturizing; blue (middle gray) = not texturizing and operator attending; red (dark gray) = not texturizing while waiting for operator.
- Field 2: State of the yarn break
  B = waiting for operator because of yarn break; op = operator re-threading or fixing yarn break.
- Field 3: State of POY creel A
  f = package on creel A full; e = package on creel A empty; op = operator changing package.
- Field 4: State of POY creel B
  f = package on creel B full; e = package on creel B empty; op = operator changing package.
- Field 5: State of DTY tubes. Number of the reserve tubes is 0 or 1.
- Field 6: State of DTY packages.
  Number of full packages ready for picking up is 0 or 1.

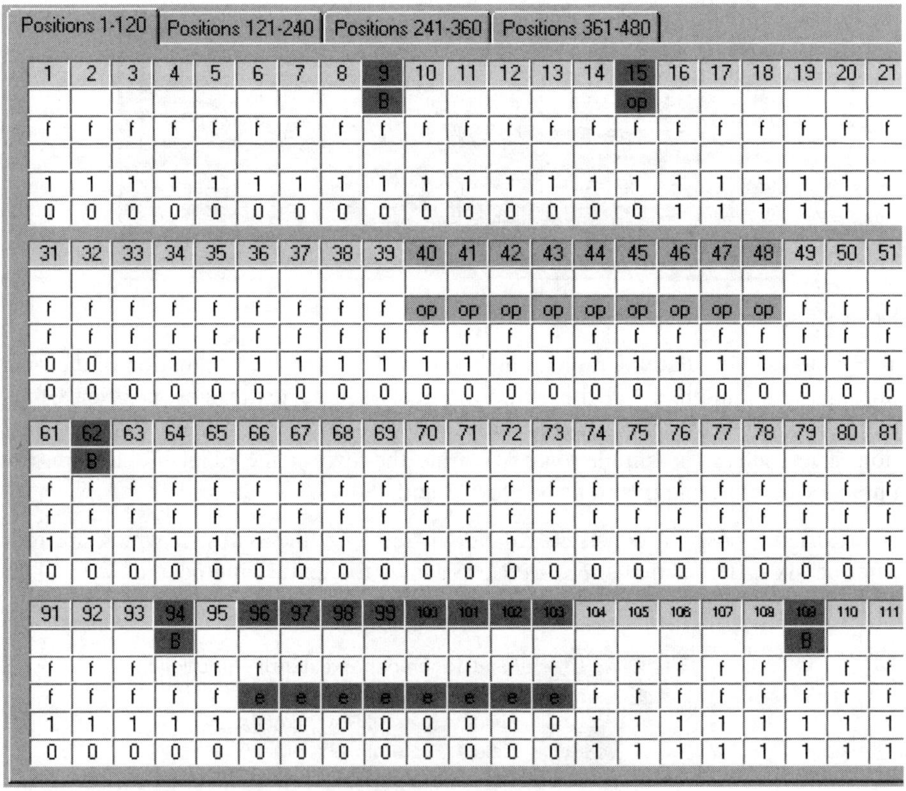

Figure C5-6: Plant animation showing states of the texturizing positions.

Figure C5-7 indicates the state of the operators. In this case, the plant is operated by 10 persons. Each operator is identified by a number. The fields with the number change its color according to the state of the pertaining operator. Blue (middle gray) means the operator is busy. Green (light gray) indicates that the operator is at work, but not busy at the moment (e.g. operator 7 and 8).

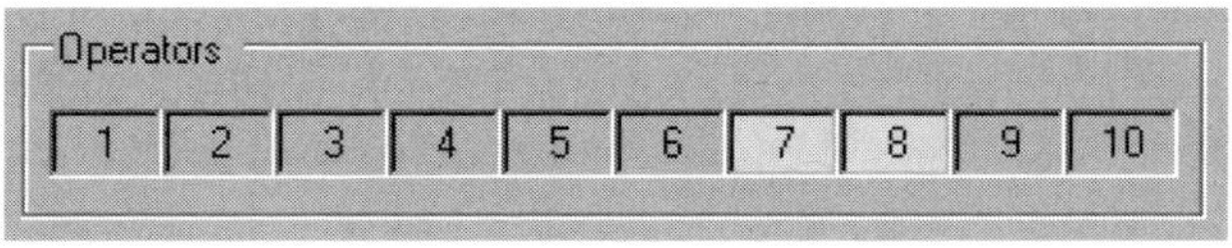

Figure C5-7: State of the operators.

5.4.6 Write each Module in Program Code

The simulation code was written in Visual Basic .NET. The setup of the program is illustrated in *Figure C5-8*. It consists of several modules that correspond to the setup, as proposed in Section D2.

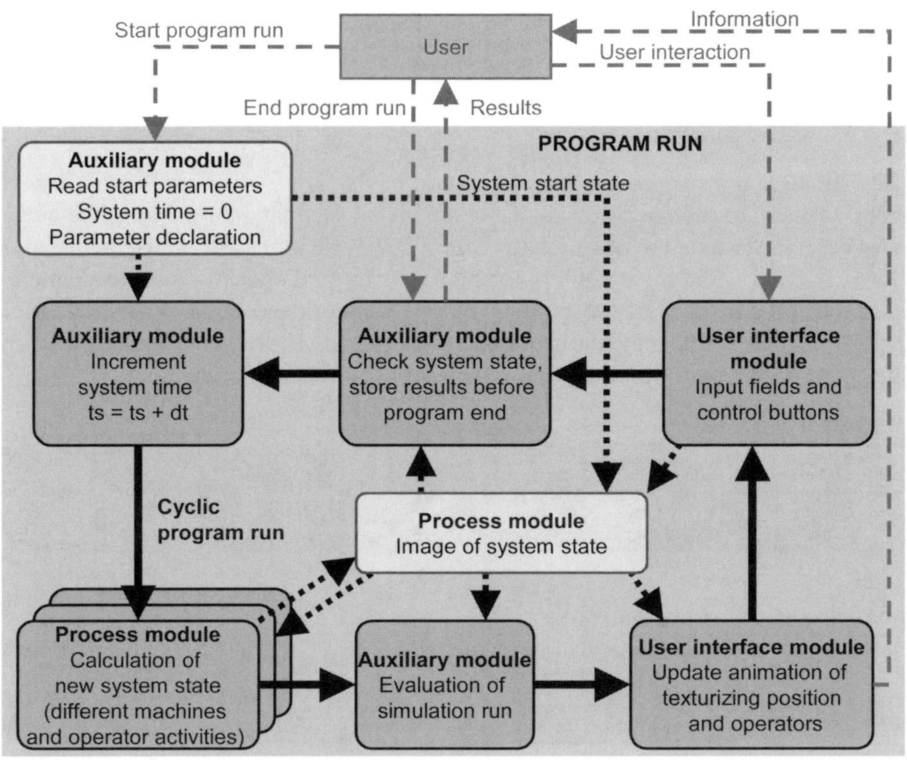

Figure C5-8: Internal structure of the simulation program.

Each of these modules is organized in submodules. The process module that calculates the new system states incorporates the following submodules:

- Yarn breaks (ends down): introduces yarn breaks and the fixing of it.
- Matching of operator and texturizing position.
- Doffing large: change of empty POY and full DTY package at the same time.
- Doffing large and new start: Starts a texturizing position that had to be shut down for lack of an available operator attending the change of empty POY and full DTY package.
- Doffing small: change of full DTY package only.
- Doffing small and new start: Starts a texturizing position that had to be shut down for lack of an available operator attending the change of DTY package.
- Preventive jobs, not done on call of the texturizing position.

The simulation program of this texturizing plant contains:

- 20 modules with in total 72 procedures
- 5,000 lines of code
- 10 forms or windows

5.4.7 Evaluation

The simulation program can support several evaluations, as defined earlier by the context diagram (*Diagram C5-5*). They are coded as submodules within the auxiliary module "Evaluation of simulation run." The evaluations are shown on different forms and may be printed out. For example, the form "EfficiencyWorkloadOnline" displays the efficiency of the whole plant and the workload of the operators during the simulation. Efficiency and workload in *Figure C5-9* are given concurrently by days; the workload of the operators is beneath the efficiency of the plant.

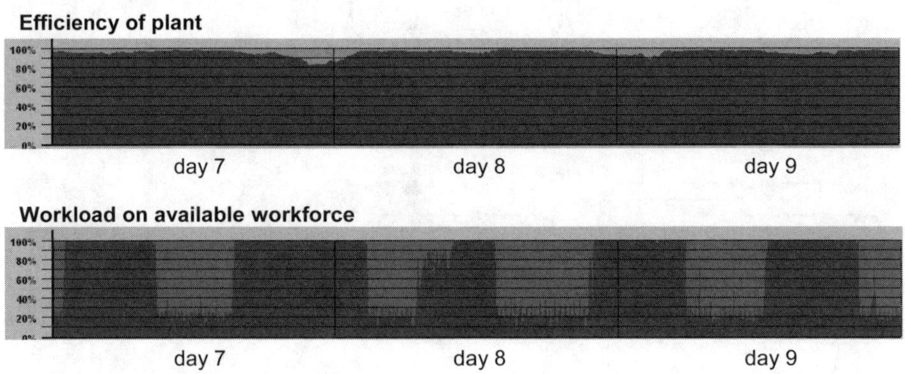

Figure C5-9: Results in Form EfficiencyWorkloadOnline, 6 operators working in assignment option 1.

The print screen in *Figure C5-9* is an extract of the window (or printout) showing the results from day 7 to day 9. The reason why the result is not shown beginning with day 1 is that the cold start of the plant from empty up to regular production takes 5 days. Thus, adding a day to assure steady-state operation, the simulation becomes useful for evaluation from day 7 on.

Figures C5-9 is the result of a simulation run with 6 operators, in the mode of assignment option 1, patrolling the whole plant. The efficiency of the plant is very good in this simulation run, while the workload of the operators is very irregular, shifting periodically from full workload down to only 20%. It can be seen that while the workload is up at 100%, the efficiency of the plant decreases slightly.

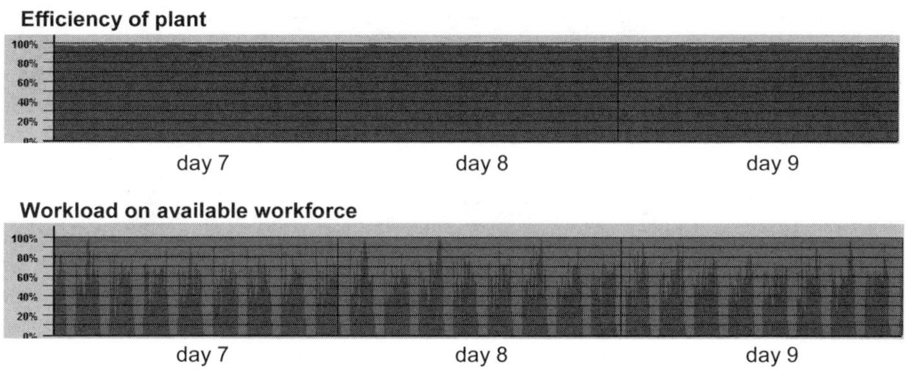

Figure C5-10: Results in Form EfficiencyWorkloadOnline, 6 operators in assignment option 2.

Figure C5-10 displays a simulation run with almost the same parameter configuration as previously. In this case, the operators patrol only a section of the plant (assignment option 2). It can be seen, that the workload of the operators is distributed in smaller packages over time and the efficiency of the plant is more constant.

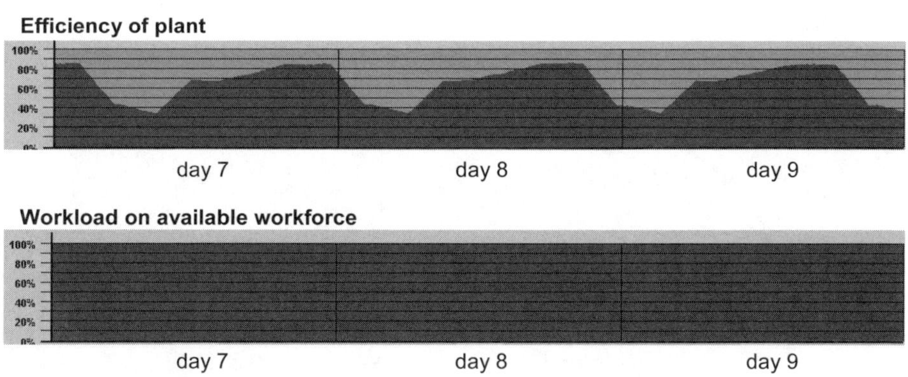

Figure C5-11: Results in Form EfficiencyWorkloadOnline, 5 operators working in option 1.

Figure C5-11 gives the results of a simulation run with 5 operators working in the mode of assignment option 1. With 5 operators, their workload is constantly on 100%. But the efficiency of the plant goes down drastically. This indicates, that 5 operators are not sufficient, when working with assignment option 1.

The simulation run resulting in *Figure C5-12* is also done with 5 operators, but working in assignment option 2 (patrolling only a section of the plant). It shows that the efficiency of the plant is again up over 95%. This indicates that assigning an operator to a specific section instead of having all operators patrolling the whole plant is more efficient.

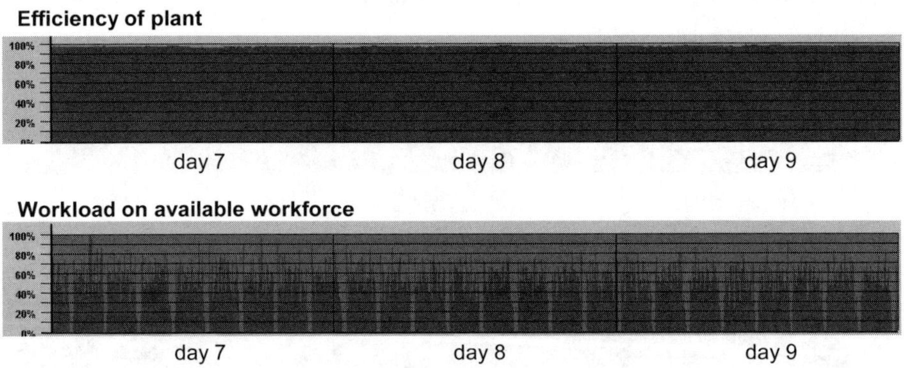

Figure C5-12: Results in Form EfficiencyWorkloadOnline, 5 operators in assignment option 2.

Figure C5-13 displays the results for the four buffer options. It shows that a second creel does not bring any advantage. The difference of the efficiency at the same number of operators is in the thousandth, and therefore negligible.

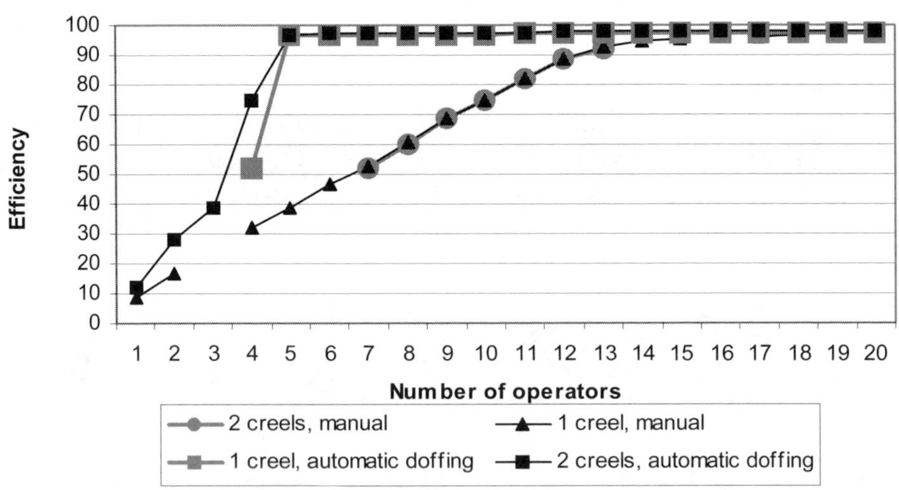

Figure C5-13: Buffer variants.

An automatic doffer, however, has a big impact. The development of the efficiency when working with 3 to 5 operators is impressive. The efficiency is about three times better. The automatic doffer may save up to 6 operators.

5.4.8 Code Example

In this chapter, two code examples of the simulation program are given. The first one shows the code of the subroutine (sub) executed when clicking on the Start button for the simulation run.

```
Private Sub cmdStarten_Click()
    For i = 1 To 14 Step 1
        If Input_Control (i) = False Then
            Call ReadIn
            Call InputValue_Control
            GoTo a
        End If
    Next i
    Call JobDuration
    Call Time_Start
    Call SimSet_Velocity
    Call InitialConditions0
    Call Grafik_Anfang
    Call Coordinates
    Call Update_Graph
    frmSimulationGraph.Visible = True
    frmSimulationGraph.tmrTimeDisplay.Enabled = True
    frmSimulationGraph.tmrTimeDisplay_r.Enabled = True
    lblInput.Visible = False
    cmdInput.Visible = False
    lblSimulation.Visible = False
    cmdStart.Visible = False
    frmInputSpecial.lblInput.Visible = False
    frmInputSpecial.cmdInput.Visible = False
End Sub
```

The second code example is the subroutine for the calculation. A part of this code is written in a select-case construct to choose the different buffer options.

```
Sub Calculate()
  'WeightPosMin is the public variable for the weight of the
  'yarn that is texturized by a position during 1 minute.
  'WeightPlantMin is the public variable for the weight of the
  'yarn texturized by the whole plant during 1 minute.
  WeightPosMin = (Tex * 0.001) * TextVelocity
  WeightPlantMin = (Tex * 0.001) * TextVelocity * NumberTexPos
  'Buffer option is assigned:
  'Buffer option = 1: 2 creels, with automatic doffing
  'Buffer option = 2: 2 creels, manual change
  'Buffer option = 3: 1 creel, with automatic doffing
  'Buffer option = 4: 1 creel, manual change
```

```
    If PoyCreel = 2 Then
        Select Case DtyBuffer
            Case 1
                Buffer = 1
            Case 0
                Buffer = 2
        End Select
    Else
        Select Case DtyBuffer
            Case 1
                Buffer = 3
            Case 0
                Buffer = 4
        End Select
    End If
End Sub
```

5.5 Results and Benefits of the Method

5.5.1 Results of the Texturizing Simulation

For this case study, 122 simulation runs were carried out. A single simulation run simulates 15 days including the start up time needed between 1 and 12 hours depending on the simulated option and the computer performance.

The results are summarized in the following statements concerning economical profit based on the standard value of the parameters (480 texturizing positions, texturizing speed of 2,500 m/min.). The benchmark of the efficiency for an acceptable option is set at 97%.

1) Automated doffing (automated change of DTY packages) is very useful. The difference of the plant efficiency is a factor 3. This makes the automated doffing a very interesting objective to pursue in development.

2) A second POY creel brings no advantage. The efficiency of the plant at the same number of operators differs less than 15%.

3) A rise of the weight of the POY packages from 17.5 kg to 24.5 kg is advantageous. The efficiency of the plant increases by 30%, respectively the workload of the operators decreases by 20%.

4) A reduction of the weight of the DTY packages from 3.5 kg down to 2.5 kg is detrimental to productivity. The efficiency of the plant drops by 30%, and the workload of the operators increases by 20%.

5) Concerning the assignment of the operators: a staggered doffing or start of the texturizing positions in sections has a big advantage.
   - The efficiency is duplicated; respectively the peak workload of the operators is cut in half.

- Flying changes of products become possible. The plant may run day and night, uninterrupted for months.

6) A high efficiency calls for spare capacity of workforce. The workload of the operators should not be set for their best efficiency, but below, taking advantage of the reserve capacity to improve productivity of the machines.

5.5.2 Benefits of the Method

System specification

POA assists in defining the investigated system. The drawing of the Flow Diagrams and subsequent State Charts allows the user to clearly determine the system. At the same time, the interfaces to the external world (other processes within the company) and the interface between the new machine and the operator are specified.

System scenarios

Based on the State Chart, a simulation model is programmed, and input parameters are defined, which need to be set for the program runs. By changing these parameters, different scenarios for the operation of the machine, as well as the starting parameters can be simulated.

Based on the Flow Diagram, the State Chart is designed and the program code developed step-by-step. Using this procedure, the resulting program code is well structured, clearly documented, and easily maintainable. The Flow Diagram and the State Chart are part of the documentation and make the source code easy to read, since they point directly to the different modules and variables therein.

Requirements for new machine

Since a second POY creel proved not cost efficient, it was not included in the final design of the machine. The further development of the machine concentrates on the automatic doffing of the DTY packages, with automatic removal and transport of the packages.

C6 REENGINEERING OF A CABLE CAR

Goals of the Case Study

- Analyze and reengineer an existing transport system.
- Assess the function and behavior of a cable car using the Flow Diagram and State Chart.
- Program a user interface and write the code for the cable car real-time control.

Figure C6-1: Car of the cable car in the fortress (Photo: ETH Zurich).

6.1 Cable Car System

During World War II, Switzerland built a number of fortresses to counter the threat of the German army. The fortress investigated in this case study was built inside a mountain on four levels, hosting 24 artillery guns, accommodation for personnel, and ammunition storage. It is accessible from the valley. The main entrance leads to a inclined tunnel with stairs and a cable car inside the mountain. When installing the cable car to connect the different levels of the fortress, little consideration was spent on the manpower required to operate the transport system and on its operational safety. The drive of the cable car was controlled by a manual switchboard in the station located on the top level of the fortress. From there, the engineer had a limited view of the top half on the track, and used a combat telephone for communication along the track.

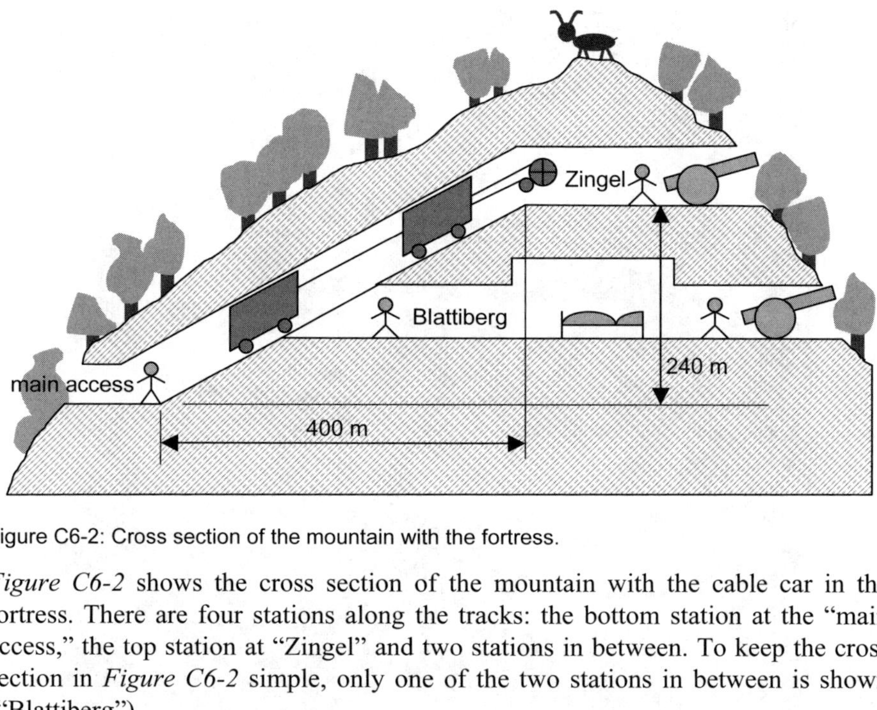

Figure C6-2: Cross section of the mountain with the fortress.

Figure C6-2 shows the cross section of the mountain with the cable car in the fortress. There are four stations along the tracks: the bottom station at the "main access," the top station at "Zingel" and two stations in between. To keep the cross section in *Figure C6-2* simple, only one of the two stations in between is shown ("Blattiberg").

During the period of the cold war, the fortress was upgraded. A second independent brake was added to the cable car drive, and the stations were equipped with emergency stop buttons. A stop signal from the moving car was transmitted by touching an overhead wire running along the track with a grounded whip.

The system was further updated later when part of the fortress was used for storage of the Swiss national gold reserve. A closed circuit television system was installed to overview the stations from the engineer's position in the top station. At this time, the emergency stop action by the overhead wire along the track turned out to be unreliable due to poor ground contact between car and rails.

When disarming the fortress after the breakdown of the communist system, the cable car was declared unsafe for further use. A key problem was now to preserve the fortress as a landmark, and to prepare it for civil use and public visitors. Because the operation of the cable car system was a prerequisite for any further public use of the fortress, a study was initiated to find a solution.

The goal was to design a concept for a remote control of the cable car to make it available to different users. The cable car should be controlled by one operator on one of the two cars, similar to an elevator. By demonstrating the feasibility of remote control operation of the cable car, an essential contribution was made to save this technical landmark for future generations.

6.2 Reengineering of a Transport Process

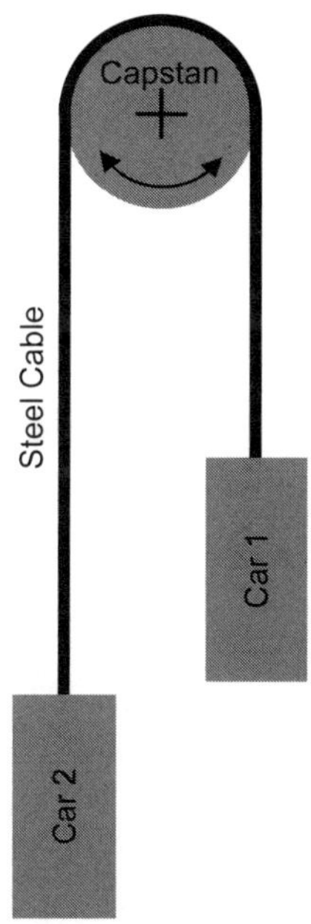

The cable car system within the fortress consists of two cars connected by a steel cable that is driven by a capstan in the top station. The cars travel on a track in opposite directions, passing each other in the middle of the distance between the terminal stations. The cars share a single track except at the point where they pass each other. There, the track splits up to enable the cars to pass.

The steel cable pulling the two cars is supported by carrier rollers located in the center of the track. The capstan, in fact a large wheel in the top station, is driven by a variable speed motor with a reduction gear. Two redundant brakes act on this wheel and on the motor shaft. The position of both cars along the track is determined by the cable; so that it is sufficient to refer to the position of car 1 to determine the position of both cars. A mechanical pointer indicates the position of the cars on the engineer's control panel by counting the revolutions of the capstan wheel.

Originally, the cable car was operated by an engineer who sat in a cockpit in the top station and had a view on the top part of the track. He would manually start the cars and operate the brake. The speed was automatically controlled by the drive in the top station; it was set for cruise between stations and reduced for approaching and passing stations. This kind of operation was feasible as long as abundant troops were deployed to the fortress.

Concept of future use

For future civil use, remote control of the cable car is a necessity. The cable car system will be controlled by a person on one of the cars, similar to an elevator. Microwave signal transmission was found to be the only feasible solution for a wireless link between the cars and the controller in the top station.

In order to conceive and check a concept of remote control, a computer simulation was made using POA. Along with this, a wireless communication system for connecting the cable cars was tested for use inside the mountain. It was discussed with the military maintenance crew of the fortress. A model of the concept was submitted to the present users of the cable car and to the authorities for approval, before starting the full-scale hardware design.

6.3 Flow Diagrams and State Charts

6.3.1 Flow Diagrams of the System

The system is first analyzed with the Flow Diagram. The processes and flows of the cable car are mapped on consecutive levels of detail. The first level of the Flow Diagram is the context diagram, which fixes the system boundaries.

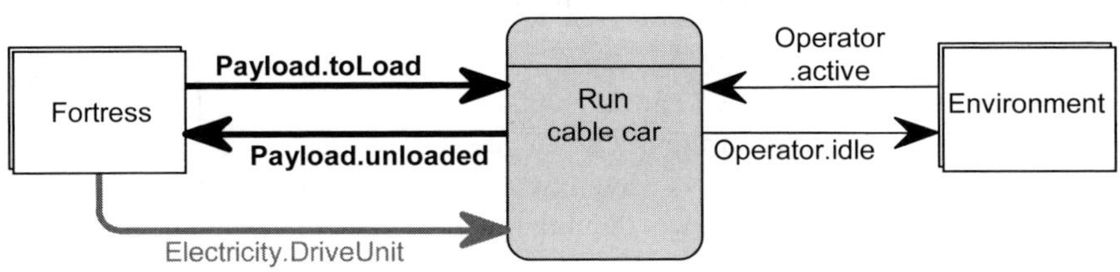

Diagram C6-1: Context diagram cable car.

In the context diagram shown in *Diagram C6-1*, the process "Run cable car" is specified within its environment. The external entities are the "Fortress" and the "Environment." Between the process and the external entities, resource flows, such as operators and payloads (boxes, pallets), are exchanged. The fortress also provides electricity to the cable car system.

The process "Run cable car" is further detailed on a child diagram. *Diagram C6-2* shows three processes: "Drive cable car" (P1), "Carry payload + operator" (P2), and "Load/unload payload" (P3). In this diagram, control processes and flows are added in order to describe the control of the cable car more in detail.

Product flows are depicted by bold colored lines, and the control flows by thin black lines. Resource flows are drawn as bold gray lines. In P3, an operator loads and unloads the payload of the cable car. In P2, the cable car transports the payload and the operator to the assigned station. P1 is the control process of the cable car and receives commands, depicted as control flows, from the operator. The operator riding on one of the cars sees where the car is and can decide on his next action. By the resource flow "CableMovement" the cars are moved. This defines the position of the cars.

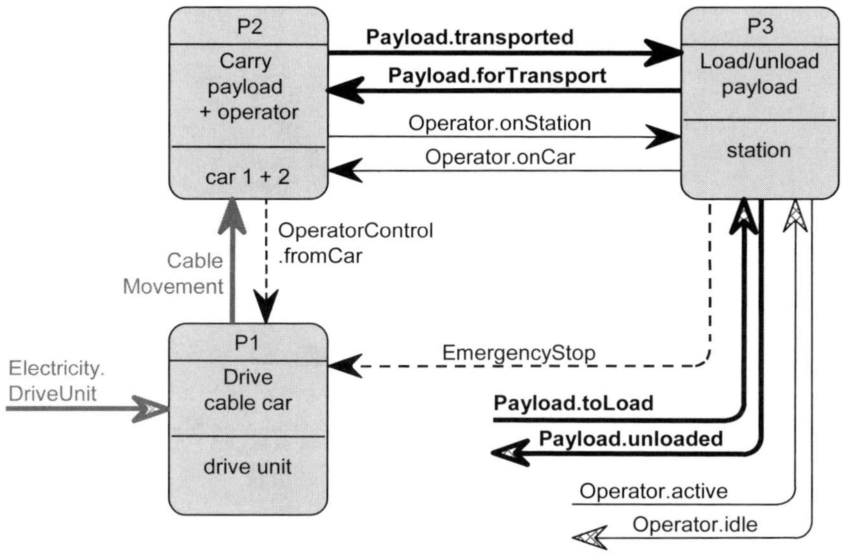

Diagram C6-2: Flow Diagram "Run cable car."

The second level of detail shows the child diagram of the control process P1 "Drive cable car" (*Diagram C6-3*). The two processes on this diagram, "Detect emergency" (P1.1) and "Move car" (P1.2), are the bottom-level processes of the Flow Diagram model. For these two processes, State Charts are modeled in order to specify their behavior.

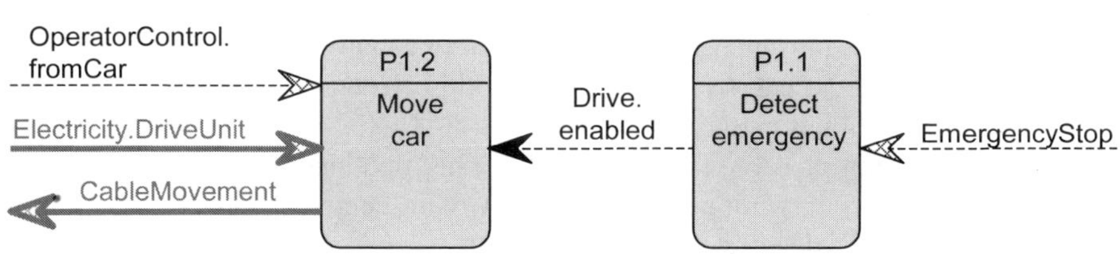

Diagram C6-3: Flow Diagram "Drive cable car."

6.3.2 State Charts of the Cable Car Drive

The drive (motor and brake combination) of the cable car, acting on the steel cable, operates in the three states: "Car 1 moving up," "Car 1 moving down," and "Stop."

This can be seen in *Diagram C6-4,* the child diagram of process P1.2 "Move car." The operation of the cable car is controlled by a console on car 1 with the push buttons "UP," "DOWN," and "STOP."

As an additional safety measure, emergency stop buttons are located in the stations. These push buttons are wired in a common emergency stop circuit. Overrunning the terminal stations is prevented by limit switches that stop and block further movement in the current direction.

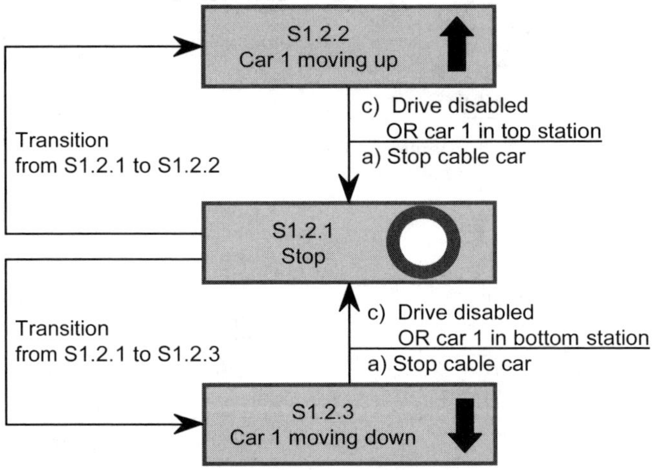

Diagram C6-4: State Chart "Move car."

By pressing the UP or DOWN button, the car starts to move in the corresponding direction. To stop the car, the stop button is pressed. This button will latch, so that it has to be pulled out for release. To simulate this pulling out on the computer screen, a release button is added to the stop button.

The transitions between the states are each triggered by a set of conditional statements. This set consists of events taking place within the system, and/or actions of the operator, all put into a single Boolean equation that finally enables the transition. As an example, if car 1 is in the upper end position, moving it down requires that the drive is enabled and a "Car 1 down" or a "Car 2 up" command is given by an operator. It is not possible to move car 1 further up since it is already in the end position. In addition, it is not possible to start any movement unless all stop buttons within the system are deactivated.

Two transitions of State Chart "Move car" (*Diagram C6-4*) are defined in *Specifications C6-1* and *C6-2*. The conditions of the transitions from state S1.2.1 to state S1.2.2 and from state S1.2.1 to state S1.2.3 are conditional statements and too large to be put on the diagram as text. Therefore, the transitions were named, and the condition and action are defined in the transition specification.

Specification C6-1: Transition from state S1.2.1 to state S1.2.2

Transition	From S1.2.1 To S1.2.2
Conditional statement	Electricity available AND (car 1 up button pressed OR car 2 down button pressed) AND car 1 not in top station AND all stop buttons released
Action	Move car 1 up

The transitions from state S1.2.2 to state S1.2.1 and from state S1.2.3 to state S1.2.1 are also conditional statements, but are short enough to be displayed on *Diagram C6-4* "Move car."

Specification C6-2: Transition from state S1.2.1 to state S1.2.3

Transition	From S1.2.1 To S1.2.3
Conditional statement	Electricity available AND (car 1 down button pressed OR car 2 up button pressed) AND car 1 not in bottom station AND all stop buttons released
Action	Move car 1 down

The State Chart for process P1.1 "Detect emergency" is shown in *Diagram C6-5* with the states "Drive enabled" and "Drive disabled."

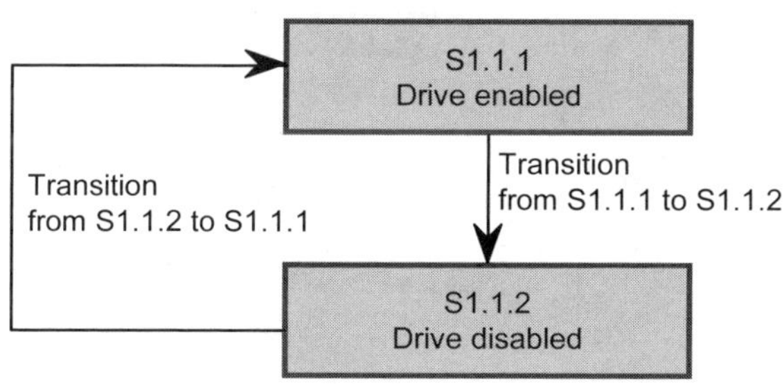

Diagram C6-5: State Chart "Detect emergency."

The transitions of *Diagram C6-5* are defined in *Specification C6-3* and *Specification C6-4* of the transition, because the conditional statements are again too large to be displayed on the diagram.

Specification C6-3: Transition from state S1.1.1 to state S1.1.2

Transition	From S1.1.1 To S1.1.2
Conditional statement	Any station stop button pressed OR any car stop button pressed
Action	Drive disabled

Specification C6-4: Transition from state S1.1.2 to state S1.1.1

Transition	From S1.1.2 To S1.1.1
Conditional statement	All station stop buttons released AND all car stop buttons released
Action	Drive enabled

6.3.3 System Hierarchy

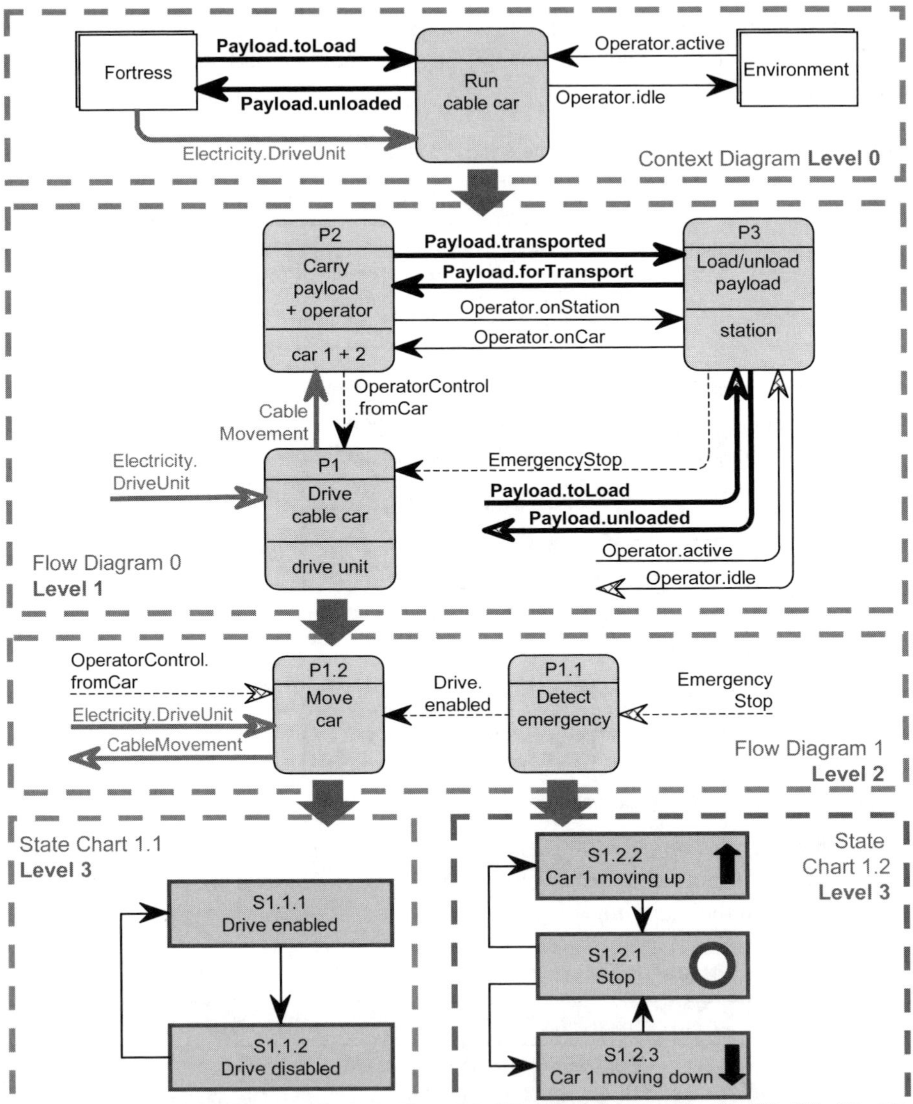

Diagram C6-6: Overview of the system architecture with different levels of detail.

To provide an overview of the model, the architecture of the cable car system is shown in *Diagram C6-6*. The top level 0 is the context diagram of the system that defines the system boundaries and the interactions with the environment. The level 1 diagram specifies the main processes of the cable car system. Level 2 details process P1 of level 1 to show the various functions. Level 3 consists of the State Charts of the processes P1.1 and P1.2 of level 2.

6.4 Transport Simulation

6.4.1 Remote Control

Originally, troops manned a cockpit at the top end of the tracks, close to the electric motor driving the cable. Persons waiting at the stations used a telephone to call the engineer on the top station for service. He used the same method to make sure that the cars were ready for departure.

For future use of the cable car system, operation of the cable car must be possible by remote control from one of the stations or one of the two cars. One single person should be able to use the cable car, for instance, when entering the fortress for the first time in the morning. The system should also ensure safe operation by multiple persons, who are waiting in the stations or riding on the cars. The basic concept of this remote control system offers two functions:

- Control of the direction of travel from at least one car. This can be accomplished, for example, by two push buttons for starting the motor drive forward or reverse.
- An emergency stop button on each car and at each station to override the start command and to disable the motor drive.

Since the motor, the gear box, the brakes, the power control switch gear, that are moving the cable, and the cars are in good shape and fulfill the requirements for safe operation, the project is limited to the two functions given above. It is split up into a hardware part, including communications, which is not treated further here, and the specification of the operator interface, including the functions of the remote control system, which are now investigated by POA.

6.4.2 User Interface

Based on the Flow Diagrams and the State Charts, the system is simulated by a program written in Visual Basic. Mouse operated command buttons are used on the screen to simulate the operator interface, and text fields on the bottom of the screen are used to indicate the system state. The screen is updated in time intervals of one second, with the option to run the simulation in real or accelerated time scale (fast motion).

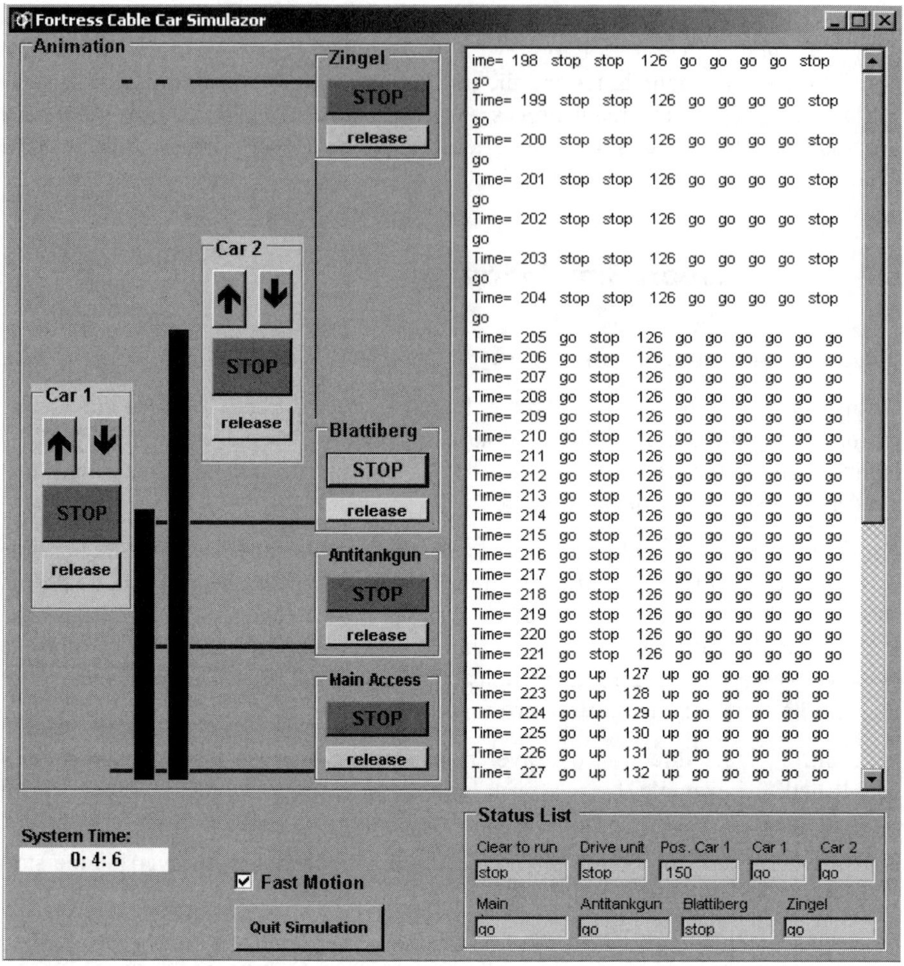

Figure C6-3: Program window for simulation program of cable car system.

The Visual Basic program is coded based on the states of the processes. The screen shot in *Figure C6-3* shows the user interface of the cable car control. The control panel displays the physical location of the cable cars in a graphical animation on the left hand side of the screen. The arrow buttons on the control panel correspond to the remote control buttons on the real cars. They initiate the start up and start down action of the cars. The pressed or released arrow buttons correspond to the states in the State Chart 1.2 “Car 1 moving up” and “Car 1 moving down.” The buttons are the conditions from the State Charts that trigger the transitions that are fulfilled if pressed or released. The cable cars can be stopped or released anytime from the car panels or from the station panels on the four stations.

Table C6-1 shows the modules of the user interface, as introduced in Section D3. The modules of the cable car real-time control are given here as examples.

Table C6-1: User interface components

User interface component	Description	Example: Cable car real-time control
Input interfaces and display	Fields where the user assigns values to variables (or parameters). A definite parameter input field gives a range of values for the input.	Fast motion selector Quit Simulation button
Output interfaces and display	Display of actual values, set values, or other values calculated by the program. They display information about the system functions and the simulation results.	System Status history
Status window	Display of the actual states of the system and parameter values at runtime.	Status list at the bottom of the screen
Control buttons	Start, stop, and other commands for simulation model. Allow the user to control the system behavior.	Stop, release, up, down
Time field	Shows the elapsed simulated time.	Simulation time
Animation	A graphical representation in motion or in changing color of the simulated object or process, for on-line check of system behavior by plausibility.	Car 1 and car 2 going up and down

The System Status history lists the status, direction, and location of cars by the time increment. The simulation control section shows the time reference of the program run. The actual System Status is depicted at the bottom of the screen.

6.4.3 Program Code

In this section, a part of the code of the Visual Basic program (VB 6) is displayed. The Public Sub (subroutine) Main is the initialization of the program, where the start parameters for the simulation are set, e.g. position of the cable cars.

Initialization

```
Public Sub Main()
  'Parameter initialization
  Time = 0
  Drive = "stop"
  Velocity = 1                'Velocity and positions in pixels
  Drive release = "stop"
  Car1_control = "stop"
  Car2_control = "ready"
  Position_Car1 = 10          'Bottom station main access
  Position_Car2 = 390         'Top station Zingel
  Position_bottom = 10
  Position_PakStation = 50
  Position_Blattibert = 100
```

```
  Position_Zingel = 390
  Stopbutton_BottomStation = "free"
  Stopbutton_PakStation = "free"
  Stopbutton_Blattiberg = "free"
  Stopbutton_Zingel = "free"
  Report = "Program Start"
  frmMask.Show              'Show form
End Sub
```

User interface

The next code example is one of the buttons of the user interface. As soon as one of the Up or Down push buttons is pressed, the state of the cable car is set accordingly. The states and the transitions are programmed with the `If-Then` construct. If the conditions are fulfilled, car 1 transitions to the state "Car 1 moving down."

```
Private Sub cmdCar1_down_Click(Index As Integer)
  'Command for click on button down
  If (Car_release = "free" And Car1_Control = "free") Then
    Car1_Control = "down"
  End If
End Sub
```

The unit of time (as real time or fast motion) is defined in the code. The code section for the drive simulation defines the direction of movement and the position of the two cars with stepwise calculations.

Animation and process module

This part of the code implements the graphical animation and shows the location of the cars at the actual point in time.

```
'Drive simulation
If drive = "up" Then          'move car 1 up
    Position_Car1 = Position_Car1 + velocity
End If
If drive = "down" Then        'move car 1 down
    Position_Car1 = Position_Car1 - velocity
End If
'derive position of car 2 from car 1:
Position_Car2 = 400 - Position_Car1
'move car symbol on screen:
frameCar_1.Top = 5520 - (13 * (Position_Car1 - 10))
frameCar_2.Top = 320 + (13 * (Position_Car1 - 10))
'end stop for movement:
If ((Position_Car1 <= Position_bottom And Drive = "down")
  Or (Position_Car1 >= Position_Zingel And drive = "up")) Then
    drive = "stop"
    Car1_control = "free"
    Car2_control = "free"
End if
```

6.5 Conclusions and Benefits of the Method

Simulation results

The simulated operation of the cable cars with the new remote control setup was successfully demonstrated and discussed with the people who operated the cable car system until its decommissioning. Two points had to be reconsidered thereafter.

- The driver should use a key switch to activate the car (1 or 2) he rides on. This ensures that only one operator at a time is in control of the system.
- The stop buttons should not be of the latching type. Any stop button accidentally pressed and forgotten would block the whole system permanently. This would require an operator to walk along the track and check all buttons. The chosen solution is to install momentary stop buttons for emergency stop only. While loading/unloading on a station, the operator would block the cars by a latching switch using his key.

These results contributed significantly to the design of the hardware of the whole system in order to put the cable car into operation again.

Program setup for remote control

POA made it possible to design and program a new remote control for the cable car in the fortress within three months. This project was carried out by a student without previous programming experience. POA allows a user to program a real-time control based on the physical structure of a system. By drawing a State Chart, the states of the individual machines or machine parts are captured and introduced into the program. This way, errors are avoided, and the control meets the specific requirements of the system.

Project costs

As a benchmark, conventional design of a given production system requires a minimum of two months worth of work for the elaboration of specification and check of concept on site. Conventional design with rapid prototyping for text and correction on the workbench needs one months worth of work. In contrast, POA based design with computer simulated control needs only one week of work for the text and corrections on screen and, therefore, it helps saving project costs.

APPENDIX

A.1 Abbreviations

a) action

A ampere

btn button

c) condition

°C degree Celsius

CASE Computer Aided Software/ Systems Engineering

CO cotton

CU Currency Unit

DD data dictionary

Dep depreciation

Dep+Int depreciation + interests

DFD Data Flow Diagram

DTY draw texturized yarn

E total energy

E_e embodied energy

E_{eA} embodied energy added

El electricity, electrical

ElComp electricity for computer

°F degree Fahrenheit

FDY fully drawn yarn

GJ gigajoule = 10^9 J

h hour

Hz hertz

Int interests

J joule

K Kelvin

kJ kilojoule

kW kilowatt

kWh kilowatt-hour

l liter

LCA Life Cycle Assessment

LD Ladder Diagram

mach machine

min minute

MJ megajoule = 1,000,000 J

MMS Modular Manufacturing Simulator

OM operation mode

PET polyester

PLC Programmable Logic Controller

POA Process Oriented Analysis

POY partially oriented yarn

RFD	Resource Flow Diagram
RPM	revolutions per minute
RT	real-time
RU	Reference Unit
s	second
SA	Structured Analysis
SC	State Chart
SM	simulation model
SU	Standard Unit
Sub	subroutine
T	temperature
TC	transfer coefficient
tex	unit for yarn mass per length: 1 tex = 1 g/1000 m
TR	room temperature
TS	temperature set point
txt	text field
UML	Unified Modeling Language
V	volt
VAT	Value Added Tax
VFD	Value Flow Diagram
W	watt
Wh	watt-hour
WIP	work in progress
WMK	workforce per kilogram product
wk	week
y	year

A.2 Glossary

The methods related to the Process Oriented Analysis are described by different authors using different terms. In this book, only one term is used. The following table lists synonyms used by other authors. The third column shows the translation in German at its synonyms.

English term	**English synonyms**	**German term with synonyms**
action		Aktion
additional information	physical reference, graphical comment	Zusatzinformation, Physikalische Referenz
child diagram	sub	Kinddiagramm
child flow	branch, subflow, flow branch	Zweigfluss, Kindfluss
child process	subprocess	Teilprozess, Subprozess
condition		Kondition, Bedingung
conditional statement		Konditionssatz, Bedingte Zuweisung/Anweisung
connection	relation, link	Verbindung, Beziehung
context diagram	top level, uppermost level	Kontext-Diagramm, oberste Ebene, Top-Diagramm
data dictionary	database, data repository	Objektverzeichnis, Datenbank
detail	explode, brake down, decompose, structure, divide, create child diagram	detaillieren, explodieren, zergliedern, aufgliedern, Kinddiagramm kreieren
detailing	decomposition, structuring, break down	Detaillierung, Zergliederung, Explosion, Verfeinerung
diagram	graph	Diagramm, grafische Darstellung
element	symbol	Element, Symbol
external entity		externe Entität
flow	interface, arrow	Fluss, Schnittstelle, Pfeil
Flow Diagram	Bubble Diagram	Flussdiagramm
interface		Schnittstelle
level		Ebene
merge		Verzweigung (Zusammenführung)
outside world	environment, external environment	Aussenwelt, Umwelt, Umsystem, Umgebung

English term	English synonyms	German term with synonyms
parent flow	trunk, flow bundle	Stammfluss, Vaterfluss, gebündelter Fluss, gruppierter Fluss
process		Prozess
	activity	Aktivität
	bubble	Blase
	function	Funktion
	transformation	Transformation
relation	relationship, connection	Beziehung, Verbindung
specification	spec	Spezifikation, Spez
split		Verzweigung (Aufteilung)
state		Zustand
State Chart	State Transition Diagram, State Diagram	Zustandsdiagramm, Zustandsübergangdiagramm
State Domain		Zustandsdomäne
State Map		Zustandsgraf
system boundary	demarcation of system	Systemgrenze, Abgrenzung
transition		Zustandsübergang, Übergang

A.3 Bibliography

Cited literature

[Booch, 1998] Booch, C., Rumbaugh, J., and Jacobson, I.: *The Unified Modeling Language Reference Manual.* Addison-Wesley, 1990

[DeMarco, 1979] DeMarco, T.: *Structured Analysis and System Specification.* Yourdon Press, 1979

[Directive, 1998] Directive 98/37/EC: *Directive 98/37/EC of the European Parliament and of the Council of 22 June 1998 on the Approximation of the Laws of the Member States Relating to Machinery.* 1998-07-23 OJ No L 207/1

[Edwards, 1993] Edwards, Keith: *Real-Time Structured Methods, System Analysis.* John Wiley & Sons, 1993

[Gane, 1977] Gane, Ch. and Sarson, T.: *Structured System Analysis: Tools Techniques.* IST, 1977

[IEC, 2003] International Electrotechnical Commission: International Standard IEC 61131-3, Programmable Controllers, Part 3: Programming Languages, 2003

[Law, 1991] Law, A. and Kelton, W.: *Simulation Modeling and Analysis.* McGraw-Hill, 1991

[Page, 1991] Page, B.: *Discrete Simulation.* Springer-Verlang 1991

[Shannon, 1975] Shannon, R.: *System Simulation – the Art and Science.* Prentice Hall, 1975

[Ward, 1985] Ward, P.T. and Mellor, St.J.: *Structured Development for Real-Time Systems.* Prentice Hall, 1987

[Yourdon, 1988] Yourdon, E.: *Modern Structured Analysis.* Prentice Hall PTR, 1988

Further literature

Constantine, L. and Yourdon, E.: *Structured Design: Fundamentals of a Discipline of Computer Program and Systems Design.* Yourdon Press, 1978

Creux, S.E.: *Strukturierte Kostenanalyse mit dem Wertfluss-Diagramm.* Dissertation ETH Zurich, 1998

Gane, Ch.: *Computer-Aided Software Engineering, the Methodologies, the Products, and the Future*. Prentice-Hall International Editions, 1990

Hatley, D. and Pribhai, I.: *Strategies for Real-Time System Specification*. Dorset House Publishing, 1987

ISO 14004: *Environmental Management Systems – General Guidelines on Systems and Support Techniques*. 2004

Kaplan, R.S. and Cooper, R.: *Cost and Effect: Using Integrated Cost Systems to Drive Profitability and Performance*. Havard Business School Press, 1997

Shelly, G., Cashman, Th., and Rosenblatt, H.: *Systems Analysis and Design*. Course Technology, Thomson Learning, 2001

Weber Marin, A.: *A Structured Model for Quantitative Analysis of Production Systems*. Dissertation ETH Zurich, 2000

Software

eM-Plant®
UGS, 5800 Granite Parkway, Suite 600, Plano, TX 75024, USA
http://www.ugs.com/products/tecnomatix/plant_design/em_plant.shtml

Modular Manufacturing Simulator
ATN Lean Manufacturing, University of Alabama, 301 Sparkman Drive, Huntsville, AL 35899, USA
http://www.atnlean.com/newsite/software.asp

POAdesigner
Institute for Manufacturing Automation, ETH Zurich, 8092 Zurich, Switzerland
http://www.texma.org

System Architect®
Telelogic AB, PO Box 4128, Kungsgatan 6, SE-203 12 Malmö, Sweden
http://www.telelogic.com/corp/products/systemarchitect/systemarchitect/overview.cfm

Visual Basic .NET®
Microsoft Corporation, One Microsoft Way, Redmond, WA 98052, USA
http://msdn.microsoft.com/vbasic/

INDEX

B

C

Other Related Titles of Interest include:

)ptimization through Intelligent Techniques
vanan, Kumaraguru College of Technology, Tamilnadu, India

90

Design, Production, Automation, and Integration
University of Toronto, Ontario Canada
37

lustrial and Systems Engineering,
ru, University of Tennessee, Knoxville, TN
99